# A FIRST COURSE IN STATISTICS

**JAMES T. McCLAVE**
University of Florida

**FRANK H. DIETRICH, II**
Northern Kentucky University

DELLEN PUBLISHING COMPANY
San Francisco and Santa Clara, California

*On the cover:* ''Zig Zag'' by John Okulick, 1982, wood and paint
California artist John Okulick's art is primarily concerned with affecting our
perceptions. His sculptured paintings, although relatively flat on a wall, give the
illusion of projecting into space. Okulick is represented by Asher-Faure, Los
Angeles, California.

© Copyright 1983 by Dellen Publishing Company, 3600 Pruneridge Avenue,
Santa Clara, California 95051

Printed in the United States of America

10 9 8 7 6 5 4 3 2 1

**Library of Congress Cataloging in Publication Data**

McClave, James T.
  A first course in statistics.

  Includes index.
  1. Statistics.    I.  Dietrich, Frank H.    II.  Title.
QA276.12.M3997    1983    519.5    82-19955

ISBN 0-89517-050-7

# A FIRST COURSE
## IN
## STATISTICS

# CONTENTS

# PREFACE

Instructors who teach a single one-semester (one-quarter) introductory statistics course find that many of the available texts have been written for a two-semester sequence. *A First Course in Statistics* is designed to cope with this problem.

This book is a modification of the second edition of our introductory text, *Statistics.* It maintains the style and flavor of *Statistics* but is greatly reduced in size. This reduction in content was achieved by deleting ''optional'' topics—that is, topics an instructor might want to cover in a two-semester sequence but not in a one-semester (one-quarter) course. For example, the optional section on counting rules was deleted from the chapter on probability (Chapter 3). This change reduces the level of difficulty for the student while still providing a good introduction to the probabilistic concepts needed for a study of statistical inference. Optional sections on the Poisson, hypergeometric, geometric, exponential, and uniform random variables were deleted, enabling us to combine coverage of the important binomial and normal probability distributions into a single chapter (Chapter 4). Material on the comparison of population means and proportions was split into two chapters, one (Chapter 7) describing the methodology for comparing two or more population means and the other (Chapter 8), the methodology for comparing two or more population proportions. This enabled us to replace the lengthy chapters on the analysis of variance and the chi square analysis of contingency tables with brief introductory (optional) sections at the ends of Chapters 7 and 8. Simple linear regression analysis is included in Chapter 9, but multiple regression analysis—a topic usually presented in the second semester of a two-semester course—has been omitted.

These changes have produced a text that is suitable for a one-semester introductory course in statistics. In addition, the text allots a modest amount of space to topics that would not usually be covered in an introductory course but might serve as a reference source for the student. Like *Statistics,* the text provides a good introduction to the basic concepts of statistics and to the important concepts and methods of statistical inference. Specific features of the text are as follows:

**1. Case Studies** (See the list of case studies on page xii.) Many important concepts are emphasized by the inclusion of case studies, which consist of brief summaries of actual applications of statistical concepts and are often drawn directly from the research literature. These case studies allow the student to see applications of important statistical concepts immediately after their introduction. The case studies also help to answer by example the often-asked questions, ''Why should I study statistics? Of what relevance is statistics to my program?'' Finally, the case studies constantly remind the student that each concept is related to the dominant theme—statistical inference.

**2.  The Use of Examples as a Teaching Device**  We have introduced and illustrated almost all new ideas by examples. Our belief is that most students will better understand definitions, generalizations, and abstractions *after* seeing an application. In most sections, an introductory example is followed by a general discussion of the procedures and techniques, and then a second example is presented to solidify the understanding of the concepts.

**3.  A Simple, Clear Style**  We have tried to achieve a simple and clear writing style. Subjects that are tangential to our objective have been avoided, even though some may be of academic interest to those well-versed in statistics. We have not taken an encyclopedic approach in the presentation of material.

**4.  Many Exercises—Labeled by Type**  The text has a large number of exercises illustrating applications in almost all areas of research. However, we believe that many students have trouble learning the mechanics of statistical techniques when problems are all couched in terms of realistic applications—the concept becomes lost in the words. Thus, the exercises at the ends of all sections are divided into two parts:

**a.  Learning the Mechanics**  These exercises are intended to be straightforward applications of the new concepts. They are introduced in a few words and are unhampered by a barrage of background information designed to make them ''practical,'' but which often detracts from instructional objectives. Thus, with a minimum of labor, the student can recheck his or her ability to comprehend a concept or a definition.

**b.  Applying the Concepts**  The mechanical exercises described above are followed by realistic exercises that allow the student to see applications of statistics across a broad spectrum. Once the mechanics are mastered, these exercises develop the student's skills at comprehending realistic problems that describe situations to which the techniques may be applied.

**5.  On Your Own . . .**  Each chapter ends with an exercise entitled ''On Your Own. . . .'' The intent of this exercise is to give the student some hands-on experience with an application of the statistical concepts introduced in the chapter. In most cases, the student is required to collect, analyze, and interpret data relating to some real application.

**6.  Where We've Been . . . Where We're Going . . .**  The first page of each chapter is a ''unification'' page. Our purpose is to allow the student to see how the chapter fits into the scheme of statistical inference. First, we briefly show how the material presented in previous chapters helps us to achieve our goal (Where we've been). Then, we indicate what the next chapter (or chapters) contributes to the overall objective (Where we're going). This feature allows us to point out that we are constructing the foundation block by block, with each chapter an important component in the structure of statistical inference. Furthermore, this feature provides a series of brief resumés of the material covered as well as glimpses of future topics.

**7. Footnotes** Although the text is designed for students with a noncalculus background, footnotes explain the role of calculus in various derivations. Footnotes are also used to inform the student about some of the theory underlying certain results. The footnotes allow additional flexibility in the mathematical and theoretical level at which the material is presented.

**8. Supplementary Materials** A study guide, a solutions manual, and a 3,000 item test bank are available.

Thanks are due to many individuals who helped in the preparation of this text. Susan L. Reiland, Jay DeVore, Roxy Peck, and Dennis Wackerly provided extremely careful line-by-line reviews of the first draft, without which the project could not have met our expectations. Many other reviewers were helpful in providing direction; they include: Charles Burrell, Western Carolina University; Donald L. Evans, Polk Community College; Frederic E. Fischer, State University of New York at Oswego; Frank Hammons, Sinclair Community College; Gordon W. Hoagland, Ricks College; Dave Hughes, University of the Pacific; William G. Koellner, Montclair State College; John T. Kontogianes, Tulsa Junior College; James R. Lackritz, California State University at San Diego; Ray Lindstrom, Northern Michigan University; Phillip McGill, Illinois Central College; David P. Nasby, Orange Coast College; James C. Nicholson, University of Texas at Arlington; Thomas A. O'Connor, University of Louisville; Don Shriner, Frostburg State College; Joseph Smith, United States Air Force Academy; Vasant B. Walkar, Miami University; Kenneth D. Wantling, Montgomery College; Abraham Weinstein, Nassau Community College; Cecilia Welna, University of Hartford. Many exercises and answers were prepared by Bill Louv, Susan L. Reiland, Nancy Shafer, and Terry Sincich. Their creativity and teaching experience are responsible for an inspiring collection of exercises.

Phyllis Niklas has our appreciation and admiration for copy editing the manuscript. Her work defies explanation; you have to see to believe the care and professionalism with which she edits.

Finally, we thank the thousands of students at the University of Florida who have helped us to form our ideas about teaching statistics. Their most common complaint seems to be that texts are written for the instructor rather than the student. We hope that this book is an exception.

# CASE STUDIES

# A FIRST COURSE
## IN
## STATISTICS

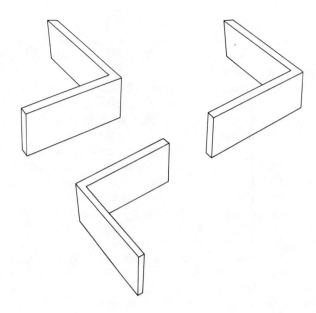

# 1

# WHAT IS STATISTICS?

**WHERE WE'RE GOING . . .**

Statistics? Is it a field of study, a group of numbers that summarize the state of our national economy, the performance of a football team, the social conditions in a particular locale, or, as the title of a popular book (Tanur et al., 1978) suggests, "a guide to the unknown"? We will attempt to answer this question in Chapter 1. Throughout the remainder of the text, we will show you how statistics can be used to interpret experimental and sample survey data. Since many jobs in government, industry, medicine, and other fields require this facility, you will see how statistics can be beneficial to you.

What does statistics mean to you? Does it bring to mind batting averages, Gallup polls, unemployment figures, numerical distortions of facts (lying with statistics!), or simply a college requirement you have to complete? We hope to convince you that statistics is a meaningful, useful science with a broad, almost limitless scope of application to business, government, and the sciences. We also want to show that statistics lie only when they are misapplied. Finally our objective is to paint a unified picture of statistics to leave you with the impression that your time was well-spent studying a subject that will prove useful to you in many ways.

Statistics means "numerical descriptions" to most people. Monthly unemployment figures, the failure rate of a particular type of steel-belted automobile tire, and the proportion of women who favor the Equal Rights Amendment all represent statistical descriptions of large sets of data collected on some phenomenon. Most often the purpose of calculating these numbers goes beyond the description of the particular set of data. Frequently, the data are regarded as a sample selected from some larger set of data. For example, a sampling of unpaid accounts for a large merchandiser would allow you to calculate an estimate of the average value of unpaid accounts. This estimate could be used as an audit check on the total value of all unpaid accounts held by the merchandiser. So, the applications of statistics can be divided into two broad areas: (1) describing large masses of data and (2) drawing conclusions (making estimates, decisions, predictions, etc.) about some set of data based on sampling. Let us examine some case studies that illustrate applications of statistics.

CASE STUDY 1.1
A SURVEY: WHERE "WOMEN'S WORK" IS DONE BY MEN

The 1980 February/March issue of *Public Opinion* describes the results of a survey of several hundred married men from each of nine countries who responded to the following question:

> In the following list, which household jobs would you say it would be reasonable that the man would often take over from his wife: washing up (doing dishes), changing baby's napkin (diaper), cleaning house, ironing, organizing meal, staying at home with sick child, shopping, none of these?

The graphs in Figure 1.1 provide an effective summary of the thousands of opinions obtained and allow for an easy comparison of attitudes across countries. The area of statistics concerned with the summarization and description of data is called descriptive statistics.

CASE STUDY 1.2
DOES JUDICIAL ACTION AFFECT THE PROBABILITY OF CONVICTION?

> Defense attorneys often remind juries that accused criminals are innocent until proven guilty. But if the judge waits to deliver his or her own version of the reminder until after the testimony is in, jurors may be more likely to render a guilty verdict than if they had heard the reminder before the trial.

This quote is from an article in the June 1980 issue of *Psychology Today** that discusses a study conducted by two social psychologists at the University of Kansas. Involved in the study were 107 student jurors, some of whom heard the judge's reminder at the start of a taped trial, some at the end, and some not at all. An analysis of the 107 student verdicts prompted the above quote.

*Reprinted from *Psychology Today* magazine. Copyright © 1980 by Ziff Davis Publishing Co.

**FIGURE 1.1**
"WOMEN'S WORK" IS RARELY DONE BY MEN IN ITALY, GERMANY . . .

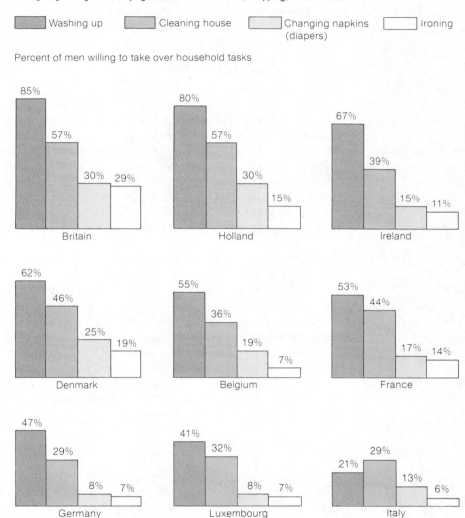

Question: In the following list, which household jobs would you say it would be reasonable that the man would often take over from his wife: washing up (doing dishes), changing baby's napkin (diaper), cleaning house, ironing, organizing meal, staying at home with sick child, shopping, none of these?

Washing up     Cleaning house     Changing napkins (diapers)     Ironing

Percent of men willing to take over household tasks

**Britain:** 85%, 57%, 30%, 29%

**Holland:** 80%, 57%, 30%, 15%

**Ireland:** 67%, 39%, 15%, 11%

**Denmark:** 62%, 46%, 25%, 19%

**Belgium:** 55%, 36%, 19%, 7%

**France:** 53%, 44%, 17%, 14%

**Germany:** 47%, 29%, 8%, 7%

**Luxembourg:** 41%, 32%, 8%, 7%

**Italy:** 21%, 29%, 13%, 6%

Note: The sample size for each country exceeded 900, except Luxembourg, where the sample was 334.
Source: Survey by the European Economic Community Commission. "Women and Men of Europe in 1978," October–November 1977, as shown in *Public Opinion*, February–March 1980, p. 37.
©Copyright American Enterprise Institute.

Note that a sample of 107 student jurors was observed and that their verdicts were used to infer that the jurors in any trial may be more likely to render a guilty verdict if they hear a reminder from the judge at the end of the trial. The branch of statistics exemplified by this study is known as inferential statistics.

Two sets of data of interest to a firm's marketing department are (1) the set of taste-preference scores given by consumers to their product and their competitors' products when all brands are clearly labeled and (2) the taste-preference scores given by the same set of consumers when all brand labels have been removed and the consumers' only means of product identification is taste. With such information the marketing department should be able to determine whether taste preference arose because of perceived physical differences in the products or as a result of the consumers' image of the brand (brand image is, of course, largely a result of a firm's marketing efforts). Such a determination should help the firm develop marketing strategies for their product.

A study using these two sets of data was conducted by Ralph Allison and Kenneth Uhl (1965) in an effort to determine whether beer drinkers could distinguish among major brands of unlabeled beer. A sample of 326 beer drinkers was randomly selected from the set of beer drinkers identified as males who drank beer at least three times a week. During the first week of the study each of the 326 participants was given a six-pack of unlabeled beer containing three major brands and was asked to taste-rate each beer on a scale from 1 (poor) to 10 (excellent). During the second week the same set of drinkers was given a six-pack containing six major brands. This time, however, each bottle carried its usual label. Again, the drinkers were asked to taste-rate each beer from 1 to 10. From a statistical analysis of the two sets of data yielded by the study, Allison and Uhl concluded that the 326 beer drinkers studied could not distinguish among brands by taste on an overall basis. This result enabled them to infer statistically that such was also the case for beer drinkers in general. Their results also indicated that brand labels and their associations did significantly influence the tasters' evaluations. These findings suggest that physical differences in the products have less to do with their success or failure in the marketplace than the image of the brand in the consumers' minds. As to the benefits of such a study, Allison and Uhl note, "to the extent that product images, and their changes, are believed to be a result of advertising . . . the ability of firms' advertising programs to influence product images can be more thoroughly examined."

This case study, like Case Study 1.2, is an example of the use of inferential statistics. Based on data collected from a sample of 326 beer drinkers, Allison and Uhl make inferences about the ability of all beer drinkers to distinguish among major brands of unlabeled beer.

These case studies provide three examples of the uses of statistics. You will note that each involves an analysis of data, either for the purpose of describing the data set (Case Study 1.1) or for making inferences about a much larger data set based on sampling (Case Studies 1.2 and 1.3). They thus provide realistic examples of the two broad areas of statistical applications.

## 1.2
## THE ELEMENTS
## OF STATISTICS

Although applications of statistics abound in almost every area of human endeavor, there are certain elements common to all statistical problems. The foundation of every statistical problem is a population:

> **DEFINITION 1.1**
> The population is a set of data that characterizes some phenomenon.

Thus, we use the term *population* to represent a set of measurements (rather than a group of people) that characterizes some phenomenon of interest to us. For example, the employment status of every person in the United States is a set of data that characterizes a certain aspect of our national health and that, consequently, is of interest to government economists and sociologists. Similarly, the lengths of the lives of Florida spiny lobsters represent a data set that characterizes the life expectancy of this important source of seafood. These data would be relevant in evaluating the growth in numbers of lobsters over time and could be beneficial in establishing minimum lobster trapping weights. Other examples of populations are the number of errors contained in each of the many pages in an accountant's ledger and the lengths of survival, after onset of the disease, of children afflicted with leukemia. Thus, we think of a population as being a large—perhaps infinitely large—collection of measurements.

The second element of a statistical problem is the sample:

> **DEFINITION 1.2**
> The sample is a set of data selected from a population.

The sample is a subset (part of) the population. The employment status of 2,000 people selected from the total of all employable persons in the United States, the lengths of the lives of twenty Florida lobsters, and the number of errors per page on 10 pages of a 100 page ledger all represent samples of the respective populations. In everyday usage, the word *sample* implies a collection of objects, for example, a sample of 1200 people from a city, a sample of ten transistors from a day's production, etc. When selecting people or objects from some group, we sometimes utilize this terminology, i.e., we use ''sample'' to refer to the selected objects rather than to the collection of measurements made on the objects. Whether we are speaking of a collection of measurements (our definition) or the collection of objects on which the measurements are made (everyday usage) becomes clear from the context of the discussion.

The usefulness of the sample is clarified by considering the third element of a statistical problem—the inference:

> **DEFINITION 1.3**
> A statistical inference is a decision, estimate, prediction, or generalization about the population based on information contained in a sample.

That is, we use the information in the smaller set of measurements (the sample) to make decisions, predictions, or generalizations about the large or whole set of measurements (the population). For example, we might use the number of accounting errors in a 10 page sample of a ledger to estimate the number of errors on all 100 pages of the ledger. Similarly, we could use a sample consisting of the employment status of 2,000 employable people to predict the national unemployment rate. Or, we might try to infer the life characteristics of the Florida lobster based on a sample of observed life lengths for twenty lobsters. In each case, we are using the information in a sample to make inferences about the corresponding population.

The preceding definitions identify three of the four elements of a statistical problem. The fourth, and perhaps the most important, is the topic of Section 1.3.

## 1.3 STATISTICS: WITCHCRAFT OR SCIENCE?

We have identified the primary objective of statistics as making inferences about a population based on information contained in a sample. However, inference-making constitutes only part of our story. We also want to measure and report the reliability of each inference made—this is the fourth element of a statistical problem.

The measure of reliability that accompanies an inference separates the science of statistics from the art of fortune-telling. A palm reader, like a statistician, may examine a sample (your hand) and make inferences about the population (your life). However, no measure of reliability can be attached to the reader's inferences. On the other hand, we will always be sure to assess the reliability of our statistical inferences. For example, if we use a sample of previous profit figures to predict a firm's future profits, we will give a bound on our prediction error. This bound is simply a number that the error of our prediction is not likely to exceed. Thus, the uncertainty of our prediction is measured by the size of the bound on the prediction errors. In general, the reliability of our statistical inferences will be discussed at length throughout this text. For now, we simply want you to realize that an inference is incomplete without a measure of its reliability.

We conclude with a summary of the elements of an inferential statistical problem:

---

FOUR ELEMENTS COMMON TO INFERENTIAL STATISTICAL PROBLEMS
1. The population of interest, with a procedure for sampling the population.
2. The sample and analysis of the information in the sample.
3. The inferences about the population, based on information contained in the sample.
4. A measure of reliability for the inference.

---

## 1.4 WHY STUDY STATISTICS?

Why study statistics? The growth in data collection associated with scientific phenomena as well as the operations of business and government (quality control, statistical auditing, forecasting, etc.) has been truly astounding over the past several decades. Published results of political, economic, and social surveys as well as increasing government emphasis on drug and product testing provide vivid evidence of the need to be able to evaluate data sets intelligently. Consequently, you will want to develop a discerning sense of rational thought that will enable you to evaluate numerical data. You may be called upon to use this ability to make intelligent decisions, inferences, and generalizations. For this reason, the study of statistics is an essential preparation for a role in modern society.

EXERCISES

**1.1.** To evaluate the current status of the dental health of school children, the American Dental Association conducted a survey to estimate the average number of cavities per child in grade school in the United States. One thousand school children from across the country were selected and the number of cavities for each was recorded.

    a. Describe the population of interest to the American Dental Association.
    b. Describe the sample.

**1.2.** Suppose that you work for a major public opinion pollster and you wish to estimate the proportion of adult citizens who think the President is doing a good job in handling the nation's economy. Clearly define the population you wish to sample.

**1.3.** A first-year chemistry student conducts an experiment to determine the amount of hydrochloric acid necessary to neutralize 2 ounces of a basic solution. The student prepares five 2 ounce portions of the solution and adds a known concentration of hydrochloric acid to each. The amount of acid necessary to achieve neutrality of the solution is recorded for each of the five portions.

    a. Describe the population of interest to the student.
    b. Describe the sample.

**1.4.** Consider a gambling game in which a die is tossed and the number on the up face is recorded. If the up face is an even number (2, 4, or 6), you win the game; if the up face is an odd number (1, 3, or 5), your opponent wins. Since you want to be sure the die is fair, you decide to toss it 100 times and record the up face on each toss before the game begins.

    a. Describe the population of interest to you.
    b. Describe the sample.
    c. Give an example of an inference you might make.

**1.5.** A manufacturer of vacuum cleaners has decided that an assembly line is operating satisfactorily if less than 2% of the cleaners produced per day are defective. If 2% or more of the cleaners are defective, the line must be shut down and proper adjustments made. To check every cleaner as it comes off the line would be costly and time-consuming. The manufacturer decides to choose thirty cleaners at random from a specific day's production and test for defects.

    a. Describe the population of interest to the manufacturer.
    b. Describe the sample.
    c. Give an example of an inference the manufacturer might make.

**1.6.** A drug company advertises a new drug which is reported to be effective in treating a particular disease, but which also produces undesirable side effects in certain patients. To determine the proportion of treated patients who develop the undesirable side effects, research physicians administer the drug to forty-five carefully selected volunteer patients and monitor their responses.

    a.   Describe the population of interest to the research physicians.
    b.   Describe the sample.
    c.   Give an example of an inference the research physicians might make.

**1.7.** An insurance company would like to determine the proportion of all medical doctors who have been involved in one or more malpractice suits. The company selects 500 doctors at random and determines the number in the sample who have ever been involved in a malpractice suit.

    a.   Describe the population of interest to the insurance company.
    b.   Describe the sample.
    c.   Give an example of an inference the insurance company might make.

**1.8.** A new teaching method has been introduced in the third grade at a local elementary school. At the end of the first year, the school board wants to evaluate the new method and determine its effectiveness. A standardized test, traditionally given at the end of the third grade, will be used to determine whether the new method is effective.

    a.   Describe the population of interest to the school board.
    b.   Describe the sample.
    c.   Give an example of an inference the school board might make.
    d.   What should accompany any inference the school board makes?

**1.9.** A manufacturer of car batteries wishes to revise the length of the battery guarantee so that only 5% of the batteries sold must be replaced while under warranty. Using a sample of their records of battery sales and replacements, the manufacturer decides to guarantee the batteries for 3 years.

    a.   Describe the population of interest to the manufacturer.
    b.   What inference was made about this population?
    c.   What information about the sample would be useful in measuring the reliability of this inference?

**1.10.** If you listened to any newscasts preceding the 1980 presidential election, you probably heard reports similar to the following: "Based on a recent public opinion poll, 40% of the voters favor Ronald Reagan. This estimate could be subject to a 3% sampling error."

    a.   Was the report based on a sample? Explain.
    b.   What was the population of interest?
    c.   What inference was made about this population?
    d.   What measure of reliability was provided? (In Chapter 6 we will learn how this measure of reliability was obtained.)

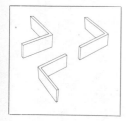

## ON YOUR OWN . . .

Scan a recent issue of a daily newspaper and look for articles that contain numerical data. The data might be a summary of the results of a public opinion poll, the results of a vote by the United States Senate, crime rates, birth or death rates, an election result, etc. For each article containing data that you find, answer the following questions:

**a.** Do the data constitute a sample or an entire population? If a sample has been taken, clearly identify both the sample and the population; otherwise, identify the population.

**b.** If a sample has been observed, does the article present an explicit (or implied) inference about the population of interest? If so, state the inference made in the article.

**c.** If an inference has been made, has a measure of reliability (statement of goodness) been included? What is it?

**REFERENCES**

Allison, R. I., & Uhl, K. P. "Influence of beer brand identification on taste perception." *Journal of Marketing Research,* Aug. 1965, 36–39.

*Careers in statistics,* American Statistical Association and the Institute of Mathematical Statistics, 1974.

Rubenstein, C. "The presumption of innocence needs prompting." *Psychology Today,* June 1980, 30.

Tanur, J. M., Mosteller, F., Kruskal, W. H., Link, R. F., Pieters, R. S., & Rising, G. R. *Statistics: A guide to the unknown.* (E. L. Lehmann, special editor.) San Francisco: Holden-Day, 1978.

" 'Women's work' is rarely done by men in Italy, Germany, . . . ." *Public Opinion,* Feb.–Mar. 1980, 37.

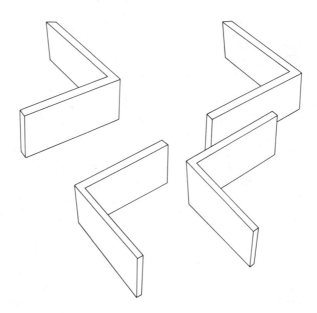

# 2
# METHODS FOR DESCRIBING SETS OF DATA

**WHERE WE'VE BEEN . . .**
By examining typical examples of the use of statistics, we identified four elements that are common to every inferential statistical problem: a population, a sample, an inference, and a measure of the reliability of the inference. The last two elements identify the goal of statistics—using sample data to make an inference (a decision, estimate, or prediction) about a population.

**WHERE WE'RE GOING . . .**
Before we make an inference, we must be able to describe a data set. Both graphical and numerical methods for describing sets of data are discussed in this chapter. As you will learn in Chapter 6, we will use some sample numerical descriptive measures to estimate the values of corresponding population descriptive measures. Therefore, our efforts in this chapter will ultimately lead to statistical inference.

Suppose we wish to evaluate the mathematical capabilities of a class of 1,000 college freshmen based on their quantitative Scholastic Aptitude Test (SAT) scores. How would you describe these 1,000 measurements? You can see that this is not an easy question to answer. The 1,000 scores provide too many bits of information for our minds to comprehend. It is clear that we need some method for summarizing the information in a data set. Methods for describing data sets are also essential for statistical inference. Most populations are large data sets. Consequently, if we are going to make descriptive statements (inferences) about a population based on information contained in a sample, we will once again need methods for describing a data set.

Two methods for describing data are presented in this chapter, one graphical and the other numerical. As you will subsequently see, both play an important role in statistics.

## 2.1
## TYPES OF DATA

Although the number of phenomena that can be measured is almost limitless, data can generally be classified as one of two types: quantitative or qualitative.

> DEFINITION 2.1
> Quantitative data are observations that are measured on a numerical scale.

The most common type of data is quantitative data, since many descriptive variables observed in nature are measured on numerical scales. Examples of quantitative data are:

1. The bacteria count in your drinking water
2. The monthly unemployment percentage
3. A person's blood pressure
4. The number of women executives in an industry

The measurements in these examples are all numerical.

All data that are not quantitative are qualitative:

> DEFINITION 2.2
> If each measurement in a data set falls into one and only one of a set of categories, the data set is called qualitative.

Qualitative data are observations that are categorical rather than numerical. Examples of qualitative data are:

1. The political party affiliations of a group of people. Each person would have one and only one political party affiliation.
2. The brand of gasoline last purchased by each person in a sample of automobile owners. Again, each measurement would fall into one and only one category.

**3.** The state in which each firm in a sample of firms in the United States has its highest yearly sales.

Notice that each of the examples has nonnumerical, or qualitative, measurements.

As you would expect, the methods used for summarizing the information in a sample of measurements depends on the type of data being collected. Qualitative data can be summarized by giving the number of observations that fall in each of the classification categories, but the description of quantitative data is more complex. Consequently, this chapter will present methods for describing quantitative data sets.

EXERCISES

Learning the mechanics

**2.1.** State whether the following types of data are qualitative or quantitative:

    a. The amount of personal life insurance on each professor at your university
    b. The flavor of ice cream each of 200 shoppers buys at a supermarket
    c. The marital status of each person living on a city block
    d. The number of years of experience possessed by each salesperson at a car dealership

**2.2.** State whether the following types of data are qualitative or quantitative:

    a. The hair color of each student in your class
    b. The length of time each of thirty patients must stay in a hospital
    c. The sex of each United States senator
    d. The religious affiliation of each of a psychologist's patients

**2.2
HISTOGRAMS:
GRAPHICAL
DESCRIPTIONS
OF
QUANTITATIVE
DATA SETS**

Before we can use the information in a sample to make inferences about a population, we need methods to summarize, or describe, a set of data. For example, the Environmental Protection Agency (EPA) performs extensive tests on all new car models to determine their mileage rating. Suppose that the 100 measurements in Table 2.1 represent the results of such tests on a particular new car model. How can we summarize the information in this rather large sample?

A visual inspection of the data indicates some obvious facts. For example, most of the mileages are in the 30's, with a smaller fraction in the 40's. But it is difficult to provide much additional information on the 100 mileage ratings without resorting to some method of summarizing the data. A graphical method for accomplishing this is displayed in Figure 2.1. This type of graph, commonly called a histogram (or relative frequency distribution), is the most popular graphical technique for depicting quantitative data. Before discussing exactly how the graph was obtained, we will extract some of the information Figure 2.1 provides.

The horizontal axis of Figure 2.1, which gives the miles per gallon for a given automobile, is divided into intervals commencing with the interval from 29.95 to 31.45 and proceeding in intervals of equal size to 43.45 to 44.95 miles per gallon. The vertical axis gives the proportion (or relative frequency) of the 100 readings that fall in each interval. Thus, you can see that 0.33, or 33%, of the owners obtained a mileage between 35.95 and 37.45. This interval contains the highest relative fre-

TABLE 2.1
EPA MILEAGE RATINGS
ON 100 CARS

| | | | | | | | | | |
|---|---|---|---|---|---|---|---|---|---|
| 36.3 | 41.0 | 36.9 | 37.1 | 44.9 | 36.8 | 30.0 | 37.2 | 42.1 | 36.7 |
| 32.7 | 37.3 | 41.2 | 36.6 | 32.9 | 36.5 | 33.2 | 37.4 | 37.5 | 33.6 |
| 40.5 | 36.5 | 37.6 | 33.9 | 40.2 | 36.4 | 37.7 | 37.7 | 40.0 | 34.2 |
| 36.2 | 37.9 | 36.0 | 37.9 | 35.9 | 38.2 | 38.3 | 35.7 | 35.6 | 35.1 |
| 38.5 | 39.0 | 35.5 | 34.8 | 38.6 | 39.4 | 35.3 | 34.4 | 38.8 | 39.7 |
| 36.3 | 36.8 | 32.5 | 36.4 | 40.5 | 36.6 | 36.1 | 38.2 | 38.4 | 39.3 |
| 41.0 | 31.8 | 37.3 | 33.1 | 37.0 | 37.6 | 37.0 | 38.7 | 39.0 | 35.8 |
| 37.0 | 37.2 | 40.7 | 37.4 | 37.1 | 37.8 | 35.9 | 35.6 | 36.7 | 34.5 |
| 37.1 | 40.3 | 36.7 | 37.0 | 33.9 | 40.1 | 38.0 | 35.2 | 34.8 | 39.5 |
| 39.9 | 36.9 | 32.9 | 33.8 | 39.8 | 34.0 | 36.8 | 35.0 | 38.1 | 36.9 |

quency, and the intervals tend to contain a smaller fraction of the measurements as the mileages get smaller or larger.

By summing the relative frequencies in the intervals 34.45–35.95, 35.95–37.45, and 37.45–38.95, you can see that 65% of the mileages are between 34.45 and 38.95. Similarly, only 2% of the cars obtained a mileage rating over 41.95. Many other summary statements can be made by further study of the histogram.

When constructing a histogram, the general rules listed in the box on page 14 should be followed.

To construct the relative frequency histogram in Figure 2.1, we must first define the measurement classes. Since the number of measurements, $n = 100$, is moderately large, we will arbitrarily choose to construct ten measurement classes. The ten classes must span the distance between the smallest measurement, 30.0, and the

FIGURE 2.1
HISTOGRAM FOR
MILES PER GALLON
DATA

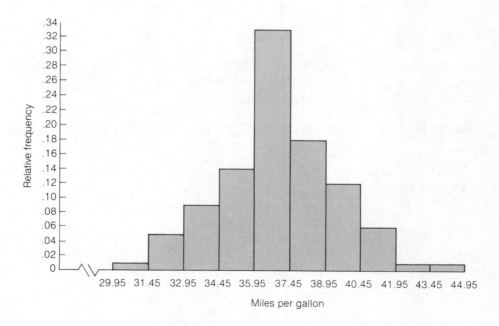

*Note that frequencies rather than relative frequencies may be used in constructing a histogram. The frequency is the actual number of measurements in each interval.

largest measurement, 44.9. Thus, each class should have a width of:

$$\frac{\text{Approximate}}{\text{class width}} = \frac{\text{Largest measurement} - \text{Smallest measurement}}{\text{Number of classes}}$$

$$= \frac{44.9 - 30.0}{10} \approx 1.50$$

Locating the lower class boundary of the first class interval at 29.95 (slightly below the smallest measurement) and adding the class width of 1.5, we find the upper class boundary to be 31.45. Adding 1.5 again, we find the upper class boundary of the

**TABLE 2.2**
**MEASUREMENT CLASSES, FREQUENCIES AND RELATIVE FREQUENCIES FOR THE CAR MILEAGE DATA**

| MEASUREMENT CLASS | FREQUENCY | RELATIVE FREQUENCY |
|---|---|---|
| 29.95–31.45 | 1 | .01 |
| 31.45–32.95 | 5 | .05 |
| 32.95–34.45 | 9 | .09 |
| 34.45–35.95 | 14 | .14 |
| 35.95–37.45 | 33 | .33 |
| 37.45–38.95 | 18 | .18 |
| 38.95–40.45 | 12 | .12 |
| 40.45–41.95 | 6 | .06 |
| 41.95–43.45 | 1 | .01 |
| 43.45–44.95 | 1 | .01 |
| Total | 100 | 1.00 |

second class to be 32.95. Continuing this process, we obtain the ten class intervals shown in Table 2.2. Note that the points that locate the class boundaries are written with a 5 in the second decimal place. This makes it impossible for any of the miles per gallon observations to fall on a class boundary (since they were given only to the nearest tenth).

The next step is to determine the class frequencies and calculate the class relative frequencies. These quantities are defined as follows:

---

**DEFINITION 2.3**

The class frequency for a given class, say class $i$, is equal to the total number of measurements that fall in that class. The class frequency for class $i$ is denoted by the symbol, $f_i$.

---

**DEFINITION 2.4**

The class relative frequency for a given class, say class $i$, is equal to the class frequency divided by the total number $n$ of measurements, i.e.,

$$\text{Relative frequency for class } i = \frac{f_i}{n}$$

---

Scan the data and count the number of measurements in each class interval. This number, the class frequency, is entered in the second column of Table 2.2. Finally, calculate the class relative frequency, the proportion of the total number of measurements that fall in each interval. The class relative frequencies for the miles per gallon data are obtained by dividing each of the class frequencies by the total number of measurements (100); these are listed in the third column of Table 2.2. As noted at the outset of this discussion, these relative frequencies were used to construct the relative frequency histogram of Figure 2.1.

By looking at a histogram (say, the relative frequency histogram in Figure 2.1), you can see two important facts. First, note the total area under the histogram, and then note the proportion of the total area that falls over a particular interval of the $x$-axis. You will see that the proportion of the total area that falls above an interval is equal to the relative frequency of measurements falling in the interval. For example, the relative frequency for the class interval 35.95–37.45 is .33. Consequently, the rectangle above the interval contains .33 of the total area under the histogram.

Second, you can imagine the appearance of the relative frequency histogram for a very large set of data (say, a population). As the number of measurements in a data set is increased, you can obtain a better description of the data by decreasing the width of the class intervals. When the class intervals become small enough, a relative frequency histogram will (for all practical purposes) appear as a smooth curve (see Figure 2.2, page 16).

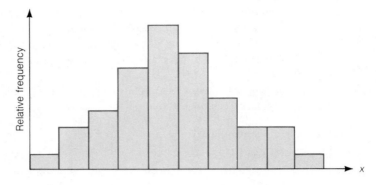

(a) Small data set

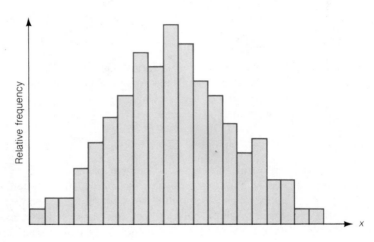

(b) Larger data set

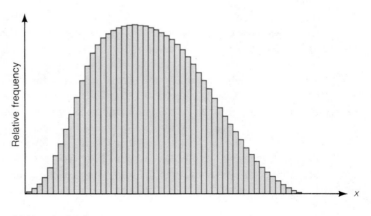

(c) Very large data set

CASE STUDY 2.1
MERCURY POISONING
AND THE DENTAL
PROFESSION

The hazards to health traced to environmental pollution constitute an area of major national concern. Mercury has been identified as one source of environmental contamination. Recognition of mercury as a hazard can be traced to Theophrastus, a Greek scientist living about 400 B.C. The physiologic effects of mercury poisoning are a matter of record with death as the ultimate possibility.

This is the introductory paragraph to an article that appeared in the November 1974 issue of the *Journal of the American Dental Association* (Miller et al., 1974).* The article discusses mercury vapor contamination levels in the air of dental offices and the resulting threat to the health of dentists and dental assistants. Such contamination might result from spills in handling mercury, the unprotected storage of scrap amalgam, or aerosols created by the use of high-speed rotary cutting instruments in removing old amalgam fillings.

The cumulative absorption of small quantities of mercury can result in serious medical problems. The constant daily exposure of dentists and their auxiliary personnel to possible mercury contamination is therefore an important concern. A level of 0.05 milligram of mercury per cubic meter of air is the largest amount considered safe for those working a 40 hour work week.

A determination of mercury vapor levels was made in sixty dental offices in San Antonio, Texas. A relative frequency histogram summarizing the information provided by these measurements is given in Figure 2.3. The histogram clearly shows that an alarming fraction ($\frac{6}{60}$, or $\frac{1}{10}$) were above the danger level of 0.05 milligram. You can see that these data have been effectively summarized and clearly indicate a need for strict policing of mercury vapor levels in dental offices.

Case Study 2.1 illustrates a shortcoming of relative frequency histograms. Suppose you wish to use the relative frequency histogram to infer the nature of the population

**FIGURE 2.3
RELATIVE FREQUENCY
HISTOGRAM**

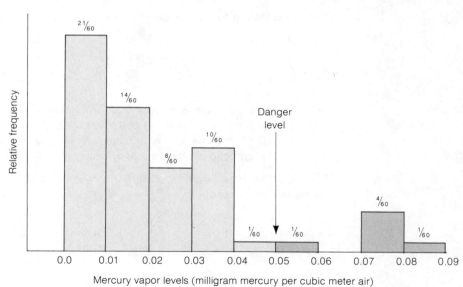

Mercury vapor levels (milligram mercury per cubic meter air)

*Copyright by the American Dental Association. Reprinted by permission.

relative frequency distribution—that is, the distribution of the mercury vapor levels in the offices of all dentists practicing in the United States. It is true that the sample and population relative frequency distributions will be similar. But, how similar? How can you explain how similar the two figures will be? Or, equivalently, how can you measure the reliability of the inference? In Section 2.3 we will explain how you can use one or more numbers (numerical descriptive measures) to describe a distribution of measurements. Further, you will see in Chapter 6 that we can use sample numerical descriptive measures to make inferences about their population counterparts and that we can measure the reliability of these inferences. In short, you will see that numerical descriptive measures are superior to graphical descriptive measures when you want to use the sample data to make inferences about the population from which the sample was selected.

EXERCISES    Learning the mechanics

**2.3.**    Construct a relative frequency histogram for the data summarized in the following relative frequency table:

| MEASUREMENT CLASS | RELATIVE FREQUENCY |
|---|---|
| 0.5– 2.5 | .10 |
| 2.5– 4.5 | .15 |
| 4.5– 6.5 | .25 |
| 6.5– 8.5 | .20 |
| 8.5–10.5 | .05 |
| 10.5–12.5 | .10 |
| 12.5–14.5 | .10 |
| 14.5–16.5 | .05 |

**2.4.**    A sample of twenty measurements is shown below:

| | | | | |
|---|---|---|---|---|
| 26 | 34 | 21 | 32 | 32 |
| 36 | 28 | 38 | 17 | 29 |
| 22 | 12 | 26 | 39 | 25 |
| 31 | 30 | 23 | 27 | 19 |

a.    Give the upper and lower boundaries for six measurement classes commencing the lower boundary of the first class at 10.5. Use a class width equal to 5. Determine the relative frequency for each of the six classes.
b.    Construct a relative frequency histogram using the results of part a.

**2.5.**    Use the data below to construct a relative frequency histogram with eight measurement classes.

| | | | | | |
|---|---|---|---|---|---|
| 5.9 | 5.3 | 1.6 | 7.4 | 9.8 | 1.7 |
| 8.6 | 1.2 | 2.1 | 4.0 | 6.5 | 7.2 |
| 7.3 | 8.4 | 8.9 | 6.7 | 9.2 | 2.8 |
| 4.5 | 6.3 | 7.6 | 9.7 | 9.4 | 8.8 |
| 3.5 | 1.1 | 4.3 | 3.3 | 3.1 | 1.3 |
| 8.4 | 1.6 | 8.2 | 6.5 | 4.1 | 3.1 |
| 1.1 | 5.0 | 9.4 | 6.4 | 7.7 | 2.7 |

**2.6.** Construct a relative frequency histogram for the following data:

| | | | | |
|---|---|---|---|---|
| 23 | 12 | 82 | 12 | 67 |
| 52 | 24 | 17 | 15 | 60 |
| 32 | 49 | 37 | 55 | 81 |
| 39 | 99 | 88 | 24 | 30 |
| 12 | 19 | 53 | 50 | 18 |
| 16 | 40 | 51 | 61 | 35 |

Applying the concepts

**2.7.** The graph below summarizes the scores obtained by 100 students on a questionnaire designed to measure aggressiveness. (Scores are integer values that range from 0 to 20. A high score indicates a high level of aggression.)

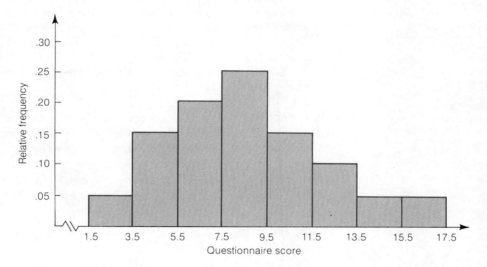

a. Which measurement class contains the highest proportion of test scores?
b. What proportion of the scores lie between 3.5 and 5.5?
c. What proportion of the scores are higher than 11.5?
d. How many students scored less than 5.5?

**2.8.** The numbers of years of experience (to the nearest half year) for the faculty of a certain statistics department are recorded as follows:

| | | | | |
|---|---|---|---|---|
| 5.0 | 4.5 | 19.5 | 1.0 | 6.5 |
| 0.5 | 4.0 | 1.0 | 2.5 | 9.0 |
| 5.5 | 5.0 | 3.5 | 4.5 | 1.5 |
| 8.0 | 21.0 | 7.0 | 13.5 | 3.5 |

a. Construct a relative frequency histogram for these data. Use six measurement classes, commencing with .45 and ending with 21.45.
b. In order to be eligible for certain benefits, a faculty member must have more

than 13 years of experience. What proportion of the statistics faculty are eligible for these benefits?

**2.9.** In the late 1970's and early 1980's gasoline prices rose at an incredible rate. The following is a sample of the prices charged (in cents per gallon) for regular grade gasoline at twenty-five gas stations in the greater Cincinnati area in January 1981:

| | | | | |
|---|---|---|---|---|
| 128.9 | 121.9 | 133.9 | 119.9 | 115.9 |
| 115.9 | 127.9 | 119.8 | 116.8 | 122.8 |
| 135.9 | 118.8 | 115.9 | 121.9 | 126.9 |
| 117.9 | 115.8 | 121.9 | 124.9 | 129.9 |
| 122.8 | 131.9 | 132.8 | 120.8 | 124.9 |

a. Construct a relative frequency histogram for these data.

b. Suppose that you were to sample twenty-five gas stations from another area of the country, say New York City or rural Kansas. Do you think a histogram for the twenty-five measurements would look similar to the one you constructed in part a? Explain.

c. Explain the correspondence between areas under the relative frequency histogram and relative frequencies.

**2.10.** Considering the climate, is it economically feasible to start an orange grove in northern Florida? If the temperature falls below 32°F, oil-burning smudge pots must be lit to keep the orange trees from freezing. Suppose a prospective grower decides that a grove would be economically feasible if the pots have to be lit an average of 15 days or less each year. The grower selects 20 years since 1900 at random and obtains the total number of days per year that the temperature fell below 32°F:

| | | | |
|---|---|---|---|
| 20 | 16 | 13 | 12 |
| 9 | 25 | 16 | 6 |
| 15 | 10 | 18 | 11 |
| 14 | 12 | 17 | 13 |
| 13 | 28 | 14 | 15 |

a. Construct a relative frequency histogram for these data.

b. Based on these sample data, estimate the proportion of years in which the pots have to be lit 15 days or less. [*Note:* We will show you how to evaluate the reliability of this estimate in Chapter 6.]

## 2.3 NUMERICAL MEASURES OF CENTRAL TENDENCY

Now that we have presented some graphical techniques for summarizing and describing data sets, we turn to numerical methods for accomplishing this objective. When we speak of a data set, we refer to either a sample or a population. If statistical inference is our goal, we will wish ultimately to use sample numerical descriptive measures to make inferences about the corresponding measures for a population.

As you will see, there are a large number of numerical methods available to describe data sets. Most of these methods measure one of two data characteristics:

**1.** The central tendency of the set of measurements, i.e., the tendency of the data to cluster or to center about certain numerical values.

**2.** The variability of the set of measurements, i.e., the spread of the data.

In this section, we will concentrate on measures of central tendency. In the next section, we will discuss measures of variability.

The most popular and best understood measure of central tendency for a quantitative data set is the arithmetic mean (or simply, the mean) of a data set:

---

DEFINITION 2.5
The mean of a set of quantitative data is equal to the sum of the measurements divided by the number of measurements contained in the data set.

---

In everyday terms, the mean is the average value of the data set.

EXAMPLE 2.1    Calculate the mean of the following five sample measurements:

$$5, \quad 3, \quad 8, \quad 5, \quad 6$$

Solution    The mean is the average of the five measurements, i.e.,

$$\frac{5 + 3 + 8 + 5 + 6}{5} = 5.4$$

Thus, the mean of this sample is 5.4.*

At this point, it will be advantageous to present some shorthand notation that will simplify calculation instructions for the mean as well as other more complicated numerical descriptive measures we will subsequently encounter. This notation is summarized in the box (page 22). Remember that such notation is used for one reason only: to avoid having to repeat the same verbal descriptions over and over again. If you mentally substitute the verbal definition of a symbol each time you read it, you will soon become accustomed to its use.

EXAMPLE 2.2    Calculate the sample mean for the 100 gas mileages given in Table 2.1.

Solution    The mean gas mileage for the 100 cars is

$$\bar{x} = \frac{36.3 + 41.0 + \cdots + 38.1 + 36.9}{100} = \frac{3,699.4}{100} = 36.994$$

*Note: In the examples given here, $\bar{x}$ is sometimes rounded to the nearest tenth, sometimes the nearest hundredth, sometimes the nearest thousandth. There is no specific rule for rounding when calculating $\bar{x}$ because $\bar{x}$ is specifically defined to be the sum of all measurements divided by $n$, i.e., it is a specific fraction. When $\bar{x}$ is used for descriptive purposes, it is often convenient to round the calculated value of $\bar{x}$ to the number of significant figures used for the original measurements. However, when $\bar{x}$ is to be used in other calculations, it may be necessary to retain more significant figures.

Given this information, you would be able to visualize a distribution of gas mileage readings centered in the vicinity of $\bar{x} = 37.0$. An examination of the relative frequency histogram (Figure 2.1) confirms that $\bar{x}$ does in fact fall near the center of the distribution.

In a practical situation, we rarely know the population mean $\mu$, but we can use the sample mean $\bar{x}$ to estimate its value. Thus, we would infer that the mean gas mileage for all cars (of the model included in the sample) is near 37.0 miles per gallon. We will show you how to evaluate the reliability of this estimate in Chapter 6.

## CASE STUDY 2.2
### HOTELS: A RATIONAL METHOD FOR OVERBOOKING

The most outstanding characteristic of the general hotel reservation system is the option of the prospective guest, without penalty, to change or cancel his reservation or even to "no-show" (fail to arrive without notice). Overbooking (taking reservations in excess of the hotel capacity) is practiced widely throughout the industry as a compensating economic measure. This has motivated our research into the problem of determining policies for overbooking which are based on some set of rational criteria.

So says Marvin Rothstein (1974) in an article that appeared in the journal for the American Institute for Decision Sciences. In this paper Rothstein introduces a method for scientifically determining hotel booking policies and applies it to the booking problems of the 133 room Sheraton Pocono Inn at Stroudsburg, Pennsylvania.

From the Sheraton Pocono Inn's records the number of reservations, walk-ins (people without reservations who expect to be accommodated), cancellations, and no-shows were tabulated for each day during the period August 1–28, 1971. The Inn's records for this period included approximately 3,100 guest histories. From the tabu-

*Note: We omit the formula for calculating the mean for grouped data (data presented in a relative frequency table). The reader interested in this special topic should consult the references included at the end of this chapter.

| TABLE 2.3 | SUNDAY | MONDAY | TUESDAY | WEDNESDAY | THURSDAY | FRIDAY | SATURDAY |
|-----------|--------|--------|---------|-----------|----------|--------|----------|
| MEAN NUMBER OF ROOM RESERVATIONS, AUGUST 1–28, 1971, 133 ROOMS | 138 | 126 | 149 | 160 | 150 | 150 | 169 |

lated data the mean or average number of room reservations per day for each of the 7 days of the week were computed. These appear in Table 2.3. In applying his booking policy decision method to the Sheraton's data, Rothstein used the means listed in Table 2.3 to help portray the Inn's demand for rooms.

The mean number of Saturday reservations during August 1–28, 1971, is 169. This may be interpreted as an estimate of $\mu$, the mean number of rooms demanded via reservations (walk-ins also contribute to the demand for rooms) on a Saturday during 1971. If the reservation data for all Saturdays during 1971 had been tabulated, $\mu$ could have been computed. But, since only August data are available, they were used to estimate $\mu$. Can you think of some problems associated with using August's data to estimate the mean for the entire year?

A median is another important measure of central tendency. In general terms, a median is the middle number when the measurements in a data set are arranged in ascending (or descending) order.

---

DEFINITION 2.6
1.  If the number $n$ of measurements in a data set is odd, the median is the middle number when the measurements are arranged in ascending (or descending) order.
2.  If the number $n$ of measurements is even, the median is the mean of the two middle measurements when the measurements are arranged in ascending (or descending) order.

---

The median is of most value in describing large data sets. If the data set is characterized by a relative frequency histogram (Figure 2.4, page 24), the median is the point on the x-axis such that half the area under the histogram lies above the median and half lies below. [*Note:* In Section 2.2 we observed that the relative frequency associated with a particular interval on the x-axis is proportional to the amount of area under the histogram that lies above the interval.]

EXAMPLE 2.3     Consider the following sample of $n = 7$ measurements:

5, 7, 4, 5, 20, 6, 2

a.  Calculate the median of this sample.
b.  Eliminate the last measurement (the 2) and calculate the median of the remaining $n = 6$ measurements.

FIGURE 2.4
LOCATION OF
THE MEDIAN

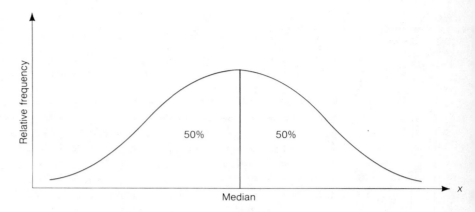

Solution

**a.** The 7 measurements in the sample arranged in ascending order are

2, 4, 5, 5, 6, 7, 20

Because the number of measurements is odd, the median is the middle measurement. Thus, the median of this sample is 5.

**b.** After removing the 2 from the set of measurements, our sample arranged in ascending order appears as follows:

4, 5, 5, 6, 7, 20

Now the number of measurements is even, and so we average the middle two measurements. The median is $(5 + 6)/2 = 5.5$.

In certain situations, the median may be a better measure of central tendency than the mean. In particular, the median is less sensitive than the mean to extremely large or small measurements. To illustrate, note that all but one of the measurements in part a of Example 2.3 center about $x = 5$. The single relatively large measurement, $x = 20$, does not affect the value of the median, 5, but it shifts the mean, $\bar{x} = 7$, to the right of most of the measurements. As another example, if you were interested in computing a measure of central tendency of the incomes of a company's employees, the mean might be misleading. If all blue- and white-collar employees' incomes are included in the data set, the high incomes of a few executives will influence the mean more than the median. Thus, the median will provide a more accurate picture of the typical income for employees of the company. Similarly, the median yearly sales for a sample of companies would locate the middle of the sales data. However, a few companies with very large yearly sales would greatly influence the mean, making it deceptively large. That is, the mean could exceed a vast majority of the sample measurements, making it a misleading measure of central tendency.

If you were to calculate the median for the 100 gas mileage readings in Table 2.1, you would find that the median, 37.0, and the mean, 36.994, are almost equal. This fact indicates that the data form an approximately symmetric distribution (compare the center figure in the box with Figure 2.1). As indicated in the box, a comparison of the mean and median gives an indication of the skewness (nonsymmetry) of a data set.

## COMPARING THE MEAN AND THE MEDIAN

If the median is less than the mean, the data set is skewed to the right:

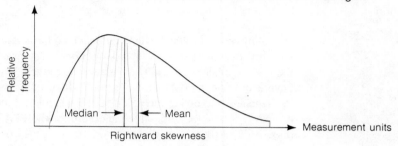

Rightward skewness

For symmetrical data sets, the mean equals the median.

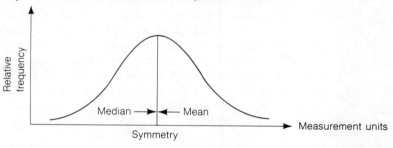

Symmetry

If the median is greater than the mean, the data set is skewed to the left:

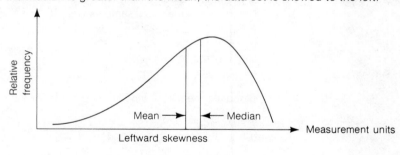

Leftward skewness

A third measure of central tendency is the **mode** of a set of measurements:

DEFINITION 2.7
The **mode** is the measurement that occurs most frequently in the data set.

Therefore, the mode shows where the data tend to concentrate.

EXAMPLE 2.4        Calculate the mode for the following ten quiz grades:

                    8,  7,  9,  6,  8,  10,  9,  9,  5,  7

Solution           Since 9 occurs most often, the mode is 9.

    The mode is of primary value in describing large data sets. When a large data set has been described using a relative frequency histogram, the mode will be located in the class containing the largest relative frequency. This is called the **modal class**. Several definitions exist for locating the position of the mode within a modal class, but the simplest is to define the mode as the midpoint of the modal class. For example, examine the relative frequency histogram for the EPA mileage per gallon readings (Figure 2.1). You can see that the modal class is the interval 35.95–37.45. The mode (the midpoint) is 36.7. Note that this measure of central tendency is very close to the mean, 36.994, and the median, 37.0.

    Because it emphasizes data concentration, the mode has applications in fields such as marketing as well as in the description of large data sets collected by state and federal agencies. For example, a retailer of men's clothing would be interested in the modal neck size and sleeve length of potential customers. A supermarket manager is interested in the cereal brand with the largest share of the market, i.e., the modal brand. The modal income class of the American worker is of interest to the Labor Department. Thus, the mode provides a useful measure of central tendency for various applications.

EXERCISES        Learning the mechanics
        **2.11.**  Calculate the mean for samples where:
            a.  $n = 20, \Sigma x = 100$
            b.  $n = 10, \Sigma x = 500$
            c.  $n = 500, \Sigma x = 10$
            d.  $n = 19, \Sigma x = 237$

        **2.12.**  Calculate the mean, median, and mode for each of the following samples:
            a.  6, −2, 4, 3, 0, 4
            b.  2, 3, 5, 3, 2, 3, 4, 3, 5, 1, 2, 3, 4
            c.  53, 50, 43, 52, 50, 47, 62, 90, 50, 41

        **2.13.**  Describe how the mean compares to the median for a distribution as follows:
            a.  Skewed to the left
            b.  Skewed to the right
            c.  Symmetric

        **2.14.**  Calculate the mean and median for each of the following samples:
            a.  0, 1, 1, 2, 2, 2, 3, 3, 4
            b.  0, 1, 1, 2, 2, 2, 3, 3, 40

c. $-100, -99, -1, 0, 1, 99, 100$
d. $-100, -99, -1, 0, 1, 2, 3$

Applying the concepts

**2.15.** A psychologist has developed a new technique intended to improve rote memory. To test the method against other standard methods, twenty high school students are selected at random and each is taught the new technique. The students are then asked to memorize a list of 100 word phrases using the technique. The following are the number of word phrases memorized correctly by the students:

| 91 | 64 | 98 | 66 | 83 |
|----|----|----|----|----|
| 87 | 83 | 86 | 80 | 93 |
| 83 | 75 | 72 | 79 | 90 |
| 80 | 90 | 71 | 84 | 68 |

a. Define the terms *mean, median,* and *mode* in the context of this problem.
b. Construct a relative frequency histogram for the data.
c. Compute the mean, median, and mode for the above data set and locate them on the histogram. Do these measures of central tendency appear to locate the center of the distribution of data?

**2.16.** Would you expect the data sets described below to possess relative frequency distributions that are symmetrical, skewed to the right, or skewed to the left? Explain.

a. The salaries of all persons employed by a large university
b. The grades on an easy test
c. The grades on a difficult test
d. The amounts of time students in your class studied last week
e. The ages of automobiles on a used car lot
f. The amounts of time spent by students on a difficult examination (maximum time is 50 minutes)

**2.17.** One index used by social scientists to measure a person's socioeconomic status is personal income. To compile a breakdown of the percentage of drinkers and abstainers by socioeconomic status, a survey was taken. The yearly incomes (in dollars) of the respondents who claim to abstain from drinking are given below.

| 12,000 | 10,000 | 10,000 | 90,800 | 9,800 |
|--------|--------|--------|--------|-------|
| 12,000 | 5,600 | 9,000 | 9,000 | 10,200 |
| 15,000 | 17,000 | 11,500 | 13,000 | 8,000 |
| 9,500 | 13,500 | 12,200 | 17,000 | 11,400 |

a. Compute the mean, median, and mode for this sample.
b. Now drop the highest value and repeat part a. What effect does dropping this large measurement have on the measures of central tendency computed in part a? Which measure of central tendency seems to be most sensitive to extremely high scores?

**2.18.** The scores for a statistics test are as follows:

| | | | |
|---|---|---|---|
| 87 | 76 | 96 | 77 |
| 94 | 92 | 88 | 85 |
| 66 | 89 | 79 | 95 |
| 50 | 91 | 83 | 88 |
| 82 | 58 | 18 | 69 |

a. Compute the mean, median, and mode for these data.

b. Which of the three measures of central tendency do you think would best represent the achievement of the class?

c. Eliminate the two lowest scores, and again compute the mean, median, and mode. Which measure of central tendency do you think is most affected by extremely low scores?

**2.19.** Ten presumably trained rats were released in a maze. Their times to escape (in seconds) are recorded below. The N's represent two rats that had still not escaped at the time of the termination of the experiment.

| | | | | |
|---|---|---|---|---|
| 100 | 38 | N | 122 | 95 |
| 116 | 56 | 135 | 104 | N |

a. Can you calculate the mean for these data? Explain.

b. Is the median a meaningful measure of central tendency for these data? Explain. Calculate the median.

**2.20.** In the National Basketball Association (NBA) there are presently a handful of superstars with contracts worth millions of dollars annually. The majority of players, however, earn less than $100,000 per year. The players' association is presently negotiating with the owners for more fringe benefits. The owners want to show that the NBA players have very lucrative salaries and do not need additional benefits.

a. What measure of central tendency of NBA salaries should the owners use to support their position? Why?

b. What measure should the players' association use? Why?

## 2.4 NUMERICAL MEASURES OF VARIABILITY

Measures of central tendency provide only a partial description of a quantitative data set. The description is incomplete without a measure of the variability, or spread, of the data set.

If you examine the two histograms in Figure 2.5, you will notice that both hypothetical data sets are symmetric, with equal modes, medians, and means. However, data set 1 in Figure 2.5(a) has measurements spread with almost equal relative frequency over the measurement classes, while data set 2 in Figure 2.5(b) has most of its measurements clustered about its center. Thus, data set 2 is *less variable* than data set 1. Consequently, you can see that we need a measure of variability as well as a measure of central tendency to describe a data set.

Perhaps the simplest measure of the variability of a quantitative data set is its range.

FIGURE 2.5

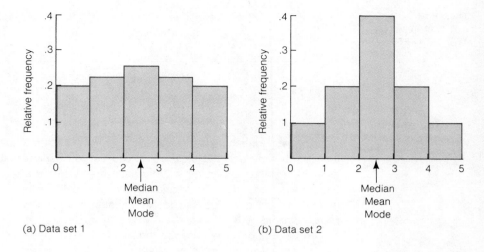

(a) Data set 1                    (b) Data set 2

DEFINITION 2.8
The range of a data set is equal to the largest measurement minus the smallest measurement.

The range is easy to compute and easy to understand, but it is a rather insensitive measure of data variation when the data sets are large. This is because two data sets can have the same range and be vastly different with respect to data variation. This phenomenon is demonstrated in Figure 2.5. Both distributions of data shown in the figure have the same range, but most of the measurements in data set 2 tend to concentrate near the center of the distribution. Consequently, the data are much less variable than the data in set 1. Thus, you can see that the range does not detect differences in data variation for some large data sets.

TABLE 2.4

| | SAMPLE 1 | SAMPLE 2 |
|---|---|---|
| MEASUREMENTS | 1, 2, 3, 4, 5 | 2, 3, 3, 3, 4 |
| MEAN | $\bar{x} = \dfrac{1 + 2 + 3 + 4 + 5}{5} = \dfrac{15}{5} = 3$ | $\bar{x} = \dfrac{2 + 3 + 3 + 3 + 4}{5} = \dfrac{15}{5} = 3$ |
| DISTANCES OF MEASUREMENT VALUES FROM $\bar{x}$ | $(1-3), (2-3), (3-3), (4-3), (5-3)$ or $-2, -1, 0, 1, 2$ | $(2-3), (3-3), (3-3), (3-3), (4-3)$ or $-1, 0, 0, 0, 1$ |

Let us see if we can find a measure of data variation that is more sensitive than the range. Consider the two samples in Table 2.4; each has five measurements (we have ordered the numbers for convenience). You will note that both samples have a mean of 3 and that we have also calculated the distance between each measurement and the

FIGURE 2.6
DOT DIAGRAMS FOR
TWO DATA SETS

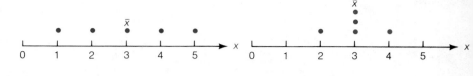

(a) Sample 1                                    (b) Sample 2

mean. What information do these distances contain? If they tend to be large in magnitude, as in sample 1, the data are spread out, or highly variable. If the distances are mostly small, as in sample 2, the data are clustered around the mean, $\bar{x}$, and therefore do not exhibit much variability. You can see that these distances, displayed graphically in Figure 2.6, provide information about the variability of the sample measurements.

The next step is to condense the information in these distances into a single numerical measure of variability. Averaging the distances from $\bar{x}$ will not help because the negative and positive distances cancel, i.e., the sum of the deviations (and thus the average deviation) is always equal to 0.

Two methods come to mind for dealing with the fact that positive and negative distances from the mean cancel. The first is to treat all the distances as though they were positive, ignoring the sign of the negative distances. We will not pursue this line of thought because the resulting measure of variability (the mean of the absolute values of the distances) is difficult to interpret. A second method of eliminating the minus signs associated with the distances is to square them. The quantity we can calculate from the squared distances will provide a meaningful description of the variability of a data set.

To use the squared distances calculated from a data set, we first calculate the sample variance:

DEFINITION 2.9
The sample variance for a sample of $n$ measurements is equal to the sum of the squared distances from the mean divided by $(n - 1)$. In symbols, using $s^2$ to represent the sample variance,

$$s^2 = \frac{\Sigma (x - \bar{x})^2}{n - 1}$$

Referring to the two samples in Table 2.4, you can calculate the variance for sample 1 as follows:

$$s^2 = \frac{(1 - 3)^2 + (2 - 3)^2 + (3 - 3)^2 + (4 - 3)^2 + (5 - 3)^2}{5 - 1}$$

$$= \frac{4 + 1 + 0 + 1 + 4}{4} = 2.5$$

2 METHODS FOR DESCRIBING SETS OF DATA

The second step in finding a meaningful measure of data variability is to calculate the **standard deviation** of the data set:

---

DEFINITION 2.10
The **sample standard deviation**, $s$, is defined as the positive square root of the sample variance, $s^2$. Thus,

$$s = \sqrt{s^2} = \sqrt{\frac{\Sigma (x - \bar{x})^2}{n - 1}}$$

---

The corresponding quantity, the population standard deviation, will be denoted by $\sigma$ (lowercase sigma). The population variance will therefore be denoted by $\sigma^2$. The symbols for the variances and standard deviations are summarized in the box.

---

$s^2 = $ Sample variance

$\sigma^2 = $ Population variance

$s = $ Sample standard deviation

$\sigma = $ Population standard deviation

---

Notice that in contrast to the variance, the standard deviation is expressed in the original units of measurement. For example, if the original measurements are in dollars, the standard deviation will be expressed in dollars. Second, you may wonder why we use the divisor $(n - 1)$ instead of $n$ when calculating the sample variance. This is because by using the divisor $(n - 1)$ you obtain a better estimate of $\sigma^2$ than you do by dividing the sum of the squared distances by $n$. Since we will ultimately want to use sample statistics to make inferences about the corresponding population parameters, $(n - 1)$ is preferred to $n$ when defining the sample variance.

EXAMPLE 2.5    Calculate the standard deviation of the following sample: 2, 3, 3, 3, 4.

Solution    For this data set, $\bar{x} = 3$. Then,

$$s = \sqrt{\frac{(2 - 3)^2 + (3 - 3)^2 + (3 - 3)^2 + (3 - 3)^2 + (4 - 3)^2}{5 - 1}}$$

$$= \sqrt{\frac{2}{4}} = \sqrt{0.5} = 0.71$$

Example 2.5 may have raised two thoughts in your mind. First, calculating $s^2$ and $s$ can be very tedious if $\bar{x}$ is a number that contains a large number of significant figures or if there are a large number of measurements in a data set. Second, we have not explained how a sample standard deviation can be used to describe the variability of a data set. Fortunately, we have an easier method for calculating $s^2$ and $s$. This method is demonstrated in Example 2.6. The interpretation of $s$ will be the subject of Section 2.5.

EXAMPLE 2.6

Calculate the sample variance, $s^2$, and the sample standard deviation, $s$, for the 100 gas mileage readings given in Table 2.1.

Solution

Recall that $\bar{x} = 36.994$. Needless to say, it would be a formidable task to calculate $\Sigma (x - \bar{x})^2$ for all 100 readings. Instead, the calculation can be done in the following manner:

---

**SHORTCUT FORMULA FOR SAMPLE VARIANCE**

$$s^2 = \frac{\left(\begin{array}{c}\text{Sum of squares of}\\ \text{sample measurements}\end{array}\right) - \dfrac{(\text{Sum of sample measurements})^2}{n}}{n - 1}$$

$$= \frac{\Sigma x^2 - \dfrac{(\Sigma x)^2}{n}}{n - 1}$$

---

**Step 1.** Calculate the sum of the squared measurements. With the aid of a calculator, we obtain

$$\Sigma x^2 = 137{,}434.38$$

**Step 2.** Calculate the square of the sum of the measurements.

$$(\Sigma x)^2 = (3{,}699.4)^2 = 13{,}685{,}560.36$$

**Step 3.** Find $\Sigma (x - \bar{x})^2 = \Sigma x^2 - [(\Sigma x)^2 / n]$, the numerator of $s^2$ in the shortcut formula.

$$\Sigma (x - \bar{x})^2 = \Sigma x^2 - \frac{(\Sigma x)^2}{n}$$

$$= 137{,}434.38 - \frac{13{,}685{,}560.36}{100} = 578.7764$$

We can now calculate $s^2$ by dividing this quantity by $(n - 1)$.

$$s^2 = \frac{578.7764}{99} = 5.85$$

$$s = +\sqrt{5.85} = 2.42$$

Note that this formula requires only the sum of the sample measurements, $\Sigma x$, and the sum of the squares of the sample measurements, $\Sigma x^2$. Be careful when you calculate these two sums. Rounding the values of $x^2$ that appear in $\Sigma x^2$ or rounding the quantity $(\Sigma x)^2 / n$, can lead to substantial errors in the calculation of $s^2$.

Learning the mechanics

**2.21.** Calculate the range, variance, and standard deviation for the following samples:

   a.  3, 0, 1, 2, 1, 1
   b.  1, 3, 4, 3, 2, 0, 1
   c.  10, 6, −4, 4, 7, 9, 10, −2, 7
   d.  0, 2, 0, 0, −1, 1, −2, 1, 0, −1, 1, −1, 0, −3, −2, −1, 0, 1

**2.22.** Calculate the variance and standard deviation for samples where:

   a.  $n = 10$, $\Sigma x^2 = 100$, $\Sigma x = 7$
   b.  $n = 40$, $\Sigma x^2 = 500$, $\Sigma x = 100$
   c.  $n = 20$, $\Sigma x^2 = 125$, $\Sigma x = 50$

**2.23.** Calculate the range, variance, and standard deviation for the following samples:

   a.  100, 102, 104, 101, 102
   b.  100, 4, 7, 96
   c.  100, 4, 7, 30

Applying the concepts

**2.24.** Consider the following sample of five measurements:

   1, 0, 2, 4, 1

   a.  Calculate the range, $s^2$, and $s$.
   b.  Add 3 to each measurement and repeat part a.
   c.  Subtract 4 from each measurement and repeat part a.
   d.  Considering your answers to parts a, b, and c, what seems to be the effect on the variability of a data set by adding the same number to or subtracting the same number from each measurement?

**2.25.** The final grades given by two professors in introductory statistics courses have been carefully examined. The students in the first professor's class had a grade-point average of 3.0 and a standard deviation of 0.2. Those in the second professor's class had grade points with an average of 3.0 and a standard deviation of 1.0. If you had a choice, which professor would you take for this course? Explain.

**2.26.** Consider the following two samples:

   Sample 1:  10, 0, 1, 9, 10, 0
   Sample 2:  0, 5, 10, 5, 5, 5

   a.  Examine both samples and identify the one that you believe has the greater variability.
   b.  Calculate the range for each sample. Does this agree with your answer to part a? Explain.
   c.  Calculate the variance for each sample. Does this agree with your answer to part a? Explain.
   d.  Which of the two, the range or the variance, provides a better measure of variability? Why?

## 2.5
## INTERPRETING THE STANDARD DEVIATION

As we have seen, if we are comparing the variability of two samples selected from a population, the sample with the larger standard deviation is the more variable of the two. Thus, we know how to interpret the standard deviation on a relative or comparative basis, but we have not explained how it provides a measure of variability for a single sample.

To better understand how the standard deviation provides a measure of variability of a data set, consider a specific data set and answer the following questions: How many measurements are within 1 standard deviation of the mean? How many measurements are within 2 standard deviations? For example, consider the 100 mileage per gallon readings given in Table 2.1.

Recall that $\bar{x} = 36.994$ and $s = 2.42$. Then

$$\bar{x} + s = 39.414 \qquad \bar{x} + 2s = 41.834$$
$$\bar{x} - s = 34.574 \qquad \bar{x} - 2s = 32.154$$

If we examine the data, we find that 68 of the 100 measurements, or 68% of the measurements, are in the interval

$$\bar{x} - s \quad \text{to} \quad \bar{x} + s$$

Similarly, we find that 96, or 96%, of the 100 measurements are in the interval

$$\bar{x} - 2s \quad \text{to} \quad \bar{x} + 2s$$

These observations identify criteria for interpreting a standard deviation that apply to any set of data, whether a population or a sample. The criteria, expressed as a mathematical theorem and as a rule of thumb, are presented in Tables 2.5 and 2.6. In the tables, we give two sets of answers to the questions of how many measurements fall within 1, 2, and 3 standard deviations of the mean. The first, which applies to *any* set of data, is derived from a theorem proved by the Russian mathematician Chebyshev. The second, which applies only to mound-shaped distributions of data, is based upon empirical evidence that has accumulated over time. The frequency histogram of a mound-shaped sample is approximately symmetric, with a clustering of measurements about the midpoint of the distribution (the mean, median, and mode should all be about the same), tailing off rapidly as we move away from the center of the histogram. Thus, the histogram will have the appearance of a mound or bell, as shown in Figure 2.7. The percentages given for the various intervals in Table 2.6 provide remarkably good approximations even when the distribution of the data is slightly skewed or asymmetric.

## TABLE 2.5
## AN AID TO INTERPRETATION OF A STANDARD DEVIATION: CHEBYSHEV'S RULE

CHEBYSHEV'S RULE

Chebyshev's rule applies to any sample of measurements, regardless of the shape of the frequency distribution:

a. It is possible that very few of the measurements will fall within 1 standard deviation of the mean ($\bar{x} - s$ to $\bar{x} + s$).

b. At least $\frac{3}{4}$ of the measurements will fall within 2 standard deviations of the mean ($\bar{x} - 2s$ to $\bar{x} + 2s$).

c. At least $\frac{8}{9}$ of the measurements will fall within 3 standard deviations of the mean ($\bar{x} - 3s$ to $\bar{x} + 3s$).

| THE EMPIRICAL RULE |
| --- |
| The Empirical rule is a rule of thumb that applies to samples with frequency distributions that are mound-shaped: |
| a. Approximately 68% of the measurements will fall within 1 standard deviation of the mean ($\bar{x} - s$ to $\bar{x} + s$). <br> b. Approximately 95% of the measurements will fall within 2 standard deviations of the mean ($\bar{x} - 2s$ to $\bar{x} + 2s$). <br> c. Essentially all the measurements will fall within 3 standard deviations of the mean ($\bar{x} - 3s$ to $\bar{x} + 3s$). |

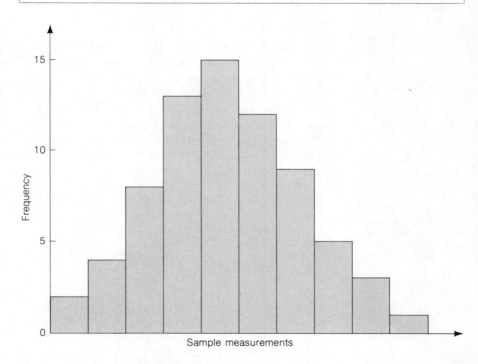

EXAMPLE 2.7

A sample of earnings per share data for thirty *Fortune* 500 companies is

| | | | | | |
| --- | --- | --- | --- | --- | --- |
| 1.97 | 0.60 | 4.02 | 3.20 | 1.15 | 6.06 |
| 4.44 | 2.02 | 3.37 | 3.65 | 1.74 | 2.75 |
| 3.81 | 9.70 | 8.29 | 5.63 | 5.21 | 4.55 |
| 7.60 | 3.16 | 3.77 | 5.36 | 1.06 | 1.71 |
| 2.47 | 4.25 | 1.93 | 5.15 | 2.06 | 1.65 |

The mean and standard deviation of these data are 3.74 and 2.20, respectively. Calculate the fraction of the thirty measurements in the intervals $\bar{x} \pm s$, $\bar{x} \pm 2s$, and $\bar{x} \pm 3s$, and compare the results with those in Tables 2.5 and 2.6.

Solution

We first form the interval

$$(\bar{x} - s, \bar{x} + s) = (3.74 - 2.20, 3.74 + 2.20) = (1.54, 5.94)$$

A check of the measurements shows that twenty-three measurements are within this 1 standard deviation interval around the mean. This represents $^{23}/_{30} \approx 77\%$ of the sample measurements.

The next interval of interest is

$$(\bar{x} - 2s, \bar{x} + 2s) = (3.74 - 4.40, 3.74 + 4.40) = (-0.66, 8.14)$$

All but two measurements are within this interval, so approximately 93% are within 2 standard deviations of $\bar{x}$.

Finally, the 3 standard deviation interval around $\bar{x}$ is

$$(\bar{x} - 3s, \bar{x} + 3s) = (3.74 - 6.60, 3.74 + 6.60) = (-2.86, 10.34)$$

All the measurements fall within 3 standard deviations of the mean.

These 1, 2, and 3 standard deviation percentages (77, 93, and 100) agree fairly well with the approximations of 68%, 95%, and 100% given by the Empirical rule (Table 2.6) for mound-shaped distributions. If you look at the frequency histogram for this data set in Figure 2.8, you will note that the distribution is not really mound-shaped, nor is it extremely skewed. Thus, we get reasonably good results from the mound-shaped approximations. Of course, we know from Chebyshev's rule (Table 2.5) that no matter what the shape of the distribution, we would expect at least 75% and 89% ($^{8}/_{9}$) of the measurements to lie within 2 and 3 standard deviations, respectively, of $\bar{x}$.

**FIGURE 2.8**
**HISTOGRAM FOR**
**EARNINGS PER SHARE**
**DATA**

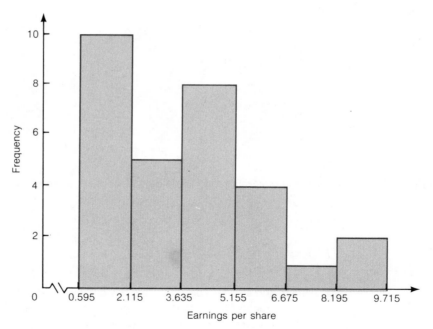

EXAMPLE 2.8

Chebyshev's rule and the Empirical rule (Tables 2.5 and 2.6) can be put to an immediate practical use as a check on the calculation of a standard deviation. Suppose you have a data set for which the smallest measurement is 20 and the largest is 80. You have calculated the standard deviation of the data set to be

$$s = 190$$

How can you use the information in Tables 2.5 and 2.6 to provide a rough check on your calculated value of $s$?

Solution

The larger the number of measurements in a data set, the greater will be the tendency for very large or very small measurements (extreme values) to appear in the data set. But from Tables 2.5 and 2.6 you know that most of the measurements (approximately 95% if the distribution is mound-shaped) will be within 2 standard deviations of the mean. And, regardless of how many measurements are in the data set, almost all of them will fall within 3 standard deviations of the mean. Consequently, we would expect the range to be equal to somewhere between 4 and 6 standard deviations, i.e., between $4s$ and $6s$. For the given data set, the range is

$$\text{Range} = \text{Largest measurement} - \text{Smallest measurement}$$
$$= 80 - 20 = 60$$

Then, if we let the range equal $6s$, we obtain

$$\text{Range} = 6s$$
$$60 = 6s$$
$$s = 10$$

Or, if we let the range equal $4s$ (see Figure 2.9), we obtain a larger (and more conservative) value for $s$, namely,

$$\text{Range} = 4s$$
$$60 = 4s$$
$$s = 15$$

FIGURE 2.9
THE RELATIONSHIP
BETWEEN THE RANGE
AND THE
STANDARD DEVIATION

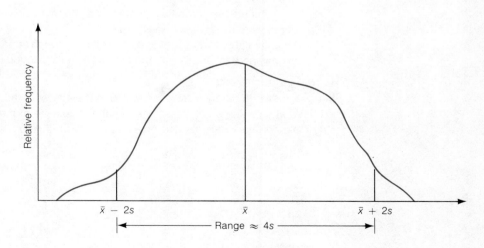

Now you can see that it does not make much difference whether you let the range equal 4s (which is more realistic for most data sets) or 6s (which is reasonable for large data sets). It is clear that your calculated value, $s = 190$, is too large, and you should check your calculations.*

Learning the mechanics

**2.27.** For any set of data, what can be said about the percentage of the measurements contained in each of the following intervals:

    a.   $\bar{x} - s$ to $\bar{x} + s$
    b.   $\bar{x} - 2s$ to $\bar{x} + 2s$
    c.   $\bar{x} - 3s$ to $\bar{x} + 3s$

**2.28.** For a set of data with a mound-shaped relative frequency distribution, what can be said about the percentage of the measurements contained in each of the following intervals:

    a.   $\bar{x} - s$ to $\bar{x} + s$
    b.   $\bar{x} - 2s$ to $\bar{x} + 2s$
    c.   $\bar{x} - 3s$ to $\bar{x} + 3s$

**2.29.** The following is a sample of 25 measurements:

| | | | | |
|---|---|---|---|---|
| 7 | 6 | 6 | 11 | 8 |
| 9 | 11 | 9 | 10 | 8 |
| 7 | 7 | 5 | 9 | 10 |
| 7 | 7 | 7 | 7 | 9 |
| 12 | 10 | 10 | 8 | 6 |

    a.   Compute $\bar{x}$, $s^2$, and $s$ for this sample.
    b.   Count the number of measurements in the intervals $\bar{x} \pm s$, $\bar{x} \pm 2s$, $\bar{x} \pm 3s$. Express the number of measurements in each interval as a percentage of the total number of measurements.
    c.   Compare the percentages found in part b to the percentages given by the Empirical rule and Chebyshev's rule.
    d.   Calculate the range, and use it to obtain a rough approximation for $s$. Does this compare favorably with the actual value for $s$ found in part a?

Applying the concepts

**2.30.** A research cardiologist is interested in the age when adult males suffer their first heart attack. The cardiologist takes a random sample of the medical records of thirty male coronary patients and obtains the following results (recorded in years):

| | | | | | |
|---|---|---|---|---|---|
| 51 | 64 | 43 | 54 | 52 | 38 |
| 45 | 70 | 75 | 71 | 49 | 42 |
| 62 | 55 | 65 | 63 | 40 | 61 |
| 49 | 57 | 58 | 67 | 53 | 54 |
| 44 | 59 | 54 | 42 | 60 | 50 |

*In the exercises and examples that follow, we will use Range $\approx 4s$, or, equivalently, $s \approx$ Range/4.

a. Compute the sample mean, variance, and standard deviation for the data.

b. What percentages of the measurements would you expect to find in the intervals $\bar{x} \pm s$, $\bar{x} \pm 2x$, and $\bar{x} \pm 3s$?

c. Count the number of measurements that actually fall in each interval of part b, and express each interval count as a percentage of the total number of measurements. Compare these to the expected percentages indicated in part b.

**2.31.** For each day of last year, the number of vehicles passing through a certain intersection was recorded by a city engineer. One objective of this study was to determine the percentage of days that more than 425 vehicles used the intersection. If the mean for the data was 375 vehicles per day and the standard deviation was 25 vehicles:

a. What can be said about the percentage of days that more than 425 vehicles used the intersection? Assume that nothing is known about the shape of the relative frequency distribution for the data.

b. What is your answer to part a if you know that the relative frequency distribution for the data is mound-shaped?

**2.32.** A buyer for a lumber company must decide whether to buy a piece of land containing 5,000 pine trees. If 1,000 of the trees are at least 40 feet tall, the buyer will purchase the land; otherwise, he will not. The owner of the land reports that the height of the trees has a mean of 30 feet and a standard deviation of 3 feet. Based on this information, what is the buyer's decision?

**2.33.** Suppose a light bulb manufacturer claims that the mean lifetime of its bulbs is 35 hours. Assume you have prior knowledge that the bulb lifetimes have a mound-shaped distribution with standard deviation 5 hours. If the manufacturer's claim is true, approximately what percentage of light bulbs will burn out in less than 20 hours? Suppose you randomly select one of the bulbs and it burns out in less than 20 hours. Do you suspect the manufacturer's claim is incorrect? Explain your answer.

**2.34.** A chemical company produces a substance composed of 98% cracked corn particles and 2% zinc phosphide for use in controlling rat populations in sugar cane fields. Production must be carefully controlled to maintain the 2% zinc phosphide because too much zinc phosphide will cause damage to the sugar cane and too little will be ineffective in controlling the rat population. Records from past production indicate that the distribution of the actual percentage of zinc phosphide present in the substance is approximately mound-shaped, with a mean of 2.0% and a standard deviation of 0.08%. If the production line is operating correctly, approximately what proportion of batches from a day's production will contain less than 1.84% zinc phosphide? Suppose one batch chosen randomly actually contains 1.80% zinc phosphide. Does this indicate that there is too little zinc phosphide in today's production? Explain your reasoning.

**2.35.** Solar energy is considered by many to be the energy of the future. A recent survey was taken to compare the cost of solar energy to the cost of gas or electric energy. Results of the survey revealed that the distribution of the amount of the monthly utility bill of a three bedroom house using gas or electric energy had a mean of $125 and a standard deviation of $10.

    a.   If nothing is known about the distribution of the amounts of monthly utility bills, what can you say about the fraction of all three bedroom homes using gas or electric energy having bills between $95 and $155?

    b.   If it is reasonable to assume that the distribution of the amounts of monthly utility bills is mound-shaped, approximately what proportion of three bedroom homes would have monthly bills less than $135?

    c.   Suppose that three houses with solar energy units had the following monthly utility bills: $101, $98, $104. Does this suggest that solar energy units might result in lower utility bills? Explain. [*Note:* We will present a statistical method in Chapter 6 for testing this conjecture.]

**2.36.**   The following data sets have been invented to demonstrate that the lower bounds given by Chebyshev's rule are appropriate. Notice that the data are contrived and would not be encountered in a real-life problem.

    a.   Consider a data set that contains ten 0's, two 1's, and ten 2's. Calculate $\bar{x}$, $s^2$, and $s$. What percentage of the measurements are in the interval $\bar{x} \pm s$? Compare this to Chebyshev's rule.

    b.   Consider a data set that contains five 0's, thirty-two 1's, and five 2's. Calculate $\bar{x}$, $s^2$, and $s$. What percentage of the measurements are in the interval $\bar{x} \pm 2s$? Compare this to Chebyshev's rule.

    c.   Consider a data set that contains three 0's, fifty 1's, and three 2's. Calculate $\bar{x}$, $s^2$, and $s$. What percentage of the measurements are in the interval $\bar{x} \pm 3s$? Compare this to Chebyshev's rule.

    d.   Draw a histogram for each of the data sets in parts a, b, and c. What do you conclude from these graphs and the answers to parts a, b, and c?

## 2.6 MEASURES OF RELATIVE STANDING

As we have seen, numerical measures of central tendency and variability describe the general nature of a data set (either a sample or a population). We may also be interested in describing the relative quantitative location of a particular measurement within a data set. Descriptive measures of the relationship of a measurement to the rest of the data are called measures of relative standing.

One measure of the relative standing of a measurement is its percentile ranking. For example, suppose you scored an 80 on an examination and you want to know how you fared in comparison with others in your class. If the instructor tells you that you scored in the 90th percentile, it means that 90% of the examination grades were less than yours and 10% were greater. Thus, if the examination scores were described by the relative frequency histogram in Figure 2.10, the 90th percentile would be located at a point such that 90% of the total area under the relative frequency histogram lies below the 90th percentile and 10% lies above. If the instructor tells you that you scored in the 50th percentile (the median of the data set), 50% of the examination grades would be less than yours and 50% would be greater.

Percentile rankings are of practical value only for large data sets. Finding them involves a process similar to the one used in finding a median. The measurements are ranked in order and a rule is selected, similar to that used in locating a median,

FIGURE 2.10
LOCATION OF 90TH
PERCENTILE FOR
EXAMINATION GRADES

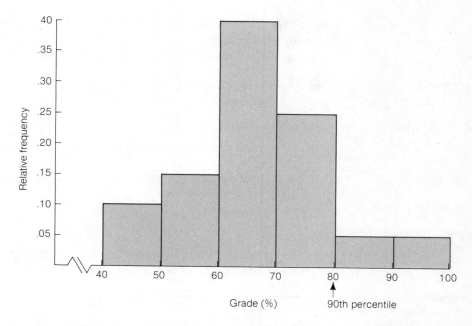

to define the location of each percentile. Since we are primarily interested in interpreting the percentile rankings of measurements (rather than finding particular percentiles for a data set), we will define the $p$th percentile of a data set as follows:

> **DEFINITION 2.11**
> For any set of $n$ measurements (arranged in ascending or descending order), the $p$th percentile is a number such that $p\%$ of the measurements fall below the $p$th percentile and $(100 - p)\%$ fall above it.

Another measure of relative standing in popular use is the $z$-score. As you can see in Definition 2.12, the $z$-score makes use of the mean and standard deviation of the data set in order to specify the relative location of a measurement:

> **DEFINITION 2.12**
> The sample $z$-score for a measurement $x$ is
>
> $$z = \frac{x - \bar{x}}{s}$$
>
> The population $z$-score for a measurement $x$ is
>
> $$z = \frac{x - \mu}{\sigma}$$

Note that the z-score is calculated by subtracting $\bar{x}$ (or $\mu$) from the measurement $x$ and then dividing the result by $s$ (or $\sigma$). The final result, the z-score, represents the distance between a given measurement $x$ and the mean expressed in standard deviations.

EXAMPLE 2.9

Suppose a sample of annual incomes for 200 steelworkers is taken. The mean and standard deviation are

$$\bar{x} = \$17,000 \qquad s = \$2,000$$

Suppose Joe Smith's annual income is $15,000. What is his sample z-score?

Solution

Joe Smith's annual income lies below the mean income of the 200 steelworkers (Figure 2.11). We compute

$$z = \frac{x - \bar{x}}{s} = \frac{\$15,000 - \$17,000}{\$2,000} = -1.0$$

which tells us that Joe Smith's annual income is 1.0 standard deviation *below* the sample mean, or, in short, his sample z-score is $-1.0$.

**FIGURE 2.11**
**ANNUAL INCOME**
**OF STEELWORKERS**

The numerical value of the z-score reflects the relative standing of the measurement. A large positive z-score implies that the measurement is larger than almost all other measurements, while a large negative z-score indicates that the measurement is smaller than almost every other measurement. If a z-score is 0 or near 0, the measurement is located near the mean of the sample or population.

We can be more specific if we know that the frequency distribution of the measurements is mound-shaped. In this case, the following interpretation of the z-scores can be given:

---

INTERPRETATION OF z-SCORES FOR MOUND-SHAPED
DISTRIBUTIONS OF DATA

**1.** Approximately 68% of the measurements will have a z-score between $-1$ and 1.
**2.** Approximately 95% of the measurements will have a z-score between $-2$ and 2.
**3.** All or almost all the measurements will have a z-score between $-3$ and 3.

---

Note that the above interpretation of z-scores is identical to that given by the Empirical rule for samples from mound-shaped distributions. The statement that a measurement falls in the interval $(\mu - \sigma)$ to $(\mu + \sigma)$ is equivalent to the statement that a measurement has a population $z$-score between $-1$ and 1, since all measurements between $(\mu - \sigma)$ and $(\mu + \sigma)$ are within 1 standard deviation of $\mu$ (see Figure 2.12).

**FIGURE 2.12**
POPULATION $z$-SCORES FOR A MOUND-SHAPED DISTRIBUTION

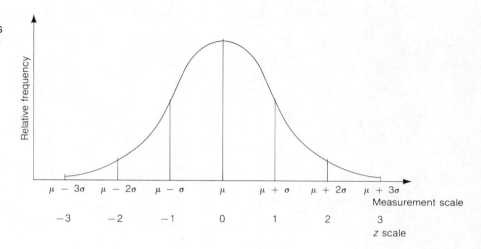

We will end this section with an example and a case study that indicate how the $z$-scores may be used to accomplish our primary objective: the use of sample information to make inferences about the population.

EXAMPLE 2.10

Suppose a female bank executive believes that her salary is low as a result of sex discrimination. To try to substantiate her belief, she collects information on the salaries of her male counterparts in the banking business. She finds that the distribution of salaries for males has a mean of $23,000 and a standard deviation of $1,000. Her salary is $19,500. Does this information substantiate her claim of sex discrimination?

Solution

The analysis might proceed as follows: First, we calculate the $z$-score for the woman's salary with respect to the sample of her male counterparts. Thus,

$$z = \frac{x - \bar{x}}{s} = \frac{\$19{,}500 - \$23{,}000}{\$1{,}000} = -3.5$$

The implication is that the woman's salary is 3.5 standard deviations *below* the mean of the male salary distribution. Furthermore, if a check of the male salary data shows that the frequency distribution is mound-shaped, we can infer that very few salaries in this distribution should have a $z$-score less than $-3$, as shown in Figure 2.13 (page 44). Therefore, a $z$-score of $-3.5$ either represents a measurement from a distribution different from the male salary distribution, or it represents a very unusual (highly improbable) measurement for the male salary distribution.

Well, which of the two situations do you think prevails? Do you think the woman's salary is simply an unusually low one in the distribution of salaries, or do you think her claim of salary discrimination is justified? Most people would probably conclude that her salary does not come from the male salary distribution. However, the careful investigator should require more information before inferring sex discrimination as the cause. We would want to know more about the sample collection technique the woman used, and more about her competence at her job. Also, perhaps other factors, such as the length of employment, should be considered in the analysis.

FIGURE 2.13
MALE SALARY
DISTRIBUTION

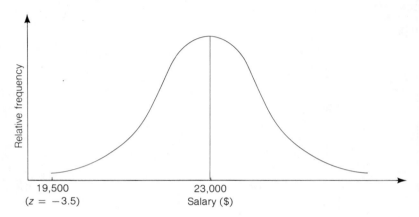

The above exemplifies an approach to statistical inference that might be called the rare event approach. An experimenter formulates some hypothesis about a population relative frequency distribution. Then a sample of measurements is drawn from the population. If the experimenter finds it unlikely that the sample came from the hypothesized distribution, the hypothesis is concluded to be false. Thus, in Example 2.10, the woman believes her salary reflects sex discrimination. She hypothesizes that her salary should be just another measurement in the distribution of her male counterparts' salaries if no discrimination exists. However, it was so unlikely that the sample (in this case, her salary) came from the male frequency distribution that she rejects that hypothesis, concluding that the distribution from which her salary was drawn is different from the distribution for the males.

This rare event approach to inference-making will be further discussed in later chapters. Proper application of the approach requires a knowledge of probability, the subject of our next chapter.

CASE STUDY 2.3
STATISTICS AND AIR
QUALITY STANDARDS

H. E. Neustadter and S. M. Sidik (1974) discuss calculational procedures for obtaining an air quality standard (AQS) to use in judging whether a company is exceeding air pollution standards. A simplification of a method they discuss is outlined below:

**1.** Obtain a large set of daily measurements from a company known to be complying with established air pollution standards.
**2.** If necessary, transform the data so that it forms a mound-shaped distribution to which the aids in Table 2.6 may be applied. For example, if the logarithm of each measurement is calculated for typical air quality data, the new set of data often forms a mound-shaped distribution.
**3.** Calculate the mean, $\bar{x}$, and standard deviation, $s$, of the mound-shaped set of data

2 METHODS FOR DESCRIBING SETS OF DATA

and use this information to obtain the AQS. For example, if it is desired to obtain a standard that would be exceeded on approximately 2.5% of the industry's operating days, calculate

$$AQS = \bar{x} + 2s$$

The number of standard deviations to be added to $\bar{x}$ can be adjusted to reflect the percentage of days a company is to be permitted to exceed the standard. As Neustadter and Sidik point out, "A major objective of air quality monitoring is often to determine compliance with air quality standards which may in part consist of a 24-hour level not to be exceeded more than once a year." They show that one should calculate the mean plus approximately 2.7 standard deviations to meet this standard.

EXERCISES

Learning the mechanics

**2.37.** Compute the $z$-score corresponding to each of the following values of $x$.

    a.   $x = 40, s = 5, \bar{x} = 30$
    b.   $x = 90, \mu = 89, \sigma = 2$
    c.   $\mu = 50, \sigma = 5, x = 50$
    d.   $s = 4, x = 20, \bar{x} = 30$
    e.   In parts a–d, state whether the $z$-score locates $x$ within a sample or a population.
    f.   In parts a–d, state whether each value of $x$ lies above or below the mean and by how many standard deviations.

**2.38.** Give the percentage of measurements in a data set that are above and below each of the following percentiles:

    a.   75th percentile
    b.   50th percentile
    c.   20th percentile
    d.   84th percentile

**2.39.** Compare $z$-scores to decide which of the following $x$ values lies the greatest distance above the mean; the greatest distance below the mean.

    a.   $x = 100, \mu = 50, \sigma = 25$
    b.   $x = 1, \mu = 4, \sigma = 1$
    c.   $x = 0, \mu = 200, \sigma = 100$
    d.   $x = 10, \mu = 5, \sigma = 3$

Applying the concepts

**2.40.** Virtually every student is concerned with his or her test score relative to the scores of the rest of the class. Suppose that you and five of your friends are taking a psychology course and the six of you received the following grades:

    83, 70, 74, 89, 97, 85

The instructor in the class tells you that the average and standard deviation for all the test scores are 85 and 5, respectively.

    a.   Calculate a $z$-score for each of the six test scores.

b. For each test score, state whether it lies above or below the mean and by how many standard deviations.

c. Do you think any of these test scores is the highest or the lowest in the class? Explain.

**2.41.** The distribution of IQ's is approximately mound-shaped, with mean 100 and variance 225. Approximately what percentage of people would you expect to have IQ's over 145? Between 70 and 115? Find the approximate value of the 16th percentile of the distribution of IQ's.

**2.42.** The distribution of scores on a nationally administered college achievement test has a median of 520 and a mean of 540.

a. Explain why it is possible for the mean to exceed the median for this distribution of measurements.

b. Suppose you are told that the 90th percentile is 660. What does this mean?

c. Suppose you are told that you scored at the 94th percentile. Interpret this statement.

**2.43.** Many firms use on-the-job training to teach their employees computer programming. Suppose you work in the Personnel Department of a firm that just finished training a group of its employees to program and you have been requested to review the performance of one of the trainees on the final test that was given to all trainees. The mean and standard deviation of the test scores are 80 and 5, respectively, and the distribution of scores is mound-shaped.

a. The employee in question scored 65 on the final test. Compute the employee's $z$-score.

b. Approximately what percentage of the trainees will have $z$-scores equal to or less than the employee of part a?

c. If a trainee were arbitrarily selected from those who had taken the final test, is it more likely that he or she would score 90 or above, or 65 or below?

**2.44.** A city librarian claims that books have been checked out an average of 7 (or more) times in the last year. You suspect he has exaggerated the checkout rate (book usage) and that the mean number of checkouts per book per year is, in fact, less than 7. Using the card catalog, we randomly select one book and find that it has been checked out 4 times in the last year. Assume that we know from previous records that the standard deviation of the number of checkouts per book per year is 1.

a. If the mean number of checkouts per book per year really is 7, what is the $z$-score corresponding to 4?

b. Considering your answer to part a, is there evidence to indicate that the librarian's claim is incorrect?

c. If you knew that the distribution of the number of checkouts were mound-shaped, would your answer to part b change? Explain.

d. If the standard deviation of the number of checkouts per book per year were 2 (instead of 1), would your answers to parts b and c change? Explain.

While it may be true in telling a story that "a picture is worth a thousand words," it is also true that pictures can be used to convey a colored and distorted message to the viewer. So the old adage applies: "Let the buyer (reader) beware." Examine relative frequency histograms and, in general, all graphical descriptions with care.

We will mention a few of the pitfalls to watch for when interpreting a chart or graph. But first we should mention the time series graph, which is often the object of distortion. This type of graph records the behavior of some variable of interest recorded over time. Examples of variables commonly graphed as time series abound: economic indices, the United States food surplus, defense spending, presidential popularity index, etc. Since time series graphs often appear in newspapers or magazines, we will use some of them to demonstrate several ways pictures are commonly distorted.

One common way to change the impression conveyed by a graph is to change the scale on the vertical axis, the horizontal axis, or both. For example, Figure 2.14 is a bar graph that shows the market share of sales for a company for each of the years 1977 to 1982. If you want to show that the change in firm A's market share over time is moderate, you should pack in a large number of units per inch on the vertical axis. That is, make the distance between successive units on the vertical scale small, as shown in Figure 2.14. You can see that a change in the firm's market share over time is barely apparent.

**FIGURE 2.14
FIRM A'S MARKET
SHARE FROM
1977 TO 1982
—PACKED
VERTICAL AXIS**

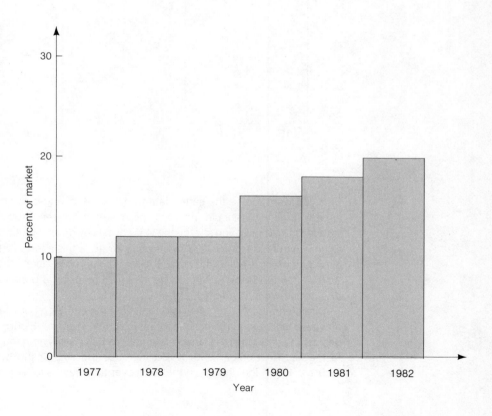

FIGURE 2.15
FIRM A'S MARKET
SHARE FROM
1977 TO 1982
—STRETCHED
VERTICAL AXIS

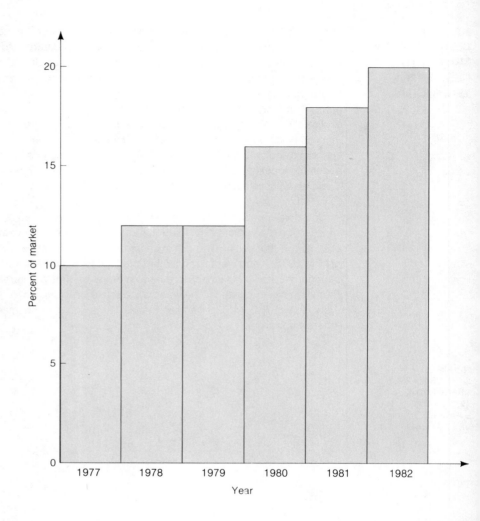

If you want to use the same data to make the changes in firm A's market share appear large, you should increase the distance between successive units on the vertical axis. That is, you stretch the vertical axis by graphing only a few units per inch, as shown in Figure 2.15. A telltale sign of stretching is a long vertical axis, but this is often hidden by starting the vertical axis at some point above 0, as shown in Figure 2.16(a). Or, the same effect can be achieved by using a broken line for the vertical axis, as shown in Figure 2.16(b).

Stretching the horizontal axis (increasing the distance between successive units) may also lead you to incorrect conclusions. For example, Figure 2.17(a) on page 50 depicts the change in the Gross National Product (GNP) from the first quarter of 1979 to the last quarter of 1980. If you increase the size of the horizontal axis, as in Figure 2.17(b), the change in the GNP over time seems to be less pronounced.

FIGURE 2.16
CHANGES IN MONEY
SUPPLY FROM
JANUARY TO
JUNE 1980

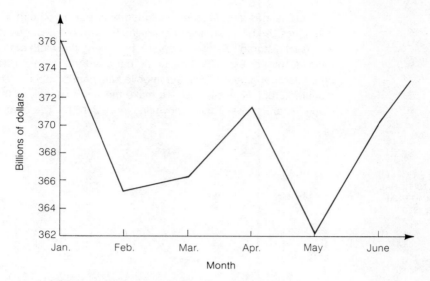

(a) Vertical axis started at a point greater than 0

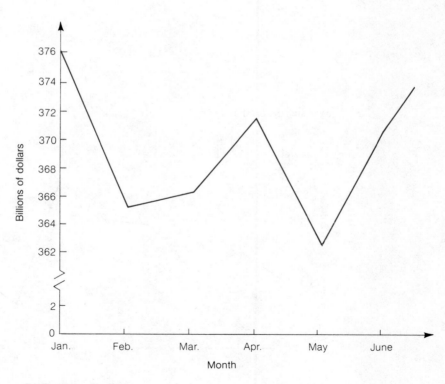

(b) Gap in vertical axis

The changes in categories indicated by a bar graph can also be emphasized or deemphasized by stretching or shrinking the vertical axis. Another method of achieving visual distortion with bar graphs is by making the width of the bars proportional to their height. For example, look at the bar chart in Figure 2.18(a), which depicts the percentage of a year's total automobile sales attributable to each of the four major manufacturers. Now suppose we make the width as well as the height grow as the market share grows. This is shown in Figure 2.18(b). The reader may tend to equate

FIGURE 2.17
GROSS NATIONAL
PRODUCT FROM
1979 TO 1980

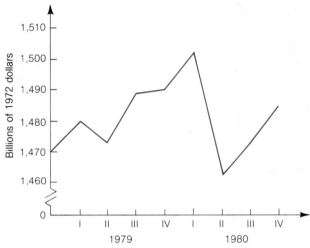

(a) Small horizontal axis

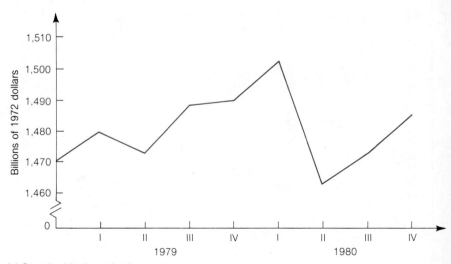

(b) Stretched horizontal axis

FIGURE 2.18
RELATIVE SHARE OF
THE AUTOMOBILE
MARKET FOR EACH OF
FOUR MAJOR
MANUFACTURERS

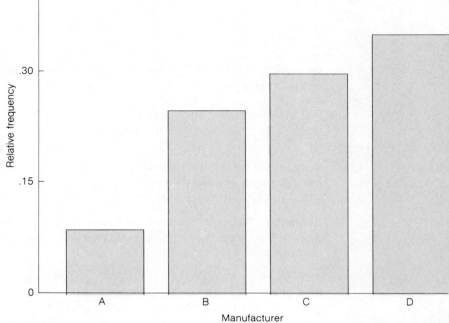

(a) Bar chart

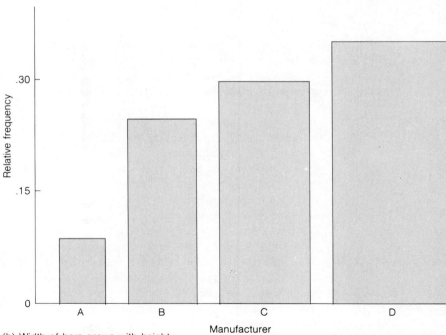

(b) Width of bars grows with height

the *area* of the bars with the relative market share of each manufacturer. In fact, the true relative market share is proportional only to the height of the bars.

We have presented only a few of the ways that graphs can be used to convey misleading pictures of phenomena. However, the lesson is clear. Examine all graphical descriptions of data with care. Particularly, check the axes and the size of the units on each axis. Ignore the visual changes and concentrate on the actual numerical changes indicated by the graph or chart.

The information in a data set can also be distorted by using numerical descriptive measures, as Example 2.11 indicates.

**EXAMPLE 2.11**    Suppose you are considering working for a small law firm that presently has a senior member and three junior members. You inquire as to the salary you could expect to earn if you join the firm. Unfortunately, you receive two answers:

**Answer A.**    The senior member tells you that an "average employee" earns $37,500.
**Answer B.**    One of the junior members later tells you that an "average employee" earns $25,000.

Which answer can you believe? The confusion exists because the phrase "average employee" has not been clearly defined. Suppose the four salaries paid are $25,000 for each of the three junior members and $75,000 for the senior member. Thus,

$$\bar{x} = \frac{3(\$25,000) + \$75,000}{4} = \frac{\$150,000}{4} = \$37,500$$

Median = $25,000

You can now see how the two answers were obtained. The senior member reported the mean of the four salaries, and the junior member reported the median. The information you received was distorted because neither person stated which measure of central tendency was being used.

Another distortion of information in a sample occurs when *only* a measure of central tendency is reported. Both a measure of central tendency and a measure of variability are needed to obtain an accurate mental image of a data set.

Suppose you want to buy a new car and are trying to decide which of two models to purchase. Since energy and economy are both important issues, you decide to purchase model A because its EPA mileage rating is 32 miles per gallon in-city, while the mileage rating for model B is only 30 miles per gallon in-city.

However, you may have acted too quickly. How much variability is associated with the ratings? As an extreme example, suppose that further investigation reveals that the standard deviation for model A mileages is 5 miles per gallon, while that for model B is only 1 mile per gallon. If the mileages form a mound-shaped distribution, then they might appear as shown in Figure 2.19. Note that the larger amount of variability associated with model A implies that more risk is involved in purchasing model A. That is, the particular car you purchase is more likely to have a mileage rating that will greatly differ from the EPA rating of 32 miles per gallon if you purchase model A, while

FIGURE 2.19
MILEAGE
DISTRIBUTIONS FOR
TWO CAR MODELS

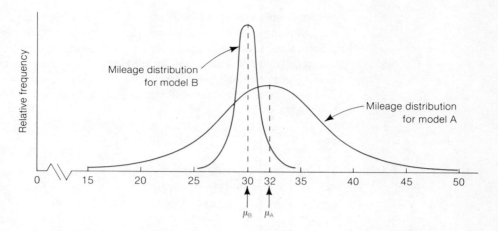

a model B car is not likely to vary from the 30 miles per gallon rating by more than 2 miles per gallon.

CASE STUDY 2.4
*CHILDREN OUT OF
SCHOOL IN AMERICA:*
MAKING AN UGLY
PICTURE LOOK WORSE

David L. Martin (1975) points out another method of distorting the truth with descriptive techniques in his article, ''Firsthand report: How flawed statistics can make an ugly picture look even worse.'' In his critique of the Children Defense Fund's (C.D.F.) 1973 report, *Children Out of School in America,* Martin quotes (italicized comments are Martin's):

•25 percent of the 16 and 17 year olds in the Portland, Me., Bayside East Housing Project were out of school. *Only eight children were surveyed; two were found to be out of school.*

•Of all the secondary school students who had been suspended more than once in census tract 22 in Columbia, S.C., 33 percent had been suspended two times and 67 percent had been suspended three or more times. *C.D.F. found only three children in that entire census tract who had been suspended; one child was suspended twice and the other two children, three or more times.*

•In the Portland Bayside East Housing Project, C.D.F. says that 50 percent of all the secondary school children who had been suspended more than once had been suspended three or more times.*The survey found two secondary school children had been suspended in that area; one of them had been suspended three or more times.*

In each of these examples the reporting of percentages instead of the numbers themselves is misleading. Any inference one would draw from the cited examples would not be reliable (we will see how to measure the reliability of estimated percentages in Chapter 6). In short, either the numbers alone should be reported instead of percentages, or, better yet, the report should state that the numbers were too small to report by region. If several regions were combined, the numbers (and percentages) would be more meaningful.

SUMMARY

Since we want to use sample data to make inferences about the population from which it is drawn, it is important for us to be able to describe the data. Graphical methods are important and useful tools for describing data sets. Our ultimate goal, however, is to use the sample to make inferences about the population. We are wary of using graphical techniques to accomplish this goal, since they do not lend themselves to a measure of the reliability for an inference. We therefore developed numerical measures to describe a data set.

These numerical methods for describing quantitative data sets can be grouped as follows:

1. Measures of central tendency
2. Measures of variability

The measures of central tendency we presented were the mean, median, and mode. The relationship between the mean and median provides information about the skewness of the frequency distribution. For making inferences about the population, the sample mean will usually be preferred to the other measures of central tendency. The range, variance, and standard deviation all represent numerical measures of variability. Of these, the variance and standard deviation are in most common use, especially when the ultimate objective is to make inferences about a population.

The mean and standard deviation may be used to make statements about the fraction of measurements in a given interval. For example, we know that at least 75% of the measurements in a data set will lie within 2 standard deviations of the mean. If the frequency distribution of the data set is mound-shaped, approximately 95% of the measurements will lie within 2 standard deviations of the mean.

Measures of relative standing provide still another dimension on which to describe a data set. The objective of these measures is to describe the location of a specific measurement relative to the rest of the data set. By doing so, you can construct a mental image of the relative frequency distribution. Percentiles and $z$-scores are important examples of measures of relative standing.

The rare event concept of statistical inference means that if the chance that a particular sample came from a hypothetical population is very small, we can conclude either that the sample is extremely rare or that the hypothesized population is not the one from which the sample was drawn. The more unlikely it is that the sample came from the hypothesized population, the more strongly we favor the conclusion that the hypothesized population is not the true one. We need to be able to assess accurately the rarity of a sample, and this requires a knowledge of probability, the subject of our next chapter.

Finally, we gave some examples that demonstrated how descriptive statistics may be used to distort the truth. You should be very critical when interpreting graphical or numerical descriptions of data sets.

**SUPPLEMENTARY EXERCISES**

Learning the mechanics

**2.45.** Construct a relative frequency distribution for the data summarized in the following table:

| MEASUREMENT CLASS | RELATIVE FREQUENCY |
|---|---|
| 0.00–0.75 | .02 |
| 0.75–1.50 | .01 |
| 1.50–2.25 | .03 |
| 2.25–3.00 | .05 |
| 3.00–3.75 | .10 |
| 3.75–4.50 | .14 |
| 4.50–5.25 | .19 |
| 5.25–6.00 | .15 |
| 6.00–6.75 | .12 |
| 6.75–7.50 | .09 |
| 7.50–8.25 | .05 |
| 8.25–9.00 | .04 |
| 9.00–9.75 | .01 |

**2.46.** Consider the following 3 measurements: 50, 70, 80. Find the $z$-score for each measurement if they are from a population with a mean and standard deviation equal to:

a. $\mu = 60$, $\sigma = 10$
b. $\mu = 60$, $\sigma = 5$
c. $\mu = 40$, $\sigma = 10$
d. $\mu = 40$, $\sigma = 100$

**2.47.** If the range of a set of data is 20, find a rough approximation to the standard deviation of the data set.

**2.48.** Calculate the mean, variance, and standard deviation of the following samples:

a. $0, -3, 4, 7, 2, 6$
b. $5, 7, 4, 8, 6, 6, 3, 9, 5, 7$
c. $12, 3, 22, 5, 19, 20, 9$

**2.49.** a. Construct a relative frequency histogram for the following set of data:

| | | | | | |
|---|---|---|---|---|---|
| 50 | 38 | 56 | 57 | 48 | 64 |
| 41 | 51 | 52 | 46 | 51 | 50 |
| 60 | 46 | 41 | 53 | 65 | 59 |
| 42 | 37 | 50 | 45 | 52 | 42 |
| 56 | 47 | 63 | 58 | 49 | 55 |

b. Calculate $\bar{x}$, $s^2$, and $s$.
c. Calculate the intervals $\bar{x} \pm s$, $\bar{x} \pm 2s$, and $\bar{x} \pm 3s$. What percentage of the measurements would you expect to fall in each interval?
d. Calculate the actual percentage of the number of measurements falling in each interval. How does this compare with your answers to part c?

**2.50.** Calculate the median, mode, and range for each of the following data sets:

a. 3, −4, 0, −6, 0, −2
b. 22, 31, 19, 25, 25, 35
c. 100, 27, 75, 80, 60, 52, 80, 83

**2.51.** Calculate the mean, variance, and standard deviation of the following samples:

a. 1.6, 2.4, 3.1, 1.8, 2.3, 4.3
b. .03, .09, .01, .04, .10, .07

**2.52.** Identify each of the following data sets as qualitative or quantitative:

a. The types of metal that are used to make automobiles
b. The amount of an anesthesia required for each of three surgical patients
c. The planting distances for certain types of garden vegetables
d. The nationality of each of twenty immigrants to the United States

Applying the concepts

**2.53.** The owner of a service station decided to conduct a survey of service records to determine the length of time (in months) between customer oil changes. A random sample of the station records produced the following times between oil changes:

| 6  | 6 | 24 | 8  | 6  |
|----|---|----|----|----|
| 6  | 6 | 16 | 6  | 12 |
| 18 | 8 | 4  | 12 | 12 |

Compute the sample mean, variance, and standard deviation.

**2.54.** In Exercise 2.53, if the individual who changed the oil in his or her car every 24 months had instead changed it every 18 months, would the sample variance increase or decrease? Why? If instead of every 24 or 18 months, the person had changed the oil every 6 months, how would the resulting sample variance compare to the sample variance when the oil was changed every 24 months? Every 18 months?

**2.55.** As a result of government pressure, automobile manufacturers in the United States are deeply involved in research to improve their products' gas mileage. One manufacturer, hoping to achieve 30 miles per gallon on one of its full-size models, measured the mileage obtained by thirty test versions of the model with the following results (rounded to the nearest mile for convenience):

| 30 | 30 | 32 | 31 | 30 | 27 |
|----|----|----|----|----|----|
| 28 | 31 | 33 | 28 | 30 | 31 |
| 29 | 31 | 27 | 33 | 31 | 30 |
| 35 | 29 | 32 | 27 | 31 | 34 |
| 30 | 27 | 26 | 29 | 28 | 30 |

a. If the manufacturer would be satisfied with a (population) mean of 30 miles per gallon, how would it react to the above test data?
b. Compute $\bar{x}$, $s^2$, and $s$ for the data set.
c. What percentage of the measurements would you expect to find in the intervals $\bar{x} \pm s$, $\bar{x} \pm 2s$, and $\bar{x} \pm 3s$?

d. Count the number of measurements that actually fall within the intervals of part c and express each interval count as a percentage of the total number of measurements. Compare these results with the results of part c.

**2.56.** A radio station claims that the amount of advertising per hour of broadcasting time has an average of 3 minutes and a standard deviation equal to 2.1 minutes. You listen to the radio station for 1 hour, at a randomly selected time, and carefully observe that the amount of advertising time is equal to 7 minutes. Does this observation appear to disagree with the radio station's claim? Explain.

**2.57.** In a study of the generality of response to pain, subjects were exposed to two kinds of pain-producing stimuli. The object of the study was to see whether the subject showed consistency in pain response. The following represent a sample of pain-responsivity scores from the study:

| | | | | |
|---|---|---|---|---|
| 10 | 13 | 20 | 15 | 13 |
| 16 | 13 | 21 | 19 | 11 |
| 12 | 16 | 11 | 15 | 16 |

a. Calculate the range for the data and use the range to calculate a very rough estimate of the sample standard deviation. Use this as a check on your arithmetic in part b.
b. Find the mean, variance, and standard deviation for the sample.
c. What proportion of the data falls within 2 standard deviations of the mean?

**2.58.** A severe drought affected several western states for 3 years. A Christmas tree farmer is worried about the drought's effect on the size of the crop of trees. To decide whether the growth of the trees has been retarded, the farmer decides to take a sample of the heights of twenty-five trees and obtains the following results (recorded in inches):

| | | | | |
|---|---|---|---|---|
| 60 | 57 | 62 | 69 | 46 |
| 54 | 64 | 60 | 59 | 58 |
| 75 | 51 | 49 | 67 | 65 |
| 44 | 58 | 55 | 48 | 62 |
| 63 | 73 | 52 | 55 | 50 |

a. Compute $\bar{x}$, $s^2$, and $s$ for these data.
b. What percentage of the tree heights would you expect to find in the intervals $\bar{x} \pm s$, $\bar{x} \pm 2s$, $\bar{x} \pm 3s$?
c. Count the number of measurements that actually fall in each interval of part b, and express each interval count as a percentage of the total number of measurements. Compare these to the results of part b.

**2.59.** The Community Attitude Assessment Scale (CAAS) measures citizens' attitudes toward fifteen life areas (e.g., education, employment, and health) on four dimensions—importance, influence, equality of opportunity, and satisfaction. In order to develop the CAAS, a number of households in each of twenty-five communities were randomly selected and sent questionnaires. Because relatively low response rates suggest that there could be a substantial but unknown opinion bias in the reported data, the percentage of the sample responding to the survey was determined in each community. The results are given below (in percent).

| | | | | |
|---|---|---|---|---|
| 21 | 14 | 18 | 20 | 14 |
| 16 | 6 | 22 | 28 | 16 |
| 26 | 14 | 13 | 15 | 25 |
| 21 | 14 | 7 | 12 | 8 |
| 15 | 14 | 21 | 22 | 10 |

a. Construct a relative frequency histogram for the data given, locating the mean, median, and mode.

b. Find the range for the data and use it to calculate a very approximate value for $s$. Use this value to check your answer to part c.

c. Calculate the variance and standard deviation for the data.

d. Find the proportion of the measurements that fall in the interval $\bar{x} \pm 2s$.

**2.60.** A small computing center has found that the number of jobs submitted per day to its computers has a distribution that is approximately mound-shaped, with a mean of 83 jobs and a standard deviation of 10.

a. On approximately what percentage of days will the number of jobs submitted be between 73 and 93?

b. On approximately what percentage of days will the number of jobs submitted be between 63 and 83?

c. On approximately what percentage of days will the number of jobs submitted be greater than 93?

**2.61.** A professor believes that if a class is allowed to work on an examination as long as desired, the times spent by the students would be approximately mound-shaped with mean 40 minutes and standard deviation 6 minutes. Approximately how long should be allotted for the examination if the professor wants almost all (say, 97.5%) of the class to finish?

**2.62.** A veterinarian was interested in determining how many animals were treated in the clinic each day. A random sample of 20 days' records produced the following results:

| | | | |
|---|---|---|---|
| 15 | 17 | 24 | 16 |
| 18 | 15 | 19 | 22 |
| 25 | 21 | 18 | 17 |
| 20 | 20 | 20 | 18 |
| 10 | 16 | 12 | 21 |

a. Calculate the sample mean, variance, and standard deviation.

b. Suppose the veterinarian had seen two additional animals on each of the 20 days. Again, calculate the sample mean, variance, and standard deviation. Compare these numbers with your results in part a. What is the effect on the sample mean, variance, and standard deviation of adding a constant to each measurement in the sample?

**2.63.** A recently hired coach of distance runners was interested in knowing how many miles America's top distance runners usually run in a week. The coach surveyed fifteen of the best distance runners with the following results (in miles):

| | | | | |
|---|---|---|---|---|
| 120 | 95 | 110 | 95 | 70 |
| 90 | 80 | ·100 | 125 | 75 |
| 85 | 100 | 115 | 130 | 90 |

a.  Find the sample mean, variance, and standard deviation for these data.
b.  Suppose each runner cut the mileage in half. Again calculate the mean, variance, and standard deviation for these data. Compare these results with those obtained in part a. What is the effect on the sample mean, variance, and standard deviation of multiplying each measurement in the sample by a constant?

**2.64.** By law a box of cereal labeled as containing 16 ounces must contain at least 16 ounces of cereal. It is known that the machine filling the boxes produces a distribution of fill weights that is mound-shaped, with mean equal to the setting on the machine and with a standard deviation equal to .03 ounce. To ensure that most of the boxes contain at least 16 ounces, the machine is set so that the mean fill per box is 16.09 ounces.

a.  What percentage of the boxes will contain less than 16 ounces if the machine is set so that $\mu = 16.09$?
b.  If the machine is set so that $\mu = 16.09$, is it likely that a randomly selected box would contain less than 16 ounces?
c.  If the machine is set so that $\mu = 16.09$, is it likely that a randomly selected box of cereal would contain as little as 16.05 ounces? Explain.

**2.65.** A study was conducted to determine the effects of cigarette smoking on the lungs. The United States Surgeon General has warned that the two most dangerous ingredients contained in cigarettes are tar and nicotine. The amount of tar (in milligrams) present in ten selected brands is recorded below. Determine the mean, range, variance, and standard deviation for the data shown.

| | | | |
|---|---|---|---|
| Winston | 19 | Viceroy | 16 |
| Marlboro | 17 | Raleigh | 16 |
| Vantage | 11 | Marlboro Lights | 12 |
| Kent Lights | 8 | Tareyton | 17 |
| Winston Lights | 12 | Parliament | 10 |

**2.66.** Most people living in metropolitan areas receive impressions of what is happening in their area primarily through their major newspapers. A study was conducted to determine whether the *Uniform Crime Report,* compiled by the United States Federal Bureau of Investigation, and the daily newspaper gave consistent information about the trend and distribution of crime in a metropolitan area. An attention score, based on the amount of space devoted to a story, was calculated for each paper's coverage of murders, assaults, robberies, etc. Suppose $\mu$, the average murder attention score of metropolitan newspapers across the country in 1980, was 60, with $\sigma = 4.5$. The *St. Louis Globe-Democrat* had a 1980 murder attention score of 69.

 a. Approximately what percentage of the newspapers had a murder attention score higher than the *Globe-Democrat* in 1980? (Make no assumptions about the nature of the distribution of scores.)
 b. Repeat part a, assuming attention scores were mound-shaped.

**2.67.** Audience sizes at concerts given by the Jacksonville Symphony over the past 2 years were recorded and found to have a sample mean of 3,125 and a sample standard deviation of 25. Calculate $\bar{x} \pm s$, $\bar{x} \pm 2s$, and $\bar{x} \pm 3s$. What fraction of the recorded audience sizes of the past 2 years would be expected to fall in each of these intervals?

**2.68.** A producer of alkaline batteries was interested in obtaining a statistical description of the shelf-life of the battery it manufactures. Twenty-five batteries were selected at random as they came off the assembly line and their shelf-lives were tested. The following are the lifetimes of the twenty-five batteries rounded to the nearest month:

| | | | | |
|---|---|---|---|---|
| 24 | 21 | 24 | 20 | 19 |
| 25 | 27 | 24 | 30 | 21 |
| 23 | 24 | 24 | 19 | 24 |
| 26 | 22 | 25 | 24 | 24 |
| 25 | 23 | 28 | 23 | 25 |

 a. Compute $\bar{x}$, $s^2$, and $s$ for this data set.
 b. What percentage of the measurements would you expect to find in the intervals $\bar{x} \pm s$, $\bar{x} \pm 2s$, and $\bar{x} \pm 3s$?
 c. Count the number of measurements that actually fall within the intervals of part b and express each interval count as a percentage of the total number of measurements. Compare these results with the results of part b.

## ON YOUR OWN . . .

We list below several sources of real-life data sets that have been obtained from Wasserman and Bernero's *Statistics Sources*. This index of data sources is very complete and is a useful reference for anyone interested in finding almost any type of data. First we list some almanacs:

 *CBS News Almanac*
 *Information Please Almanac*
 *World Almanac and Book of Facts*

United States government publications are also rich sources of data:

 *Agricultural Statistics*
 *Digest of Educational Statistics*

*Handbook of Labor Statistics*
*Housing and Urban Development Yearbook*
*Social Indicators*
*Uniform Crime Reports for the United States*
*Vital Statistics of the United States*
*Business Conditions Digest*
*Economic Indicators*
*Monthly Labor Review*
*Survey of Current Business*
*Bureau of the Census Catalog*

Many data sources are published on an annual basis:

*Commodity Yearbook*
*Facts and Figures on Government Finance*
*Municipal Yearbook*
*Standard and Poor's Corporation, Trade and Securities: Statistics*

Some sources contain data that are international in scope:
*Compendium of Social Statistics*
*Demographic Yearbook*
*United Nations Statistical Yearbook*
*World Handbook of Political and Social Indicators*

Utilizing the data sources listed above, sources suggested by your instructor, or your own resourcefulness, find one real-life quantitative data set that stems from an area of particular interest to you.

**a.** Describe the data set using a relative frequency histogram.
**b.** Find the mean, median, variance, standard deviation, and range of the data set.
**c.** Use Tables 2.5 and 2.6 to describe the distribution of this data set. Count the actual number of observations that fall within 1, 2, and 3 standard deviations of the mean of the data set and compare these counts with the description of the data set you developed in part b.

**REFERENCES**

Huff, D. *How to lie with statistics.* New York: Norton, 1954.

Martin, D. L. "Children out of school: Firsthand report: how flawed statistics can make an ugly picture look even worse." *The American School Board Journal,* 1975, *162,* 57–59.

Mendenhall, W. *Introduction to probability and statistics.* 6th ed. Boston: Duxbury, 1983. Chapter 3.

Miller, S. L., Domey, R. G., Elston, S. F. A., & Milligan, G. "Mercury vapor levels in the dental office: a survey." *Journal of the American Dental Association,* Nov. 1974, *89,* 1084–1091.

Neustadter, H. E., & Sidik, S. M. "On evaluating compliance with air pollution levels 'not to be exceeded more than once a year'." *Journal of Air Pollution Control Association,* 1974, *24*(6), 559–563.

Rothstein, M. "Hotel overbooking as a Markovian sequential decision process." *Decision Sciences,* July 1974, *5,* 389–405.

Wasserman, P., & Bernero, J. *Statistics sources.* 5th ed. Detroit: Gale Research Company, 1978.

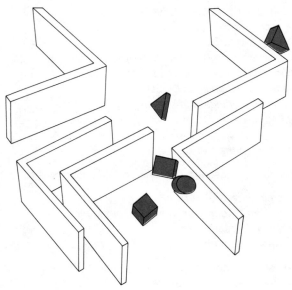

# 3
# PROBABILITY

**WHERE WE'VE BEEN . . .**

We've identified inference, from a sample to a population, as the goal of statistics. To reach this goal, we must be able to describe a set of measurements. The use of graphical and numerical methods for describing data sets and for phrasing inferences was the topic of Chapter 2.

**WHERE WE'RE GOING . . .**

Now that we know how to phrase an inference about a population, we turn to the problem of making the inference. What is it that permits us to make the inferential jump from sample to population and then to give a measure of reliability for the inference? As you will subsequently see, the answer is *probability*. This chapter is devoted to a study of probability—what it is and some of the basic concepts of the theory that surrounds it.

You will recall that statistics is concerned with decisions about a population based on sample information. Understanding how this is accomplished will be easier if you understand the relationship between population and sample. This understanding is enhanced by reversing the statistical procedure of making inferences from sample to population. In this chapter we assume the population *known* and calculate the chances of obtaining various samples from the population. Thus, probability is the "reverse" of statistics: In probability we use the population information to infer the probable nature of the sample.

Probability plays an important role in decision-making. To illustrate, suppose you have an opportunity to invest in an oil exploration company. Past records show that for ten out of ten previous oil drillings (a sample of the company's experiences), all ten resulted in dry wells. What do you conclude? Do you think the chances are better than 50–50 that the company will hit a producing well? Should you invest in this company? We think your answer to these questions will be an emphatic "No." If the company's exploratory prowess is sufficient to hit a producing well 50% of the time, a record of ten dry wells out of ten drilled is an event that is just too improbable. Do you agree?

As another illustration, suppose you are playing poker with what your opponents assure you is a well-shuffled deck of cards. In three consecutive 5 card hands, the person on your right is dealt 4 aces. Based on this sample of three deals, do you think the cards are being adequately shuffled? Again, we think your answer will be "No" and that you will reach this conclusion because dealing three hands of 4 aces is just too improbable assuming that the cards were properly shuffled.

Note that the decisions concerning the potential success of the oil drilling company and the decision concerning the card shuffling were both based on probabilities, namely the probabilities of certain sample results. Both situations were contrived so that you could easily conclude that the probabilities of the sample results were small. Unfortunately, the probabilities of many observed sample results are not so easy to evaluate intuitively. For these cases we will need the assistance of a theory of probability.

## 3.1
## EVENTS, SAMPLE SPACES, AND PROBABILITY

We will commence the discussion of probability with simple examples that are easily described, thus eliminating any discussion that could be distracting. With the aid of simple examples, important definitions are introduced and the notion of probability is more easily developed.

Suppose a coin is tossed once and the up face of the coin is recorded. This is an **observation**, or **measurement**. Any process of making an observation is called an **experiment**. Our definition of experiment is broader than that used in the physical sciences, where you would picture test tubes, microscopes, etc. Other practical examples of statistical experiments are recording whether a customer prefers one of two brands of electronic calculators, recording a voter's opinion on an important political issue, measuring the amount of dissolved oxygen in a polluted river, observing the closing price of a stock, counting the number of errors in an inventory, and observing the fraction of insects killed by a new insecticide. This list of statistical experiments could be continued ad infinitum, but the point is that our definition of experiment is very broad.

DEFINITION 3.1

An **experiment** is the process of making an observation or taking a measurement.

Consider another simple experiment consisting of tossing a die and observing the number on the up face of the die. The six basic possible outcomes to this experiment are:

1. Observe a 1    4. Observe a 4
2. Observe a 2    5. Observe a 5
3. Observe a 3    6. Observe a 6

Note that if this experiment is conducted once, you can observe one and only one of these six basic outcomes. Also, these possibilities cannot be decomposed into more basic outcomes. These very basic possible outcomes to an experiment are called **simple events**.

DEFINITION 3.2

A **simple event** is the most basic outcome of an experiment.

EXAMPLE 3.1

Two coins are tossed and the up faces of both coins are recorded. List all the simple events for this experiment.

Solution

Even for a seemingly trivial experiment, we must be careful when listing the simple events. At first glance the basic outcomes seem to be Observe two heads, Observe two tails, Observe one head and one tail. However, further reflection reveals that the last of these, Observe one head and one tail, can be decomposed into Head on coin 1, Tail on coin 2 and Tail on coin 1, Head on coin 2.* Thus, the simple events are as follows:

1. Observe *HH*    3. Observe *TH*
2. Observe *HT*    4. Observe *TT*

(where *H* in the first position means ''Head on coin 1,'' *H* in the second position means ''Head on coin 2,'' etc.).

We will often wish to refer to the collection of all the simple events of an experiment. This collection will be called the **sample space** of the experiment. For example, there

---

*Even if the coins are identical in appearance, there are, in fact, two distinct coins. Thus, the designation of one coin as ''coin 1'' and the other as ''coin 2'' is legitimate in any case.

are six simple events in the sample space associated with the die tossing experiment. The sample spaces for the experiments discussed thus far are shown in Table 3.1.

---

**DEFINITION 3.3**
The **sample space** of an experiment is the collection of all its simple events.

---

**TABLE 3.1**
**EXPERIMENTS AND THEIR SAMPLE SPACES**

Experiment:   Observe the up face on a coin

Sample space:   1.   Observe a head
                2.   Observe a tail

This sample space can be represented in set notation as a set containing two simple events
   $S$:   $\{H, T\}$
where $H$ represents the simple event Observe a head and $T$ represents the simple event Observe a tail.

---

Experiment:   Observe the up face on a die

Sample space:   1.   Observe a 1
                2.   Observe a 2
                3.   Observe a 3
                4.   Observe a 4
                5.   Observe a 5
                6.   Observe a 6

This sample space can be represented in set notation as a set of six simple events
   $S$:   $\{1, 2, 3, 4, 5, 6\}$

---

Experiment:   Observe the up faces on two coins

Sample space:   1.   Observe $HH$
                2.   Observe $HT$
                3.   Observe $TH$
                4.   Observe $TT$

This sample space can be represented in set notation as a set of four simple events
   $S$:   $\{HH, HT, TH, TT\}$

---

Just as graphs are useful in describing sets of data, a pictorial method for presenting the sample space and its simple events will often be useful. Figure 3.1 (page 66) shows such a representation for each of the experiments in Table 3.1. In each case, the sample space is shown as a closed figure, labeled $S$, containing a set of points, called sample points, with each point representing one simple event. Note that the number of sample points in a sample space $S$ is equal to the number of simple events associated with the respective experiment: two for the coin toss, six for the die toss, and four for the two-coin toss. These graphical representations are called Venn diagrams.

FIGURE 3.1
VENN DIAGRAMS FOR
THE THREE
EXPERIMENTS
FROM TABLE 3.1

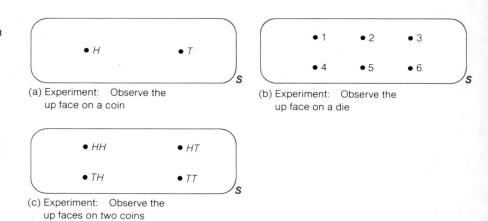

(a) Experiment:   Observe the
up face on a coin

(b) Experiment:   Observe the
up face on a die

(c) Experiment:   Observe the
up faces on two coins

Now that we have defined simple events as the basic outcomes of the experiment and the sample space as the collection of all the simple events, we are prepared to discuss the probabilities of simple events. You have undoubtedly used the term *probability* and have some intuitive idea about its meaning. Probability is generally used synonymously with "chance," "odds," and similar concepts. We will begin our treatment of probability using these informal concepts and then solidify what we mean later. For example, if a fair coin is tossed, we might reason that both the simple events, Observe a head and Observe a tail, have the same chance of occurring. Thus, we might state that "the probability of observing a head is 50%," or "the odds of seeing a head are 50—50." Both these statements are based on an informal knowledge of probability.

The probability of a simple event is a number that measures the likelihood that the event will occur when the experiment is performed. This number is usually taken to be the relative frequency of the occurrence of a simple event in a very long series of repetitions of an experiment. Or, when this information is not available, we select the number based upon experience. For example, if we are assigning probabilities to the two simple events in the coin toss experiment (Observe a head and Observe a tail), we might reason that if we toss a balanced coin a very large number of times, the simple events Observe a head and Observe a tail will occur with the same relative frequency of .5. Thus, the probability of each simple event is .5.

In other cases we may choose the probability based on general information about the experiment. For example, if the experiment is observing whether a venture succeeds or fails (the simple events), we may assess the probability of success by considering the personnel managing the venture, the success of similar ventures, and any other information deemed pertinent. If we finally decide that the venture has an 80% chance of succeeding, we assign a probability of .8 to the simple event Success. We hope that .8 is a reasonably accurate measure of the likelihood of the occurrence of the simple event Success. If it is not, we may be misled on any decisions based on this probability or based on any calculations in which it appears.

No matter how you assign the probabilities to simple events, the probabilities assigned must obey two rules:

> **1.** All simple event probabilities *must* lie between 0 and 1.
> **2.** The probabilities of all the simple events within a sample space must sum to 1.

Although the probabilities of simple events are often of interest in their own right, it is usually probabilities of collections of simple events that are important. Example 3.2 demonstrates this point.

**EXAMPLE 3.2**

A fair die is tossed and the up face is observed. If the face is even, you win $1. Otherwise, you lose $1. What is the probability that you win?

**Solution**

Recall that the sample space for this experiment contains six simple events, namely

$$S: \quad \{1, 2, 3, 4, 5, 6\}$$

Since the die is balanced, we assign a probability of $\frac{1}{6}$ to each of the simple events in this sample space. An even number will occur if one of the simple events, Observe a 2, Observe a 4, or Observe a 6, occurs. A collection of simple events such as this will be called an event, and we will denote this event by the letter $A$. Since the event $A$ contains three simple events—all with probability $\frac{1}{6}$—and since no simple events can occur simultaneously, we reason that the probability of $A$ is the sum of the probabilities of the simple events in $A$. Thus, the probability of $A$ is $\frac{1}{6} + \frac{1}{6} + \frac{1}{6} = \frac{1}{2}$. This implies that *in the long run* you will win $1 half the time and lose $1 half the time.

Figure 3.2 is a Venn diagram depicting the sample space associated with a die toss experiment and the event $A$, Observe an even number. The event $A$ is represented by the closed figure inside the sample space $S$. This closed figure $A$ contains all the simple events that comprise it.

**FIGURE 3.2**
DIE TOSS EXPERIMENT WITH EVENT $A$, OBSERVE AN EVEN NUMBER

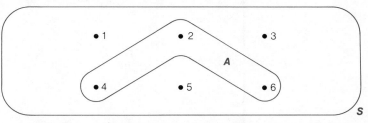

How do you decide which simple events belong to the set associated with an event $A$? Test each simple event in the sample space $S$. If event $A$ occurs when a particular simple event occurs, then that simple event is in the event $A$. For example, the event $A$, Observe an even number, in the die toss experiment will occur if the simple event Observe a 2 occurs. By the same reasoning, the simple events Observe a 4 and Observe a 6 are also in event $A$.

To summarize, we have demonstrated that an event can be defined in words or it can be defined as a specific set of simple events. This leads us to the following general definition of an event:

---

**DEFINITION 3.4**
An **event** is a specific collection of simple events.

---

EXAMPLE 3.3

Consider the experiment of tossing two coins. Suppose the coins are *not* balanced and the correct probabilities associated with the simple events are as follows:

| SIMPLE EVENT | PROBABILITY |
|:---:|:---:|
| *HH* | $\frac{4}{9}$ |
| *HT* | $\frac{2}{9}$ |
| *TH* | $\frac{2}{9}$ |
| *TT* | $\frac{1}{9}$ |

[*Note:* The necessary properties for assigning probabilities to simple events are satisfied.]

Consider the events

A: {Observe exactly one head}

B: {Observe at least one head}

Calculate the probability of A and the probability of B.

Solution

Event A contains the simple events *HT* and *TH*. Since two or more simple events cannot occur at the same time, we can easily calculate the probability of event A by summing the probabilities of the two simple events. Thus, the probability of observing exactly one head (event A), denoted by the symbol $P(A)$, would be

$$P(A) = P(\text{Observe } HT) + P(\text{Observe } TH)$$

$$= \frac{2}{9} + \frac{2}{9} = \frac{4}{9}$$

Similarly, since B contains the simple events *HH*, *HT*, and *TH*,

$$P(B) = \frac{4}{9} + \frac{2}{9} + \frac{2}{9} = \frac{8}{9}$$

This example leads us to a general procedure for finding the probability of an event A, namely:

---

The probability of an event A is calculated by summing the probabilities of the simple events in A.

---

Thus, we can summarize the steps for calculating the probability of any event:*

---

**STEPS FOR CALCULATING PROBABILITIES OF EVENTS**
**1.** Define the experiment, i.e., describe the process used to make an observation and the type of observation that will be recorded.
**2.** List the simple events.
**3.** Assign probabilities to the simple events.
**4.** Determine the collection of simple events contained in the event of interest.
**5.** Sum the simple event probabilities to get the event probability.

---

**EXAMPLE 3.4**

A bank wishes to divide its service windows into two groups corresponding to the type of customer account: commercial or personal. One problem facing the bank is deciding how to apportion the service windows to the categories of service. At this stage of our study, we do not have the tools to solve this problem, but we can say that one of the important factors affecting the solution is the proportion of the two types of customers that enter the bank at a particular time. To illustrate, what is the probability that an incoming customer will have a commercial account? What is the probability that the next two customers will have commercial accounts? What is the probability for the general case of *k* customers? Explain how you might attempt to solve this problem. [*Note:* For this example we use the term *customer* to refer only to people who seek teller service.]

**Solution**

**Step 1.** Define the experiment. The experiment corresponding to the entrance of a single customer is identical in underlying structure to the coin tossing experiment illustrated in Figure 3.1(a). A customer, who possesses either a commercial account (call this a head) or a personal account (call this a tail), is observed and the type of account is recorded.

Experiment: Observe the type of account a single customer possesses.

**Step 2.** List the simple events. There are only two possible outcomes of the experiment. These simple events are:

Simple events: 1. *C*: {Customer possesses a commercial account}
2. *D*: {Customer possesses a personal account}

**Step 3.** Assign probabilities to the simple events. The difference between this problem and the coin tossing problem becomes apparent when we attempt to assign probabilities to the two simple events. What probability should we assign to the simple event *C*? Some people might say .5, as for the coin tossing experiment, but you can see that finding $P(C)$, the probability of simple event *C*, is not so easy. Suppose that a check of the bank's records showed that 10% of their accounts were personal. Then, at first glance, it would appear that $P(C)$ is .10. But this may not be correct, because

*A thorough treatment of this topic can be found in the classical text by W. Feller (1968).

the probability will depend upon how frequently the two types of customers use the accounts—i.e., the proportion of customers per week, on the average, that seek banking service. So, the important point to note is that this is a case where equal probabilities are not assigned to the simple events. How can we find these probabilities? A good procedure might be to monitor the system for a period of time, and ask incoming customers which type of service they desire. Then the proportions of the two types of customers could be used to approximate the probabilities of the two simple events.

We could then continue with steps 4 and 5 to calculate any probability of interest for this experiment with two simple events.

The experiment, assessing the service requirements of two customers, is identical to the experiment of Example 3.3, tossing two coins, except that the probabilities of the simple events are not the same. We will learn how to find the probabilities of the simple events for this experiment, or for the general case of $k$ customers, in Section 3.5.

For the experiments discussed thus far, listing the simple events has been easy. For more complex experiments, the number of simple events may be so large that listing them is impractical. Sometimes these large numbers of simple events can be counted with the assistance of a computer or by using combinatorial mathematics (see the references). Or, we might calculate the probability of an event using the method discussed in Sections 3.2–3.5, which involves forming a composition of events and then applying two rules of probability.

CASE STUDY 3.1
COMPARING
SUBJECTIVE
PROBABILITY
ASSESSMENTS
WITH RELATIVE
FREQUENCIES

Preston and Baratta (1948) performed an experiment with the objective of comparing how an individual's subjective assessment of the probability of an event compares with the known probability (relative frequency of occurrence) of the event. The individuals selected for the experiment ranged from undergraduates with no training in probability theory to professors of mathematics and statistics with a "substantial acquaintance with probability theory." Each individual participated in a game in which he or she bet part of an initial stake on one of seven different outcomes of a combination card–dice game. The probabilities of these seven different events, known only to the experimenters, were .01, .05, .25, .50, .75, .95, and .99. From the amount the subject is willing to bet, the individual's subjective probabilities can be determined and then can be compared with the actual probabilities of the events.

In Figure 3.3 we reproduce the authors' figure depicting the average subjective probability assessed by the experimental subjects, compared to the true probabilities. Some of the conclusions reached were:

**1.** Events with probabilities that were less than .25 were subjectively overestimated.
**2.** Events with probabilities that were more than .25 were subjectively underestimated.
**3.** These conclusions were the same for both the probabilistically naive and sophisticated subjects.

For events with probabilities that are not clearly defined (such as the probability of rain tomorrow), it is important to have information on general tendencies in subjec-

FIGURE 3.3
OBSERVED
RELATIONSHIP
BETWEEN TRUE AND
SUBJECTIVE
PROBABILITIES

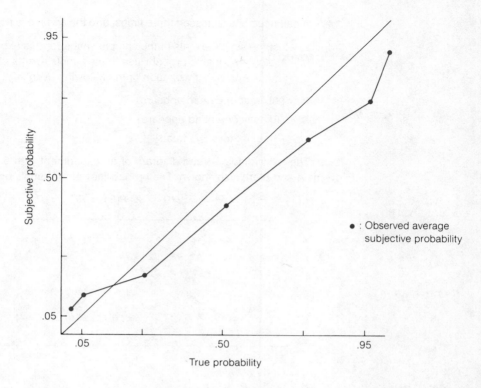

tively evaluating the probabilities. Of course, much more evidence has been collected on the subjective evaluation of probabilities since the Preston and Baratta article (not all of which corroborate their conclusions), but it remains an interesting evaluation of a person's ability to evaluate probabilities subjectively.

EXERCISES

Learning the mechanics

3.1.  The sample space for an experiment contains 5 simple events with probabilities as shown in the following table:

| SIMPLE EVENTS | PROBABILITIES |
|---|---|
| 1 | .15 |
| 2 | .10 |
| 3 | .30 |
| 4 | .25 |
| 5 | .20 |

Find the probability of each of the following events:

  A:   {Either 1, 2, or 3 occurs}

  B:   {Either 1, 3, or 5 occurs}

  C:   {4 does not occur}

**3.2.** A balanced coin is tossed three times, and the up face is noted after each toss.

    a.   List the simple events in the sample space for this experiment.

    b.   Assign reasonable probabilities to the simple events.

    c.   Find the probability of each of the following events:

   *A:*   {At least one head appears}

   *B:*   {Exactly one head appears}

   *C:*   {The first toss is a head}

**3.3.** The following is a Venn diagram of an experiment with 6 simple events. The events *A* and *B* are also shown. The probabilities of the simple events are as follows:

$$P(1) = P(2) = P(4) = 2/9 \qquad P(3) = P(5) = P(6) = 1/9$$

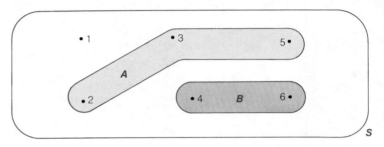

    a.   Find *P(A)*.

    b.   Find *P(B)*.

    c.   Find the probability the events *A* and *B* occur *simultaneously*.

**3.4.** Two fair dice are tossed, and the up face on each die is recorded.

    a.   List the 36 simple events contained in the sample space.

    b.   Find the probability of observing each of the following events:

   *A:*   {A 3 appears on each of the two dice}

   *B:*   {The sum of the numbers is even}

   *C:*   {The sum of the numbers is equal to 7}

   *D:*   {A 5 appears on at least one of the dice}

   *E:*   {The sum of the numbers is 10 or more}

**3.5.** Consider the experiment composed of one roll of a fair die, followed by one toss of a fair coin. List the simple events. Assign a logical probability to each simple event. Determine the probability of observing each of the following events:

   *A:*   {6 on the die; *H* on the coin}

   *B:*   {Even number on the die; *T* on the coin}

   *C:*   {Even number on the die}

   *D:*   {*T* on the coin}

**3.6.** Two marbles are randomly drawn from a box containing two blue marbles and three red marbles. Determine the probability of observing each of the following events:

A: {Two blue marbles are drawn}

B: {A red and a blue marble are drawn}

C: {Two red marbles are drawn}

**3.7.** Simulate the experiment described in Exercise 3.6 using any five identically shaped objects, two of which are one color and three are another. Mix the objects, randomly draw two, record the results, and then replace the objects. Repeat the experiment a large number of times (at least 100). Calculate the proportion of times events A, B, and C occur. How do these proportions compare with the probabilities that you calculated in Exercise 3.6? Should these proportions equal the probabilities? Explain.

Applying the concepts

**3.8.** A hospital reports that two patients have been admitted who have contracted Legionnaire's disease. Suppose our experiment consists of observing whether the patients survive or die as a result of the disease. The simple events and probabilities of their occurrence are as follows (where S in the first position means that patient 1 survives, D in the first position means that patient 1 dies, etc.):

| SIMPLE EVENTS | PROBABILITIES |
|---|---|
| SS | .81 |
| SD | .09 |
| DS | .09 |
| DD | .01 |

Find the probabilities of each of the following events:

A: {Both patients survive the disease}

B: {At least one patient dies}

C: {Exactly one patient survives the disease}

**3.9.** The corporations in the highly competitive razor blade industry do a tremendous amount of advertising each year. Corporation G gave a supply of the three top name brands, G, S, and W, to a consumer and asked him to use them and rank them in order of preference. The corporation was, of course, hoping the consumer would prefer its brand and rank it first, thereby giving them some material for a consumer interview advertising campaign. If the consumer did not prefer one blade more than any other, but was still required to rank the blades, what is the probability that:

a. The consumer ranked brand G first?

b. The consumer ranked brand G last?

c. The consumer ranked brand G last and brand W second?

d. The consumer ranked brand W first, brand G second, and brand S third?

**3.10.** An individual's genetic makeup is determined by the genes obtained from each parent. For every genetic trait, each parent possesses a gene pair; and each contributes one-half of this gene pair, with equal probability, to their offspring, forming a new gene pair. The offspring's traits (eye color, baldness, etc.) come from this new gene pair, where each gene in this pair possesses some characteristic.

For the particular gene pair that determines eye color, each gene trait may be one of two types: dominant brown (*B*) or recessive blue (*b*). A person possessing the gene pair *BB* or *Bb* has brown eyes, while the gene pair *bb* produces blue eyes.

a. Suppose that both parents of an individual are brown-eyed, each with a gene pair of the type *Bb*. What is the probability that a randomly selected child of this couple will have blue eyes? [*Hint:* Construct the sample space for the experiment.]

b. If one parent has brown eyes, type *Bb,* and the other has blue eyes, what is the probability that a randomly selected child of this couple will have blue eyes?

c. Suppose that one parent is brown-eyed, type *BB*. What is the probability that a child has blue eyes?

**3.11.** Three people play a game called "odd man out." In this game, each player flips a fair coin until the outcome (heads or tails) for one of the players is not the same as the other two players'. This player is then the odd man out and loses the game. Find the probability that the game ends (i.e., either exactly one of the coins will fall heads or exactly one of the coins will fall tails) after only one toss by each player. Suppose that one of the players, hoping to reduce his chances of being the odd man, uses a two-headed coin. Will this ploy be successful? Solve by listing the simple events in the sample space.

**3.12.** The breakdown of workers in a particular state according to their political affiliation and type of job held is as follows:

| | | POLITICAL AFFILIATION | | |
| | | Republican | Democrat | Independent |
| --- | --- | --- | --- | --- |
| TYPE OF JOB | White-collar | 12% | 12% | 6% |
| | Blue-collar | 23% | 43% | 4% |

Suppose a worker is selected at random within the state and the worker's political affiliation and type of job are noted.

a. List all simple events for this experiment.

b. What is the set of all simple events called?

c. Let *A* be the event that the worker is a white-collar worker. Find $P(A)$.

d. Let *B* be the event that the worker is a Republican. Find $P(B)$.

e. Let *C* be the event that the worker is a Democrat. Find $P(C)$.

f.   Let $D$ be the event that the worker is a white-collar worker and a Democrat. Find $P(D)$.

**3.13.**   One regulation that concerns high school students who participate in varsity track is that they may enter at most three events during a meet. Suppose a star senior at a certain school can compete equally well in the following five events: high jump, triple jump, long jump, 220 yard dash, and 100 yard dash.

a.   List the different ways this track star can choose three events in which to participate at the upcoming track meet.

b.   Unknown to the senior, an opposing team will set state track records in the 100 yard dash and the triple jump. If the three events in which to participate are selected at random, what is the probability that the track star will complete in both events in which a record will be set?

**3.14.**   Before placing a person in a highly skilled position, a company gives the applicants a series of three examinations. The first is a physical examination, and each applicant is classified as satisfactory or unsatisfactory. The other two are verbal and quantitative examinations, and the scores are used to classify each applicant as high, medium, or low in each area. Thus, each individual will receive a health score, a verbal score, and a quantitative score.

a.   List the different sets of classifications that can result from this battery of examinations.

b.   If all applicants who take the examinations were equally qualified, and all the variation in test scores was caused by random variation due to the types of test questions, what is the probability that an applicant receives the lowest classification on all three examinations?

c.   If an applicant scores in the highest category on at least two of the three examinations, the applicant will get a position. What is the probability that a randomly selected applicant will get a position?

**3.2
COMPOUND
EVENTS**

An event can often be viewed as a composition of two or more other events. Such events are called **compound events**; they can be formed (composed) in two ways.

---

**DEFINITION 3.5**
The **union** of two events $A$ and $B$ is the event that occurs if either $A$ or $B$ or both occur on a single performance of the experiment. We will denote the union of events $A$ and $B$ by the symbol $A \cup B$.

**DEFINITION 3.6**
The **intersection** of two events $A$ and $B$ is the event that occurs if both $A$ and $B$ occur on a single performance of the experiment. We will write $AB$ for the intersection of events $A$ and $B$.

---

**EXAMPLE 3.5**    Consider the die toss experiment. Define the following events:

   *A*:  {Toss an even number}

   *B*:  {Toss a number less than or equal to 3}

**a.** Describe $A \cup B$ for this experiment.
**b.** Describe $AB$ for this experiment.
**c.** Calculate $P(A \cup B)$ and $P(AB)$ assuming the die is fair.

**Solution**

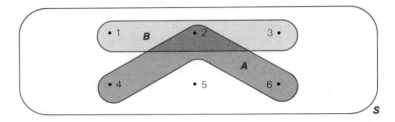

**a.** The union of *A* and *B* is the event that occurs if we observe either an even number, a number less than or equal to 3, or both on a single throw of the die. Consequently, the simple events in the event $A \cup B$ are those for which *A* occurs, *B* occurs, or both *A* and *B* occur. Testing the simple events in the entire sample space, we find that the collection of simple events in the union of *A* and *B* is

$$A \cup B = \{1, 2, 3, 4, 6\}$$

**b.** The intersection of *A* and *B* is the event that occurs if we observe **both** an even number and a number less than or equal to 3 on a single throw of the die. Testing the simple events to see which imply the occurrence of **both** events *A* and *B*, we see that the intersection contains only one simple event:

$$AB = \{2\}$$

In other words, the intersection of *A* and *B* is the simple event Observe a 2.

**c.** Recalling that the probability of an event is the sum of the probabilities of the simple events of which the event is composed, we have

$$P(A \cup B) = P(1) + P(2) + P(3) + P(4) + P(6)$$

$$= \frac{1}{6} + \frac{1}{6} + \frac{1}{6} + \frac{1}{6} + \frac{1}{6} = \frac{5}{6}$$

and

$$P(AB) = P(2) = \frac{1}{6}$$

Unions and intersections also can be defined for more than two events. For example, the event $A \cup B \cup C$ represents the union of three events, *A*, *B*, and *C*. This event, which includes the set of simple events in *A*, *B*, or *C*, will occur if any one or more of the events A, B, or C occurs. Similarly, the intersection ABC is the event that all three of the events A, B, and C occur. Therefore, ABC is the set of simple events that are in all three of the events A, B, and C.

EXAMPLE 3.6    Refer to Example 3.5 and define the event

C:    {Toss a number greater than 1}

Find the simple events in

**a.**  $A \cup B \cup C$
**b.**  $ABC$

where

A:    {Toss an even number}
B:    {Toss a number less than or equal to 3}

Solution    **a.**  Event C contains the simple events corresponding to tossing a 2, 3, 4, 5, or 6, and event B contains the simple events 1, 2, and 3. Therefore, the event that either A, B, or C occurs contains all six simple events in S, i.e., those corresponding to tossing a 1, 2, 3, 4, 5, or 6.

**b.**  You can see that you will observe all of the events A, B, and C only if you observe a 2. Therefore, the intersection ABC contains the single simple event Toss a 2.

3.3
COMPLEMENTARY
EVENTS

A very useful concept in the calculation of event probabilities is the notion of complementary events:

---
DEFINITION 3.7
The complement of an event A is the event that A does not occur, i.e., the event consisting of all simple events that are not in event A. We will denote the complement of A by A'.

---

An event A is a collection of simple events, and the simple events included in A' are those that are not in A. Figure 3.4 demonstrates this. You will note from the

FIGURE 3.4
VENN DIAGRAM OF
COMPLEMENTARY
EVENTS

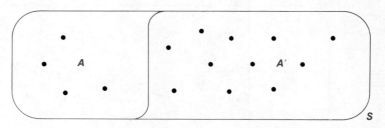

figure that all simple events in *S* are included in *either A* or *A'*, and that *no* simple event is in both *A* and *A'*. This leads us to conclude that the probabilities of an event and its complement must sum to 1:

> The sum of the probabilities of complementary events equals 1, i.e.,
>
> $$P(A) + P(A') = 1$$

In many probability problems it will be easier to calculate the probability of the complement of the event of interest rather than the event itself. Then, since

$$P(A) + P(A') = 1$$

we can calculate $P(A)$ by using the relationship

$$P(A) = 1 - P(A')$$

EXAMPLE 3.7

Consider the experiment of tossing two fair coins. Calculate the probability of event *A*:  {Observing at least one head} by using the complementary relationship.

Solution

We know that the event *A*:  {Observing at least one head} consists of the simple events

$$A: \quad \{HH, HT, TH\}$$

The complement of *A* is defined as the event that occurs when *A* does not occur. Therefore,

$$A': \quad \{\text{Observe no heads}\} = \{TT\}$$

This complementary relationship is shown in Figure 3.5. Assuming the coins are balanced,

$$P(A') = P(TT) = \frac{1}{4}$$

and

$$P(A) = 1 - P(A') = 1 - \frac{1}{4} = \frac{3}{4}$$

**FIGURE 3.5**
**COMPLEMENTARY EVENTS IN THE TOSS OF TWO COINS**

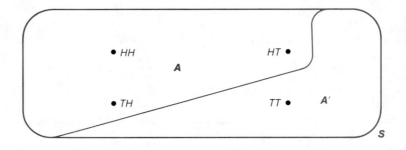

**EXAMPLE 3.8**          A fair coin is tossed ten times and the up face is recorded after each toss. What is the probability of the event A:  {Observe at least one head}?

**Solution**          We will solve this problem by following the five steps for calculating probabilities of events (see Section 3.1).

**Step 1.** Define the experiment. The experiment is to record the results of the ten tosses of the coin.

**Step 2.** List the simple events. A simple event consists of a particular sequence of ten heads and tails. Thus, one simple event is

  *HHTTTHTHTT*

which denotes head on first toss, head on second toss, tail on third toss, etc. Others would be *HTHHHTTTTT* and *THHTHTHTTH*. There is obviously a very large number of simple events—too many to list. It can be shown that there are $2^{10} = 1,024$ simple events for this experiment.

**Step 3.** Assign probabilities. Since the coin is fair, each sequence of heads and tails has the same chance of occurring and therefore all the simple events are equally likely. Then

$$P(\text{Each simple event}) = \frac{1}{1,024}$$

**Step 4.** Determine the simple events in event A. A simple event is in A if at least one H appears in the sequence of ten tosses. However, if we consider the complement of A, we find that

  A':  {No heads are observed in ten tosses}

Thus, A' contains only one simple event

  A':  {*TTTTTTTTTT*}      *and*      $P(A') = \dfrac{1}{1,024}$

**Step 5.** Since we know the probability of the complement of A, we use the relationship for complementary events:

$$P(A) = 1 - P(A') = 1 - \frac{1}{1,024} = \frac{1,023}{1,024} = .999$$

That is, we are virtually certain of observing at least one head in ten tosses of the coin.

**EXERCISES**          Learning the mechanics
**3.15.** A fair coin is tossed three times and the events A and B are defined as follows:

  A:  {At least one head is observed}

  B:  {The number of heads observed is odd}

  a.  Identify the simple events in the events, A, B, A ∪ B, A', and AB.
  b.  Find $P(A)$, $P(B)$, $P(A \cup B)$, $P(A')$, and $P(AB)$ by summing the probabilities of the appropriate simple events.

**3.16.** A pair of fair dice is tossed. Define the following events:

   *A*:   {You will roll a 7}

(i.e., the sum of the numbers of dots on the upper faces of the two dice is equal to 7)

   *B*:   {At least one of the two dice is showing a 4}

   a.   Identify the simple events in the events *A*, *B*, *AB*, *A* ∪ *B*, and *A'*.
   b.   Find $P(A)$, $P(B)$, $P(AB)$, $P(A \cup B)$, and $P(A')$ by summing the probabilities of the appropriate simple events.

**3.17.** The Venn diagram portrays an experiment that contains 5 simple events. Two events, *A* and *B*, are also shown. The probabilities of the simple events are: $P(1) = 1/10$, $P(2) = 2/10$, $P(3) = 3/10$, $P(4) = 1/10$, $P(5) = 3/10$. Find $P(A')$, $P(B')$, $P(A'B)$, $P(AB)$, $P(A \cup B)$, $P(A \cup B')$, $P[(AB)']$, and $P[(A \cup B)']$.

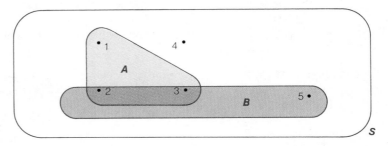

Applying the concepts

**3.18.** A buyer for a large metropolitan department store must choose two firms from the four available to supply the store's fall line of men's slacks. The buyer has not dealt with any of the four firms before and considers their products equally attractive. Unknown to the buyer, two of the four firms are having serious financial problems that may result in their not being able to deliver the fall line of slacks as soon as promised. The four firms are identified as $G_1$ and $G_2$ (firms in good financial condition) and $P_1$ and $P_2$ (firms in poor financial condition). Simple events identify the pairs of firms selected. If the probability of the buyer selecting a particular firm from among the four is the same for each firm, the following are the simple events and their probabilities for this buying experiment:

| SIMPLE EVENTS | PROBABILITIES |
|:---:|:---:|
| $G_1G_2$ | $1/6$ |
| $G_1P_1$ | $1/6$ |
| $G_1P_2$ | $1/6$ |
| $G_2P_1$ | $1/6$ |
| $G_2P_2$ | $1/6$ |
| $P_1P_2$ | $1/6$ |

We will define the following events:

   *A*:   {At least one of the selected firms is in good financial condition}

   *B*:   {Firm $P_1$ is selected}

a. Define the event $AB$ as a specific collection of simple events.
b. Define the event $A \cup B$ as a specific collection of simple events.
c. Define the event $A'$ as a specific collection of simple events.
d. Find $P(A)$, $P(B)$, $P(AB)$, $P(A \cup B)$, and $P(A')$ by summing the probabilities of the appropriate simple events.

**3.19.** A firm's accounting department claims the firm will make a profit in the first quarter of the year in one of the following ranges, with probability as noted:

| PROFIT RANGE ($) | PROBABILITY |
|---|---|
| Under 75,000 | .10 |
| 75,000–99,999 | .15 |
| 100,000–124,999 | .25 |
| 125,000–149,999 | .35 |
| 150,000–174,999 | .10 |
| 175,000 or over | .05 |

Define the following events:

A: {The firm makes $99,999 or less}

B: {The firm makes over $149,999}

C: {The firm makes between $100,000 and $149,999}

Find $P(AB)$, $P(BC)$, $P(C')$, $P(A \cup C)$, and $P(B \cup C)$.

**3.20.** One game that is very popular in many American casinos is roulette. Roulette is played by spinning a ball on a circular wheel that has been divided into thirty-eight arcs of equal length; these bear the numbers 00, 0, 1, 2, . . . , 35, 36. The number of the arc on which the ball comes to rest is the outcome of one play of the game. The numbers are also colored in the following manner:

Red     1, 3, 5, 7, 9, 12, 14, 16, 18, 19, 21, 23, 25, 27, 30, 32, 34, 36

Black    2, 4, 6, 8, 10, 11, 13, 15, 17, 20, 22, 24, 26, 28, 29, 31, 33, 35

Green   00, 0

Players may place bets on the table in a variety of ways, including bets on odd, even, red, black, high, low, etc. Define the following events:

A: {Outcome is an odd number}

(00 and 0 are not considered odd or even)

B: {Outcome is a black number}

C: {Outcome is a low number (1–18)}

a. Define the event $AB$ as a specific set of simple events.
b. Define the event $A \cup B$ as a specific set of simple events.
c. Find $P(A)$, $P(B)$, $P(AB)$, $P(A \cup B)$, and $P(C)$ by summing the probabilities of the appropriate simple events.

d. Define the event $ABC$ as a specific set of simple events.

e. Find $P(ABC)$ by summing the probabilities of the simple events given in part d.

f. Define the event $(A \cup B \cup C)$ as a specific set of simple events.

g. Find $P(A \cup B \cup C)$ by summing the probabilities of the simple events given in part f.

## 3.4 CONDITIONAL PROBABILITY

The event probabilities we have been discussing give the relative frequencies of the occurrences of the events when the experiment is repeated a very large number of times. They are called unconditional probabilities because no special conditions are assumed other than those that define the experiment.

Sometimes we may wish to alter the probability of an event when we have additional knowledge that might affect its outcome. This probability is called the conditional probability of the event. For example, we have shown that the probability of observing an even number (event $A$) on a toss of a fair die is $\frac{1}{2}$. However, suppose you are given the information that on a particular throw of the die the result was a number less than or equal to 3 (event $B$). Would you still believe that the probability of observing an even number on that throw of the die is equal to $\frac{1}{2}$? If you reason that making the assumption that $B$ has occurred reduces the sample space from six simple events to three simple events (namely, those contained in event $B$), the reduced sample space is as shown in Figure 3.6.

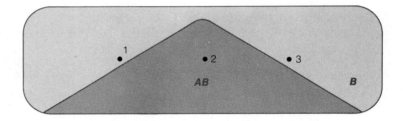

**FIGURE 3.6**

REDUCED SAMPLE SPACE FOR THE DIE TOSS EXPERIMENT —GIVEN THAT EVENT $B$ HAS OCCURRED

Since the only even number of the three numbers in the reduced sample space $B$ is the number 2 and since the die is fair, we conclude that the probability that $A$ occurs given that $B$ occurs is one in three, or $\frac{1}{3}$. We will use the symbol $P(A|B)$ to represent the probability of event $A$ given that event $B$ occurs. For the die toss example

$$P(A|B) = \frac{1}{3}$$

To get the probability of event $A$ given that event $B$ occurs, we proceed as follows: We divide the probability of the part of $A$ that falls within the reduced sample space $B$, namely $P(AB)$, by the total probability of the reduced sample space, namely $P(B)$. Thus, for the die toss example where event $A$: {Observe an even number} and event $B$: {Observe a number less than or equal to 3}, we find

$$P(A|B) = \frac{P(AB)}{P(B)} = \frac{P(2)}{P(1) + P(2) + P(3)} = \frac{\frac{1}{6}}{\frac{3}{6}} = \frac{1}{3}$$

This formula for $P(A|B)$ is true in general:

> To find the conditional probability that event $A$ occurs given that event $B$ occurs, divide the probability that *both* $A$ and $B$ occur by the probability that $B$ occurs, that is,
>
> $$P(A|B) = \frac{P(AB)}{P(B)} \quad \text{(we assume that } P(B) \neq 0)$$

EXAMPLE 3.9

Many medical researchers have conducted experiments to examine the relationship between cigarette smoking and cancer. Let $A$ represent the event that an individual smokes and let $C$ represent the event that an individual develops cancer. Thus, $AC$ is the simple event that an individual smokes and develops cancer, $AC'$ is the simple event that an individual smokes and does not develop cancer, etc. Assume that the probabilities associated with the four simple events are as shown below for a certain section of the United States:

| SIMPLE EVENTS | PROBABILITIES |
|:---:|:---:|
| $AC$ | .15 |
| $AC'$ | .25 |
| $A'C$ | .10 |
| $A'C'$ | .50 |

How can these simple event probabilities be used to examine the relationship between smoking and cancer?

Solution

One method of determining whether these probabilities indicate that smoking and cancer are related is to compare the conditional probability that an individual acquires cancer given that he or she smokes with the conditional probability that an individual acquires cancer given that he or she does not smoke. That is, we will compare $P(C|A)$ with $P(C|A')$. The calculations are as follows:

$$P(C|A) = \frac{P(AC)}{P(A)}$$

where the event $A$, a person smokes, contains two simple events, $AC$ (a person smokes and develops cancer) and $AC'$ (a person smokes and does not develop cancer). Remembering that the probability of an event is the sum of the probabilities of its simple events, we obtain

$$P(A) = P(AC) + P(AC')$$
$$= .15 + .25 = .40$$

then

$$P(C|A) = \frac{P(AC)}{P(A)} = \frac{.15}{.40} = .375$$

The conditional probability that a nonsmoking individual develops cancer is calculated in a similar manner:

$$P(C|A') = \frac{P(A'C)}{P(A')} = \frac{.10}{.60} = .167$$

where

$$P(A') = P(A'C) + P(A'C') = .10 + .50 = .60, \text{ or } P(A') = 1 - P(A) = 1 - .40 = .60$$

(either way we get the same answer).

Comparing the two conditional probabilities, we see that the probability that a smoker develops cancer, .375, is more than twice the probability that a nonsmoker develops cancer, .167. This does not imply that smoking *causes* cancer, but it does suggest a very pronounced link between smoking and cancer.

EXAMPLE 3.10    The investigation of consumer product complaints by the Federal Trade Commission (FTC) has generated much interest by manufacturers in the quality of their products. A manufacturer of an electromechanical kitchen aid conducted an analysis of a large number of consumer complaints and found that they fell into the six categories shown in Table 3.2. If a consumer complaint is received, what is the probability that the cause of the complaint was product appearance given that the complaint originated prior to the end of the guarantee period?

**TABLE 3.2**
**DISTRIBUTION OF PRODUCT COMPLAINTS**

| | REASON FOR COMPLAINT | | | |
| | Electrical | Mechanical | Appearance | TOTALS |
| --- | --- | --- | --- | --- |
| DURING GUARANTEE PERIOD | 18% | 13% | 32% | 63% |
| AFTER GUARANTEE PERIOD | 12% | 22% | 3% | 37% |
| TOTALS | 30% | 35% | 35% | 100% |

Solution    Let $A$ represent the event that the cause of a particular complaint was product appearance and let $B$ represent the event that the complaint occurred prior to the termination of the guarantee period. Checking Table 3.2, you can see that $(18 + 13 + 32) = 63\%$

of the complaints occur prior to the termination of the guarantee time. Hence, $P(B) = .63$. The percentage of complaints that were caused by appearance and occurred prior to the termination time (the event $AB$) is 32%. Therefore, $P(AB) = .32$.

Using these probability values, we can calculate the conditional probability $P(A|B)$ that the cause of a complaint is appearance given that the complaint occurred prior to the termination of the guarantee time:

$$P(A|B) = \frac{P(AB)}{P(B)} = \frac{.32}{.63} = .51$$

Consequently, you can see that slightly more than half of the complaints that occurred prior to guarantee time were due to scratches, dents, or other imperfections in the surface of the kitchen devices.

**CASE STUDY 3.2**
PURCHASE PATTERNS AND THE CONDITIONAL PROBABILITY OF PURCHASING

In his doctoral dissertation, Alfred A. Kuehn (1958) examined sequential purchase data to gain some insight into consumer brand switching. He analyzed the frozen orange juice purchases of approximately 600 Chicago families during 1950–1952. The data were collected by the *Chicago Tribune* Consumer Panel. Kuehn was interested in determining the influence of a consumer's last four orange juice purchases on the next purchase. Thus, sequences of five purchases were analyzed.

Table 3.3 contains a summary of the data collected for Snow Crop brand orange juice and part of Kuehn's analysis of the data. In the column labeled "Previous Purchase Pattern" an S stands for the purchase of Snow Crop by a consumer and an

**TABLE 3.3**
OBSERVED APPROXIMATE CONDITIONAL PROBABILITY OF PURCHASING SNOW CROP, GIVEN THE FOUR PREVIOUS BRAND PURCHASES

| PREVIOUS PURCHASE PATTERN S = Snow Crop O = Other brand | SAMPLE SIZE | FREQUENCY | OBSERVED APPROXIMATE CONDITIONAL PROBABILITY OF PURCHASE P{Purchase\|Previous purchase pattern} |
|---|---|---|---|
| SSSS | 1,047 | 844 | .806 |
| OSSS | 277 | 191 | .690 |
| SOSS | 206 | 137 | .665 |
| SSOS | 222 | 132 | .595 |
| SSSO | 296 | 144 | .486 |
| OOSS | 248 | 137 | .552 |
| SOOS | 138 | 78 | .565 |
| OSOS | 149 | 74 | .497 |
| SOSO | 163 | 66 | .405 |
| OSSO | 181 | 75 | .414 |
| SSOO | 256 | 78 | .305 |
| OOOS | 500 | 165 | .330 |
| OOSO | 404 | 77 | .191 |
| OSOO | 433 | 56 | .129 |
| SOOO | 557 | 86 | .154 |
| OOOO | 8,442 | 405 | .048 |

O stands for the purchase of a brand other than Snow Crop. Thus, for example, SSSO is used to represent the purchase of Snow Crop three times in a row followed by the purchase of some other brand of frozen orange juice. The column labeled "Sample Size" lists the number of occurrences of the purchase sequences in the first column. The column labeled "Frequency" lists the number of times the associated purchase sequence in the first column led to the next purchase (i.e., the fifth purchase in the sequence) being Snow Crop.

The column labeled "Observed Approximate Conditional Probability of Purchase" contains the relative frequency with which each sequence of the first column led to the next purchase being Snow Crop. These relative frequencies, which give approximate conditional probabilities, are computed for each sequence of the first column by dividing the frequency of the sequence by the sample size of the sequence. For example, .806 is the approximate conditional probability that the next purchase will be Snow Crop given that the previous four purchases were also Snow Crop.

An examination of the approximate conditional probabilities in the fourth column indicates that both the most recent brand purchased and the number of times a brand is purchased have an effect on the next brand purchased. It appears that the influence on the next brand of orange juice purchased by the second most recent purchase is not as strong as the most recent purchase, but is stronger than the third most recent purchase. In general, it appears that the probability of a particular consumer purchasing Snow Crop the next time he or she buys orange juice is inversely related to the number of consecutive purchases of another brand he or she made since last purchasing Snow Crop and is directly proportional to the number of Snow Crop purchases among the four purchases.

Kuehn conducts a more formal statistical analysis of these data, which we will not pursue here. We simply want you to see that probability is a basic tool for making inferences about populations using sample data.

EXERCISES

Learning the mechanics

**3.21.** Two fair coins are tossed and the events $A$ and $B$ are defined as follows:

> $A$: {At least one head appears}
>
> $B$: {Exactly one head appears}

Find $P(A)$, $P(B)$, $P(AB)$, $P(A|B)$, and $P(B|A)$.

**3.22.** A sample space contains 6 simple events and events $A$, $B$, and $C$ as shown at the top of page 87. The probabilities of the simple events are: $P(1) = .25$, $P(2) = .05$, $P(3) = .30$, $P(4) = .10$, $P(5) = .10$, $P(6) = .20$. Use the Venn diagram and the probabilities of the simple events to find:

a. $P(A)$, $P(B)$, and $P(C)$
b. $P(AB)$, $P(AC)$, and $P(BC)$
c. $P(A|B)$, $P(B|C)$, $P(C|A)$, and $P(C|A')$

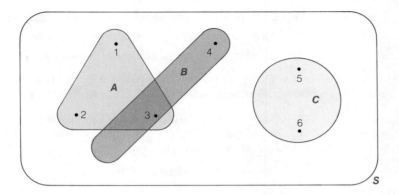

**3.23.** A box contains two white, two red, and two blue poker chips. Two chips are randomly chosen without replacement and their colors are noted. Define the events

       *A*:   {Both chips are of the same color}

       *B*:   {Both chips are red}

       *C*:   {At least one chip is red}

Find $P(B|A)$, $P(B|A')$, $P(B|C)$, $P(A|C)$, and $P(C|A')$.

### Applying the concepts

**3.24.** A soap manufacturer has decided to market two new brands. An analysis of current market conditions and a review of the firm's past successes and failures with new brands have led the manufacturer to believe that the simple events and the probabilities of their occurrence in this marketing experiment are as follows (where *S* means the brand succeeds and *F* means the brand fails in the first year):

| SIMPLE EVENTS | PROBABILITIES |
|---|---|
| *SS* | .16 |
| *SF* | .24 |
| *FS* | .24 |
| *FF* | .36 |

Define the following events:

       *A*:   {Both new brands are successful in the first year}

       *B*:   {At least one new brand is successful in the first year}

    a.   Find $P(A)$, $P(B)$, and $P(AB)$.

    b.   Find $P(A|B)$ and $P(B|A)$.

**3.25.** Six people apply for two identical positions in a company. Four are minority applicants and the remainder are nonminority. Define the following events:

A:  {Both persons selected are nonminority candidates}

B:  {Both persons selected are minority candidates}

C:  {At least one of the persons selected will be a minority candidate}

If all of the applicants are equally qualified and the choice is essentially a random selection of two applicants from the six available, find:

a.  $P(A)$    b.  $P(B)$    c.  $P(C)$    d.  $P(B|C)$

e.  Assume that the minority candidates are numbered 1, 2, 3, 4 for purposes of identification. Define the event

D:  {Minority candidate 1 is selected}

Find $P(D|C)$.

**3.26.**  A fast-food restaurant chain with 700 outlets in the United States describes the geographic location of its restaurants with the following table of percentages:

|  |  | REGION | | | |
|---|---|---|---|---|---|
|  |  | NE | SE | SW | NW |
| POPULATION OF CITY | Under 10,000 | 5% | 6% | 3% | 0 |
|  | 10,000–100,000 | 10% | 15% | 12% | 5% |
|  | Over 100,000 | 25%. | 4% | 5% | 10% |

A restaurant is to be chosen at random from the 700 to test market a new style of chicken.

a.  Given the restaurant chosen is in a city with population over 100,000, what is the probability that it is located in the northeast?

b.  Given the restaurant chosen is in the southeast, what is the probability that it is located in a city with population under 10,000?

c.  If the restaurant selected is located in the southwest, what is the probability that the city it is in has a population of 100,000 or less?

d.  If the restaurant selected is located in the northwest, what is the probability that the city it is in has a population of 10,000 or more?

**3.27.**  There are several methods of typing, or classifying, human blood. The most common procedure types blood into the general classifications of A, B, O, or AB. A method which is not as well known examines phosphoglucomutase (PGM) and classifies the blood into one of three main categories, 1-1, 2-1, or 2-2. Suppose that a certain geographical region of the United States has the PGM percentages as shown in the following table:

|  |  | 1-1 | 2-1 | 2-2 |
|---|---|---|---|---|
| RACE | White | 50.4% | 35.1% | 4.5% |
|  | Black | 6.7% | 2.9% | 0.4% |

A person is to be chosen at random from this geographical region.

    a.   What is the probability that a black person is chosen?

    b.   Given that a black is chosen, what is the probability he or she is PGM type 1-1?

    c.   Given that a white is chosen, what is the probability he or she is PGM type 1-1?

**3.28.**   The following table of percentages describes the 1,000 apartment units in a large suburban apartment complex:

| | | APARTMENT SIZE | | |
| | | One bedroom | Two bedroom | Three bedroom |
| --- | --- | --- | --- | --- |
| LOCATION WITHIN BUILDING | First floor | 8% | 30% | 10% |
| | Second floor | 22% | 20% | 10% |

The manager of the complex is considering installing new carpets in all the apartments. Before doing so, he wants to wear-test the brand in which he is interested for 6 months in one of the 1,000 apartments. He plans to choose one apartment at random from the 1,000 and install a test carpet.

    a.   What is the probability that he will choose a first floor, two bedroom apartment?

    b.   Given that he chooses a second floor apartment, what is the probability that the apartment has three bedrooms?

    c.   Given that he chooses a one bedroom apartment, what is the probability that the apartment is on the second floor?

    d.   Given that the apartment selected is on the first floor, what is the probability that it has two or three bedrooms?

**3.5
PROBABILITIES
OF UNIONS AND
INTERSECTIONS**

Since unions and intersections of events are themselves events, we can always calculate their probabilities by adding the probabilities of the simple events that compose them. However, when the probabilities of certain events are known, it is easier to use one or both of two rules to calculate the probability of unions and intersections. How and why these rules work will be illustrated by example.

EXAMPLE 3.11

A loaded (unbalanced) die is tossed and the up face is observed. The following two events are defined:

    $A$:   {Observe an even number}

    $B$:   {Observe a number less than 3}

Suppose that $P(A) = .4$, $P(B) = .2$, and $P(AB) = .1$. Find $P(A \cup B)$. [*Note:*  Assuming that we would know these probabilities in a practical situation is not very realistic, but the example will illustrate a point.]

Solution

By studying the Venn diagram in Figure 3.7, we can obtain information that will help us find $P(A \cup B)$. We can see that

$$P(A \cup B) = P(1) + P(2) + P(4) + P(6)$$

Also, we know that

$$P(A) = P(2) + P(4) + P(6) = .4$$
$$P(B) = P(1) + P(2) = .2$$
$$P(AB) = P(2) = .1$$

FIGURE 3.7
VENN DIAGRAM FOR
DIE TOSS

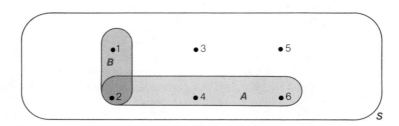

If we add the probabilities of the simple events that comprise events $A$ and $B$, we find

$$P(A) + P(B) = \overbrace{\underbrace{P(2) + P(4) + P(6)}_{P(A)}} + \overbrace{\underbrace{P(1) + P(2)}_{P(B)}}$$

$$= \overbrace{\underbrace{P(1) + P(2) + P(4) + P(6)}_{P(A \cup B)}} + \overbrace{\underbrace{P(2)}_{P(AB)}}$$

Thus, by subtraction,

$$P(A \cup B) = P(A) + P(B) - P(AB)$$
$$= .4 + .2 - .1 = .5$$

By studying the Venn diagram in Figure 3.8, you can see that the method used in Example 3.11 may be generalized to find the union of two events for any experiment.

FIGURE 3.8
VENN DIAGRAM OF
UNION

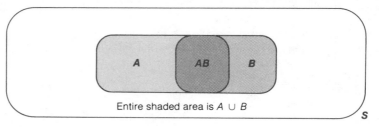

Entire shaded area is $A \cup B$

The probability of the union of two events, $A$ and $B$, can always be obtained by summing $P(A)$ and $P(B)$ and subtracting $P(AB)$. Note that we must subtract $P(AB)$ because the simple event probabilities in $AB$ have been included twice—once in $P(A)$ and once in $P(B)$.

The formula for calculating the probability of the union of two events, often called the additive rule of probability, is given in the box.

---

ADDITIVE RULE OF PROBABILITY

The probability of the union of events $A$ and $B$ is the sum of the probabilities of events $A$ and $B$ minus the probability of the intersection of events $A$ and $B$, i.e.,

$$P(A \cup B) = P(A) + P(B) - P(AB)$$

---

EXAMPLE 3.12    Hospital records show that 12% of all patients are admitted for surgical treatment, 16% are admitted for obstetrics, and 2% receive both obstetrics and surgical treatment. If a new patient is admitted to the hospital, what is the probability that the patient will be admitted either for surgery, obstetrics, or both?

Solution    Consider the following events:

$A$:  {A patient admitted to the hospital receives surgical treatment}

$B$:  {A patient admitted to the hospital receives obstetrics treatment}

Then, from the information above,

$$P(A) = .12 \qquad P(B) = .16$$

and the probability of the event that a patient receives both obstetrics and surgical treatments is

$$P(AB) = .02$$

The event that a patient admitted to the hospital receives either surgical treatment, obstetrics treatment, or both is the union, $A \cup B$. The probability of $A \cup B$ is given by the additive rule of probability:

$$P(A \cup B) = P(A) + P(B) - P(AB)$$
$$= .12 + .16 - .02 = .26$$

Thus, 26% of all patients admitted to the hospital receive either surgical treatment, obstetrics treatment, or both.

A very special relationship exists between the events $A$ and $B$ when $AB$ contains no simple events. In this case, we call the events $A$ and $B$ mutually exclusive events.

DEFINITION 3.8
Events *A* and *B* are mutually exclusive if *AB* contains no simple events.

**FIGURE 3.9**
**VENN DIAGRAM OF**
**MUTUALLY EXCLUSIVE**
**EVENTS**

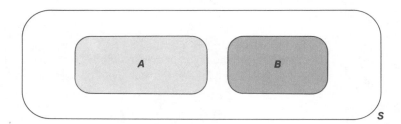

Figure 3.9 shows a Venn diagram of two mutually exclusive events. The events *A* and *B* have no simple events in common, i.e., *A* and *B* cannot occur simultaneously, and $P(AB) = 0$. Thus, we have the following important relationship:

If two events *A* and *B* are mutually exclusive, the probability of the union of *A* and *B* equals the sum of the probabilities of *A* and *B*, i.e.,

$$P(A \cup B) = P(A) + P(B)$$

EXAMPLE 3.13

Consider the experiment of tossing two balanced coins. Find the probability of observing *at least* one head.

Solution

Define the events

    *A*: {Observe at least one head}
    *B*: {Observe exactly one head}
    *C*: {Observe exactly two heads}

Note that

$$A = B \cup C$$

and that *BC* contains no simple events (see Figure 3.10). Thus, *B* and *C* are mutually exclusive, so that

$$P(A) = P(B \cup C) = P(B) + P(C)$$

$$= \frac{1}{2} + \frac{1}{4} = \frac{3}{4}$$

FIGURE 3.10
VENN DIAGRAM FOR
COIN TOSS
EXPERIMENT

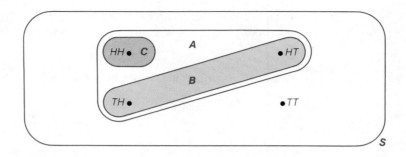

Although this example is very simple, the concept of writing events with verbal descriptions that include the phrases "at least" or "at most" as unions of mutually exclusive events is a very useful one. This enables us to find the probability of the event by adding the probabilities of the mutually exclusive events.

The second rule of probability, which will help us find the probability of the intersection of two events, is illustrated by Example 3.14.

EXAMPLE 3.14

An agriculturalist, who is interested in planting wheat next year, is concerned with the following events:

$A$:  {The production of wheat will be profitable}
$B$:  {A serious drought will occur}

Based on available information, the agriculturalist believes that the probability is .01 that production of wheat will be profitable *assuming* a serious drought will occur in the same year, and that the probability is .05 that a serious drought will occur. That is,

$$P(A|B) = .01$$
$$P(B) = .05$$

Based on the information provided, what is the probability that a serious drought will occur *and* that a profit will be made? That is, find $P(AB)$.

Solution

As you will see, we have already developed a formula for finding the probability of an intersection of two events. Recall that the conditional probability of $A$ given $B$ is

$$P(A|B) = \frac{P(AB)}{P(B)}$$

Multiplying both sides of this equation by $P(B)$, we obtain a formula for the probability of the intersection of events $A$ and $B$. This is often called the multiplicative rule of probability and is given by

$$P(AB) = P(A|B)P(B)$$

Thus,

$$P(AB) = (.01)(.05) = .0005$$

The probability that a serious drought occurs *and* the production of wheat is profitable is only .0005. As we might intuitively expect, this intersection is a very rare event.

---

MULTIPLICATIVE RULE OF PROBABILITY

$$P(AB) = P(A|B)P(B) = P(B|A)P(A)$$

---

Intersections often contain only a few simple events, and then the probability of an intersection is easy to calculate by summing the appropriate simple event probabilities. However, the formula for calculating intersection probabilities plays a very important role, particularly in an area of statistics known as Bayesian statistics.(More complete discussions of Bayesian statistics are contained in the references at the end of the chapter.)

EXAMPLE 3.15

Consider the experiment of tossing a fair coin twice and recording the up face on each toss. The following events are defined:

$A$:  {First toss is a head}

$B$:  {Second toss is a head}

Does *knowing* that event $A$ has occurred affect the probability that $B$ will occur?

Solution

Intuitively the answer should be no, since what occurs on the first toss should in no way affect what occurs on the second toss. Let us check our intuition. Recall the sample space for this experiment:

1.  Observe $HH$       3.  Observe $TH$
2.  Observe $HT$       4.  Observe $TT$

Each of these simple events has a probability of $\frac{1}{4}$. Thus,

$$P(B) = P(HH) + P(TH) \qquad \text{and} \qquad P(A) = P(HH) + P(HT)$$

$$= \frac{1}{4} + \frac{1}{4} = \frac{1}{2} \qquad\qquad\qquad = \frac{1}{4} + \frac{1}{4} = \frac{1}{2}$$

Now, what is $P(B|A)$?

$$P(B|A) = \frac{P(AB)}{P(A)} = \frac{P(HH)}{P(A)}$$

$$= \frac{\frac{1}{4}}{\frac{1}{2}} = \frac{1}{2}$$

We can now see that $P(B) = \frac{1}{2}$ and $P(B|A) = \frac{1}{2}$. Knowing that the first toss resulted in a head does not affect the probability that the second toss will be a head. The probability is ½ whether or not we know the result of the first toss. When this occurs, we say that the two events $A$ and $B$ are independent.

---

**DEFINITION 3.9**

Events $A$ and $B$ are independent if the occurrence of $B$ does not alter the probability that $A$ has occurred, i.e.,

$$P(A|B) = P(A)$$

When events $A$ and $B$ are independent it will also be true that

$$P(B|A) = P(B)$$

Events that are not independent are said to be dependent.

---

EXAMPLE 3.16

Consider the experiment of tossing a fair die and let

    $A$:  {Observe an even number}
    $B$:  {Observe a number less than or equal to 4}

Are events $A$ and $B$ independent?

Solution

The Venn diagram for this experiment is shown in Figure 3.11. We first calculate

$$P(A) = P(2) + P(4) + P(6) = \frac{1}{2}$$

$$P(B) = P(1) + P(2) + P(3) + P(4) = \frac{4}{6} = \frac{2}{3}$$

$$P(AB) = P(2) + P(4) = \frac{2}{6} = \frac{1}{3}$$

Now assuming $B$ has occurred, the conditional probability of $A$ given $B$ is

$$P(A|B) = \frac{P(AB)}{P(B)} = \frac{\frac{1}{3}}{\frac{2}{3}} = \frac{1}{2} = P(A)$$

**FIGURE 3.11**

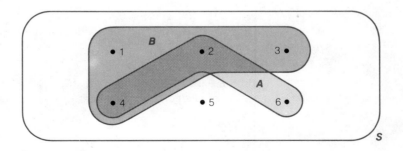

Thus, assuming that event $B$ occurs does not alter the probability of observing an even number—it remains $\frac{1}{2}$. Therefore, the events $A$ and $B$ are independent. Note that if we calculate the conditional probability of $B$ given $A$, our conclusion is the same.

$$P(B|A) = \frac{P(AB)}{P(A)} = \frac{\frac{1}{3}}{\frac{1}{2}} = \frac{2}{3} = P(B)$$

**EXAMPLE 3.17**

Refer to the consumer product complaint study in Example 3.10. The percentages of complaints of various types in the pre- and post-guarantee periods are shown in Table 3.2. Define the following events:

A: {Cause of complaint is product appearance}

B: {Complaint occurred during the guarantee term}

Are $A$ and $B$ independent events?

**Solution**

Events $A$ and $B$ are independent if $P(A|B) = P(A)$. We calculated $P(A|B)$ in Example 3.10 to be .51, and from Table 3.2 we can see that

$$P(A) = .32 + .03 = .35$$

Therefore, $P(A|B)$ is not equal to $P(A)$, and $A$ and $B$ are not independent events.

We will make three final points about independence. The first is that the property of independence, unlike the mutually exclusive property, cannot be shown on or gleaned from a Venn diagram and you cannot trust your intuition. In general, the only way to check for independence is by performing the calculations of the probabilities in the definition.

The second point concerns the relationship between the mutually exclusive and independence properties. Suppose that events $A$ and $B$ are mutually exclusive, as shown in Figure 3.9. Are these events independent or dependent? That is, does the assumption that $B$ occurs alter the probability of the occurrence of $A$? It certainly does, because if we assume that $B$ has occurred, it is impossible for $A$ to have occurred simultaneously. Thus, mutually exclusive events are dependent events.

The third point is that the probability of the intersection of independent events is very easy to calculate. Referring to the formula for calculating the probability of an intersection, we find

$$P(AB) = P(B)P(A|B)$$

Thus, since $P(A|B) = P(A)$ when $A$ and $B$ are independent, we have the following useful rule:

---

If events $A$ and $B$ are independent, the probability of the intersection of $A$ and $B$ equals the product of the probabilities of $A$ and $B$, i.e.,

$$P(AB) = P(A)P(B)$$

---

In the die toss experiment, we showed in Example 3.16 that the events $A$: {Observe an even number} and $B$: {Observe a number less than or equal to 4} are independent if the die is fair. Thus,

$$P(AB) = P(A)P(B) = \left(\frac{1}{2}\right)\left(\frac{2}{3}\right) = \frac{1}{3}$$

This agrees with the result

$$P(AB) = P(2) + P(4) = \frac{2}{6} = \frac{1}{3}$$

that we obtained in the example.

EXAMPLE 3.18

In Example 3.4, a bank considered the problem of apportioning its service windows according to two types of accounts: personal or commercial. In the example, we discussed the problem of finding the probability that one, two, or, in general, $k$ customers arriving at the bank possessed a commercial account. We are now ready to find the probability that both of two customers arriving at the bank possess commercial accounts. Suppose that a study of arriving customers showed that 20% of arriving customers intend to utilize their commercial accounts at the bank.

**a.** If two customers arrive at the bank, what is the probability that they will both utilize a commercial account?

**b.** If $k$ customers arrive at the bank, what is the probability that all will utilize commercial accounts?

Solution

**a.** Let $C_1$ be the event that customer 1 will utilize a commercial account and let $C_2$ be a similar event for customer 2. The event that *both* customers will utilize commercial accounts is the intersection $C_1C_2$. Then, since it is not unreasonable to assume that the service requirements of the customers would be independent of one another, the probability that both will utilize commercial accounts is

$$P(C_1C_2) = P(C_1)P(C_2)$$
$$= (.2)(.2) = (.2)^2 = .04$$

**b.** Let $C_i$ represent the event that the $i$th customer will utilize a commercial account. Then the event that all three of three arriving customers will utilize a commercial account is the intersection of the event $C_1C_2$ (from part a) with the event $C_3$. Assuming independence of the events, $C_1$, $C_2$, and $C_3$, we have

$$P(C_1C_2C_3) = P(C_1C_2)P(C_3)$$
$$= (.2)^2(.2) = (.2)^3 = .008$$

Noting the pattern, you can see that the probability that all $k$ out of $k$ arriving customers will utilize a commercial account is the probability of $C_1C_2 \cdots C_k$, or

$$P(C_1C_2 \cdots C_k) = (.2)^k \qquad \text{for} \quad k = 1, 2, 3, \ldots$$

Learning the mechanics

**3.29.** Three fair coins are tossed and the following events are defined:

  $A$: {Observe at least one head}

  $B$: {Observe exactly two heads}

  $C$: {Observe exactly two tails}

  $D$: {Observe at most one head}

a. Sum the probabilities of the appropriate simple events to find: $P(A)$, $P(B)$, $P(C)$, $P(D)$, $P(AB)$, $P(AD)$, $P(CB)$, and $P(BD)$.

b. Use the formulas of this section and your answers to part a to calculate: $P(A \cup B)$, $P(D \cup A)$, $P(B \cup C)$, $P(B|A)$, $P(A|D)$, and $P(C|B)$.

c. Are events $A$ and $B$ independent? Mutually exclusive?

d. Are events $A$ and $D$ independent? Mutually exclusive?

e. Are events $B$ and $C$ independent? Mutually exclusive?

**3.30.** Two fair dice are tossed and the following events are defined:

  $A$: {Sum of the numbers showing is odd }

  $B$: {Sum of the numbers showing is 8, 9, 11, *or* 12}

a. Are events $A$ and $B$ independent? Why?

b. Are events $A$ and $B$ mutually exclusive? Why?

**3.31.** Two events, $A$ and $B$, are mutually exclusive with $P(A) = .2$ and $P(B) = .3$.

a. Find $P(A \cup B)$.

b. Find $P(A|B)$.

c. Find $P(B|A)$.

d. Are $A$ and $B$ independent? Why?

**3.32.** For two events, $A$ and $B$, $P(A) = 1/2$ and $P(B) = 1/3$.

a. If $A$ and $B$ are independent, find $P(AB)$, $P(A|B)$, and $P(A \cup B)$.

b. If $A$ and $B$ are dependent, with $P(A|B) = 3/5$, find $P(AB)$ and $P(B|A)$.

Applying the concepts

**3.33.** Each of a random sample of fifty people was asked to name his or her favorite soft drink. The responses are shown below:

| | |
|---|---|
| Pepsi-Cola | 18 |
| Coca-Cola | 16 |
| Mr. Pibb | 4 |
| Seven-Up | 6 |
| Sprite | 4 |
| Nehi Orange | 1 |
| Dr Pepper | 1 |

Suppose a person is selected at random from the survey. Let $A$ be the event that the person preferred a soft drink bottled by the Coca-Cola Company (Coca-Cola, Mr.

Pibb, or Sprite). Let $B$ be the event that the person did *not* choose a cola (either Pepsi-Cola or Coca-Cola). Find the following:

a. $P(A)$      b. $P(B)$      c. $P(A \cup B)$

d. $P(AB)$      e. $P(AB')$      f. $P(A|B)$

g. Are $A$ and $B$ independent events? Why?

h. Are $A$ and $B$ mutually exclusive events? Why?

**3.34.** A county welfare agency employs thirty welfare workers who interview prospective food stamp recipients. Periodically, the supervisor selects, at random, the forms completed by two workers to audit for illegal deductions. Unknown to the supervisor, six of the workers have regularly been giving illegal deductions to applicants.

    a. What is the probability that the first worker chosen has been giving illegal deductions?

    b. Given that the first worker chosen has been giving illegal deductions, what is the probability that the second worker chosen has also been giving illegal deductions?

    c. What is the probability that neither of the two workers chosen has been giving illegal deductions?

**3.35.** The percentages of all the teenagers, aged 14 to 19, in two small communities fell in the following six community-delinquency categories:

|  |  | COMMUNITY A | B |
|---|---|---|---|
| | Nondelinquents | 28% | 42% |
| DELINQUENCY | First offenders | 5% | 15% |
| | Repeat offenders | 7% | 3% |

Suppose a teenager was chosen at random from one of the two communities and the following events are defined:

$A$: {Teenager chosen is from community A}

$B$: {Teenager chosen is from community B}

$C$: {Teenager chosen is not delinquent}

$D$: {Teenager chosen has committed at least one crime}

Find the following probabilities:

a. $P(C')$      b. $P(A \cup D)$      c. $P(C|B)$      d. $P(D|A)$

**3.36.** A local YMCA has a membership of 1,000 people and operates facilities that include both a running track and an indoor swimming pool. Before setting up the new schedule of operating hours for the two facilities, the manager would like to know how many members regularly use each facility. A survey of the membership indicates that 65% use the running track, 45% use the swimming pool, and 5% use neither.

a. Construct a Venn diagram to describe the population.

b. If one member is chosen at random, what is the probability that the member uses both the track and the pool?

c. If a member is chosen at random, what is the probability that the member uses only the track?

**3.37.** The probability that a certain electronics component fails when first used is .10. If it does not fail immediately, the probability that it lasts for 1 year is .99. What is the probability that a new component will last 1 year?

**3.38.** Consider an experiment that consists of drawing 1 card from a standard 52 card playing deck and define the following events:

$A$:  {Drawing a black card}

$B$:  {Drawing a face card (J, Q, K)}

$C$:  {Drawing a heart or spade}

$D$:  {Drawing a diamond}

a. Are $A$ and $B$ mutually exclusive? Are they independent? Explain.

b. Are $A$ and $C$ independent? Explain.

c. Are $C$ and $D$ mutually exclusive? Are they independent? Explain.

**3.39.** Some strings of Christmas tree lights are wired in series; thus, if one bulb fails, the entire string goes out. Suppose the probability of an individual bulb failing during a certain period of time is .05. What is the probability that a string of ten lights goes out during that period of time? What assumption did you make concerning the light bulbs? [*Hint:* The complement of the event, At least one of the ten lights fails, is the event, None of the ten lights fails.]

**3.40.** A microwave oven manufacturer claims that only 10% of the ovens it makes will need repair in the first year. Suppose three recent customers are independently selected.

a. If the manufacturer is correct, what is the probability that at least two of the three ovens will need repair in the first year?

b. If at least two of the three customers' ovens need repair in the first year, what inference may be made about the manufacturer's claim?

**3.41.** Psychologists tend to believe that there is a relationship between aggressiveness and order of birth. To test this belief, a psychologist chose 500 elementary school students at random and administered to each a test designed to measure the student's aggressiveness. Each student was classified according to one of four categories. The percentages of students falling in the four categories are shown below:

|  | FIRSTBORN | NOT FIRSTBORN |
|---|---|---|
| AGGRESSIVE | 15% | 15% |
| NOT AGGRESSIVE | 25% | 45% |

a. If one student is chosen at random from the 500, what is the probability that the student is firstborn?

b. What is the probability that the student is aggressive?

c. What is the probability that the student is aggressive, given the student was firstborn?

d. If

    *A*:  {Student chosen is aggressive}

    *B*:  {Student chosen is firstborn}

are *A* and *B* independent events? Explain.

**3.42.** A popular dice game, called "craps," is played in the following manner: A player starts by rolling two dice. If the result is a 7 or 11, the player wins. For any other sum appearing on the dice, the player continues to roll the dice until that outcome reoccurs (in which case, the player wins) or until a 7 or 11 occurs (in which case, the player loses). If on any roll the outcome is 2 (snake-eyes), the game is over and the player loses.

a. What is the probability that a player wins the game on the first roll of the dice (assume the dice are balanced)?

b. What is the probability that a player loses the game on the first roll of the dice?

c. If the player throws a total of 3 on the first roll, what is the probability that the game ends on the next roll?

**3.43.** Despite penicillin and other antibiotics, bacterial pneumonia still kills thousands of Americans every year. Just recently, the United States Food and Drug Administration (FDA) approved the use of a new antipneumonia vaccine, called Pneumovax. It is designed especially for elderly or debilitated patients, who are usually the most vulnerable to bacterial pneumonia. Field trials proved the new vaccine to be 90% effective in stimulating the production of antibodies to pneumonia-producing bacteria (i.e., 90% successful in preventing a person exposed to pneumonia-producing bacteria from acquiring the disease). Suppose the probability of an elderly or debilitated person being exposed to these bacteria is .40 (whether inoculated or not), and after being exposed, the probability of each person contracting bacterial pneumonia if not inoculated with the vaccine is .95. Find the probability that an elderly or debilitated person inoculated with this new vaccine acquires pneumonia. What is the probability if this person has not been inoculated?

**3.6
PROBABILITY AND
STATISTICS:
AN EXAMPLE**

We have introduced a number of new concepts in the preceding sections, and this makes the study of probability a particularly arduous task. It is therefore very important to establish clearly the connection between probability and statistics, which we will do in the remaining chapters. However, we present one brief example in this section so that you can begin to understand why some knowledge of probability is important in the study of statistics.

Suppose a psychologist is researching the hypothesis that rats who have been trained will pass on at least part of the training to their offspring. To test the hypothesis, three offspring (no two with the same parents) of trained rats are randomly selected and subjected to a training test. It is known from many previous experiments that the relative frequency distribution of the scores for untrained rats is mound-shaped, with a mean of 60 and a standard deviation of 10. Suppose all three of the trained rats' offspring score more than 70 on the test. What can the research psychologist conclude?

FIGURE 3.12
RELATIVE FREQUENCY
DISTRIBUTION FOR
TRAINING
TEST SCORES

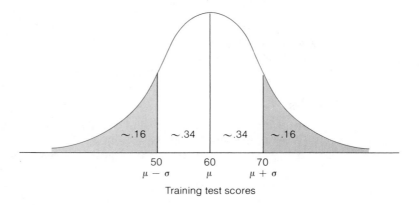

Training test scores

The relative frequency distribution of the scores for untrained rats is shown in Figure 3.12. If the distribution is mound-shaped and approximately symmetric about the mean, we can conclude that approximately 16% of untrained rats will score more than 70 on the test (see Table 2.6). Now, define the events

$A_1$:  {Offspring 1 scores more than 70}

$A_2$:  {Offspring 2 scores more than 70}

$A_3$:  {Offspring 3 scores more than 70}

We want to find $P(A_1A_2A_3)$, the probability that all three offspring score more than 70 on the training test.

Since the offspring are selected so that they have different parents, it may be plausible to assume that the events $A_1$, $A_2$, and $A_3$ are independent. That is,

$$P(A_2|A_1) = P(A_2)$$

In words, knowing that the first offspring scores more than 70 on the test does not affect the probability that the second offspring scores more than 70. With the assumption of independence, we can calculate the probability of the intersection by multiplying the individual probabilities:

$$P(A_1A_2A_3) = P(A_1)P(A_2)P(A_3)$$
$$\approx (.16)(.16)(.16) = .004096$$

Thus, the probability that the research psychologist will observe all three offspring score more than 70 is only about .004 *if the offspring are untrained.* If this event were to occur, the psychologist might conclude that it lends credence to the theory that the offspring inherit some of the parents' training, *since it is so unlikely to occur if they are untrained.* Such a conclusion would be an application of the rare event approach to statistical inference, and you can see that the basic principles of probability play an important role.

**3.7
RANDOM
SAMPLING**

How a sample is selected from a population is of vital importance in statistical inference because the probability of an observed sample will be used to infer the characteristics of the sampled population. To illustrate, suppose you deal yourself 4 cards from a deck of 52 cards and all 4 cards are aces. Do you conclude that your deck is an ordinary bridge deck, containing only 4 aces, or do you conclude that the deck is stacked with more than 4 aces? It depends on how the cards were drawn. If the 4 aces were always placed at the top of a standard bridge deck, drawing 4 aces is not unusual—it is certain. On the other hand, if the cards are thoroughly mixed, drawing 4 aces in a sample of 4 cards is highly improbable. The point, of course, is that in order to use the observed sample of 4 cards to draw inferences about the population (the deck of 52 cards), you need to know how the sample was selected from the deck.

One of the simplest and most frequently employed sampling procedures produces what is known as a random sample.

---

**DEFINITION 3.10**
If *n* elements are selected from a population in such a way that every set of *n* elements in the population has an equal probability of being selected, the *n* elements are said to be a random sample.*

---

EXAMPLE 3.19

Suppose a lottery consists of ten tickets (this number is small to simplify our example). One ticket stub is to be chosen and the corresponding ticket holder will receive a generous prize. How would you select this ticket stub so that the prize will be awarded fairly?

Solution

If the prize is to be awarded fairly, it seems reasonable to require that each ticket stub has the same probability of being drawn. That is, each stub should have a probability of $\frac{1}{10}$ of being selected. A method to achieve the objective of equal selection probabilities is to *thoroughly mix* the ten stubs and *blindly* pick one of the stubs. If this procedure were repeatedly used, each time replacing the selected stub, a particular stub should be chosen approximately $\frac{1}{10}$ of the time in a long series of draws. This method of sampling is known as random sampling.

If a population is not too large and the elements can be numbered on slips of paper,

---

*Strictly speaking, this is a simple random sample. There are many different types of random samples. The simple random sample is the most common.

poker chips, etc., you can physically mix the slips of paper or chips and remove $n$ elements from the total. The numbers that appear on the chips selected would indicate the population elements to be included in the sample. Such a procedure will not guarantee a random sample, because it is often difficult to achieve a thorough mix, but it provides a reasonably good approximation to random sampling.

EXAMPLE 3.20

Suppose you wish to randomly sample 5 (we will keep the number in the sample small to simplify our example) from a population of 100,000 households. Give a procedure for selecting this random sample.

Solution

Since there are 100,000 households, it is not feasible to select a random sample by numbering slips of paper (or poker chips, etc.), mixing them, and choosing 5. Instead, we will enlist the aid of Table I in the Appendix.

First, we number the households in the population from 1 to 100,000. Then, we turn to a page of Table I, say the first page. (A partial reproduction of the first page of Table I is shown in Figure 3.13.) Now, randomly select a starting number, say the random number appearing in the third row, second column. This number is 48360. Proceed down the second column to obtain the remaining four random numbers. The five selected random numbers are shaded in Figure 3.13. Using the first five digits to represent the households from 1 to 99,999 and the number 00000 to represent household 100,000, you can see that the households numbered

| 48,360 | 93,093 | 39,975 | 6,907 | 72,905 |

should be included in your sample.

FIGURE 3.13
PARTIAL
REPRODUCTION OF
TABLE I IN THE
APPENDIX

| ROW \ COLUMN | 1 | 2 | 3 | 4 | 5 | 6 |
|---|---|---|---|---|---|---|
| 1 | 10480 | 15011 | 01536 | 02011 | 81647 | 91646 |
| 2 | 22368 | 46573 | 25595 | 85393 | 30995 | 89198 |
| 3 | 24130 | 48360 | 22527 | 97265 | 76393 | 64809 |
| 4 | 42167 | 93093 | 06243 | 61680 | 07856 | 16376 |
| 5 | 37570 | 39975 | 81837 | 16656 | 06121 | 91782 |
| 6 | 77921 | 06907 | 11008 | 42751 | 27756 | 53498 |
| 7 | 99562 | 72905 | 56420 | 69994 | 98872 | 31016 |
| 8 | 96301 | 91977 | 05463 | 07972 | 18876 | 20922 |
| 9 | 89579 | 14342 | 63661 | 10281 | 17453 | 18103 |
| 10 | 85475 | 36857 | 53342 | 53988 | 53060 | 59533 |
| 11 | 28918 | 69578 | 88231 | 33276 | 70997 | 79936 |
| 12 | 63553 | 40961 | 48235 | 03427 | 49626 | 69445 |
| 13 | 09429 | 93969 | 52636 | 92737 | 88974 | 33488 |
| 14 | 10365 | 61129 | 87529 | 85689 | 48237 | 52267 |
| 15 | 07119 | 97336 | 71048 | 08178 | 77233 | 13916 |

Table I in the Appendix is just one example of a table of random numbers. Most samplers use such a table to obtain random samples. Random number tables are constructed in such a way that every number occurs with (approximately) equal probability. Further, the occurrence of any one number in a position is independent of any of the other numbers that appear in the table. To use a table of random numbers, number the *N* elements in the population from 1 to *N*. Then turn to Table I and select a starting number in the table. Proceeding from this number either across the rows or down the column, remove and record *n* numbers from the table. Use only the necessary number of digits in each random number to identify the element to be included in the sample.

## SUMMARY

We have developed some of the basic tools of probability to enable us to assess the probability of various sample outcomes given a specific population structure. Although many of the examples we presented were of no practical importance, they accomplished their purpose if you now understand the concepts and definitions necessary for a basic understanding of probability.

In the next several chapters we will present probability models that can be used to solve practical problems. You will see that for most applications, we will need to make inferences about unknown aspects of these probability models, i.e., we will need to apply inferential statistics to the problem.

## SUPPLEMENTARY EXERCISES

Learning the mechanics

**3.44.** A fair die is tossed, and the up face is noted. If the number is even, the die is tossed again; if the number is odd, a fair coin is tossed. Define the events:

A:  {A head appears on the coin}

B:  {The die is tossed only one time}

a. List the simple events in the sample space.
b. Give the probability for each of the simple events.
c. Find $P(A)$ and $P(B)$.
d. Identify the simple events in $A'$, $B'$, $AB$, and $A \cup B$.
e. Find $P(A')$, $P(B')$, $P(AB)$, $P(A \cup B)$, $P(A|B)$, and $P(B|A)$.
f. Are *A* and *B* mutually exclusive events? Independent events? Why?

**3.45.** The following Venn diagram shows a sample space containing 6 simple events and 3 events, *A, B,* and *C:*

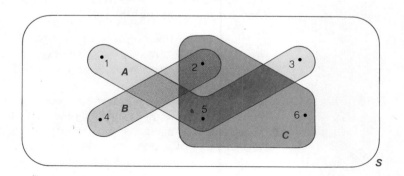

The probabilities of the simple events are

$$P(1) = .4 \qquad P(2) = .1 \qquad P(3) = .1$$
$$P(4) = .2 \qquad P(5) = .1 \qquad P(6) = .1$$

a. Find $P(AB)$, $P(BC)$, $P(A \cup C)$, $P(A \cup B \cup C)$, $P(B')$, $P(A'B)$, $P(B|C)$, and $P(B|A)$.

b. Are $A$ and $B$ independent? Mutually exclusive? Why?

c. Are $B$ and $C$ independent? Mutually exclusive? Why?

**3.46.** Two events, $A$ and $B$, are independent, with $P(A) = .3$ and $P(B) = .1$.

a. Are $A$ and $B$ mutually exclusive? Why?

b. Find $P(A|B)$ and $P(B|A)$.

c. Find $P(A \cup B)$.

**3.47.** A balanced die is thrown once. If a 4 appears, a ball is drawn from urn 1, otherwise a ball is drawn from urn 2. Urn 1 contains four red, three white, and three black balls. Urn 2 contains six red and four white balls.

a. Find the probability that a red ball is drawn.

b. Find the probability that urn 1 was used given that a red ball was drawn.

**3.48.** Consider an experiment that consists of simultaneously flipping a nickel, dime, and quarter. List the simple events. Let $A$ be the event that the up face on the dime is heads. Let $B$ be the event of observing at least two heads. Let $C$ be the event that the outcomes (on the up faces) on all three coins are the same.

a. Identify the simple events in $BC$.

b. Identify the simple events in $A \cup B$.

c. Find $P(BC)$ by summing the probabilities of the simple events in $BC$.

d. Find $P(A \cup B)$ by summing the probabilities of the simple events in $A \cup B$.

e. Find $P(B \cup C)$ by summing the probabilities of the simple events in $B \cup C$.

f. Find $P(B \cup C)$ using the additive rule of probability.

Applying the concepts

**3.49.** In college basketball games a player may be afforded the opportunity to shoot two consecutive foul shots (free throws).

a. Suppose a particular player who scores on 80% of his foul shots has been awarded two free throws. If the two throws are considered independent, what is the probability that the player scores on both shots? Exactly one? Neither shot?

b. Suppose a particular player who scores on 80% of his first attempted foul shots has been awarded two free throws, and the outcome on the second shot is dependent on the outcome of the first shot. In fact, if this player makes the first shot he makes 90% of the second shots, and if he misses the first shot, he makes 70% of the second shots. In this case, what is the probability that the player scores on both shots? Exactly one? Neither shot?

c. In parts a and b, we considered two ways of *modeling* the probability a basketball player scores on two consecutive foul shots. Which model do you feel is a more realistic attempt to explain the outcome of shooting foul shots, i.e., do you think two consecutive foul shots are independent or dependent? Explain.

**3.50.** Two companies, A and B, package and market a chemical substance and claim .15 of the total weight of the substance is sodium. However, a careful survey of 4,000 packages (half from each company) indicated the actual proportion varied around .15, with the following results:

|  |  | PROPORTION OF SODIUM | | | |
|---|---|---|---|---|---|
|  |  | Less than .100 | .100–.149 | .150–.199 | Over .200 |
| CHEMICAL BRAND | A | 25% | 10% | 10% | 5% |
|  | B | 5% | 5% | 10% | 30% |

Suppose a package is chosen at random from the 4,000 packages. If

$A$: {Package chosen was brand A}

$B$: {Package chosen was brand B}

$C$: {Package chosen contained less than .100 sodium}

$D$: {Package chosen contained between .100 and .149 sodium}

$E$: {Package chosen contained between .150 and .199 sodium}

$F$: {Package chosen contained over .200 sodium}

then describe the characteristics of a package portrayed by the following events:

a. $A \cup B$    b. $B \cup F$    c. $AD$

d. $EB$    e. $(AC) \cup (AD)$

**3.51.** Refer to Exercise 3.50. Find the probabilities of the following events by summing the probabilities of the appropriate simple events.

a. $A, B, C, D, E, F$    b. $A \cup B$    c. $BC$

d. $AF$    e. $AB$    f. $C \cup D$

g. $(AC) \cup (AD)$

**3.52.** A survey of the faculty of a large university yielded the following breakdown according to sex and marital status:

|  |  | MARITAL STATUS | | | |
|---|---|---|---|---|---|
|  |  | Married | Single | Divorced | Widowed |
| SEX | Male | 60% | 12% | 8% | 5% |
|  | Female | 10% | 3% | 2% | 0 |

Suppose a faculty member is chosen at random. If

> $A$:   {Faculty member chosen is female}
>
> $B$:   {Faculty member chosen is male}
>
> $C$:   {Marital status of chosen faculty member is married}
>
> $D$:   {Marital status of chosen faculty member is single}
>
> $E$:   {Marital status of chosen faculty member is widowed}
>
> $F$:   {Marital status of chosen faculty member is divorced}

then describe the characteristics of a faculty member portrayed by the following events:

| | | |
|---|---|---|
| a.  $A \cup B$ | b.  $BF$ | c.  $C \cup D$ |
| d.  $EF$ | e.  $F \cup B$ | f.  $E \cup F$ |

**3.53.**   Refer to Exercise 3.52. Find the probability of the following events by summing the probabilities of the appropriate simple events:

| | | |
|---|---|---|
| a.  $A, B, C, D, E, F$ | b.  $A \cup B$ | c.  $BF$ |
| d.  $C \cup D$ | e.  $EF$ | f.  $F \cup B$ |

**3.54.**   In the game of parcheesi each player rolls a pair of dice on each turn. In order to begin the game you must throw a 5 on at least one of the dice, or a total of 5 on the two dice. What is the probability that you can begin the game on your first turn? The second turn? The third turn? The $n$th turn?

**3.55.**   Entomologists are often interested in studying the effect of chemical attractants (pheromones) on insects. One common technique is to release several insects equidistant from the pheromone being studied and from a control substance. If the pheromone has an effect, more insects will travel toward it rather than toward the control. Otherwise, the insects are equally likely to travel in either direction. Suppose the pheromone under study has no effect so that it is equally likely that an insect will move toward either the pheromone or the control.

>   a.   If five insects are released, what is the probability that all five travel toward the pheromone?
>
>   b.   Exactly four?

**3.56.**   The probability of the union of three events, $A$, $B$, and $C$, is given by the expression (proof omitted):

$$P(A \cup B \cup C) = P(A) + P(B) + P(C) - P(AB) - P(AC) - P(BC) + P(ABC)$$

>   a.   Sketch a Venn diagram showing the events $A$, $B$, and $C$ and their intersections. Let the areas in the diagram corresponding to events $A$, $B$, $C$, $AB$, $AC$, $BC$, and $ABC$ represent their probabilities. Then use the diagram to justify the formula for $P(A \cup B \cup C)$.
>
>   b.   A national poll of biostatisticians was conducted to ascertain their professional responsibilities. An analysis of the responses gave the following dis-

tribution of professional responsibilities:

A:  {Research}                          40%
B:  {Professional consultation}         64%
C:  {Data collection and analysis}      36%

Suppose 10% are involved in all three activities; 15% are involved in both A and C; and 17% are involved in both A and B. Using this information, find the percentage of all biostatisticians that are involved in both B and C (i.e., in professional consultation and in data collection and analysis). Assume that $P(A \cup B \cup C) = 1$.

**3.57.** A manufacturer of 35 mm cameras knows that a shipment of thirty cameras sent to a large discount store contains six defective cameras. The manufacturer also knows that the store will choose two of the cameras at random, test them, and accept the shipment if neither is defective.

    a.  What is the probability that the first camera chosen by the store will be defective?
    b.  Given that the first camera chosen passed inspection, what is the probability that the second camera chosen will fail inspection?
    c.  What is the probability that the shipment will be accepted?

**3.58.** The probability that a mini-computer salesperson sells a computer to a prospective customer on the first visit to the customer is .4. If the salesperson fails to make the sale on the first visit, the probability that the sale will be made on the second visit is .65. The salesperson never visits a prospective customer more than twice. What is the probability that the salesperson will make a sale to a particular customer?

**3.59.** Seventy-five percent of all women who submit to pregnancy tests are really pregnant. A certain pregnancy test gives a false positive result with probability .02 and a valid positive result with probability .99. If a particular woman's test is positive, what is the probability that she really is pregnant? [*Hint:* If A is the event that a woman is pregnant and B is the event that the pregnancy test is positive, then B is the union of the two mutually exclusive events AB and A'B. Also, the probability of a "false positive result" may be written as $P(B|A') = .02$.]

**3.60.** Blackjack, a favorite game of gamblers, is played by a dealer and at least one opponent and uses a 52 card bridge deck. At the outset of the game, 2 cards are dealt to the player and 2 cards to the dealer. Drawing an ace and a face card is called "blackjack." If the dealer draws it, he or she automatically wins. If the dealer does not draw a blackjack and the player does, the player wins.

    a.  What is the probability that the dealer will draw a blackjack?
    b.  What is the probability that the player wins with a blackjack?

**3.61.** A small brewery has two bottling machines. Machine A produces 75% of the bottles and machine B produces 25%. One out of every twenty bottles filled by A is rejected for some reason, while one out of every thirty bottles from B is rejected. What proportion of bottles is rejected? What is the probability that a bottle comes from machine A, given that it is accepted?

**3.62.** A clinical psychologist is asked to view tapes in which each of six experimental subjects is discussing his or her recent dreams. Three of the six subjects have previously been classified as "high-anxiety" individuals, and the other three as "low-anxiety." The psychologist is told only that there are three of each type and is asked to select the three high-anxiety subjects.

    a.   List all possible outcomes (simple events) for this experiment.

    b.   Assuming the psychologist guesses at the classifications of the subjects, assign probabilities to the simple events.

    c.   Find the probability that the psychologist guesses all classifications correctly.

    d.   Find the probability that the psychologist guesses at least two of the three high-anxiety subjects correctly.

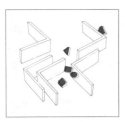

**ON YOUR OWN . . .**

Obtain a standard deck of 52 playing cards (the kind commonly used for bridge, poker, or solitaire). An experiment will consist of drawing one card at random from the deck of cards and recording which card was observed. This will be simulated by shuffling the deck thoroughly and observing the top card. Consider the following two events:

    $A$:   {Card observed is a heart}

    $B$:   {Card observed is an ace, king, queen, or jack}

**a.**   Find $P(A)$, $P(B)$, $P(AB)$, and $P(A \cup B)$.

**b.**   Conduct the experiment ten times and record the observed card each time. Be sure to return the observed card each time and thoroughly shuffle the deck before making the draw. After 10 cards have been observed, calculate the proportion of observations that satisfy event $A$, event $B$, event $AB$, and event $A \cup B$. Compare the observed proportions with the true probabilities calculated in part a.

**c.**   Conduct the experiment forty more times to obtain a total of fifty observed cards. Now, calculate the proportion of observations that satisfy event $A$, event $B$, event $AB$, and event $A \cup B$. Compare these proportions with those found in part b and the true probabilities found in part a.

**d.**   Conduct the experiment fifty more times to obtain a total of 100 observations. Compare the observed proportions for the 100 trials with those found previously. What comments do you have concerning the different proportions found in parts b, c, and d as compared to the true probabilities found in part a? How do you think the observed proportions and true probabilities would compare if the experiment were conducted 1,000 times? 1,000,000 times?

**REFERENCES**

Feller, W. *An introduction to probability theory and its applications,* 3d ed. Vol. I. New York: Wiley, 1968. Chapters 1, 3, 4, and 5.

Kuehn, A. ''An Analysis of the Dynamics of Consumer Behavior and Its Implications for Marketing Management.'' Unpublished doctoral dissertation, Graduate School of Industrial Administration, Carnegie Institute of Technology, 1958.

Mendenhall, W., Scheaffer, R. L., & Wackerly, D. *Mathematical statistics with applications.* 2d ed. North Scituate, Mass.: Duxbury, 1980. Chapter 2.

Parzen, E. *Modern probability theory and its applications.* New York: Wiley, 1960. Chapters 1 and 2.

Preston, M. G., & Baratta, P. ''An experimental study of the auction-value of an uncertain outcome.'' *American Journal of Psychology,* 1948, *61,* 183–193.

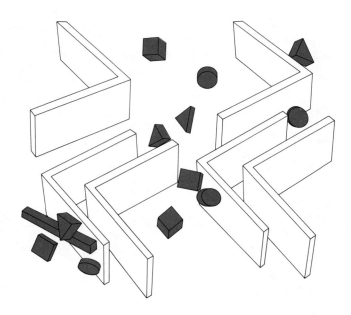

# 4

# RANDOM VARIABLES AND PROBABILITY DISTRIBUTIONS

**WHERE WE'VE BEEN . . .**

By illustration we indicated in Chapter 3 how probability would be used to make an inference about a population from data contained in an observed sample. We also noted that probability would be used to measure the reliability of the inference.

**WHERE WE'RE GOING . . .**

Most experimental events in Chapter 3 were events described in words and denoted by capital letters. In real life, most sample observations are numerical—in other words, numerical data. In this chapter, we will learn that data are observed values of random variables. We will study several important random variables and will learn how to find the probabilities of specific numerical outcomes.

You may have noticed that many of the examples of experiments in Chapter 3 generated quantitative (numerical) observations. The unemployment rate, the percentage of voters favoring a particular candidate, the cost of textbooks for any school term, and the amount of pesticide in the discharge waters of a chemical plant are all examples of numerical measurements of some phenomenon. Thus, most experiments have simple events that correspond to values of some numerical variable.

---

DEFINITION 4.1

A random variable is a rule that assigns one (and only one) numerical value to each simple event of an experiment.*

---

EXAMPLE 4.1

Three potential voters are asked whether they are in favor of a bond for building a fine arts center. Each response is recorded as Yes or No. Define a random variable that could be of interest in this experiment.

Solution

In this experiment, the simple events are *not* numerical in nature. In fact, the eight simple events listed in Table 4.1 are sequences of Yes's and No's. It is doubtful that anyone would be interested in the exact sequence of Yes's and No's, each corresponding to one of the voters polled. One random variable of interest is the number of voters in favor of the bond issue. Thus, the value 3 would be assigned to the first simple event, 2 to the second, etc., as shown in Table 4.1. Note that only one numerical value is assigned to each simple event, although several simple events have the same numerical value of the random variable.

**TABLE 4.1
POSSIBLE OUTCOMES
OF VOTER SAMPLING**

|   | VOTER 1 | VOTER 2 | VOTER 3 | RANDOM VARIABLE (NUMBER OF YES'S) |
|---|---------|---------|---------|-----------------------------------|
| 1 | Yes | Yes | Yes | 3 |
| 2 | Yes | Yes | No  | 2 |
| 3 | Yes | No  | Yes | 2 |
| 4 | No  | Yes | Yes | 2 |
| 5 | Yes | No  | No  | 1 |
| 6 | No  | Yes | No  | 1 |
| 7 | No  | No  | Yes | 1 |
| 8 | No  | No  | No  | 0 |

The term *random variable* is more meaningful than just the term *variable,* because the adjective *random* indicates that the experiment may result in one of the several possible values of the variable, according to the random outcome of the experiment. For example, if the experiment is to count the number of customers who use the drive-up window of a bank each day, the random variable (the number of customers) will vary from day to day, partly because of the random phenomena that influence whether

---

*By *experiment,* we mean an experiment that yields random outcomes (as defined in Chapter 3).

customers use the drive-up window. Thus, the possible values of this random variable range from zero to the maximum number of customers the window could possibly serve in a day.

## 4.1
## TWO TYPES OF
## RANDOM
## VARIABLES

Dividing one unit of probability among the simple events in a sample space and consequently assigning probabilities to the values of a random variable is not always as easy as the examples in Chapter 3 might lead you to believe. If the number of simple events is finite, that is, if they can be completely listed, the job is relatively easy. However, some experiments result in an infinite number of sample points, in which case assignment of probabilities will be more difficult. In fact, we will have to use different types of probability models depending on the number of values that a random variable can assume.

EXAMPLE 4.2

A panel of ten wine experts are asked to taste a new white wine and assign a rating of 0, 1, 2, or 3. A score is then obtained by adding together the ratings of the ten experts. How many values can this random variable assume?

Solution

A simple event is a sequence of ten numbers associated with the rating of each expert. The random variable assigns a score to each one of these simple events by adding the ten numbers together. Thus, the smallest score is 0 (all ten ratings are 0) and the largest score is 30 (all ten ratings are 3). Since every integer between 0 and 30 is a possible score, the random variable $x$ can assume thirty-one values.

This is an example of a discrete random variable, since there is a finite number of distinct possible values. Whenever all the possible values a random variable can assume can be listed (or counted), the random variable is discrete.

EXAMPLE 4.3

Suppose the Environmental Protection Agency (EPA) takes readings once a month on the amount of pesticide in the discharge water of a chemical company. If the amount of pesticide exceeds the maximum level set by the EPA, the company is forced to take corrective action and may be subject to penalty. Consider the random variable:

Number, $x$, of months before the company's discharge exceeds the EPA's maximum level

What values can $x$ assume?

Solution

The company's discharge of pesticide may exceed the maximum allowable level on the first month of testing, the second month of testing, etc. It is possible that the company's discharge will *never* exceed the maximum level. Thus, the set of possible values for the number of months until the level is first exceeded is the set of all positive integers:

1, 2, 3, 4, . . .

If we can list the values of a random variable $x$, even though the list is never-ending, we call the list countable and the corresponding random variable discrete. Thus, the number of months until the company's discharge first exceeds the limit is a discrete random variable.

EXAMPLE 4.4    Refer to Example 4.3. A second random variable of interest is the amount, $x$, of pesticide (in milligrams per liter) found in the monthly sample of discharge waters from the chemical company. What values can this random variable assume?

Solution    Unlike the variable, *Number* of months before the company's discharge exceeds the EPA's maximum level, the set of all possible values for the *amount* of discharge *cannot* be listed, i.e., is not countable. The possible values for the amount, $x$, of pesticide would correspond to the points on the line interval between 0 and the largest possible value the amount of the discharge could attain, the maximum number of milligrams that could occupy one liter of volume. (Practically, the interval would be much smaller, say, between 0 and 500 milligrams per liter.) When the values of a random variable are not countable, but instead correspond to the points on some line interval, we call it a continuous random variable. Thus, the amount of pesticide in the chemical plant's discharge waters is a continuous random variable.

---

**DEFINITION 4.2**
Random variables that can assume a countable number of values are called discrete.

**DEFINITION 4.3**
Random variables that can assume values corresponding to any of the points contained in one or more intervals on a line are called continuous.

---

Examples of discrete random variables are:

**1.** The number of sales made by a salesperson in a given week:  $x = 0, 1, 2, \ldots$ [*Note:* Theoretically, $x$ could become very large.]
**2.** The number of students in a sample of 500 who favor an increase in student activities and, correspondingly, an increase in student activity fees: $x = 0, 1, 2, \ldots,$ 499, 500.
**3.** The number of students applying to medical schools this year:  $x = 0, 1, 2, \ldots$ [*Note:* Theoretically, $x$ could become very large.]
**4.** The number of errors on a page of an accountant's ledger:  $x = 0, 1, 2, \ldots$
**5.** The number of customers waiting to be served in a restaurant at a particular time: $x = 0, 1, 2, \ldots$

Note that each of the examples of discrete random variables begins with the words "The number of . . . ." This is very common, since the discrete random variables most frequently observed are counts.

Examples of continuous random variables are:

**1.** The length of time between arrivals at a hospital clinic: $0 \leq x < \infty$ (infinity).
**2.** For a new apartment complex, the length of time from completion until a specified number of apartments are rented: $0 \leq x < \infty$.
**3.** The amount of carbonated beverage loaded into a 12 ounce can in a can filling operation: $0 \leq x \leq 12$.
**4.** The depth at which a successful oil drilling venture first strikes oil: $0 \leq x \leq c$, where $c$ is the maximum depth obtainable.
**5.** The weight of a food item bought in a supermarket: $0 \leq x \leq 500$. [*Note:* Theoretically, there is no upper limit on $x$, but it is unlikely that it would exceed 500 pounds.]

In the succeeding sections, we will explain how to construct probability models for both discrete and continuous random variables. Then we will describe in detail the properties of two random variables that are often encountered in the real world, and we will see how we can apply their probability distributions to solve some practical problems.

EXERCISES

Applying the concepts

**4.1.** Classify the following random variables according to whether they are discrete or continuous.

  a. The number of words spelled correctly by a student on a spelling test
  b. The number of sofas a furniture store sells monthly
  c. The length of time an employee is late for work
  d. The amount of liquid waste a plant purifies daily
  e. The price per gallon of unleaded gas at a local service station
  f. The starting salary of a college graduate

**4.2.** Identify the following random variables as discrete or continuous:

  a. The number of coin flips until a head is observed
  b. The heart rate (number of beats per minute) of an American male
  c. The time for a rat to negotiate a maze
  d. The yearly inflation rate in the United States
  e. The pressure (pounds per square inch) necessary to break a steel cable
  f. The number of students enrolled in a certain university

**4.3.** Identify the following random variables as discrete or continuous:

  a. The number of violent crimes committed per month in your community
  b. The length of time needed for you to recover from the common cold
  c. The amount of precipitation observed per day in your community
  d. The number of commercial aircraft accidents that occur per month in the United States
  e. Your blood pressure
  f. Your weight

**4.2
PROBABILITY
DISTRIBUTIONS
FOR DISCRETE
RANDOM
VARIABLES**

A complete description of a discrete random variable requires that we specify the possible values the random variable can assume and the probability associated with each value. To illustrate, consider Example 4.5.

EXAMPLE 4.5

Recall the experiment of tossing two coins (Chapter 3), and let $x$ be the number of heads observed. Find the probability associated with each value of the random variable $x$, assuming the two coins are fair.

**FIGURE 4.1
VENN DIAGRAM FOR
THE TWO-COIN TOSS
EXPERIMENT**

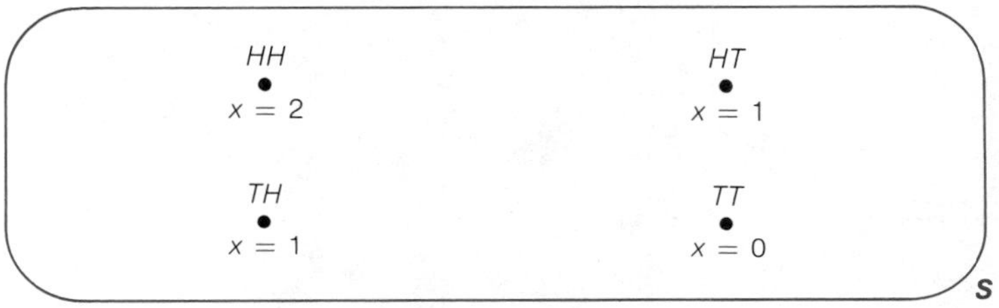

Solution

Recall from Chapter 3 that the sample space and simple events for this experiment are as shown in Figure 4.1, and the probability associated with each of the four simple events is $\frac{1}{4}$. The random variable $x$ can assume values 0, 1, 2. Then, identifying the probabilities of the simple events associated with each of these values of $x$, we have

$$P(x = 0) = P(TT) = \frac{1}{4}$$

$$P(x = 1) = P(TH) + P(HT) = \frac{1}{4} + \frac{1}{4} = \frac{1}{2}$$

$$P(x = 2) = P(HH) = \frac{1}{4}$$

Thus, we now know the values the random variable can assume, and we know how the probability is distributed over these values. This completely describes the random variable and will be referred to as the probability distribution. This probability distribution is given on page 118 in the form of a table (Table 4.2) and as a graph (Figure 4.2). Denoting the probability of $x$ by the symbol $p(x)$, we have $p(0) = \frac{1}{4}$, $p(1) = \frac{1}{2}$, and $p(2) = \frac{1}{4}$.

Despite the fact that the probabilities $p(0)$, $p(1)$, and $p(2)$ are concentrated at the points $x = 0$, 1, and 2, respectively, we represent the probabilities in Figure 4.2 as rectangles similar to the rectangles that are used in a relative frequency histogram. This emphasizes the relationship between probability distributions and relative frequency histograms, and we will use the same representation when we approximate certain types of probabilities.

**TABLE 4.2**
**PROBABILITY**
**DISTRIBUTION**
**FOR COIN TOSS**
**EXPERIMENT:**
**TABULAR FORM**

| $x$ | $p(x)$ |
|-----|--------|
| 0   | $\frac{1}{4}$ |
| 1   | $\frac{1}{2}$ |
| 2   | $\frac{1}{4}$ |

**FIGURE 4.2**
**PROBABILITY**
**DISTRIBUTION**
**FOR COIN TOSS**
**EXPERIMENT:**
**GRAPHICAL FORM**

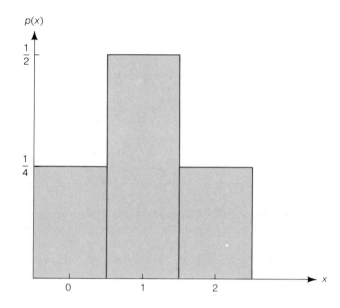

We could also present the probability distribution for $x$ as a formula, but this would unnecessarily complicate a very simple example. We will give the formula for the probability distribution of one common discrete random variable later in this chapter.

---

DEFINITION 4.4
The probability distribution of a discrete random variable is a graph, table, or formula that specifies the probability associated with each possible value the random variable can assume.

---

Two requirements must be satisfied by all probability distributions for discrete random variables:

REQUIREMENTS FOR THE PROBABILITY DISTRIBUTION OF A
DISCRETE RANDOM VARIABLE $x$

$$p(x) \geq 0 \quad \text{for all values of } x$$

$$\sum p(x) = 1$$

Where the summation* of $p(x)$ is over all possible values of $x$.

Example 4.5 illustrates how the probability distribution for a discrete random variable can be derived, but for many practical situations, the task is much more difficult. Fortunately, many experiments and associated discrete random variables observed in nature possess identical characteristics. Thus, you might observe a random variable in a psychology experiment that would possess the same probability distribution as a random variable observed in an engineering experiment or a social sample survey. We classify random variables according to type of experiment, derive the probability distribution for each of the different types, and then use the appropriate probability distribution when a particular type of random variable is observed in a practical situation. The probability distributions for most commonly occurring discrete random variables have already been derived. This fact simplifies the problem of finding the probability distributions for random variables.

EXERCISES

Learning the mechanics

**4.4.** Consider the following probability distribution:

| $x$ | −4 | 0 | 1 | 3 |
|------|------|------|------|------|
| $p(x)$ | .1 | .3 | .4 | .2 |

   a.  List the values that $x$ may assume.
   b.  What value of $x$ is most probable?
   c.  What is the probability that $x$ is greater than 0?
   d.  What is the probability that $x = -2$?

**4.5.** A discrete random variable $x$ can assume five possible values: 2, 3, 5, 8 and 10. Its probability distribution is shown below:

| $x$ | 2 | 3 | 5 | 8 | 10 |
|------|------|------|------|------|------|
| $p(x)$ | .20 | .10 | | .25 | .15 |

   a.  What is $p(5)$?
   b.  What is the probability that $x$ equals 2 or 10?
   c.  What is $P(x \leq 8)$?

*Unless otherwise indicated, summations will always be over all possible values of $x$.

4.6. Toss three fair coins and let $x$ equal the number of heads observed.

    a. Identify the simple events associated with this experiment and assign a value of $x$ to each simple event.
    b. Calculate $p(x)$ for each value of $x$.
    c. Construct a probability histogram for $p(x)$.
    d. What is $P(x = 2$ or $x = 3)$?

4.7. A fair die is tossed twice and $x$, the sum of the up faces, is recorded.

    a. Give the probability distribution for $x$ in tabular form.
    b. Find $P(x \geq 8)$.
    c. Find $P(x < 8)$.
    d. What is the probability that $x$ is odd? Even?
    e. What is $P(x = 7)$?

Applying the concepts

4.8. A real estate agent has five homes available to show to a potential buyer. Unknown to the realtor, the buyer would be willing to buy any of three of the homes and would refuse to buy either of the other two. The real estate agent will show the potential buyer homes, randomly selected one at a time, until a sale is concluded. Let $x$ be the number of homes the real estate agent must show. Find the probability distribution for $x$. (Remember, we assume that the realtor selects the homes to be shown in a random order. This assumption could be reasonable if the realtor has no idea which homes the potential buyer prefers.)

4.9. Every human possesses two sex chromosomes. A copy of one or the other (equally likely) is contributed to an offspring. Males have one X chromosome and one Y chromosome. Females have two X chromosomes. If a couple has three children, what is the probability that they have at least one boy? [*Hint:* Define the random variable $z$ as the number of male offspring, and find the probability distribution of $z$.]

4.10. Recent studies have found that 49% of all American households possess a gun of some kind, and of these gun owners, 61% favor gun control. Three households are chosen at random, and $x$, the number of households possessing a gun while also favoring gun control, is counted.

    a. Calculate $p(x)$ for $x = 0, 1, 2, 3$.
    b. Graph $p(x)$.
    c. Find the probability that at least one of the three households possesses a gun and favors gun control.

4.11. Suppose that you have studied very hard for a history test and a sociology test that you must take tomorrow. You are very confident that you will pass both tests, that there is very little chance that you will fail one test and pass the other, and that there is virtually no chance that you will fail both tests. Let $x$ be the number of tests that you pass.

a.  Give an example of a probability distribution for x that reflects your beliefs.

b.  Suppose that you had not studied for the tests. Would the probability distribution given in part a be appropriate? Explain.

**4.12.** Each of the following is a possible probability distribution for the market demand for the number of new four bedroom homes a building contractor will build next summer.

| PROBABILITY DISTRIBUTION I | | PROBABILITY DISTRIBUTION II | |
|---|---|---|---|
| $x$ = Number of homes | $p(x)$ | $y$ = Number of homes | $p(y)$ |
| 0 | .05 | 2 | .03 |
| 1 | .13 | 3 | .05 |
| 2 | .18 | 4 | .06 |
| 3 | .25 | 5 | .13 |
| 4 | .20 | 6 | .16 |
| 5 | .11 | 7 | .23 |
| 6 | .06 | 8 | .15 |
| 7 | .02 | 9 | .12 |
| | | 10 | .07 |

a.  Which probability distribution is more appropriate if the general rate of new homes being built is rather slow? Why?

b.  For each probability distribution, find the probability that the contractor will be requested to build more than five homes.

c.  For each probability distribution, find the probability that the contractor will be requested to build at most three homes.

## 4.3 EXPECTED VALUES OF DISCRETE RANDOM VARIABLES

If a discrete random variable x were observed a very large number of times and if the data generated were arranged in a relative frequency distribution, the relative frequency distribution would be indistinguishable from the probability distribution for the random variable. Thus, the probability distribution for a random variable is a theoretical model for the relative frequency distribution of a population. To the extent that the two distributions are equivalent (and we will assume they are), the probability distribution for x possesses a mean $\mu$ and a variance $\sigma^2$ that are identical to the corresponding descriptive measures for the population. The purpose of this section is to explain how you can find the mean value for a random variable. We will illustrate the procedure with an example.

Examine the probability distribution for x (the number of heads observed in the toss of two coins) in Figure 4.3 (page 122). Try to locate the mean of the distribution intuitively. We may reason that the mean $\mu$ of this distribution is equal to 1 as follows:

in a large number of experiments, we should observe 0 heads $\frac{1}{4}$ of the time, 1 head $\frac{1}{2}$ of the time, and 2 heads $\frac{1}{4}$ of the time. Thus, the average number of heads is

$$\mu = 0\left(\frac{1}{4}\right) + 1\left(\frac{1}{2}\right) + 2\left(\frac{1}{4}\right)$$

$$= 0 \quad + \frac{1}{2} \quad + \frac{1}{2} = 1$$

**FIGURE 4.3**
**PROBABILITY**
**DISTRIBUTION FOR**
**A TWO-COIN TOSS**

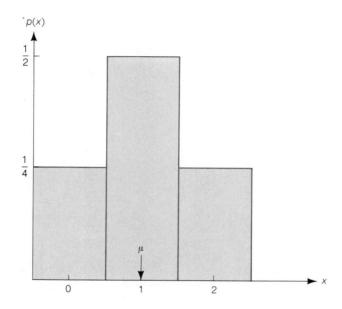

Note that to get the population mean of the random variable $x$, we multiply each possible value of $x$ by its probability $p(x)$, and then we sum this product over all possible values of $x$. Another term often used as a substitute for the mean of $x$ is the expected value of $x$, denoted $E(x)$.

---

DEFINITION 4.5

The expected value of a discrete random variable $x$ is

$$\mu = E(x) = \sum xp(x)$$

---

EXAMPLE 4.6

Suppose you work for an insurance company and you sell a $10,000 whole life insurance policy at an annual premium of $290. Actuarial tables show that the probability of death during the next year for a person of your customer's age, sex, health, etc., is .001. What is the expected gain (amount of money made by the company) for a policy of this type?

**Solution**

The experiment is to observe whether the customer survives the upcoming year. The probabilities associated with the two simple events, Live and Die, are .999 and .001, respectively. The random variable you are interested in is the gain, $x$, which can assume the following values:

| GAIN $x$ | SIMPLE EVENT | PROBABILITY |
|---|---|---|
| $290 | Customer lives | .999 |
| $290 - $10,000 | Customer dies | .001 |

If the customer lives, the company gains the $290 premium as profit. If the customer dies, the gain is negative because the company must pay $10,000, for a net "gain" of $(290 - 10,000)$. The expected gain is therefore

$$\mu = E(x) = \sum xp(x)$$
$$= (290)(.999) + (290 - 10,000)(.001)$$
$$= 290(.999 + .001) - 10,000(.001)$$
$$= 290 - 10 = \$280$$

In other words, if the company were to sell a very large number of 1 year $10,000 policies to customers possessing the characteristics described above, it would (on the average) net $280 per sale in the next year.

We want to measure the variability as well as the central tendency of a probability distribution. The population variance $\sigma^2$ is defined as the average squared distance of the $x$ measurements from the population mean $\mu$. This quantity is also called the expected value of the squared distance from the mean, i.e., $\sigma^2 = E[(x - \mu)^2]$:

---

**DEFINITION 4.6**
The variance of a discrete random variable $x$ is

$$\sigma^2 = E[(x - \mu)^2] = \sum (x - \mu)^2 p(x)$$

---

**DEFINITION 4.7**
The standard deviation of a discrete random variable is equal to the square root of the variance, i.e., to $\sigma = \sqrt{\sigma^2}$.

---

Thus, to calculate the average of $(x - \mu)^2$, we multiply all possible values of $(x - \mu)^2$ by $p(x)$ and then sum over all possible $x$ values. *

---

*It can be shown that $E[(x - \mu)^2] = E(x^2) - \mu^2$, where $E(x^2) = \sum x^2 p(x)$. Note the similarity between this expression and the shortcut formula $\sum (x - \bar{x})^2 = \sum x^2 - \left(\sum x\right)^2 / n$ given in Chapter 2.

EXAMPLE 4.7    Medical research has shown that a certain type of chemotherapy is successful 70% of the time when used to treat skin cancer. Suppose five skin cancer patients are treated with this type of chemotherapy and let $x$ equal the number of successful cures out of the five. The probability distribution for the number $x$ of successful cures out of five is given in the table:

| $x$ | $p(x)$ |
|-----|--------|
| 0 | .002 |
| 1 | .029 |
| 2 | .132 |
| 3 | .309 |
| 4 | .360 |
| 5 | .168 |

a.  Find $\mu = E(x)$.
b.  Find $\sigma = \sqrt{E[(x - \mu)^2]}$.
c.  Graph $p(x)$. Locate $\mu$ and the interval $\mu \pm 2\sigma$ on the graph. Explain how $\mu$ and $\sigma$ can be used to describe $p(x)$.

Solution    a.  Applying the formula

$$\mu = E(x) = \sum xp(x)$$

$$= 0(.002) + 1(.029) + 2(.132) + 3(.309) + 4(.360) + 5(.168)$$

$$= 3.50$$

b.  Now we calculate the variance of $x$:

$$\sigma^2 = E[(x - \mu)^2] = \sum (x - \mu)^2 p(x)$$

$$= (0 - 3.5)^2(.002) + (1 - 3.5)^2(.029) + (2 - 3.5)^2(.132)$$

$$+ (3 - 3.5)^2(.309) + (4 - 3.5)^2(.360) + (5 - 3.5)^2(.168)$$

$$= 1.05$$

Thus, the standard deviation is

$$\sigma = \sqrt{\sigma^2} = \sqrt{1.05} = 1.02$$

c.  The graph of $p(x)$ is shown in Figure 4.4. Note that the mean $\mu$ and the interval $\mu \pm 2\sigma$ are shown on the graph. We can use $\mu$ and $\sigma$ to describe the probability distribution $p(x)$ in the same way that we used $\bar{x}$ and $s$ to describe a relative frequency distribution in Chapter 2. Note particularly that $\mu = 3.5$ locates the center of the probability distribution. If five skin cancer patients receive the chemotherapy treatment, we expect the number $x$ that are cured to be near 3.5. Similarly, $\sigma = 1.02$ measures the spread of the probability distribution $p(x)$. Since this distribution is a theoretical relative frequency distribution that is moderately mound-shaped (see Figure 4.4), we expect (see Tables 2.5 and 2.6) at least 75% and, more likely, near 95% of observed $x$ values to fall in the interval $\mu \pm 2\sigma$, i.e., between 1.46 and 5.54. Compare this with the actual probability that $x$ falls in the interval $\mu \pm 2\sigma$. From Figure 4.4 you can see that this probability includes the sum of $p(x)$ for all values of $x$ except $p(0) = .002$ and $p(1) = .029$. Therefore, 96.9% of the probability distribution lies within 2 standard deviations of the mean. This percentage is consistent with Tables 2.5 and 2.6.

**FIGURE 4.4**
**GRAPH OF $p(x)$ FOR**
**EXAMPLE 4.7**

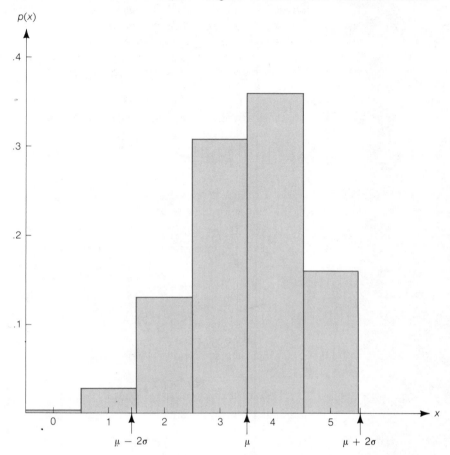

The June 1, 1977, business section of the Orlando, Florida, *Sentinel Star* featured the following headline: "Red Lobster to Fight Tax Claim." According to the *Sentinel Star,* the Red Lobster Inns of America, a national seafood chain, had decided to take the state of Florida to court. The dispute concerned the 4% sales tax levied on most purchases in the state and mainly focused on the state's "bracket collection system." According to the bracket system, a merchant must collect 1¢ for sales between 10¢ and 25¢, 2¢ for sales from 26¢ to 50¢, 3¢ for sales between 51¢ and 75¢, and 4¢ for sales between 76¢ and 99¢. Red Lobster contended that if this system is followed, merchants will always collect more than 4%. That is, if a sale were made for $10.41, 4% will be collected on the $10, but more than 4% will be collected on the 41¢. This, they contend, would amount to more than 4% on the total sale and therefore is not consistent with the 4% tax required by law.

Concrete evidence supplied by the state of Florida tax records does indeed support the contention that the amount of tax collected using the bracket system exceeds the 4% specified by law. It appears that the state sales tax receipts exceeded expected revenue (based on 4%) by $9.5 million in a single year.

What percent sales tax should the state expect to receive using the bracket system for computing the tax? (As noted, the tax on the whole dollar portion of the sale will be 4%.) Using the formula for calculating expected values, you can show (see Exercise 4.109) that the expected percent tax paid on the cents portion of a sale is 4.6%.

Learning the mechanics

**4.13.** Consider the following probability distribution:

| $x$ | 0 | 10 | 20 | 30 |
|------|-----|-----|-----|-----|
| $p(x)$ | .5 | .2 | .1 | .2 |

a. Find $\mu = E(x)$.
b. Find $\sigma^2 = E[(x - \mu)^2]$.
c. Find $\sigma$.

**4.14.** Consider the following probability distribution:

| $x$ | 1 | 2 | 3 | 5 | 10 |
|------|-----|-----|-----|-----|-----|
| $p(x)$ | .1 | .3 | .2 | .1 | .3 |

a. Find the mean of the distribution.
b. Find the variance.
c. Find the standard deviation.

**4.15.** Consider the following probability distribution:

| $x$ | −4 | −3 | −2 | −1 | 0 | 1 | 2 | 3 | 4 |
|------|-----|-----|-----|-----|-----|-----|-----|-----|-----|
| $p(x)$ | .05 | .07 | .10 | .15 | .26 | .15 | .10 | .07 | .05 |

a. Calculate $\mu$, $\sigma^2$, and $\sigma$.
b. Graph $p(x)$. Locate $\mu$, $\mu - 2\sigma$, and $\mu + 2\sigma$ on the graph.
c. What is the probability that $x$ is in the interval, $\mu \pm 2\sigma$?

**4.16.** Consider the following probability distributions:

| $x$ | 0 | 1 | 2 |
|------|-----|-----|-----|
| $p(x)$ | .3 | .4 | .3 |

| $y$ | 0 | 1 | 2 |
|------|-----|-----|-----|
| $p(y)$ | .1 | .8 | .1 |

a. Use your intuition to find the mean for each distribution. How did you arrive at your choice?
b. Which distribution appears to be more variable? Why?
c. Calculate $\mu$ and $\sigma^2$ for each distribution. Compare these answers to your answers in parts a and b.

Applying the concepts

**4.17.** A rehabilitation officer at a county jail questioned each inmate to determine how many previous convictions, $x$, each had prior to the one for which he or she was now serving. The relative frequencies corresponding to $x$ are given in the following probability distribution:

| $x$ | 0 | 1 | 2 | 3 | 4 |
|------|-----|-----|-----|-----|-----|
| $p(x)$ | .16 | .53 | .20 | .08 | .03 |

If we can regard the relative frequencies as the approximate values for $p(x)$, find the expected number of previous convictions for an inmate.

**4.18.** A hospital research laboratory purchases rats from a distributor for use in experiments. The distributor sells four different strains of rats, each at different prices, and fills requests with one of the four strains depending on availability. The laboratory would like to estimate how much it will have to spend on rats in the next year. The following table lists the four strains, price per fifty rats for each strain, and the probability of the purchase of each strain:

| STRAIN | PRICE PER 50 RATS | PROBABILITY |
|--------|-------------------|-------------|
| A | $50.00 | $\frac{1}{10}$ |
| B | $75.00 | $\frac{2}{5}$ |
| C | $87.50 | $\frac{3}{10}$ |
| D | $100.00 | $\frac{1}{5}$ |

a. If $x$ is the price of a shipment of fifty rats, what is the expected price?
b. What is the variance of the price?
c. Graph $p(x)$. Locate $\mu$ and $\mu \pm 2\sigma$ on the graph. What proportion of the time does $x$ fall in this interval?

**4.19.** A company's marketing and accounting departments have determined that if the company markets its newly developed line of party favors, the following probability distribution will describe the contribution of the new line to the firm's profit during the next 6 months:

| Profit contribution | $p$(Profit contribution) |
|---------------------|--------------------------|
| −$5,000* | .3 |
| $10,000 | .4 |
| $30,000 | .3 |

*A negative contribution is a loss.

The company has decided it should market the new line of party favors if the expected contribution to profit for the next 6 months is over $10,000. Based on the probability distribution, will the company market the new line?

**4.20.** On a particular busy holiday weekend, a national airline has many requests for standby flights at half of the usual one-way air fare. However, past experience has shown that these passengers have only about a one in five chance of getting on the standby flight. When they fail to get on a flight as a standby, their only other choice is to fly first class on the next flight out. Suppose that the usual one-way air fare to a certain city is $70 and the cost of flying first class is $90. Should a passenger who wishes to fly to this city opt to fly as a standby? [*Hint:* Find the expected cost of the trip for a person flying standby.]

**4.21.** Odds makers try to predict which football teams will win and by how much (the *spread*). If the odds makers do this accurately, adding the spread to the underdog's score should make the final score a tie. Suppose a bookie will give you $6 for every $1 you risk if you pick the winners in three ballgames (adjusted by the spread). Thus, for every $1 bet, you will either lose $1 or gain $5. What is the bookie's expected earnings per dollar wagered?

**4.22.** A buyer for a large department store wants to project the required inventory for men's ten-speed bicycles next spring. Experience has shown that demand ($x$) has approximately the probability distribution shown in the table.

| DEMAND $x$ = Number of bicycles | $p(x)$ |
|:---:|:---:|
| 40 | .01 |
| 60 | .06 |
| 80 | .16 |
| 100 | .24 |
| 120 | .23 |
| 140 | .15 |
| 160 | .10 |
| 180 | .05 |

a. Find $E(x)$.     b.   Find $E[(x - \mu)^2]$.      c.   Find $\sigma$.

d.   Graph $p(x)$ and locate $\mu$ and the interval $\mu \pm 2\sigma$ on the graph.

e.   Based on the above information, how many bicycles would you order for next spring? Explain the reasoning behind your decision.

**4.4
THE BINOMIAL
RANDOM
VARIABLE**

Many experiments result in dichotomous responses, i.e., responses for which there exist two possible alternatives, such as Yes—No, Pass—Fail, Defective—Nondefective, Male—Female, etc. A simple example of such an experiment is the coin tossing experiment. A coin is tossed a number of times, say ten. Each toss results in one of two outcomes, Head or Tail, and the probability of observing each of these two outcomes remains the same for each of the ten tosses. Ultimately, we are interested in the probability distribution of $x$, the number of heads observed. Many other experiments are equivalent to tossing a coin (either balanced or unbalanced) a fixed number $n$ of times and observing the number $x$ of times that one of the two possible outcomes occurs. Random variables that possess these characteristics are called binomial random variables.

Public opinion and consumer preference polls (e.g., the Gallup and Harris polls) frequently yield observations on binomial random variables. For example, suppose a sample of 100 students is selected from a large student body and each person is asked whether he or she favors (a Head) or opposes (a Tail) some particular campus issue. Ultimately, we are interested in $x$, the number of people in the sample who favor the issue. If each student is randomly selected from the student body and if the (unknown) proportion of students favoring the issue is $p$, then observing whether a student favors or is opposed to the issue is analogous to tossing an unbalanced coin. The chance that any randomly selected student favors the issue is $p$; the probability that he or she opposes the issue is $(1 - p)$. Sampling 100 students is analogous to tossing the coin 100 times. Thus, you can see that opinion polls that record the number of people who favor some particular issue are real-life equivalents of coin tossing experiments.

The experiment we have been describing is called a binomial experiment and is identified by the following characteristics:

---

CHARACTERISTICS OF A BINOMIAL RANDOM VARIABLE
**1.** The experiment consists of $n$ identical trials.
**2.** There are only two possible outcomes on each trial. We will denote one outcome by $S$ (for Success) and the other by $F$ (for Failure).
**3.** The probability of $S$ remains the same from trial to trial. This probability will be denoted by $p$, and the probability of $F$ will be denoted by $q$. Note that $q = 1 - p$.
**4.** The trials are independent.
**5.** The binomial random variable $x$ is the number of $S$'s in $n$ trials.

---

EXAMPLE 4.8    For each of the following examples, decide whether $x$ is a binomial random variable.

**a.** Suppose a university scholarship committee must select two students to receive a scholarship for the next academic year. The committee receives ten applications for the scholarships—six from male students and four from female students. Suppose the applicants are all equally qualified, so that the selections are randomly made. Let $x$ be the number of female students who receive a scholarship.
**b.** Before marketing a new product on a large scale, many companies will conduct a consumer-preference survey to determine whether the product is likely to be successful. Suppose a company develops a new diet soda and then conducts a taste-preference survey with 100 randomly chosen consumers stating their preference among the new soda and the two leading sellers. Let $x$ be the number of the 100 who choose the new brand over the two others.

**c.** Some surveys are conducted using a method of sampling other than simple random sampling (defined in Chapter 3). For example, suppose a television cable company is trying to decide whether to establish a branch in a particular city. The company plans to conduct a survey to determine the fraction of households in the city that would use the cable television service. The sampling method is to choose a city block at random and then to survey every household on that block. This sampling technique is called cluster sampling. Suppose ten blocks are so sampled, producing a total of 124 household responses. Let $x$ be the number of the 124 households that would use the television cable service.

Solution

**a.** In checking the binomial characteristics, a problem arises with independence (characteristic 4 in the box). Given that the first student selected is female, the probability the second chosen is female is $\frac{3}{9}$. On the other hand, given that the first selection is a male student, the probability the second is female is $\frac{4}{9}$. Thus, the conditional probability of a Success (choosing a female student to receive a scholarship) on the second trial (selection) depends upon the outcome of the first trial, and the trials are therefore dependent. Since the trials are *not independent,* this is not a binomial random variable.

**b.** Surveys that produce dichotomous responses and use random sampling techniques are classical examples of binomial experiments. In our example, each randomly selected consumer either states a preference for the new diet soda or does not. The sample of 100 consumers is a very small proportion of the totality of potential consumers, so the response of one would be, for all practical purposes, independent of another. Thus, $x$ is a binomial random variable.

**c.** This example is a survey with dichotomous responses (Yes or No to the cable service), but the sampling method is not simple random sampling. Again, the binomial characteristic of independent trials would very probably not be satisfied. The responses of households within a particular block almost surely would be dependent, since households within a block tend to be similar with respect to income, race, and general interests. Thus, the binomial model would not be satisfactory for $x$ if the cluster sampling technique were employed.

EXAMPLE 4.9

The Heart Association claims that only 10% of adults over 30 years of age in the United States can pass the minimum fitness requirements established by the President's Physical Fitness Commission. Suppose that four adults are randomly selected and each is given the fitness test. Let $x$ be the number of the four who pass the minimum requirements. Find the probability distribution for $x$, assuming the Heart Association's claim is true.

Solution

Recall that a probability distribution describes a discrete random variable by assigning probabilities to each of its values. In this example, the possible values of $x$, the number of four adults who pass the minimum requirements, are 0, 1, 2, 3, 4. Furthermore, if you check the characteristics of this experiment, you can see that it is a binomial experiment.

Let us first consider the event $x = 0$, i.e., the event that none of the four tested adults passes the test. You can see that the event $x = 0$ is equivalent to the simple event

$FFFF$

where $F$ in the first position implies adult 1 fails, $F$ in the second position implies adult 2 fails, etc. Since the trials are independent in this binomial experiment (knowing whether adult 1 passes should not affect the probability that adult 2 passes), we can find the probability of an intersection by multiplying the probabilities of the events. Thus,

$$P(x = 0) = P(FFFF) = P(F)P(F)P(F)P(F)$$
$$= (.9)(.9)(.9)(.9) = (.9)^4 = .6561$$

The event $x = 1$ implies that one of the four adults passes the physical fitness test and three fail it. The following list of simple events contains all simple events that imply $x = 1$:

$SFFF, \quad FSFF, \quad FFSF, \quad FFFS$

where $S$ in the first position corresponds to adult 1 passing the test, $S$ in the second position corresponds to adult 2 passing the test, etc. Note that each of these simple events will have the same probability, $(.1)(.9)^3$, where .1 corresponds to the one adult who passes the test and $(.9)^3$ corresponds to the three who fail it. Remembering from Chapter 3 that we obtain the probability of an event by summing the probabilities of the simple events of which it is composed, we get

$$P(x = 1) = 4[(.1)(.9)^3] = .2916$$

The event $x = 2$ implies that two adults pass the test and two fail it, and consists of the following six simple events:

$SSFF, \quad SFSF, \quad SFFS, \quad FSSF, \quad FSFS, \quad FFSS$

Each of the above simple events has probability $(.1)^2(.9)^2$, so that

$$P(x = 2) = 6[(.1)^2(.9)^2] = .0486$$

Similarly,

$$P(x = 3) = 4[(.1)^3(.9)] = .0036$$
$$P(x = 4) = (.1)^4 = .0001$$

The complete probability distribution is given in Table 4.3 and shown in Figure 4.5.

**TABLE 4.3**
**PROBABILITY DISTRIBUTION FOR PHYSICAL FITNESS EXAMPLE: TABULAR FORM**

| $x$ | $p(x)$ |
|---|---|
| 0 | .6561 |
| 1 | .2916 |
| 2 | .0486 |
| 3 | .0036 |
| 4 | .0001 |

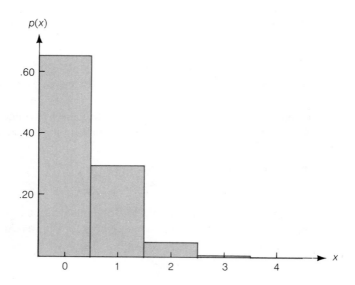

FIGURE 4.5
PROBABILITY
DISTRIBUTION FOR
PHYSICAL FITNESS
EXAMPLE:
GRAPHICAL FORM

Before we give a formula for $p(x)$, we must introduce a new notation. The symbol $n!$ is to be read "$n$ factorial," and is calculated by

$$n! = n(n - 1)(n - 2) \cdots 3 \cdot 2 \cdot 1$$

We define $0! = 1$. Thus, for example, $4! = 4 \cdot 3 \cdot 2 \cdot 1 = 24$.

Using factorial notation, we write the formula for a binomial probability distribution with $n = 4$ and $p = .2$:

$$p(x) = \frac{4!}{x!(4 - x)!} (.1)^x(.9)^{4-x}$$

Then, for $x = 2$, we have

$$p(2) = \frac{4!}{2!(4 - 2)!} (.1)^2(.9)^{4-2} = \frac{4 \cdot 3 \cdot 2 \cdot 1}{(2 \cdot 1)(2 \cdot 1)} (.1)^2(.9)^2$$

$$= 6(.1)^2(.9)^2 = .0486$$

which agrees with our simple event calculation. Note that the first part of the formula, $4!/x!(4 - x)!$, counts the number of simple events that result in $x$ adults passing the physical fitness test. The second part of the formula, $(.1)^x(.9)^{4-x}$, is the probability assigned to each simple event that has $x$ adults passing and $(4 - x)$ failing. When we multiply the *number* of simple events by the *probability* assigned to each simple event, we get the probability that $x$ adults pass the test. We will see that this formula can be generalized to give the probability distribution of any binomial random variable.

Note that $\binom{n}{x}$ is shorthand for $n!/x!(n - x)!$. This is the number of simple events that have $x$ successes and $(n - x)$ failures, and $p^x q^{n-x}$ is the probability assigned to each simple event that has $x$ successes and $(n - x)$ failures. The product of these

THE BINOMIAL PROBABILITY DISTRIBUTION

$$p(x) = \binom{n}{x} p^x q^{n-x} \qquad (x = 0, 1, 2, \ldots, n)$$

where

$p$ = Probability of a success on a single trial

$q = 1 - p$

$n$ = Number of trials

$x$ = Number of successes in $n$ trials

$$\binom{n}{x} = \frac{n!}{x!(n - x)!}$$

two quantities, $\binom{n}{x} p^x q^{n-x}$, is the probability that $x$ successes and $(n - x)$ failures are observed.

EXAMPLE 4.10      Refer to Example 4.9. Calculate $\mu$ and $\sigma$, the mean and standard deviation, respectively, of the number of the four adults who pass the test.

Solution      From Section 4.3 we know that the mean of a discrete probability distribution is

$$\mu = \sum xp(x)$$

Referring to Table 4.3, the probability distribution for the number $x$ who pass the fitness test, we find

$$\mu = 0(.6561) + 1(.2916) + 2(.0486) + 3(.0036) + 4(.0001)$$
$$= .4 = 4(.1) = np$$

The relationship $\mu = np$ will hold in general for a binomial random variable.

The variance is

$$\sigma^2 = \sum (x - \mu)^2 p(x) = \sum (x - .4)^2 p(x)$$
$$= (0 - .4)^2(.6561) + (1 - .4)^2(.2916) + (2 - .4)^2(.0486)$$
$$\quad + (3 - .4)^2(.0036) + (4 - .4)^2(.0001)$$
$$= .104976 + .104976 + .124416 + .024336 + .001296$$
$$= .36 = 4(.1)(.9) = npq$$

The relationship $\sigma^2 = npq$ will hold in general for a binomial random variable.

Finally, the standard deviation of the number who pass the fitness test is

$$\sigma = \sqrt{\sigma^2} = \sqrt{.36} = .6$$

We emphasize that you need not use the expectation summation rules to calculate $\mu$ and $\sigma^2$ for a binomial random variable. You can find them easily using the formulas $\mu = np$ and $\sigma^2 = npq$.

---

MEAN, VARIANCE, AND STANDARD DEVIATION FOR A BINOMIAL RANDOM VARIABLE

Mean: $\mu = np$

Variance: $\sigma^2 = npq$

Standard deviation: $\sigma = \sqrt{npq}$

---

As we demonstrated in Chapter 2, the mean and standard deviation provide measures of the central tendency and variability, respectively, of a distribution. Thus, we can use $\mu$ and $\sigma$ to obtain a rough visualization of the probability distribution for $x$ when the calculation of the probabilities is too tedious. To illustrate the use of the binomial probability distribution, consider Example 4.11.

EXAMPLE 4.11

A poll of twenty voters is taken in a large city. The purpose is to determine $x$, the number in favor of a particular candidate for mayor. Suppose that (unknown to us) 60% of all the city's voters favor this candidate.

a.  Find the mean and standard deviation of $x$.
b.  Find the probability that $x$ is less than or equal to ten ($x \leq 10$).
c.  Find the probability that $x$ exceeds twelve ($x > 12$).
d.  Find the probability that $x$ equals eleven ($x = 11$).
e.  Graph the probability distribution of $x$ and locate the interval $\mu - 2\sigma$ to $\mu + 2\sigma$ on the graph.

Solution

a.  Given that the sample of twenty was randomly selected from a large number of voters, it is likely that $x$, the number of the twenty who favor the candidate, is a binomial random variable. The value of $p$ is the fraction of the total voters who favor the candidate, i.e., $p = .6$. Therefore, we calculate the mean and variance:

$$\mu = np = 20(.6) = 12$$
$$\sigma^2 = npq = 20(.6)(.4) = 4.8$$

The standard deviation is then

$$\sigma = \sqrt{4.8} = 2.2$$

**b.** Calculating binomial probabilities when $n$ is large is a formidable task. For example, to find the probability that $x \leq 10$, we would calculate

$$P(x \leq 10) = p(0) + p(1) + p(2) + \cdots + p(10)$$

$$= \sum_{x=0}^{10}{}^{*}p(x) = \sum_{x=0}^{10} \binom{20}{x}(.6)^x(.4)^{20-x}$$

We can avoid these tedious calculations by making use of cumulative binomial probability tables (Table II in the Appendix). Part of Table II is shown in Figure 4.6. The entries in Table II are the cumulative sums

$$P(x \leq k) = p(0) + p(1) + p(2) + \cdots + p(k)$$

for values of $k = 0, 1, 2, \ldots , (n - 1)$. Observe that the bottom row of the table, the one corresponding to $k = n$, is omitted. This is because the sum of $p(x)$ from $x = 0$ to $x = n$ is always equal to 1; i.e., $P(x \leq n) = 1$ for any binomial random variable.

**FIGURE 4.6**
**PARTIAL REPRODUCTION OF TABLE II IN THE APPENDIX**

**h.** $n = 20$

| k \ p | 0.01 | 0.05 | 0.10 | 0.20 | 0.30 | 0.40 | 0.50 | 0.60 | 0.70 | 0.80 | 0.90 | 0.95 | 0.99 |
|---|---|---|---|---|---|---|---|---|---|---|---|---|---|
| 0 | .818 | .358 | .122 | .012 | .001 | .000 | .000 | .000 | .000 | .000 | .000 | .000 | .000 |
| 1 | .983 | .736 | .392 | .069 | .008 | .001 | .000 | .000 | .000 | .000 | .000 | .000 | .000 |
| 2 | .999 | .925 | .677 | .206 | .035 | .004 | .000 | .000 | .000 | .000 | .000 | .000 | .000 |
| 3 | 1.000 | .984 | .867 | .411 | .107 | .016 | .001 | .000 | .000 | .000 | .000 | .000 | .000 |
| 4 | 1.000 | .997 | .957 | .630 | .238 | .051 | .006 | .000 | .000 | .000 | .000 | .000 | .000 |
| 5 | 1.000 | 1.000 | .989 | .804 | .416 | .126 | .021 | .002 | .000 | .000 | .000 | .000 | .000 |
| 6 | 1.000 | 1.000 | .998 | .913 | .608 | .250 | .058 | .006 | .000 | .000 | .000 | .000 | .000 |
| 7 | 1.000 | 1.000 | 1.000 | .968 | .772 | .416 | .132 | .021 | .001 | .000 | .000 | .000 | .000 |
| 8 | 1.000 | 1.000 | 1.000 | .990 | .887 | .596 | .252 | .057 | .005 | .000 | .000 | .000 | .000 |
| 9 | 1.000 | 1.000 | 1.000 | .997 | .952 | .755 | .412 | .128 | .017 | .001 | .000 | .000 | .000 |
| 10 | 1.000 | 1.000 | 1.000 | .999 | .983 | .872 | .588 | .245 | .048 | .003 | .000 | .000 | .000 |
| 11 | 1.000 | 1.000 | 1.000 | 1.000 | .995 | .943 | .748 | .404 | .113 | .010 | .000 | .000 | .000 |
| 12 | 1.000 | 1.000 | 1.000 | 1.000 | .999 | .979 | .868 | .584 | .228 | .032 | .000 | .000 | .000 |
| 13 | 1.000 | 1.000 | 1.000 | 1.000 | 1.000 | .994 | .942 | .750 | .392 | .087 | .002 | .000 | .000 |
| 14 | 1.000 | 1.000 | 1.000 | 1.000 | 1.000 | .998 | .979 | .874 | .584 | .196 | .011 | .000 | .000 |
| 15 | 1.000 | 1.000 | 1.000 | 1.000 | 1.000 | 1.000 | .994 | .949 | .762 | .370 | .043 | .003 | .000 |
| 16 | 1.000 | 1.000 | 1.000 | 1.000 | 1.000 | 1.000 | .999 | .984 | .893 | .589 | .133 | .016 | .000 |
| 17 | 1.000 | 1.000 | 1.000 | 1.000 | 1.000 | 1.000 | 1.000 | .996 | .965 | .794 | .323 | .075 | .001 |
| 18 | 1.000 | 1.000 | 1.000 | 1.000 | 1.000 | 1.000 | 1.000 | .999 | .992 | .931 | .608 | .264 | .017 |
| 19 | 1.000 | 1.000 | 1.000 | 1.000 | 1.000 | 1.000 | 1.000 | 1.000 | .999 | .988 | .878 | .642 | .182 |

*The value of $x$ below the $\Sigma$ symbol, $x = 0$, is the first member, or lower limit, of the summation. The value of $x$ above the $\Sigma$ symbol, $x = 10$, is the last member, or upper limit, of the summation. Thus, $\sum_{x=0}^{10} p(x) = p(0) + p(1) + \cdots + p(9) + p(10)$. We will include these limits when the summation extends over only some of the possible values of $x$.

To find $P(x \leq 10)$ for $n = 20$ and $p = .6$, we first find the column corresponding to $p = .6$ and then the row corresponding to $k = 10$. The recorded value, shaded in Figure 4.6, is

$$P(x \leq 10) = .245$$

**c.** To find the probability

$$P(x > 12) = p(13) + p(14) + \cdots + p(19) + p(20) = \sum_{x=13}^{20} p(x)$$

we use the fact that for all probability distributions, $\sum p(x) = 1$. Therefore,

$$P(x > 12) = 1 - \{p(0) + p(1) + \cdots + p(12)\}$$

$$= 1 - P(x \leq 12) = 1 - \sum_{x=0}^{12} p(x)$$

Consulting Table II, we find the entry in row $k = 12$, column $p = .6$ to be .584. Thus,

$$P(x > 12) = 1 - .584 = .416$$

**d.** To find the probability that exactly eleven voters favor the candidate, recall that the entries in Table II are cumulative probabilities and use the relationship

$$P(x = 11) = [p(0) + p(1) + \cdots + p(10) + p(11)]$$
$$- [p(0) + p(1) + \cdots + p(9) + p(10)]$$
$$= P(x \leq 11) - P(x \leq 10)$$

Then

$$P(x = 11) = .404 - .245 = .159$$

**e.** The probability distribution for $x$ is shown in Figure 4.7. Note that

$$\mu - 2\sigma = 12 - 2(2.2) = 7.6$$
$$\mu + 2\sigma = 12 + 2(2.2) = 16.4$$

The interval $\mu - 2\sigma$ to $\mu + 2\sigma$ is shown in Figure 4.7. The probability that $x$ falls in the interval, $\mu \pm 2\sigma$, i.e., $P\{x = 8, 9, 10, \ldots, 16\} = P(x \leq 16) - P(x \leq 7) = .984 - .021 = .963$. Note that this probability is very close to the .95 given by the Empirical rule.

**CASE STUDY 4.2**
**A SURVEY OF CHILDREN'S POLITICAL KNOWLEDGE**

Children's images of political leaders in Britain, France, and the United States were studied by Fred I. Greenstein (1975). Data were collected by means of interviews with small samples of children in the three countries "in order to examine various standard assumptions about political culture and socialization among the three nations, as well as black–white differences in the United States." During one phase of the study, twenty-five black children from the United States were asked to name the President of their country. This represents a binomial experiment with $n = 25$ trials and $p$ equal

FIGURE 4.7
THE BINOMIAL
PROBABILITY
DISTRIBUTION FOR *x*
IN EXAMPLE 4.11

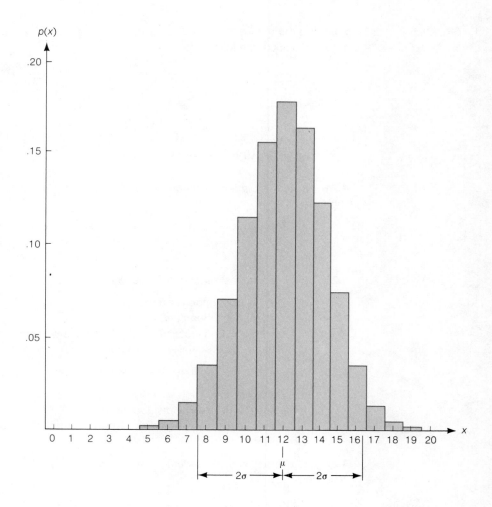

to the proportion of all black children who could correctly name the President at that time (1969–1970). One objective of the experiment was to obtain an estimate of the value of *p*.

Of the sample of twenty-five black children, twenty-four correctly identified Richard Nixon as President. The implication of this result is that the proportion of all black children who could have made a correct identification must have been quite high. In fact, if the true proportion were equal to .8, the probability that at least twenty-four out of twenty-five would correctly identify the President is only .027 (you can verify this by using Table II in the Appendix). Thus, unless the observed outcome represents a rare event, the proportion of all black children who could have correctly identified the President was probably in excess of .8 at that time.

In Chapter 6 we will develop a more systematic approach for making inferences about proportions.

Learning the mechanics

**4.23.** Compute the following:

a.  $3!$

b.  $0!$

c.  $\dfrac{5!}{2!\,(5-2)!}$

d.  $(.2)^5$

e.  $\dbinom{6}{3}$

f.  $\dbinom{4}{0}$

g.  $\dbinom{4}{3}(.1)^3(.9)^1$

**4.24.** If $x$ is a binomial random variable, compute $p(x)$ for each of the following cases:

a.  $n = 4,\quad x = 2,\quad p = .5$

b.  $n = 3,\quad x = 3,\quad q = .9$

c.  $n = 2,\quad x = 0,\quad p = .3$

d.  $n = 4,\quad x = 3,\quad p = .4$

e.  $n = 3,\quad x = 1,\quad q = .7$

f.  $n = 2,\quad x = 1,\quad p = .3$

**4.25** If $x$ is a binomial random variable with $n = 6$ and $p = .5$:

a.  Display $p(x)$ in tabular form.
b.  Compute the mean and variance of $x$.
c.  Graph $p(x)$ and locate $\mu$ and the interval $\mu \pm 2\sigma$ on the graph.
d.  What is the probability that $x$ falls within the interval $\mu \pm 2\sigma$?

**4.26.** If $x$ is a binomial random variable with $n = 5$ and $p = .3$:

a.  Calculate the value of $p(x)$, $x = 0, 1, 2, 3, 4, 5$, using the formula for a binomial probability distribution.
b.  Using your answers to part a, give the probability distribution for $x$ in tabular form.

**4.27.** If $x$ is a binomial random variable, calculate $\mu$, $\sigma^2$, and $\sigma$ for each of the following:

a.  $n = 20,\quad p = .2$

b.  $n = 20,\quad p = .8$

c.  $n = 100,\quad p = .5$

d.  $n = 100,\quad p = .2$

e.  $n = 50,\quad p = .2$

f.  $n = 500,\quad p = .01$

**4.28.** If $x$ is a binomial random variable, use Table II, Appendix, to find the following probabilities:

a.  $P(x = 2)$ for $n = 10,\quad p = .1$
b.  $P(x \le 5)$ for $n = 15,\quad p = .4$
c.  $P(x > 1)$ for $n = 5,\quad p = .5$
d.  $P(x < 10)$ for $n = 25,\quad p = .8$
e.  $P(x \ge 10)$ for $n = 15,\quad p = .9$
f.  $P(x = 2)$ for $n = 20,\quad p = .6$

**4.29.** If $x$ is a binomial random variable with $n = 10$ and $p = .5$:

a. Find $\mu$, $\sigma^2$, and $\sigma$.

b. Using Table II, Appendix, graph the probability distribution of $x$.

c. Locate the interval $\mu \pm 2\sigma$ on the graph. What is the probability that $x$ is in the interval, $\mu \pm 2\sigma$?

## Applying the concepts

**4.30.** A large southern university has determined from past records that the probability a student who registers for fall classes will have his or her schedule rejected (due to overfilled classrooms, clerical error, etc.) is .2.

a. Suppose that 25,000 students register for fall classes, and $x$ is the number of students who have their schedules rejected. Is $x$ a binomial random variable? Explain.

b. Suppose that a random sample of 20 students is selected from the total of 25,000, and $x$ is the number of these students who have their schedules rejected. Is $x$ a binomial random variable? Explain.

c. Suppose that you sample the results of the first 1,000 students who register next fall and record $x$, the number of rejected registrations. Is $x$ a binomial random variable? Explain.

d. For the random variables in parts a, b, and c that you identified as being binomial, find $\mu$, $\sigma^2$, and $\sigma$.

**4.31.** Suppose that 60% of all people who are eligible for jury duty in a large Florida city are in favor of capital punishment. How does this affect the composition of a jury in a murder trial? Suppose that a jury of twelve is to be randomly selected from among all the prospective jurors in this city.

a. What is the expected number of jurors who favor capital punishment?

b. What is the probability that none of the twelve jurors selected favors capital punishment?

c. If the jury were really selected at random, would you be surprised if none of the jurors favored capital punishment? Explain.

**4.32.** A particular system in a space vehicle must work properly in order for the space ship to gain reentry into the earth's atmosphere. One component of the system operates successfully only 85% of the time. To increase the reliability of the system, four of the components will be installed in such a way that the system will operate successfully if at least one component is working successfully. What is the probability that the system will fail? Assume the components operate independently.

**4.33.** A basketball star has been successful on 80% of his foul shots during his career. As time runs out in a big game, the star's team is losing by 1 point. The star misses a desperation shot at the buzzer, but is fouled and awarded two foul shots. Assuming that this represents a binomial experiment, what is the probability that the game goes into overtime? Why might this not be a binomial experiment? [*Note:* A successful foul shot is worth 1 point, and the game will go into overtime only if the score is tied.]

4.34. Over the years, a physician has found that one out of every ten diabetics receiving insulin develops antibodies against the hormone, thus requiring a more costly form of medication.

    a.  Find the probability that, of the next five diabetic patients the physician treats, none will develop antibodies against insulin.

    b.  Find the probability that at least one will develop antibodies.

    c.  What assumptions are needed for solving this problem?

4.35. A problem of great concern to a manufacturer is the cost of repair and replacement required under a product's guarantee agreement. Assume it is known that 10% of all electronic pocket calculators purchased are returned for repair while their guarantee is still in effect. If a firm purchased twenty-five pocket calculators for its salespeople, what is the probability that five or more of these calculators will need repair while their guarantees are still in effect?

4.36. Experiments with animals are often conducted to determine whether certain chemicals are linked to cancer. Suppose that when a chemical is ingested in large doses by experimental rats, 24% of the rats contract thyroid tumors.

    a.  If each rat in a group of 200 ingests a large dose of the chemical, how many would be expected to be free of thyroid tumors?

    b.  Within what limits would we expect the number of rats free of thyroid tumors to fall? [*Hint:* Use Chebyshev's rule to assist in establishing the limits.]

4.37. An accountant believes that 10% of the company's invoices contain errors. To check this theory the accountant randomly samples twenty-five invoices and finds that seven contain errors.

    a.  If the accountant's theory is correct, what is the probability that of the twenty-five invoices written, seven or more contain errors?

    b.  What assumptions do you have to make to solve this problem using the methodology of this section?

    c.  If these assumptions are satisfied and if the sample of twenty-five invoices produces seven that contain errors, do you think that more than 10% of the company's invoices contain errors? Explain.

4.38. Suppose you are a purchasing officer for a large company. You have purchased five million electrical switches and have been guaranteed by the supplier that the shipment will contain no more than 0.1% defectives. To check the shipment, you randomly sample 500 switches, test them, and find that four are defective. If the switches are as represented, calculate $\mu$ and $\sigma$ for this sample of 500. Based on this evidence, do you think the supplier has complied with the guarantee? Explain. [*Hint:* Calculate $\mu$ and $\sigma$ for this binomial random variable with $p = .001$ to see if a value of $x$ as large as 4 is probable.]

4.39. An experiment is to be conducted to see whether an acclaimed psychic has extrasensory perception (ESP). Five different cards are shuffled and one is chosen at random. The psychic will then try to identify which card was drawn without seeing it. The experiment is to be repeated twenty times and $x$, the number of correct decisions, is recorded. (Assume the twenty trials are independent.)

a. If the psychic is guessing, i.e., if the psychic does not possess ESP, what is the value of $p$, the probability of a correct decision on each trial?

b. If the psychic is guessing, what is the expected number of correct decisions in twenty trials?

c. If the psychic is guessing, what is the probability of six or more correct decisions in twenty trials?

d. Suppose that the psychic makes six correct decisions in twenty trials. Is there evidence to indicate that the psychic is *not* guessing and actually has ESP? Explain.

**4.40.** After a costly study, a market analyst claims that 12% of all consumers in a particular sales region prefer a certain noncarbonated beverage. To check the validity of this figure, you decide to conduct a survey in the region. You randomly sample $n = 400$ consumers and find that $x = 31$ prefer the beverage.

a. Compute $\mu$ and $\sigma$ for $x$ assuming the market analyst's claim is correct.

b. Based on a sample of 400, is it likely that you will observe a value of $x \leq 31$ if the market analyst's claim is correct? Explain.

c. Do the results of your survey agree with the 12% estimate given by the market analyst? Explain.

**4.41.** A new drug has been synthesized which is designed to reduce a person's blood pressure. Twenty randomly selected hypertensive patients receive the new drug. Suppose eighteen or more of the patients' blood pressures drop.

a. Suppose that the probability that a hypertensive patient's blood pressure drops if he or she is *untreated* is .5. Then what is the probability of observing eighteen or more blood pressure drops in a random sample of twenty treated patients if the new drug is in fact ineffective in reducing blood pressure?

b. Considering this probability (part a), do you think you have observed a rare event or do you conclude that the drug is effective in reducing hypertension?

**4.42.** A literature professor decides to give a twenty question true–false quiz to determine who has read an assigned novel. She wants to choose the passing grade such that the probability of passing someone who is purely guessing is less than .05. What score should constitute the lowest passing grade?

**4.43.** Most firms utilize sampling plans to control the quality of manufactured items ready for shipment or the quality of items that have been purchased. To illustrate the use of a sampling plan, suppose you are shipping electrical fuses in lots, each containing 10,000 fuses. The plan specifies that you will randomly sample twenty-five fuses from each lot and accept (or ship) the lot if $x$, the number of defective fuses in the sample, is less than three. If $x \geq 3$, you will reject the lot and hold it for a complete reinspection.

a. What is the probability of accepting a lot ($x = 0$, 1, or 2) if the actual fraction defective in the lot is:

(i)  1      (ii)  .8      (iii)  .5      (iv)  .2      (v)  .05      (vi)  0

b. Construct a graph showing $P(A)$, the probability of lot acceptance, as a function of lot fraction defective, $p$. This graph is called the **operating characteristic curve** for the sampling plan.

c. Suppose the sampling plan called for sampling $n = 25$ fuses and accepting a lot if $x \leq 3$. Calculate the quantities specified in part a and construct the operating characteristic curve for this sampling plan. Compare this curve with the curve obtained in part b. (Note how the curve characterizes the ability of the plan to screen bad lots from shipment.)

## 4.5 CONTINUOUS PROBABILITY DISTRIBUTIONS

The graphical form of the probability distribution for a continuous random variable $x$ will be a smooth curve that might appear as shown in Figure 4.8. This curve, a function of $x$, is denoted by the symbol $f(x)$ and is variously called a probability density function, a frequency function, or a probability distribution.

The areas under a probability distribution correspond to probabilities for $x$. For example, the area $A$ beneath the curve between the two points $a$ and $b$, as shown in Figure 4.8, is the probability that $x$ assumes a value between $a$ and $b$ ($a < x < b$). Because there is no area over a point. say $x = a$, it follows that (according to our model) the probability associated with a particular value of $x$ is equal to 0, i.e.,

**FIGURE 4.8**
**A PROBABILITY DISTRIBUTION $f(x)$ FOR A CONTINUOUS RANDOM VARIABLE $x$**

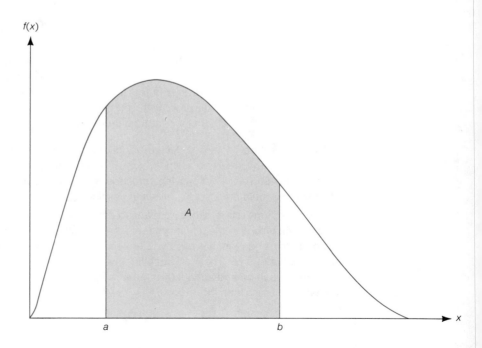

$P(x = a) = 0$ and $P(a < x < b) = P(a \le x \le b)$, i.e., the probability is the same regardless of whether you include the endpoints of the interval. Also, because areas over intervals represent probabilities, it follows that the total area under a probability distribution, the probability assigned to all values of $x$, should equal 1. Note that probability distributions for continuous random variables will possess different shapes depending on the relative frequency distributions of real data that the probability distributions are supposed to model.

The areas under most probability distributions are obtained by the use of the calculus* or other numerical methods. Because this is often a difficult procedure, we will give the areas for some of the most common probability distributions in tabular form in the Appendix. Then, to find the area between two values of $x$, say $x = a$ and $x = b$, you will simply have to consult the appropriate table.

For each of the continuous random variables presented in this chapter, we will give the formula for the probability distribution along with its mean and standard deviation. These two numbers, $\mu$ and $\sigma$, will enable you to make some approximate probability statements about a random variable even when you do not have access to a table of areas under the probability distribution.

## 4.6
## THE NORMAL
## DISTRIBUTION

One of the most commonly observed continuous random variables has a bell-shaped probability distribution, as shown in Figure 4.9. It is known as a normal random variable and its probability distribution is called a normal distribution.

**FIGURE 4.9**
**A NORMAL**
**PROBABILITY**
**DISTRIBUTION**

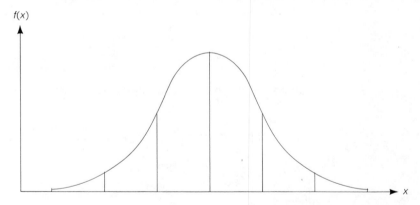

You will see during the remainder of this text that the normal distribution plays a very important role in the science of statistical inference. In addition, many phenomena generate random variables with probability distributions that are very well approximated by a normal distribution. For example, the error made in measuring a person's blood pressure may be a normal random variable, and the probability distribution for the yearly rainfall in a certain region might be approximated by a normal probability distribution. The normal distribution might also provide an accurate model for the

*Students with knowledge of the calculus should note that the probability that $x$ assumes a value in the interval $a < x < b$ is $P(a < x < b) = \int_a^b f(x)\, dx$, assuming the integral exists. Similar to the requirements for a discrete probability distribution, we require $f(x) \ge 0$ and $\int_{-\infty}^{\infty} f(x)\, dx = 1$.

probability distribution of the weights of loads of produce shipped to a supermarket. You can determine the adequacy of the normal approximation to an existing population of data by comparing the relative frequency distribution of a sample of the data (200 or 300 measurements, or more) to the normal probability distribution. Tests to detect a disagreement between a set of data and the assumption of normality are available, but they are beyond the scope of this book.

The normal distribution is perfectly symmetric about its mean $\mu$, as can be seen in the examples in Figure 4.10. Its spread is determined by the value of its standard deviation, $\sigma$.

FIGURE 4.10
SEVERAL NORMAL
DISTRIBUTIONS,
WITH DIFFERENT
MEANS AND STANDARD
DEVIATIONS

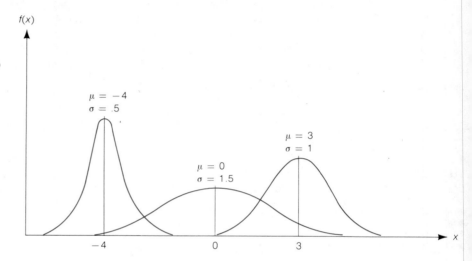

The formula for the normal probability distribution is shown below:

PROBABILITY DISTRIBUTION FOR A NORMAL RANDOM VARIABLE $x$

$$f(x) = \frac{1}{\sigma\sqrt{2\pi}} e^{-(1/2)[(x-\mu)/\sigma]^2}$$

where

$\mu$ = Mean of the normal random variable $x$

$\sigma$ = Standard deviation

$\pi$ = 3.1416 . . .

$e$ = 2.71828 . . . .

Note that the mean $\mu$ and standard deviation $\sigma$ appear in this formula, so that no separate formulas for $\mu$ and $\sigma$ are necessary. To graph the normal curve we will have to know the numerical values of $\mu$ and $\sigma$.

Computing the area over intervals under the normal probability distribution is a difficult task.* Consequently, we will use the computed areas listed in Table III of the Appendix. Since there is an infinitely large number of normal curves—one for each pair of values for $\mu$ and $\sigma$—we have formed a single table that will apply to any of these normal curves. This is done by constructing the table of areas as a function of the $z$-score (presented in Section 2.6). The population $z$-score for a measurement was defined as the *distance* between the measurement and the population mean, divided by the population standard deviation. Thus, the $z$-score gives the distance between a measurement and the mean in units equal to the standard deviation. In symbolic form, the $z$-score for the measurement $x$ is

$$z = \frac{x - \mu}{\sigma}$$

Note that when $x = \mu$, we obtain $z = 0$.

To illustrate the use of Table III, suppose we know that the length of time, $x$, between charges of a pocket calculator has a normal distribution, with a mean of 50 hours and a standard deviation of 15 hours. If we were to observe the length of time that elapses before the need for the next charge, what is the probability that this measurement will assume a value between 50 and 70 hours? This probability is the area under the normal probability distribution between 50 and 70, as shown in the shaded area, $A$, of Figure 4.11.

**FIGURE 4.11**
**NORMAL FREQUENCY FUNCTION:**
$\mu = 50$, $\sigma = 15$

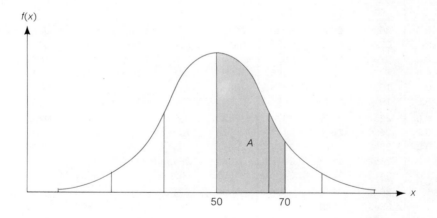

The first step in finding the area $A$ is to calculate the $z$-score corresponding to the measurement 70. We calculate

$$z = \frac{x - \mu}{\sigma} = \frac{70 - 50}{15} = \frac{20}{15} = 1.33$$

*The student with knowledge of the calculus should note that there is not a closed-form expression for $P(a < x < b) = \int_a^b f(x)\, dx$ for the normal probability distribution. The value of this definite integral can be obtained to any desired degree of accuracy by approximation procedures. For this reason, it is tabulated for the user.

Thus, the measurement, 70, is 1.33 standard deviations above the mean, $\mu = 50$. The second step is to refer to Table III (a partial reproduction of this table is shown in Figure 4.12). Note that $z$-scores are listed in the left-hand column of the table. To find the area corresponding to a $z$-score of 1.33, we first locate the value 1.3 in the left-hand column. Since this column lists $z$ values to one decimal place only, we refer to the top row of the table to get the second decimal place, .03. Finally, we locate the number where the row labeled $z = 1.3$ and the column labeled .03 meet. This number represents the area between the mean, $\mu$, and the measurement that has a $z$-score of 1.33:

$$A = .4082$$

Or, the probability that the calculator operates between 50 and 70 hours before needing a charge is .4082.

**FIGURE 4.12**
**REPRODUCTION OF PART OF TABLE III IN THE APPENDIX**

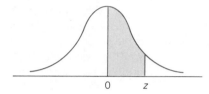

| z | .00 | .01 | .02 | .03 | .04 | .05 | .06 | .07 | .08 | .09 |
|---|-----|-----|-----|-----|-----|-----|-----|-----|-----|-----|
| 0.0 | .0000 | .0040 | .0080 | .0120 | .0160 | .0199 | .0239 | .0279 | .0319 | .0359 |
| 0.1 | .0398 | .0438 | .0478 | .0517 | .0557 | .0596 | .0636 | .0675 | .0714 | .0753 |
| 0.2 | .0793 | .0832 | .0871 | .0910 | .0948 | .0987 | .1026 | .1064 | .1103 | .1141 |
| 0.3 | .1179 | .1217 | .1255 | .1293 | .1331 | .1368 | .1406 | .1443 | .1480 | .1517 |
| 0.4 | .1554 | .1591 | .1628 | .1664 | .1700 | .1736 | .1772 | .1808 | .1844 | .1879 |
| 0.5 | .1915 | .1950 | .1985 | .2019 | .2054 | .2088 | .2123 | .2157 | .2190 | .2224 |
| 0.6 | .2257 | .2291 | .2324 | .2357 | .2389 | .2422 | .2454 | .2486 | .2517 | .2549 |
| 0.7 | .2580 | .2611 | .2642 | .2673 | .2704 | .2734 | .2764 | .2794 | .2823 | .2852 |
| 0.8 | .2881 | .2910 | .2939 | .2967 | .2995 | .3023 | .3051 | .3078 | .3106 | .3133 |
| 0.9 | .3159 | .3186 | .3212 | .3238 | .3264 | .3289 | .3315 | .3340 | .3365 | .3389 |
| 1.0 | .3413 | .3438 | .3461 | .3485 | .3508 | .3531 | .3554 | .3577 | .3599 | .3621 |
| 1.1 | .3643 | .3665 | .3686 | .3708 | .3729 | .3749 | .3770 | .3790 | .3810 | .3830 |
| 1.2 | .3849 | .3869 | .3888 | .3907 | .3925 | .3944 | .3962 | .3980 | .3997 | .4015 |
| 1.3 | .4032 | .4049 | .4066 | .4082 | .4099 | .4115 | .4131 | .4147 | .4162 | .4177 |
| 1.4 | .4192 | .4207 | .4222 | .4236 | .4251 | .4265 | .4279 | .4292 | .4306 | .4319 |
| 1.5 | .4332 | .4345 | .4357 | .4370 | .4382 | .4394 | .4406 | .4418 | .4429 | .4441 |

**EXAMPLE 4.12**

Suppose you have a normal random variable $x$ with $\mu = 50$ and $\sigma = 15$. Find the probability that $x$ will fall within the interval $30 < x < 50$.

**Solution**

The solution to this example can be seen from Figure 4.13. Note that both $x = 30$ and $x = 70$ lie the same distance from the mean, $\mu = 50$; $x = 30$ lies below the mean

FIGURE 4.13
NORMAL PROBABILITY
DISTRIBUTION:
$\mu = 50$, $\sigma = 15$

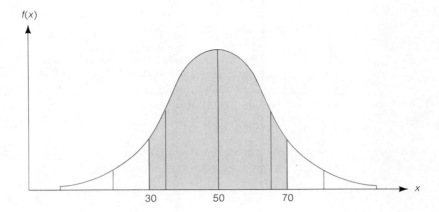

and $x = 70$ lies above it. Then, because the normal curve is symmetric about the mean, the area representing the probability that $x$ falls between $x = 30$ and $\mu = 50$ is equal to the area representing the probability that it falls between $\mu = 50$ and $x = 70$. The probability of observing a value between 50 and 70 (from Table III) is .4082 (obtained in the previous discussion).

Because $x = 30$ lies to the left of the mean, the corresponding $z$-score should be negative and of the same magnitude as the $z$-score corresponding to $x = 70$. Checking, we obtain

$$z = \frac{x - \mu}{\sigma} = \frac{30 - 50}{15} = -1.33$$

In finding areas (probabilities) under the normal curve, it is easier to show the location of the $z$-scores rather than the corresponding values of $x$. For example, the $z$-scores corresponding to $x = 30$ and $x = 70$ are located on the distribution of $z$-scores at the points shown in Figure 4.14. The distribution of $z$-scores, known as a standard normal distribution, will always have a mean equal to 0 and a standard deviation equal to 1.

FIGURE 4.14
A DISTRIBUTION
OF $z$-SCORES
(A STANDARD NORMAL
DISTRIBUTION)

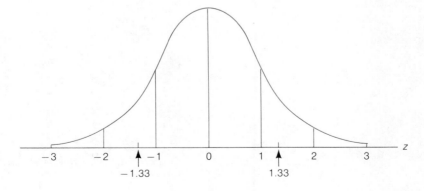

EXAMPLE 4.13    Use Table III to determine the area to the right of the z-score 1.64 for the standard normal distribution, i.e., find $P(z > 1.64)$.

Solution    The probability that a normal random variable will fall more than 1.64 standard deviations to the right of its mean is indicated in Figure 4.15. Because the normal distribution is symmetric, half of the total probability (.5) lies to the right of the mean and half to the left. Therefore, the desired probability is

$$P(z > 1.64) = .5 - A$$

where $A$ is the area between $\mu = 0$ and $z = 1.64$, as shown in the figure. Referring to Table III, we find that the area $A$ corresponding to $z = 1.64$ is .4495. So,

$$P(z > 1.64) = .5 - A = .5 - .4495 = .0505$$

**FIGURE 4.15**
**STANDARD NORMAL**
**DISTRIBUTION:**
$\mu = 0, \sigma = 1$

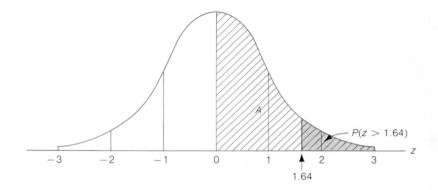

EXAMPLE 4.14    Find the total area to the right of $z = -.74$ for the standard normal distribution. This area is $P(z > -.74)$.

Solution    The standard normal distribution is shown in Figure 4.16, with the area to the right of $-.74$, $P(z > -.74)$, shaded. Note that we have divided the total shaded area corresponding to $P(z > -.74)$ into two parts: the area to the left of $z = 0$, $A_1$, and the area to the right of $z = 0$, $A_2$. Whenever the desired area overlaps the mean, it is necessary to make this division and to find the areas separately in Table III. The area $A_2$ is easy to

**FIGURE 4.16**
**STANDARD NORMAL**
**DISTRIBUTION:**
$\mu = 0, \sigma = 1$

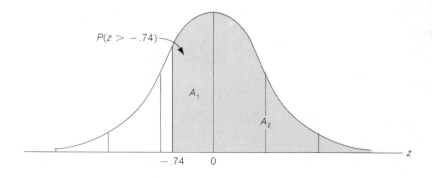

find, since it is all the area to the right of the mean. Thus, $A_2 = .5$. The area $A_1$ is the area between $z = 0$ and $z = -.74$. This area, which is equivalent to the tabulated area (Table III) between $z = 0$ and $z = .74$ is $A_1 = .2704$. Then the total area, $A$, to the right of $z = -.74$ is the sum of the areas $A_1$ and $A_2$:

$$P(z > -.74) = A_1 + A_2 = .2704 + .5 = .7704$$

EXAMPLE 4.15

Find the total area to the right of $z = 1.96$ and to the left of $z = -1.96$. To put this in probabilistic terminology, find the probability that a normal random variable lies more than 1.96 standard deviations away from the mean.

Solution

The requested probability $P(z > 1.96$ or $z < -1.96)$ is the sum of the two areas $A_1$ and $A_2$ shown in Figure 4.17. Because the normal distribution is symmetric, the areas lying to the right of $z = 1.96$ and to the left of $z = -1.96$ must be equal. Therefore, $A_1 = A_2$.

Checking Table III, we find the area corresponding to $z = 1.96$ to be .4750. This is the area between $z = 0$ and $z = 1.96$. Therefore,

$$A_2 = .5 - .4750 = .0250$$

And, because of the symmetry of the normal distribution,

$$P(z > 1.96 \text{ or } z < -1.96) = A_1 + A_2 = .0250 + .0250 = .0500$$

**FIGURE 4.17**
**STANDARD NORMAL**
**DISTRIBUTION:**
$\mu = 0, \sigma = 1$

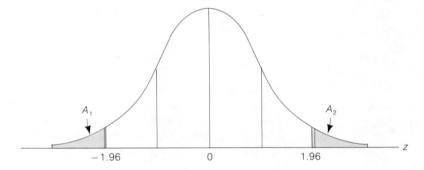

EXAMPLE 4.16

Suppose an automobile manufacturer introduces a new model, which has an advertised mean in-city mileage of 27 miles per gallon. Although such advertisements never report any measure of variability from car to car, suppose you write the manufacturer for the details of the tests, and you find that the standard deviation is 3 miles per gallon.

The information leads you to formulate a probability model for the random variable $x$, the in-city mileage for this car model. You believe that the probability distribution of $x$ can be approximated by a normal distribution, with a mean of 27 and a standard deviation of 3.

**a.** If you were to buy this model of automobile, what is the probability that you would purchase one that averages less than 20 miles per gallon for in-city driving?
**b.** Suppose you purchase one of these new models and it does get less than 20 miles per gallon for in-city driving. Should you conclude that your probability model is incorrect?

FIGURE 4.18
NORMAL PROBABILITY
DISTRIBUTION FOR $x$
IN EXAMPLE 4.16:
$\mu = 27$ MILES
PER GALLON,
$\sigma = 3$ MILES
PER GALLON

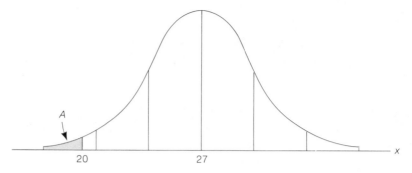

Solution

**a.** The probability model proposed for $x$, the in-city mileage, is shown in Figure 4.18. We are interested in finding the area, $A$, to the left of 20, since this area corresponds to the probability that a measurement chosen from this distribution falls below 20. Or, in other words, if this model is correct, the area $A$ represents the fraction of cars that can be expected to get less than 20 miles per gallon for in-city driving. To find $A$, we first calculate the $z$ value corresponding to $x = 20$. That is,

$$z = \frac{x - \mu}{\sigma} = \frac{20 - 27}{3} = -\frac{7}{3} = -2.33$$

Since Table III gives only areas to the right of the mean (and because the normal distribution is symmetric about its mean), we look up 2.33 in Table III and find that the corresponding area is .4901. This is equal to the area between $z = 0$ and $z = -2.33$, so we find

$$A = .5 - .4901 = .0099 \approx .01$$

According to this probability model, you should have only about a 1% chance of purchasing a car of this make with an in-city mileage under 20 miles per gallon.

**b.** Now you are asked to make an inference based on a sample—the car you purchased. You are getting less than 20 miles per gallon for in-city driving. What do you infer? We think you will agree that one of two possibilities is true:

The probability model is correct, and you simply were unfortunate to have purchased one of the cars in the 1% that get less than 20 miles per gallon in the city.
The probability model is incorrect. That is, if the manufacturer meant that the in-city mileage for the cars has a normal distribution with a mean equal to 27 and $\sigma = 3$, the claim is false.

You have no way of knowing which possibility is the correct one, but certainly the evidence points to the second one. We are again relying on the rare event approach to statistical inference that we introduced earlier. The sample (one measurement in this case) was so unlikely to have been drawn from the proposed probability model that it casts serious doubt on the model. We would be inclined to believe that the model is somehow in error. Perhaps the assumption of a normal distribution is unwarranted, or the mean of 27 is an overestimate, or the standard deviation of 3 is an underestimate, or some combination of these errors was made. At any rate, the form of the actual probability model certainly merits further investigation.

**CASE STUDY 4.3**
GRADING ON THE
CURVE

How did teachers ever suppose that the statistician's bell-shaped curve ought somehow to be imposed on the results of their work? The famous curve was developed to describe the distribution of natural phenomena. If we weigh 10,000 grains of corn, or measure the heights of men and women, or their ability to learn nonsense syllables, the findings will cluster around some central value and taper at both ends. The bell-shaped curve is descriptive of raw or unselected phenomena, and then only if vast numbers of cases are used. When the teacher receives his pupils, their distribution with respect to some characteristics may follow the "normal curve." But, having received his charges, the skillful teacher sets out as fast as he can to destroy the "natural" state of affairs.

The above is the introduction to Clyde W. Bresee's (1976) article about "grading on the curve." His main point is that many teachers who consistently grade on the curve are assuming the measures of learning (test scores and the like) will always form a normal or bell-shaped distribution. These teachers may even use $z$-scores to determine grades and the corresponding areas under the normal curve (those in Table III) to obtain the percentages of students who will receive each grade. The implication is that a student's performance will be measured only in relation to the other students in the class. Bresee relates the following anecdote:

After the completion of a particularly successful unit, a teacher was heard to say, "I don't know how I'll grade this thing because they all did so well." This is a sad and dangerous statement because this teacher is on the verge of undoing a month's or even a year's work. Here is a teacher so indoctrinated by an erroneous concept that he questions his own accomplishment even in the face of clear evidence. But question it he must, if he has been trained to grade "on the curve." In a class of 20, for example, the only possible explanation for 15 A's is that he is a weak teacher or a "soft grader," or both.

The caveat presented by Bresee can be more generally applied. The normal distribution is widely used and in many situations it provides an adequate approximation to reality. However, before each new application of the normal distribution, the situation should be carefully studied, and the user must be confident that all attendant assumptions are satisfied. When using the normal distribution to "grade on a curve," Bresee worries that the day may come when a supervisor admonishes a teacher, "Last year we sent you 20 children to teach. We spent a year's time and upwards of $20,000 on them, and all we have to show for it is the same old bell-shaped curve!"

EXERCISES

Learning the mechanics

**4.44.** Find the area under the standard normal probability distribution between the following pairs of $z$-scores:

    a.  $z = 0$ and $z = 2.00$       b.  $z = 0$ and $z = 1.50$
    c.  $z = 0$ and $z = 3.06$       d.  $z = 0$ and $z = 0.50$
    e.  $z = -2.00$ and $z = 0$     f.  $z = -1.50$ and $z = 0$
    g.  $z = -3.06$ and $z = 0$     h.  $z = -0.50$ and $z = 0$

**4.45.** Find each of the following:

    a.  $P(0 \le z \le 2.00)$     b.  $P(-1.50 < z < 0)$
    c.  $P(0 \le z < 3.06)$     d.  $P(-.50 < z < 0)$

**4.46.** Find each of the following:

    a.  $P(-1 \le z \le 1)$       b.  $P(-2 \le z \le 2)$
    c.  $P(-2.03 < z \le 1.72)$     d.  $P(-1.50 < z < 0.46)$
    e.  $P(z \ge -2.16)$         f.  $P(z < 2.57)$

**4.47.** Find each of the following:

    a.  $P(z > 1)$           b.  $P(z < -2)$
    c.  $P(1.62 \le z \le 2.15)$     d.  $P(-1.92 \le z < -.67)$
    e.  $P(z \ge 0)$           f.  $P(-2.76 < z < 1.00)$

**4.48.** Find a $z$-score, call it $z_0$, such that:

    a.  $P(z \ge z_0) = .5$       b.  $P(z \ge z_0) = .025$
    c.  $P(z \le z_0) = .025$     d.  $P(z \ge z_0) = .0228$
    e.  $P(0 \le z \le z_0) = .4803$    f.  $P(z < z_0) = .0401$

**4.49.** Find a $z$-score, call it $z_0$, such that:

    a.  $P(z > z_0) = .9850$       b.  $P(z < z_0) = .9850$
    c.  $P(-z_0 \le z \le z_0) = .95$     d.  $P(-z_0 \le z \le z_0) = .90$
    e.  $P(-z_0 \le z \le z_0) = .6826$     f.  $P(-z_0 \le z \le z_0) = .9950$

**4.50.** Find each of the following:

    a.  $P(z \le 1.72)$         b.  $P(-2.01 \le z \le 0.91)$
    c.  $P(z \ge 2.90)$         d.  $P(z \ge -1.70)$
    e.  $P(-1.96 \le z \le -1.26)$    f.  $P(z \le 4.00)$
    g.  $P(z \le -3.20)$        h.  $P(-1.65 \le z \le 1.65)$

**4.51.** Find a z-score, call it $z_0$, such that:

a. $P(z \leq z_0) = .0013$
b. $P(z \geq z_0) = .0013$
c. $P(z \leq z_0) = .9515$
d. $P(-z_0 \leq z \leq z_0) = .6528$
e. $P(-2 \leq z \leq z_0) = .8804$
f. $P(z_0 \leq z \leq 2.6) = .0312$

**4.52.** Give the z-score for a measurement from a normal distribution for the following:

a. One standard deviation above the mean
b. One standard deviation below the mean
c. Equal to the mean
d. Two and one-half standard deviations below the mean
e. Three standard deviations above the mean

**4.53.** Suppose that the random variable $x$ is best described by a normal distribution with $\mu = 30$ and $\sigma = 5$. Find the z-score that corresponds to each of the following $x$ values:

a. $x = 25$
b. $x = 30$
c. $x = 37.5$
d. $x = 10$
e. $x = 50$
f. $x = 32$

**4.54.** A random variable $x$ is normally distributed with $\mu = 20$ and $\sigma = 4$. Determine the distance, in units of standard deviations, between each of the following values of $x$ and the mean, $\mu = 20$:

a. $x = 10$
b. $x = 20$
c. $x = 25$
d. $x = 15$
e. $x = 0$
f. $x = 60$

**4.55.** Suppose that $x$ is a normally distributed random variable with $\mu = 10$ and $\sigma = 2$. Find each of the following:

a. $P(10 \leq x \leq 12)$
b. $P(6 \leq x \leq 10)$
c. $P(13 \leq x \leq 16)$
d. $P(7.8 \leq x \leq 12.6)$
e. $P(x \geq 13.24)$
f. $P(x \geq 7.62)$

**4.56.** Suppose that $x$ is a normally distributed random variable with $\mu = 50$ and $\sigma = 10$. Find each of the following:

a. $P(x \leq 50)$
b. $P(x \leq 35.6)$
c. $P(40.7 \leq x \leq 75.8)$
d. $P(22.9 \leq x \leq 33.2)$
e. $P(x \geq 25.3)$
f. $P(x \leq 25.3)$

**4.57.** Suppose that $x$ is a normally distributed random variable with $\mu = 40$ and $\sigma = 9$. Find a value of the random variable, call it $x_0$, such that:

a. $P(x \geq x_0) = .5$
b. $P(x < x_0) = .025$
c. $P(x > x_0) = .10$
d. $P(x > x_0) = .95$
e. 10% of the values of $x$ are less than $x_0$
f. 80% of the values of $x$ are less than $x_0$
g. 1% of the values of $x$ are greater than $x_0$

**4.58.** Suppose that $x$ is a normally distributed random variable with mean 100 and standard deviation 8. Draw a rough graph of the distribution of $x$. Locate $\mu$ and the interval $\mu \pm 2\sigma$ on the graph. Find the following probabilities:

a.  $P(\mu - 2\sigma \leq x \leq \mu + 2\sigma)$    b.  $P(x \geq \mu + 2\sigma)$
c.  $P(x \leq 92)$                    d.  $P(92 \leq x \leq 116)$
e.  $P(92 \leq x \leq 96)$       f.  $P(76 \leq x \leq 124)$

Applying the concepts

**4.59.** The amount of oxygen dissolved in rivers and streams depends upon the water temperature and upon the amounts of decaying organic matter from natural processes or human disturbances that are present in the water. The Council on Environmental Quality (CEQ) considers a dissolved oxygen content of less than 5 milligrams per liter of water to be undesirable because it is unlikely to support aquatic life. Suppose that an industrial plant discharges its waste into a river and that the downstream daily oxygen content measurements are normally distributed with a mean equal to 6.3 milligrams per liter and a standard deviation of .6 milligram per liter.

    a.  What percentage of the days would the dissolved oxygen content in the river be considered undesirable by the CEQ?
    b.  Within what limits would we expect the dissolved oxygen content to fall?

**4.60.** The scores on a test designed to measure elementary school teachers' attitudes toward handicapped students are normally distributed with a mean score equal to 67 and a standard deviation equal to 10.8.

    a.  If an elementary school teacher is chosen at random, what is the probability that he or she would score above 95 on the test?
    b.  A program has been developed that is designed to improve teachers' attitudes toward handicapped students. One teacher who has completed the program is chosen at random and scores above 95 on the test. Would you conclude that the mean test score for teachers completing the program is higher than that of other teachers? Why?

**4.61.** The pulse rate per minute of the adult male population between 18 and 25 years of age in the United States is known to have a normal distribution with a mean of 72 beats per minute and a standard deviation of 9.7. If the requirements for military service state that anyone with a pulse rate over 100 is medically unsuitable for service, what proportion of the males between 18 and 25 years of age would be declared unfit because their pulse rates are too high?

**4.62.** Testing has shown that a new washing machine has a length of life that is normally distributed with mean equal to 5.10 years and a standard deviation equal to 1.54 years. If the washers are guaranteed for 2 years, what percentage will fail before the end of the guarantee time?

**4.63.** A network television department that sells commercial time to advertisers claims the number of homes reached per afternoon by a certain weekday soap opera has a mean of 4.5 million homes and a standard deviation of .5 million. Assume that the number of homes reached per afternoon by this program has a normal distribution.

a.  Assuming the department's claim is true, for what fraction of the weekday afternoons does this program reach fewer than 3.5 million homes?

b.  For what fraction of weekday afternoons does this program reach between 3.1 and 5.6 million homes?

c.  If the department's claim is correct, what is the probability that this program will reach fewer than 3.5 million homes on each of two randomly selected weekday afternoons?

d.  A television rating service randomly selects two weekday afternoons and finds 3.2 and 3.4 million homes, respectively, were reached by this program on the two days. Using these findings and your answer to part c, what can be said about the television department's claim?

**4.64.**  A machine used to regulate the amount of dye dispensed for mixing shades of paint can be set so that it discharges an average of $\mu$ milliliters of dye per can of paint. The amount of dye discharged is known to have a normal distribution with a standard deviation of .4 milliliter. If more than 6 milliliters of dye are discharged when making a particular shade of blue paint, the shade is unacceptable. Determine the setting for $\mu$ so that only 1% of the cans of paint will be unacceptable.

**4.65.**  A physical fitness association is including the mile run in their secondary school fitness test for boys. The time for this event for boys in secondary school is approximately normally distributed with a mean of 450 seconds and a standard deviation of 40 seconds. If the association wants to designate the fastest 10% as "excellent," what time should the association set for this criterion?

**4.66.**  The board of examiners that administers the real estate brokers' examination in a certain state found that the mean score on the test was 435 and the standard deviation was 72. If the board wants to set the passing score so that only the best 30% of all applicants pass, what is the passing score? Assume the scores are normally distributed.

**4.67.**  The distribution of the demand (in number of units per unit time) for a product can often be approximated by a normal probability distribution. For example, a bakery has determined that the number of loaves of its white bread demanded daily has a normal distribution with mean 7,200 loaves and standard deviation 300 loaves. Based on cost considerations, the company has decided that its best strategy is to produce a sufficient number of loaves so that it will fully supply demand on 94% of all days.

a.  How many loaves of bread should the company produce?

b.  Based on the production in part a, on what percentage of days will the company be left with more than 500 loaves of unsold bread?

4.7

APPROXIMATING
A BINOMIAL
DISTRIBUTION
WITH A NORMAL
DISTRIBUTION

When a binomial random variable can assume a large number of values, the calculation of its probabilities may become very tedious. To contend with this problem, we
provide tables in the Appendix to give the probabilities for some values of $n$ and $p$,
but these tables are by necessity incomplete. In particular, the binomial probability
table (Table II) can be used only for $n = $ 5, 10, 15, 20, or 25. To deal with this
limitation, we seek approximation procedures for calculating the probabilities associated with a binomial probability distribution.

When $n$ is large, a normal probability distribution may be used to provide a good
approximation to the probability histogram of a binomial random variable. To show
how this approximation works, we refer to Example 4.11, in which we used the binomial distribution to model the number $x$ of twenty voters who favor a candidate.
We assumed that 60% of all the eligible voters favored the candidate. The mean
and standard deviation of $x$ were found to be $\mu = 12$ and $\sigma = 2.2$. The binomial
distribution for $n = 20$ and $p = .6$ is shown in Figure 4.19 and the approximat-

**FIGURE 4.19**
BINOMIAL
DISTRIBUTION FOR
$n = 20$, $p = .6$ AND
NORMAL DISTRIBUTION
WITH $\mu = 12$, $\sigma = 2.2$

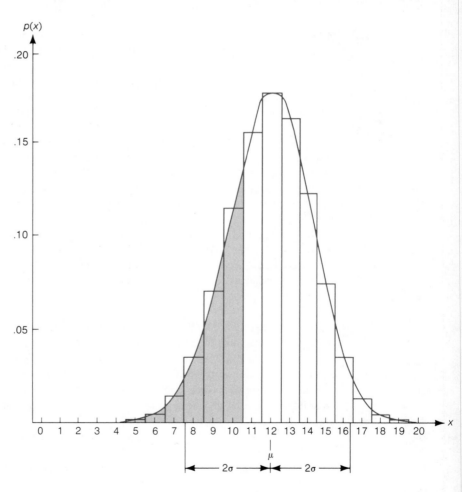

ing normal distribution with mean $\mu = 12$ and standard deviation $\sigma = 2.2$ is superimposed.

As part of Example 4.11, we used Table II to find the probability that $x \leq 10$. This probability, which is equal to the sum of the areas contained in the rectangles (shown in Figure 4.19) that correspond to $p(0), p(1), p(2), \ldots, p(10)$, was found to equal .245. The portion of the approximating normal curve that would be used to approximate the area $p(0) + p(1) + \cdots + p(10)$ is shaded in Figure 4.19. Note that this shaded area lies to the left of 10.5 (not 10), so we may include all of the probability in the rectangle corresponding to $p(10)$. The $z$-score corresponding to this value of $x$, 10.5, is

$$z = \frac{x - \mu}{\sigma} = \frac{10.5 - 12}{2.2} = -.68$$

From Table III of the Appendix we find that the area between $z = 0$ and $z = -.68$ is .2517. Thus, the approximating normal probability is the area to the left of $x = 10.5$, or

$$P(x \leq 10) \approx .5 - .2517 = .2483$$

You can see that this approximation yields a value that differs only slightly from the exact value, .245.

You may be wondering how large $n$ should be before the normal distribution provides an adequate approximation to the binomial. We will, as a rule of thumb, require that the interval $\mu \pm 3\sigma$ lie completely within the range of values for $x$, i.e., within the interval from 0 to $n$. When this condition is *not* satisfied, the binomial probability distribution will be skewed (to the right or left, depending upon the value of $p$) and the symmetric normal curve will provide a poor approximation to it. In the example above, $\mu \pm 3\sigma = 12 \pm 3(2.2) = 12 \pm 6.6 = (5.4, 18.6)$. This lies within the interval from 0 to 20, so the normal approximation should be adequate.

---

THE SAMPLE SIZE NECESSARY FOR THE NORMAL DISTRIBUTION
TO PROVIDE A GOOD APPROXIMATION TO THE
BINOMIAL PROBABILITY DISTRIBUTION
The sample size should be large enough so that the interval $\mu \pm 3\sigma$ (where $\mu = np$ and $\sigma = \sqrt{npq}$) lies completely within the range of the values of $x$, i.e., within the interval, 0 to $n$.

---

EXAMPLE 4.17

An integral part of many modern electronic appliances (television sets, tape recorders, pocket calculators, etc.) is solid-state circuitry. Since this circuitry is mass-produced (stamped) by machine, many of these products have become relatively inexpensive. A problem with anything that is mass-produced is quality control, and the manufacturing process must somehow be monitored to make certain the fraction of defective items produced is kept at an acceptable level.

One method of dealing with this problem is lot acceptance sampling, in which a sample of the items produced is selected, and each item in the sample is carefully tested. The lot of items is then accepted (inferring there are few defectives in the entire lot) or rejected (inferring there is an unacceptable fraction of defectives in the entire lot), based upon the number of defectives in the sample. For example, suppose a manufacturer of solid-state circuits for television sets chooses 200 stamped circuits from the day's production and determines $x$, the number of defective circuits in the sample. If the manufacturer is willing to accept the production of up to 6% defectives:

**a.** Find the mean and standard deviation of $x$, assuming the true proportion of defectives is .06.

**b.** Use the normal approximation to determine the probability that twenty or more defectives are observed in the sample of 200 circuits, i.e., that $x \geq 20$.

**Solution**

**a.** The random variable $x$ is binomial, with $n = 200$ and the fraction defective $p = .06$. Thus,

$$\mu = np = 200(.06) = 12$$
$$\sigma = \sqrt{npq} = \sqrt{200(.06)(.94)} = \sqrt{11.28} = 3.36$$

Note that

$$\mu \pm 3\sigma = 12 \pm 3(3.36) = 12 \pm 10.08 = (1.92, 22.08)$$

lies completely within the range from 0 to 200, so that the normal probability distribution should provide an adequate approximation to this binomial distribution.

**b.** To find the approximating area corresponding to $x \geq 20$, refer to Figure 4.20. Note that we want to include all of the binomial probability histogram from 20 to 200, inclusive. But in order to include the entire rectangle corresponding to $x = 20$, we must begin the approximating area at $x = 19.5$. Thus, our $z$ value is

$$z = \frac{x - \mu}{\sigma} = \frac{19.5 - 12}{3.36} = \frac{7.5}{3.36} = 2.23$$

Referring to Table III in the Appendix, we find that the area to the right of the mean corresponding to $z = 2.23$ (see Figure 4.21) is .4871. So, the area $A$ is

**FIGURE 4.20**
**NORMAL APPROXIMATION TO THE BINOMIAL DISTRIBUTION WITH** $n = 200$, $p = .06$

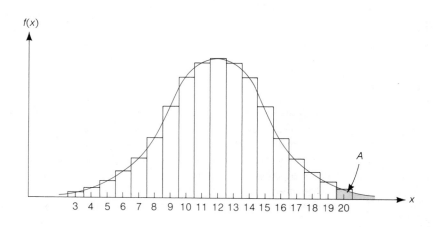

$f(x)$

3  4  5  6  7  8  9  10 11  12 13 14 15 16 17 18 19 20

$x$

$A$

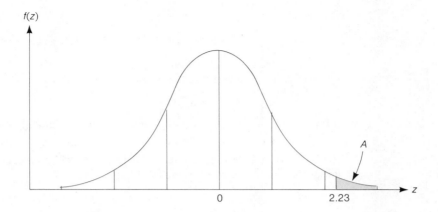

FIGURE 4.21
STANDARD NORMAL
DISTRIBUTION

$A = .5 - .4871 = .0129$

Thus, the normal approximation to the binomial probability is

$P(x \geq 20) \approx .0129$

In other words, the probability is extremely small that twenty or more defectives will be observed in a sample of 200 circuits, if in fact the true fraction defective is .06. If the manufacturer observes $x \geq 20$, the likely reason is that the process is producing more than the acceptable 6% defectives. The lot acceptance sampling procedure is another example of using the rare event approach to make inferences.

The following are useful aids when using the normal probability distribution to approximate the binomial probability distribution:

---

**AIDS IN USING NORMAL PROBABILITIES TO APPROXIMATE BINOMIAL PROBABILITIES**

**1.** Find $\mu = np$ and $\sigma = \sqrt{npq}$. Check to make certain that the interval $\mu \pm 3\sigma$ lies completely within the interval from 0 to $n$.

**2.** Sketch the probability rectangles you wish to approximate and sketch the approximating normal curve. Label the mean of the normal curve.

**3.** Shade the area of interest. Remember that when you compute $z$ values, the $x$ values that locate this area on the approximating normal curve will always end in .5. You will be able to see this in your sketch (item 2).

**4.** Use the method discussed in Section 4.6 for finding normal areas to obtain the approximating normal probability.

---

EXERCISES

**Learning the mechanics**

**4.68.** Assume that $x$ is a binomial random variable with $n = 25$ and $p = .4$. Use Table II in the Appendix and the normal approximation to find the exact and approximate values, respectively, for the following probabilities:

    a.  $P(x \leq 10)$    b.  $P(x \leq 12)$    c.  $P(x = 8)$    d.  $P(9 \leq x \leq 17)$

**4.69.** Assume that $x$ is a binomial random variable with $n = 50$ and $p = .4$. Use a normal approximation to find the following:

    a.  $P(x \leq 40)$    b.  $P(25 \leq x \leq 30)$    c.  $P(x \geq 20)$

**4.70.** What conditions must be satisfied in order for the normal distribution to provide a good approximation to the binomial probability distribution?

**4.71.** Calculate the approximate probability that the binomial random variable $x$ is larger than 50 for each of the following situations:

    a.  $n = 100$,  $p = .6$    b.  $n = 100$,  $p = .4$

    c.  $n = 200$,  $p = .1$    d.  $n = 200$,  $p = .2$

    e.  $n = 200$,  $p = .3$

## Applying the concepts

**4.72.** Of all adults in a major city in the United States, 80% favor an increased emphasis on the basics of education—reading, writing, and arithmetic. From the total number of adults in the city, 150 are randomly selected.

    a.  What is the probability that at least 110 of the 150 adults surveyed favor an increased emphasis on the basics of education?

    b.  What is the probability that more than 125 of the adults surveyed favor an increased emphasis on the basics of education?

**4.73.** It is against the law to discriminate against job applicants because of race, religion, sex, or age. Of the individuals who apply for an accountant's position in a large corporation, 40% are over 45 years of age. If the company decides to choose fifty of a very large number of applicants for closer credential screening, claiming that the selection will be random and not age-biased, what is the approximate probability that fewer than fifteen of those chosen are over 45 years of age? (Assume that the applicant pool is large enough so that $x$, the number over 45 in the sample, has a binomial probability distribution.)

**4.74.** An advertising agency was hired to introduce a new product. It claimed that after its campaign, 30% of all consumers were familiar with the product. To check the claim, the manufacturer of the product surveyed 2,000 consumers. Of this number, 527 consumers had learned about the product through sources attributable to the campaign. What is the approximate probability that as few as 527 (i.e., 527 or less) would have learned about the product if the campaign was really 30% effective?

**4.75.** A manufacturer of pencils randomly chooses 5 gross (1 gross = 12 dozen) from each day's production and inspects for defects (chips, cracks, etc.). The manufacturer is willing to tolerate up to $\frac{1}{10}$ defectives in the production process. If the process fraction defective is more than $\frac{1}{10}$, the process is considered out of control.

    a.  Assuming that 10% of all pencils produced are defective, what is the approximate probability of observing eighty or more defectives in a day's sample?

b.   Is it likely that the process is in control if 100 defective pencils are observed in a day's sample? Explain.

**4.76.**  A recent study involving attrition rates at a major university has shown that 43% of all incoming freshmen do not graduate within 4 years of entrance.

a.   If 200 freshmen are randomly sampled this year, and their progress through college is followed, what is the approximate probability that no less than half will graduate within the next 4 years?

b.   What is the approximate probability that the number of sampled freshmen graduating within 4 years will be between forty and eighty?

**4.77.**  To check on the effectiveness of a new production process, 700 photoflash devices were randomly selected from a large number that had been produced. If the process actually produces 6% defectives, what is the approximate probability that:

a.   More than fifty defectives appear in the sample of 700?

b.   The number of defectives in the sample of 700 is forty-five or less?

**4.78.**  Due to a recent drought, conditions were favorable for the growth of a mold that produces the cancer-causing substance, aflatoxin. Researchers estimated that the mold affected 45% of the corn crop, thus making the corn unfit as feed for livestock.

a.   If a random sample of 500 ears of corn is taken, what is the approximate probability that fewer than 200 ears will be affected, assuming the estimate of 45% is correct?

b.   Suppose a sample of 500 ears was actually taken and only 190 ears were affected. What, if anything, would you conclude?

**4.79.**  A credit card company claims that 80% of all clothing purchases in excess of $10 are made with credit cards. A random check of 100 clothing purchases in excess of $10 showed that seventy-three had been made with credit cards. If 80% of all clothing purchases in excess of $10 are made with credit cards and if $x$ is the number in a sample of 100 that make credit card purchases, approximate:

a.   $P(x \leq 73)$      b.   $P(75 \leq x \leq 85)$

**4.80.**  The Department of Labor is interested in the fraction of the employed work force in the United States who feel in danger of losing their jobs during the next year. A random sample of 100 members of the work force is taken and $x$, the number who feel their job is in danger, is observed.

a.   What type of random variable is $x$? Justify your answer.

b.   Assuming 30% of the work force feels insecure about their job, what are the mean and standard deviation of the random variable $x$?

c.   What is the approximate probability that no more than fifteen workers will feel that their job is in danger?

**4.81.**  The median time a patient waits to see a doctor in a large clinic is 20 minutes. On a day when 150 patients visited the clinic, what is the approximate probability:

a. That more than half will have to wait more than 20 minutes?

b. That more than eighty-five will have to wait more than 20 minutes?

c. That more than sixty but less than ninety will have to wait more than 20 minutes?

**4.82.** The percentage of fat in the bodies of American men is an approximate normal random variable with mean equal to 15% and standard deviation equal to 2%.

a. If these values were used to describe the body fat of men in the United States Army and if 20% or more body fat is characterized as obese, what is the approximate probability that a random sample of 10,000 United States Army men will contain fewer than fifty who would be characterized as obese?

b. If the Army actually were to check the percentage of body fat for a random sample of 10,000 men and if only thirty contained 20% (or higher) body fat, would you conclude that the Army was successful in reducing the percentage of obese men below the percentage in the general population? Explain your reasoning.

## SUMMARY

Many of the events of interest in the real world are numerical. The variables that generate these data, called **random variables,** may be either **discrete** or **continuous.** A discrete random variable is one that can assume a countable number of values. In contrast, a continuous random variable is one that can assume any one of the infinitely large number of values associated with the points in a line interval.

A table, graph, or formula that gives the probability associated with each value that a discrete random variable can assume is its **probability distribution.** In contrast, the **probability function for a continuous random variable** $x$ gives the probability that $x$ will fall within an interval. This is usually a smooth curve that models a population relative frequency distribution and is constructed so that the total area under the curve is equal to 1. The area under the distribution over the interval $a \leq x \leq b$ is equal to the probability that a randomly selected value of $x$ will fall within that interval.

One of the most useful discrete random variables is one that satisfies the characteristics of a binomial experiment. This experiment and its associated **binomial probability distribution** provide a good model for the relative frequency distributions of many discrete types of data.

Of even greater importance in statistics is the continuous normal random variable and its **normal probability distribution.** Many relative frequency distributions of data can be approximated by this distribution. But more important, you will learn in the following chapters that the normal distribution plays a primary role in statistical inference.

## SUPPLEMENTARY EXERCISES

Learning the mechanics

**4.83** Consider the following discrete probability distribution:

| $x$ | 10 | 12 | 18 | 20 |
|---|---|---|---|---|
| $p(x)$ | .2 | .2 | .1 | .5 |

a. Calculate $\mu$, $\sigma^2$, and $\sigma$.
b. What is the probability that $x < 15$?
c. Calculate $\mu \pm 2\sigma$.
d. What is the probability that $x$ is in the interval $\mu \pm 2\sigma$?

**4.84.** Suppose that $x$ is a binomial random variable. Find $p(x)$ for each of the following combinations of $x$, $n$, and $p$.

a. $x = 1$, $n = 3$, $p = .1$
b. $x = 4$, $n = 20$, $p = .3$
c. $x = 0$, $n = 2$, $p = .4$
d. $x = 4$, $n = 5$, $p = .5$
e. $n = 15$, $x = 12$, $p = .9$
f. $n = 10$, $x = 8$, $p = .6$

**4.85.** Suppose that $x$ is a binomial random variable with $n = 20$ and $p = .6$. Find:

a. $P(x = 14)$      b. $P(x \le 10)$      c. $P(x > 10)$
d. $P(8 \le x \le 17)$      e. $P(8 < x < 17)$      f. $\mu$, $\sigma^2$, and $\sigma$
g. What is the probability that $x$ is in the interval, $\mu \pm 2\sigma$?

**4.86.** Calculate the area under the standard normal probability distribution between the following pairs of $z$-scores:

a. 0 and 1.96      b. $-1.96$ and 1.96      c. $-1.65$ and 1.65
d. $-3.00$ and $-1.75$      e. 0.96 and 2.78      f. $-1.52$ and 2.20

**4.87.** Find the following probabilities:

a. $P(z \le .6)$      b. $P(z \ge .6)$
c. $P(z \ge -2.5)$      d. $P(-2.75 \le z \ge -.25)$
e. $P(-2.75 \le z \le .25)$      f. $P(z \le -.93)$

**4.88.** Find a $z$-score, call it $z_0$, such that:

a. $P(z \le z_0) = .5199$      b. $P(z \ge z_0) = .3300$
c. $P(z \le z_0) = .5$      d. $P(-z_0 \le z \le z_0) = .8882$
e. $P(z \ge z_0) = .8508$      f. $P(z \ge z_0) = .0091$

**4.89.** The random variable $x$ has a normal distribution with $\mu = 80$ and $\sigma = 10$. Find the following probabilities:

a. $P(x \le 75)$      b. $P(x \ge 90)$      c. $P(60 \le x \le 70)$
d. $P(x > 75)$      e. $P(x = 75)$      f. $P(x \le 105)$

**4.90.** The random variable $x$ has a normal distribution with $\mu = 75$ and $\sigma^2 = 81$. Find a value of $x$, call it $x_0$, such that:

a. $P(x \ge x_0) = .5$      b. $P(x \le x_0) = .9911$
c. $P(x \le x_0) = .0028$      d. $P(x \ge x_0) = .0228$
e. $P(x \le x_0) = .1003$      f. $P(x \ge x_0) = .7995$

**4.91.** Assume that $x$ is a binomial random variable with $n = 50$ and $p = .6$. Use the normal probability distribution to approximate the following probabilities:

    a.  $P(x \le 35)$        b.  $P(10 \le x \le 30)$      c.  $P(x \ge 40)$
    d.  $P(20 \le x \le 33)$    e.  $P(x = 30)$           f.  $P(x \le 24 \text{ or } x \ge 36)$

Applying the concepts

**4.92.** Due to pollution, it is thought that 5% of the fish found in a particular river contain a level of mercury that is harmful to humans. To test this theory, environmentalists sample ten fish from the river and analyze each for the presence of a dangerous amount of mercury.

    a.  If the 5% contamination figure is correct, what is the probability that none of the ten fish contain a dangerous level of the substance?
    b.  What is the probability that no more than two of the fish contain a dangerous level of the substance?

**4.93.** Many minor operations at a hospital can properly be performed during the same day the patient is admitted. A hospital serving a large metropolitan area has found that in the past, 20% of newly admitted patients needing an operation are scheduled for same-day surgery. Suppose that ten patients are randomly selected from those admitted to the hospital for surgery over the past year. If $x$, the number in the sample of ten who receive same-day surgery, possesses a binomial probability distribution:

    a.  What is the probability that exactly five of these patients have same-day surgery?
    b.  What is the probability that at most one has same-day surgery?
    c.  If the ten patients are selected from the admissions on a single given day, is it reasonable to expect $x$ to possess the characteristics of a binomial random variable, that is, is this a binomial experiment? Explain.

**4.94.** The head librarian of a large library claims that 60% of the books in the library have been published since 1960.

    a.  If twenty books are chosen at random from the library, what is the probability that at least fifteen were published since 1960, assuming the claim is true?
    b.  What is the probability that fewer than nine books chosen were published after 1960?

**4.95.** An advertisement for laundry soap claims that the soap is preferred over all others by 30% of American women.

    a.  Assuming this claim is true, what is the probability that fewer than four women in a random sample of twenty-five prefer the advertiser's brand?
    b.  What is the probability that the number of women preferring the brand takes a value in the interval $3 \le x \le 13$?
    c.  If a sample of twenty-five American women was taken and only three preferred the brand, what conclusions would you draw? Why?

**4.96.** Suppose it is known that 5% of all radios produced by a manufacturer have defective tuning mechanisms. Your store receives a large shipment of the radios from which you choose ten radios to inspect. You have decided not to accept the shipment if you discover one or more defective radios. Before inspecting the ten radios, what is the probability that you will not accept the shipment?

**4.97.** An important function in any business is long-range planning. Additions to a firm's physical plant, for example, cannot be achieved overnight; their construction must be planned years in advance. Anticipating a substantial growth in sales over the next 5 years, a printing company is planning today for the warehouse space it will need 5 years hence. It obviously cannot be certain exactly how many square feet of storage space, $x$, it will need in 5 years, but the company can project its needs using a probability distribution such as the following:

| $x$ | 10,000 | 15,000 | 20,000 | 25,000 | 30,000 | 35,000 |
|------|--------|--------|--------|--------|--------|--------|
| $p(x)$ | .05 | .15 | .35 | .25 | .15 | .05 |

What is the expected number of square feet of storage space the printing company will need in 5 years?

**4.98.** An employee of a firm has an option to invest $1,000 in the company's bonds. At the end of 1 year, the company will buy back the bonds at a price determined by its profits for the year. From past years, the company predicts it will buy the bonds back at the following prices with the associated probabilities ($x$ = Price paid for bonds):

| $x$ | $0 | $500 | $1,000 | $1,500 | $2,000 |
|------|-----|------|--------|--------|--------|
| $p(x)$ | .01 | .22 | .30 | .22 | .25 |

a. What is the probability the employees will receive $1,000 or less for the investment?
b. What is the expected price paid for the bonds?
c. What is the employee's expected profit?
d. Find $\sigma^2$ and $\sigma$ for this probability distribution.

**4.99.** The probability that a person responds to a mailed questionnaire is .4.

a. What is the probability that of twenty questionnaires, more than twelve will be returned?
b. How many questionnaires should be mailed if you want to be reasonably certain that at least 100 will be returned?

**4.100.** The state highway patrol has determined that one out of every six calls for help originating from roadside call boxes is a hoax. Five calls for help have been received and five tow trucks dispatched.

a. What is the probability that none of the calls was a hoax?
b. What is the probability that only three of the callers really needed assistance?
c. What assumptions do you have to make in order to solve this problem?

d. If the highway patrol answers 10,000 calls for help next year and each call costs the patrol about $30 (labor, gas, etc.), approximately how much money will be wasted answering false alarms?

**4.101.** A sales manager has determined that a salesperson makes a sale to 70% of the retailers visited.

a. If the salesperson visits five retailers today and twenty tomorrow, what is the probability that he or she makes exactly four sales today *and* more than ten tomorrow?

b. If the salesperson visits five retailers today and twenty tomorrow, what is the probability that he or she makes more than fourteen sales out of the twenty-five visits?

c. If the salesperson visits four retailers today and five tomorrow, what is the probability that in these two days he or she will make exactly two sales?

**4.102.** The owner of construction company A makes bids on jobs so that if awarded the job, company A will make a $10,000 profit. The owner of construction company B makes bids on jobs so that if awarded the job, company B will make a $15,000 profit. Each company describes the probability distribution of the number of jobs the company is awarded per year as follows:

| COMPANY A | | COMPANY B | |
|---|---|---|---|
| $x$ | $p(x)$ | $x$ | $p(x)$ |
| 2 | .05 | 2 | .15 |
| 3 | .15 | 3 | .30 |
| 4 | .20 | 4 | .30 |
| 5 | .35 | 5 | .20 |
| 6 | .25 | 6 | .05 |

a. Find the expected number of jobs each will be awarded in a year.

b. What is the expected profit for each company?

c. Find the variance and standard deviation of the distribution of number of jobs awarded per year for each company.

d. Graph $p(x)$ for both companies A and B. For each company, what proportion of the time will $x$ fall in the interval $\mu \pm 2\sigma$?

**4.103.** A large cigarette manufacturer has determined that the probability of a new brand of cigarettes obtaining a large enough market share to make production profitable is .3. Over the next 3 years, this manufacturer will introduce one new brand a year.

a. What is the probability that at least one new brand will obtain sufficient market share to make its production profitable?

b. What is the probability that all three new brands will obtain sufficient market share?

c. What assumptions do you have to make in order to solve this problem?

**4.104.** If it is known that 5% of the finished products coming off an assembly line are defective, what is the probability that exactly one of the next four products coming off the line is defective? What assumptions do you have to make to solve this problem using the methodology of this chapter?

**4.105.** Suppose you are an airport manager. In looking over your records for the past year, you note that 60% of the time the 8:10 PM flight from Atlanta is 20 or more minutes late.

      a.   If you assume that the probability of the 8:10 PM flight being 20 or more minutes late on each day during the upcoming 5 day period is .6, what is the probability that the plane will be 20 or more minutes late exactly three times in the next 5 days?
      b.   What is the probability that the plane will be late at least three times in the next 5 days?

**4.106.** In recent years, the use of the telephone as a data collection instrument for public opinion polls has been steadily increasing. However, one of the major factors bearing on the extent to which the telephone will become an acceptable data collection tool in the future is the refusal rate, i.e., the percentage of the eligible subjects actually contacted who refuse to take part in the poll. Suppose that past records indicate a refusal rate of 20% in a large city. A poll of twenty-five city residents is to be taken and $x$ is the number of residents contacted by telephone who refuse to take part in the poll.

      a.   Find the mean and variance of $x$.
      b.   Find $P(x \leq 5)$.
      c.   Find $P(x > 10)$.

**4.107.** The efficacy of insecticides is often measured by the dose necessary to kill a certain percentage of insects. Suppose that a certain dose of a new insecticide is supposed to kill 80% of the exposed insects. To test the claim, twenty-five insects are put in contact with the insecticide.

      a.   If the insecticide really kills 80% of the exposed insects, what is the probability that fewer than fifteen die?
      b.   If you observed such a result, what would you conclude about the new insecticide? Explain your logic.

**4.108.** A physical fitness specialist claims that the probability is greater than .5 that an average adult male can improve his physical condition by spending 5 minutes per day on a certain exercise program. To test the claim, fifteen randomly selected adult males follow the program for a specified amount of time. Maximal oxygen uptake is measured before and after the program for each male and serves as the criterion for assessing physical condition. If the program is not really beneficial (i.e., the probability of improvement is only .5), what is the probability that eleven or more of the fifteen men have improved maximal oxygen uptake? Suppose eleven or more showed increased maximal oxygen uptake. Assuming the specialist's exercise program is ineffective, would you regard $x \geq 11$ as a rare event or would you conclude that, in actuality, the probability of improvement exceeds $p = \frac{1}{2}$ and that the program is effective?

**4.109.** [*Warning:* This exercise is realistic, but the computations involved are tedious.] Refer to Case Study 4.1—the Red Lobster sales tax problem. Let $x = 0, 1, 2, \ldots, 99$ be the number of cents (exceeding whole dollars) involved in a sale, and assume that the sales tax is assessed using the bracket system listed in the table.

| Values of $x$ | Tax (¢) |
|---|---|
| 0, 1, . . . , 9 | 0 |
| 10, 11, . . . , 25 | 1 |
| 26, 27, . . . , 50 | 2 |
| 51, 52, . . . , 75 | 3 |
| 76, 77, . . . , 99 | 4 |

Suppose that $x$ has a probability distribution $p(x) = .01, x = 0, 1, 2, \ldots, 99$ (an assumption that might be fairly accurate for restaurant sales). Find the expected value of the percentage of tax paid on the cents portion of a sale. [*Hint:* You can write the percent tax—call it $y$—for each value of $x$. You also know the probabilities associated with each value of $x$ and, consequently, each value of $y$. Then the expected percentage of tax paid is $E(y) = \Sigma\, y p(y)$.]

**4.110.** For a student to graduate, a high school requires that each student demonstrate competence in mathematics by scoring 70% or above on a mathematics achievement test. The scores of those students taking the test for the first time are normally distributed with a mean of 77% and a standard deviation of 7.3%. What percentage of students who take the test for the first time will pass the test?

**4.111.** Farmers often sell fruits and vegetables at roadside stands during the summer. One such roadside stand has a daily demand for tomatoes that is approximately normally distributed with a mean equal to 125 tomatoes per day and a standard deviation equal to 30 tomatoes per day.

    a.  If there are 90 tomatoes available to be sold at the roadside stand at the beginning of a day, what is the probability that they will all be sold?
    b.  If there are 200 tomatoes available to be sold, what is the probability that 50 or more will not be sold that day?
    c.  How many tomatoes must be available on any given day so that there will be only a 10% chance that all of the tomatoes will be sold?

**4.112.** A firm believes the internal rate of return for its proposed investment can best be described by a normal distribution with mean 20% and standard deviation 3%. What is the probability that the internal rate of return for the investment will be the following?

    a.  Greater than 26% or less than 14%
    b.  At least 16%
    c.  More than 28.5%

**4.113.** The probability distribution of the number of people per month who open a savings account in a large banking system is approximately normally distributed with $\mu = 1{,}280$ and $\sigma = 265$.

    a.   If the bank gives a $10 gift to each new account holder, what portion of the time will the bank's payout for gifts exceed $15,000 per month?

    b.   What is the bank's mean monthly payout for gifts?

**4.114.** The length of time required to assemble a photoelectric cell is normally distributed with mean equal to 18.1 minutes and $\sigma = 1.3$ minutes. What is the probability that it will require more than 20 minutes to assemble a cell?

**4.115.** A local track club has decided to sponsor a 10,000 meter road race. From past results of races across the state, it is known that the length of time to complete the race has an approximate normal distribution with a mean of 49 minutes and a standard deviation of 8 minutes. The club has decided that everyone who completes the race will receive a T-shirt. Those who run between 34 and 60 minutes will also receive a medal, and those finishing under 34 minutes will receive a plaque.

    a.   What proportion of racers would you expect to receive a medal and a T-shirt?

    b.   What proportion of racers would you expect to receive a plaque and a T-shirt?

**4.116.** A woman has made a daily record of the length of time it takes to travel from her house to her place of work. The distribution is approximately normal, with a mean of 20 minutes and a standard deviation of 2.1 minutes.

    a.   If the woman leaves for work 15 minutes before she is to start working, what is the probability that she will make it on time?

    b.   How long before she is to start working should she leave so that she is late only 10% of the time?

**4.117.** The amount of coal produced per day by a coal mine has a normal distribution with a mean of 60.8 tons and a standard deviation of 7.9 tons.

    a.   On any given day, what is the probability the mine produces less than 50 tons of coal?

    b.   Between 62 and 66 tons?

**4.118.** Suppose the present value of a risky investment is approximately normally distributed with mean $10,000 and standard deviation $4,000.

    a.   What is the probability that the present value of the investment is less than $1,000?

    b.   Greater than $20,000?

**4.119.** After extensive testing of a new low-tar cigarette, a tobacco company concluded that the number of milligrams of tar yielded by each cigarette had a probability distribution that was approximately normal with $\mu = 8$ and $\sigma = 1.9$ milligrams.

a.  What is the probability that one of the new low-tar cigarettes will yield more than 10 milligrams of tar?

b.  If two of the new low-tar cigarettes are chosen at random and tested, what is the probability that both will yield less than 6 milligrams of tar?

**4.120.** Eighty 10 pound bags of sugar were randomly sampled from the stock of a local sugar wholesaler and weighed. The following are the results rounded to the nearest tenth of a pound:

| | | | | | | | |
|------|------|------|------|------|------|------|------|
| 10.2 | 10.2 | 9.6  | 9.9  | 10.0 | 10.0 | 10.0 | 10.1 |
| 10.4 | 9.9  | 9.6  | 10.0 | 9.9  | 10.1 | 9.9  | 9.9  |
| 9.8  | 10.0 | 10.4 | 10.0 | 9.7  | 9.9  | 10.2 | 9.8  |
| 9.9  | 10.3 | 10.0 | 10.0 | 9.8  | 9.8  | 10.1 | 9.9  |
| 10.1 | 9.9  | 10.2 | 9.8  | 10.0 | 10.1 | 9.5  | 10.0 |
| 10.0 | 10.1 | 10.1 | 10.1 | 10.2 | 10.5 | 9.9  | 10.1 |
| 9.5  | 9.9  | 9.7  | 9.9  | 9.9  | 10.1 | 9.9  | 10.3 |
| 10.1 | 10.1 | 9.8  | 10.0 | 10.0 | 10.0 | 9.7  | 10.0 |
| 10.0 | 10.2 | 10.0 | 10.1 | 10.3 | 9.8  | 10.0 | 10.2 |
| 9.6  | 9.9  | 9.8  | 10.0 | 9.8  | 10.2 | 10.0 | 9.9  |

a.  Construct a relative frequency histogram for these data and suggest a continuous probability distribution that could be used to approximate the weight distribution of the wholesaler's stock of 10 pound bags of sugar.

b.  Using the histogram, estimate $\mu$ and $\sigma$ for the weight distribution of the wholesaler's stock of 10 pound bags of sugar.

c.  Calculate $\bar{x}$ and $s$ for the sample of eighty weights. How do these estimates of $\mu$ and $\sigma$ compare to your answer in part b?

**4.121.** A company has a lump-sum incentive plan for salespeople that is dependent upon their level of sales. If they sell less than $100,000 per year, they receive a $1,000 bonus; from $100,000 to $200,000, they receive $5,000; and above $200,000, they receive $10,000. If the annual sales per salesperson has approximately a normal distribution with $\mu = \$180,000$ and $\sigma = \$50,000$:

a.  Find $p_1$, the proportion of salespeople who receive a $1,000 bonus.

b.  Find $p_2$, the proportion of salespeople who receive a $5,000 bonus.

c.  Find $p_3$, the proportion of salespeople who receive a $10,000 bonus.

d.  What is the mean value of the bonus payout for the company? [*Hint:* See the definition for the expected value of a random variable.]

**4.122.** It is quite common for the standard deviation of a random variable to increase proportionally as the mean increases. When this occurs, the coefficient of variation,

$$CV = \frac{\sigma}{\mu}$$

the ratio of $\sigma$ to $\mu$, is the proportionality constant. To illustrate, the error (in dollars) in assessing the value of a house increases as the house increases in value. Suppose that long experience with assessors in a given region has shown that the coefficient of variation is .08 and that the probability distribution of assessed valuations on the same house by many different assessors is approximately normal with a mean we will call the "true value" of the house. Suppose the true value of your house is $50,000 and it is being assessed for taxation purposes. What is the probability the assessor will assess your house in excess of $55,000?

**4.123.** The net weight per package of a certain brand of corn chips is listed as 10 ounces. The weight actually delivered to each package by an automated machine is a normal random variable with mean 10.5 ounces and standard deviation .2 ounce. Suppose 100 packages are chosen at random and the net weights are ascertained. Let $x$ be the number of the 100 selected packages that contain at least 10 ounces of corn chips. Then $x$ is a binomial random variable with $n = 100$ and $p = $ Probability that a randomly selected package contains at least 10 ounces. What is the probability that they all contain at least 10 ounces of corn chips? What is the probability that at least 90% of the packages contain 10 ounces or more?

**4.124.** An admissions officer for a law school indicates that 35% of the applicants meet all ten requirements and 95% meet at least eight of the ten requirements.

    a. If a random sample of 300 applicants is taken, what is the approximate probability that fewer than 250 will fail to meet all ten requirements?

    b. What is the approximate probability that more than 280 will meet at least eight of the ten requirements?

**4.125.** Sixteen percent of the American black population is known to suffer from sickle-cell anemia. If 1,000 American black people are sampled at random, what is the approximate probability:

    a. That more than 175 have the disease?

    b. That fewer than 140 have the disease?

    c. That the number of people in the sample with the disease is between 130 and 180?

**4.126.** A loan officer in a large bank has been assigned to screen sixty loan applications during the next week.

    a. If her past record indicates that she turns down 20% of the applicants, what is the approximate probability that forty-one or more of the sixty applications will be approved?

    b. What is the approximate probability that between forty-five and fifty of the applications will be approved?

**4.127.** Contrary to our intuition, very reliable decisions concerning the proportion of a large group of consumers favoring a particular product or a particular social issue can be based upon relatively small samples. For example, suppose the target population of consumers contains 50,000,000 people and we wish to decide whether the proportion of consumers, $p$, in the population that favor some product (or issue) is as large as some value, say .2. Suppose you randomly select a sample as small as 1,600 from the 50,000,000 and you observe the number $x$ of consumers in the sample who favor the new product. Assuming that $p = .2$, find the mean and standard deviation of $x$. Suppose that 400 (or 25%) of the sample of 1,600 consumers favor the new product. Why might this sample result lead you to conclude that $p$ (the proportion of consumers favoring the product in the population of 50,000,000) is at least as large as .2? [*Hint:* Find the values of $\mu$ and $\sigma$ for $p = .2$, and use them to decide whether the observed value of $x$ is unusually large.]

**4.128.** Golf balls that do not meet a manufacturer's shape specifications are referred to as being "out of round" and may be sold as rejects. Assume that 10% of the balls produced by a particular machine are out of round. What is the approximate probability that of the next 200 balls produced by the machine, twenty-five or more are out of round?

**4.129.** The median income of residents in a certain community is $20,000. If 1,000 people are randomly sampled from the population of residents, let $x$ equal the number whose incomes are less than $20,000. Compute approximate values for the following:

   a.   $P(x > 500)$        b.   $P(x \leq 480)$        c.   $P(475 \leq x \leq 525)$

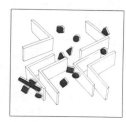

**ON YOUR OWN . . .**

For large values of $n$ the computational effort involved in working with the binomial probability distribution is considerable. Fortunately, in many instances the normal distribution provides a good approximation to the binomial distribution. This exercise was designed to enable you to demonstrate to yourself how well the normal distribution approximates the binomial distribution.

**a.**   Let the random variable $x$ have a binomial probability with $n = 10$ and $p = .5$. Using the binomial distribution, find the probability that $x$ takes on a value in each of the following intervals: $\mu \pm \sigma$, $\mu \pm 2\sigma$, and $\mu \pm 3\sigma$.

**b.**   Find the probabilities requested in part a using a normal approximation to the given binomial distribution.

**c.**   Determine the magnitude of the difference between each of the three probabilities as determined by the binomial distribution and by the normal approximation.

**d.**   Letting $x$ have a binomial distribution with $n = 20$ and $p = .5$, repeat parts a, b, and c. Notice that the probability estimates provided by the normal distribution are more accurate for $n = 20$ than for $n = 10$.

**e.**   Letting $x$ have a binomial distribution with $n = 20$ and $p = .01$, repeat parts a, b, and c. Notice that the probability estimates provided by the normal distribution are very poor in this case. Explain why this occurs.

**REFERENCES**

Bresee, C. W. "On 'grading on the curve.'". *The Clearing House*, Nov. 1976, *50*, 108–110.

Greenstein, F. I. "The benevolent leader revisited: children's images of political leaders in three democracies." *American Political Science Review*, Dec. 1975, *69*: 1371–1398.

Hogg, R. V., & Craig, A. T. *Introduction to mathematical statistics.* 4th ed. New York: Macmillian, 1978. Chapters 1 and 3.

Mendenhall, W., Scheaffer, R. L., & Wackerly, D. *Mathematical statistics with applications.* 2d ed. North Scituate, Mass.: Duxbury, 1980. Chapters 3 and 4.

Mood, A. M., Graybill, F. A., & Boes, D. C. *Introduction to the theory of statistics.* 3d ed. New York: McGraw-Hill, 1963. Chapter 3.

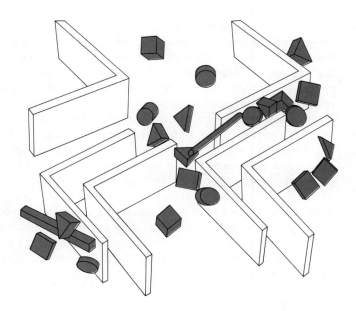

# 5

# SAMPLING DISTRIBUTIONS

**WHERE WE'VE BEEN . . .**

We have learned in earlier chapters that the objective of most statistical investigations is inference—that is, making decisions or predictions about a population based on information in a sample. To actually make the decision, we use the sample data to compute sample statistics, such as the sample mean or variance. The knowledge of random variables and their probability distributions enables us to construct theoretical models of populations.

**WHERE WE'RE GOING . . .**

Because sample measurements are observed values of random variables, the value that you compute for a sample statistic will vary in a random manner from sample to sample. In other words, since sample statistics are random variables, they therefore possess probability distributions that are either discrete or continuous, as discussed in Chapter 4. These probability distributions, called *sampling distributions* because they characterize the distribution of values of the various statistics over a very large number of samples, are the topic of this chapter. In particular, you will learn why many sampling distributions tend to be approximately normal, and you will see how sampling distributions can be used to evaluate the reliability of inferences made using the statistics.

In Chapter 4 we assumed that we knew the probability distribution of a random variable, and based on this knowledge, we were able to compute the mean, variance, and probabilities associated with the random variable. However, in most practical applications, this information will not be available. To illustrate, in Example 4.11 we calculated the probability that the binomial random variable $x$, the number of twenty polled voters who favor a particular mayoral candidate, assumed specific values. To do this, it was necessary to assume some value for $p$, the proportion of all voters who favor the candidate. Thus, for the purposes of illustration, we assumed $p = .6$ when, in all likelihood, the exact value of $p$ would be unknown. In fact, the probable purpose of taking the poll is to estimate $p$. Similarly, when we modeled the in-city gas mileage of a certain automobile model, we used the normal probability distribution with an *assumed* mean and standard deviation of 27 and 3 miles per gallon, respectively. In most situations, the true mean and standard deviation are unknown quantities that would have to be estimated. Numerical quantities that describe probability distributions are called parameters. Thus, $p$, the probability of a success in a binomial experiment, and $\mu$ and $\sigma$, the mean and standard deviation of a normal distribution, are examples of parameters.

---

DEFINITION 5.1

A parameter is a numerical descriptive measure of a population.

---

We have also discussed the sample mean, $\bar{x}$, sample variance, $s^2$, sample standard deviation, $s$, etc., which are numerical descriptive measures calculated from the sample. We will often use the information contained in these sample statistics to make inferences about the parameters of a population.

---

DEFINITION 5.2

A sample statistic is a quantity calculated from the observations in a sample.

---

Note that the term statistic refers to a sample quantity, and the term parameter refers to a population quantity.

Before we can show you how to use sample statistics to make inferences about population parameters, we need to be able to evaluate their properties. Does one sample statistic contain more information than another about a population parameter? On what basis should we choose the "best" statistic for making inferences about a parameter? The purpose of this chapter is to answer these questions.

## 5.1
## WHAT IS A
## SAMPLING
## DISTRIBUTION?

If we want to estimate a parameter of a population—say, the population mean $\mu$—there are a number of sample statistics that could be used for the estimate. Two possibilities are the sample mean $\bar{x}$ and the sample median $m$. Which of these do you think will provide a better estimate of $\mu$?

Before answering this question, consider the following example: Toss a fair die and let $x$ equal the number of dots showing on the up face. Suppose the die is tossed three times, producing sample measurements 2, 2, 6. The sample mean is $\bar{x} = 3.33$ and the sample median is $m = 2$. Since the population mean of $x$ is $\mu = 3.5$, you can see that for this sample of three measurements, the sample mean $\bar{x}$ provides an estimate that falls closer to $\mu$ than does the sample median [see Figure 5.1 (a)].

FIGURE 5.1
COMPARING THE
SAMPLE MEAN ($\bar{x}$) AND
SAMPLE MEDIAN ($m$)
AS ESTIMATORS OF THE
POPULATION MEAN ($\mu$) :
TWO SAMPLES, EACH
CONSISTING OF
THREE TOSSES
OF A DIE

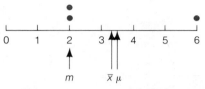

(a) Sample 1:  $\bar{x}$ is closer than $m$ to $\mu$

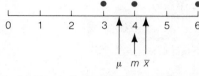

(b) Sample 2:  $m$ is closer than $\bar{x}$ to $\mu$

Now suppose we toss the die three more times and obtain the sample measurements 3, 4, 6. The mean and median of this sample are $\bar{x} = 4.33$ and $m = 4$, respectively. This time $m$ is closer to $\mu$ [see Figure 5.1(b)].

This simple example illustrates an important point: Neither the sample mean nor the sample median will *always* fall closer to the population mean. Consequently, we cannot compare these two sample statistics, or, in general, any two sample statistics, on the basis of their performance for a single sample. Instead, we need to recognize that sample statistics are themselves random variables, because different samples can lead to different values for the sample statistics. As random variables, sample statistics must be judged and compared on the basis of their probability distributions, i.e., the collection of values and associated probabilities of each statistic that would be obtained if the sampling experiment were repeated a very large number of times. We will illustrate this concept with an example.

Suppose it is known that in a certain part of Canada the daily high temperature recorded for all past months of January has a mean of $\mu = 10°F$ and a standard deviation of $\sigma = 5°F$. Consider an experiment consisting of randomly selecting twenty-five daily high temperatures from the records of past months of January and calculating the sample mean $\bar{x}$. If this experiment were repeated a very large number of times, the value of $\bar{x}$ would vary from sample to sample. For example, the first sample of twenty-five temperature measurements might have a mean $\bar{x} = 9.8$, the second sample a mean $\bar{x} = 11.4$, the third sample a mean $\bar{x} = 10.5$, etc. If the sampling experiment were repeated a very large number of times, the resulting histogram of sample means would be approximately the probability distribution of $\bar{x}$. If $\bar{x}$ is a good estimator of $\mu$, we would expect the values of $\bar{x}$ to cluster around $\mu$ as shown in Figure 5.2 (page 176). This probability distribution is called a **sampling distribution** because it is generated by repeating a sampling experiment a very large number of times.

FIGURE 5.2
SAMPLING
DISTRIBUTION FOR $\bar{x}$
BASED ON A SAMPLE
OF $n = 25$
MEASUREMENTS

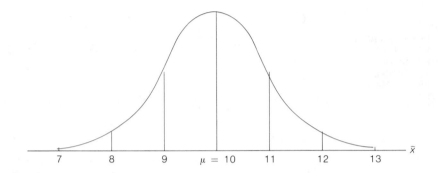

In actual practice, the sampling distribution of a statistic is obtained mathematically or (at least approximately) by simulating the sample on a computer using a procedure similar to that just described.

If $\bar{x}$ has been calculated from a sample of $n = 25$ measurements selected from a population with mean $\mu = 10$ and standard deviation $\sigma = 5$, the sampling distribution (Figure 5.2) provides all the information you may wish to know about its behavior. For example, the probability that you will draw a sample of twenty-five measurements and obtain a value of $\bar{x}$ in the interval $9 \le \bar{x} \le 10$ will be the area under the sampling distribution over that interval. Note that you need not know the actual value of $\mu$ in order to use the information that a sampling distribution provides about the distance between $\bar{x}$ and $\mu$. All you need to know is the position of the sampling distribution relative to $\mu$.

Since the properties of a statistic are typified by its sampling distribution, it follows that to compare two statistics, you compare their sampling distributions. For example, if you have two statistics, $A$ and $B$, for estimating the same parameter (for purposes of illustration, suppose the parameter is the population variance $\sigma^2$) and if their sampling distributions are as shown in Figure 5.3, you would choose statistic $A$ in preference to statistic $B$. You would make this choice because the sampling distribution for statistic $A$ centers over $\sigma^2$ and has less spread (variation) than the sampling distribution for statistic $B$. When you draw a single sample in a practical sampling situation, the probability is higher that statistic $A$ will fall nearer $\sigma^2$.

Remember that in practice we will not know the numerical value of the unknown parameter $\sigma^2$, so we will not know whether statistic $A$ or statistic $B$ is closer to $\sigma^2$ for particular samples. We have to rely on our knowledge of the theoretical sampling distributions to choose the best sample statistic, and then use it sample after sample.

FIGURE 5.3
TWO SAMPLING
DISTRIBUTIONS FOR
ESTIMATING THE
POPULATION
VARIANCE, $\sigma^2$

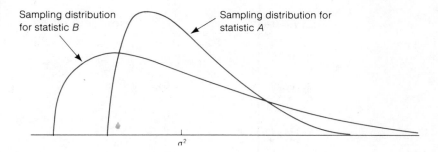

The procedure for finding the sampling distribution for a statistic is demonstrated in Example 5.1.

EXAMPLE 5.1

Consider a population consisting of the measurements 0, 3, and 12, and described by the probability distribution shown in the table:

| $x$ | $p(x)$ |
|-----|--------|
| 0   | $\frac{1}{3}$ |
| 3   | $\frac{1}{3}$ |
| 12  | $\frac{1}{3}$ |

A random sample of $n = 3$ measurements is selected from the population.

**a.** Find the sampling distribution of the sample mean $\bar{x}$.
**b.** Find the sampling distribution of the sample median $m$.

Solution

Every possible sample of $n = 3$ measurements is listed in Table 5.1 (page 178) along with the sample mean and median. Also, because any one sample is as likely to be selected as any other (random sampling), the probability of observing any particular sample is $1/27$.

**a.** From Table 5.1, you can see that $\bar{x}$ can assume the values 0, 1, 2, 3, 4, 5, 6, 8, 9, and 12. Because $\bar{x} = 0$ occurs in only one sample, $P(\bar{x} = 0) = 1/27$. Similarly, $\bar{x} = 1$ occurs in three samples, (0, 0, 3), (0, 3, 0), and (3, 0, 0). Therefore, $P(\bar{x} = 1) = 3/27 = 1/9$. Calculating the probabilities of the remaining values of $\bar{x}$ and arranging them in a table, we obtain the following probability distribution:

PROBABILITY
DISTRIBUTION OF
THE SAMPLE MEAN $\bar{x}$,
EXAMPLE 5.1

| $\bar{x}$ | $p(\bar{x})$ |
|-----------|--------------|
| 0   | $\frac{1}{27}$ |
| 1   | $\frac{3}{27}$ |
| 2   | $\frac{3}{27}$ |
| 3   | $\frac{1}{27}$ |
| 4   | $\frac{3}{27}$ |
| 5   | $\frac{6}{27}$ |
| 6   | $\frac{3}{27}$ |
| 8   | $\frac{3}{27}$ |
| 9   | $\frac{3}{27}$ |
| 12  | $\frac{1}{27}$ |

TABLE 5.1

| POSSIBLE SAMPLES | $\bar{x}$ | $m$ | PROBABILITY |
|---|---|---|---|
| 0, 0, 0 | 0 | 0 | $\frac{1}{27}$ |
| 0, 0, 3 | 1 | 0 | $\frac{1}{27}$ |
| 0, 0, 12 | 4 | 0 | $\frac{1}{27}$ |
| 0, 3, 0 | 1 | 0 | $\frac{1}{27}$ |
| 0, 3, 3 | 2 | 3 | $\frac{1}{27}$ |
| 0, 3, 12 | 5 | 3 | $\frac{1}{27}$ |
| 0, 12, 0 | 4 | 0 | $\frac{1}{27}$ |
| 0, 12, 3 | 5 | 3 | $\frac{1}{27}$ |
| 0, 12, 12 | 8 | 12 | $\frac{1}{27}$ |
| 3, 0, 0 | 1 | 0 | $\frac{1}{27}$ |
| 3, 0, 3 | 2 | 3 | $\frac{1}{27}$ |
| 3, 0, 12 | 5 | 3 | $\frac{1}{27}$ |
| 3, 3, 0 | 2 | 3 | $\frac{1}{27}$ |
| 3, 3, 3 | 3 | 3 | $\frac{1}{27}$ |
| 3, 3, 12 | 6 | 3 | $\frac{1}{27}$ |
| 3, 12, 0 | 5 | 3 | $\frac{1}{27}$ |
| 3, 12, 3 | 6 | 3 | $\frac{1}{27}$ |
| 3, 12, 12 | 9 | 12 | $\frac{1}{27}$ |
| 12, 0, 0 | 4 | 0 | $\frac{1}{27}$ |
| 12, 0, 3 | 5 | 3 | $\frac{1}{27}$ |
| 12, 0, 12 | 8 | 12 | $\frac{1}{27}$ |
| 12, 3, 0 | 5 | 3 | $\frac{1}{27}$ |
| 12, 3, 3 | 6 | 3 | $\frac{1}{27}$ |
| 12, 3, 12 | 9 | 12 | $\frac{1}{27}$ |
| 12, 12, 0 | 8 | 12 | $\frac{1}{27}$ |
| 12, 12, 3 | 9 | 12 | $\frac{1}{27}$ |
| 12, 12, 12 | 12 | 12 | $\frac{1}{27}$ |

**b.** In Table 5.1 you can see that the median $m$ can assume one of the three values 0, 3, or 12. The value $m = 0$ occurs in seven different samples. Therefore, $P(m = 0) = \frac{7}{27}$. Similarly, $m = 3$ occurs in thirteen samples and $m = 12$ occurs in seven samples. Therefore, the probability distribution for the median $m$ is as follows:

PROBABILITY
DISTRIBUTION OF THE
SAMPLE MEDIAN $m$,
EXAMPLE 5.1

| $m$ | $p(m)$ |
|---|---|
| 0 | $\frac{7}{27}$ |
| 3 | $\frac{13}{27}$ |
| 12 | $\frac{7}{27}$ |

Example 5.1 demonstrates the procedure for finding the exact sampling distribution of a statistic when the number of different samples that could be selected from the population is relatively small. In the real world, populations are often generated by random variables that can assume a very large number of values; the samples are difficult (or impossible) to enumerate. When this situation occurs, we may choose to obtain the approximate sampling distribution for a statistic by simu-

lating the sampling over and over again and recording the proportion of times different values of the statistic occur. Example 5.2 illustrates this procedure.

EXAMPLE 5.2   Suppose we perform the following experiment over and over again: Take a random sample of eleven measurements from a uniform probability distribution. A graph of this distribution is shown in Figure 5.4. Calculate the two sample statistics.

$$\bar{x} = \text{Sample mean} = \frac{\Sigma x}{11}$$

$m = \text{Median} = \text{Sixth sample measurement when the eleven measurements}$
$\text{are arranged in ascending order}$

Obtain approximations to the sampling distributions of $\bar{x}$ and $m$.

Solution   We use a computer to generate 1,000 samples, each with $n = 11$ observations. Then, we compute $\bar{x}$ and $m$ for each sample. Our goal is to obtain approximations to the sampling distributions of $\bar{x}$ and $m$ and to find out which sample statistic ($\bar{x}$ or $m$) contains more information about $\mu$. [*Note:* The mean of the uniform probability distribution (Figure 5.4) is $\mu = .5$.] The first ten of the 1,000 samples generated are presented in Table 5.2 (page 181). For example, the first computer-generated sample from the uniform distribution (arranged in ascending order) contained the following measurements: .125, .138, .139, .217, .419, .506, .516, .757, .771, .786, .919. The sample means, $\bar{x}$, and median, $m$, computed for this sample are

$$\bar{x} = \frac{.125 + .138 + \cdots + .919}{11} = .481$$

$m = \text{Sixth ordered measurement} = .506$

FIGURE 5.4
UNIFORM FREQUENCY
FUNCTION FROM
0 TO 1

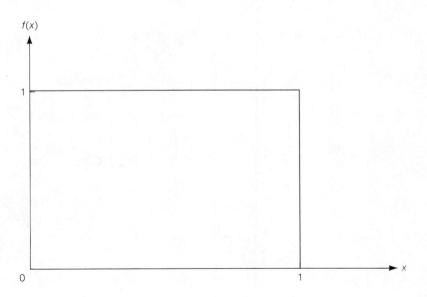

FIGURE 5.5

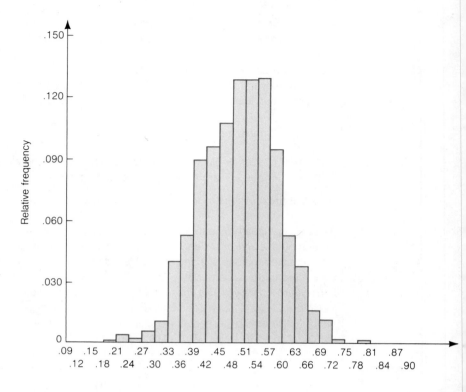

(a) Sampling distribution of $\bar{x}$ (based on 1,000 samples of $n = 11$ measurements)

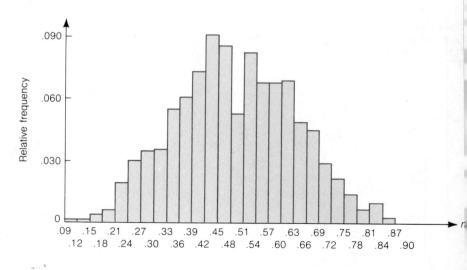

(b) Sampling distribution of $m$ (based on 1,000 samples of $n = 11$ measurements)

**TABLE 5.2**
**FIRST TEN SAMPLES**
**OF** $n = 11$
**MEASUREMENTS FROM**
**A UNIFORM**
**DISTRIBUTION**

| SAMPLE | MEASUREMENTS | | | | | | | | | | |
|---|---|---|---|---|---|---|---|---|---|---|---|
| 1 | .217 | .786 | .757 | .125 | .139 | .919 | .506 | .771 | .138 | .516 | .419 |
| 2 | .303 | .703 | .812 | .650 | .848 | .392 | .988 | .469 | .632 | .012 | .065 |
| 3 | .383 | .547 | .383 | .584 | .098 | .676 | .091 | .535 | .256 | .163 | .390 |
| 4 | .218 | .376 | .248 | .606 | .610 | .055 | .095 | .311 | .086 | .165 | .665 |
| 5 | .144 | .069 | .485 | .739 | .491 | .054 | .953 | .179 | .865 | .429 | .648 |
| 6 | .426 | .563 | .186 | .896 | .628 | .075 | .283 | .549 | .295 | .522 | .674 |
| 7 | .643 | .828 | .465 | .672 | .074 | .300 | .319 | .254 | .708 | .384 | .534 |
| 8 | .616 | .049 | .324 | .700 | .803 | .399 | .557 | .975 | .569 | .023 | .072 |
| 9 | .093 | .835 | .534 | .212 | .201 | .041 | .889 | .728 | .466 | .142 | .574 |
| 10 | .957 | .253 | .983 | .904 | .696 | .766 | .880 | .485 | .035 | .881 | .732 |

The relative frequency histograms for $\bar{x}$ and $m$ for the 1,000 samples of size $n = 11$ are shown in Figure 5.5.

You can see that the values of $\bar{x}$ tend to cluster around $\mu$ to a greater extent than do the values of $m$. Thus, on the basis of the observed sampling distributions, we conclude that $\bar{x}$ contains more information about $\mu$ than $m$ does—at least for samples of $n = 11$ measurements from the uniform distribution.

As noted earlier, many sampling distributions can be derived mathematically, but the theory necessary to do this is beyond the scope of this text. Consequently, when we need to know the properties of a statistic, we will present its sampling distribution and describe its properties. Several of the important properties we look for in sampling distributions are discussed in the next section.

**EXERCISES**

[*Note:*  *Starred (\*) exercises require the use of a computer.*]

**5.1**   The following probability distribution describes a population of measurements that can assume values of 0, 2, 4, and 6, each of which occurs with the same relative frequency:

| $x$ | 0 | 2 | 4 | 6 |
|---|---|---|---|---|
| $p(x)$ | $\frac{1}{4}$ | $\frac{1}{4}$ | $\frac{1}{4}$ | $\frac{1}{4}$ |

a.   List all of the different samples of $n = 2$ measurements that can be selected from this population.
b.   Calculate the mean for each different sample listed in part a.
c.   If a sample of $n = 2$ measurements is randomly selected from the population, what is the probability that any particular sample will be selected?
d.   Assume that a random sample of $n = 2$ measurements is selected from the population. List the different values of $\bar{x}$ found in part b and find the probability of each. Then give the sampling distribution of the sample mean, $\bar{x}$, in tabular form.
e.   Construct a probability histogram for the sampling distribution of $\bar{x}$.

**5.2.** Simulate sampling from the population (Exercise 5.1) by marking the values of $x$, one on each of four identical coins (or poker chips, etc.). Place the coins (marked 0, 2, 4, and 6) into a bag, randomly select one, and observe its value. Replace this coin, draw a second coin, and observe its value. Finally, calculate the mean, $\bar{x}$, for this sample of $n = 2$ observations randomly selected from the population (Exercise 5.1). Replace the coins, mix, and, using the same procedure, select a sample of $n = 2$ observations from the population. Record the numbers and calculate $\bar{x}$ for this sample. Repeat this sampling process until you acquire 100 values of $\bar{x}$. Construct a relative frequency distribution for these 100 sample means. This distribution will be an approximation to the exact sampling distribution of $\bar{x}$ found in part e of Exercise 5.1. Compare the two distributions. The distribution obtained in this exercise will not be exactly the same as the exact sampling distribution (Exercise 5.1, part e).

If you were to repeat the sampling procedure, drawing two coins not 100 times but 10,000 times, the relative frequency distribution for the 10,000 sample means would be almost identical to the sampling distribution of $\bar{x}$ found in Exercise 5.1, part e.

**5.3.** Consider the population described by the following probability distribution:

| $x$ | 1 | 2 | 3 | 4 | 5 |
|---|---|---|---|---|---|
| $p(x)$ | .2 | .3 | .2 | .2 | .1 |

The random variable $x$ is observed twice. If these observations are independent, verify that the different samples of size two and their probabilities are as follows:

| SAMPLE | PROBABILITY | SAMPLE | PROBABILITY | SAMPLE | PROBABILITY |
|---|---|---|---|---|---|
| 1, 1 | .04 | 3, 1 | .04 | 5, 1 | .02 |
| 1, 2 | .06 | 3, 2 | .06 | 5, 2 | .03 |
| 1, 3 | .04 | 3, 3 | .04 | 5, 3 | .02 |
| 1, 4 | .04 | 3, 4 | .04 | 5, 4 | .02 |
| 1, 5 | .02 | 3, 5 | .02 | 5, 5 | .01 |
| 2, 1 | .06 | 4, 1 | .04 | | |
| 2, 2 | .09 | 4, 2 | .06 | | |
| 2, 3 | .06 | 4, 3 | .04 | | |
| 2, 4 | .06 | 4, 4 | .04 | | |
| 2, 5 | .03 | 4, 5 | .02 | | |

a. Find the sampling distribution of the sample mean, $\bar{x}$.
b. What is the probability that $\bar{x}$ is 4.5 or larger?
c. Would you expect to observe a value of $\bar{x}$ equal to 4.5 or larger? Explain.

*5.4. In Example 5.2 we used the computer to generate 1,000 samples, each containing $n = 11$ observations, from a uniform distribution over the interval from 0 to 1. For this exercise, generate 500 samples, each containing $n = 15$ observations, from this population.

a.   Calculate the sample mean for each sample. To approximate the sampling distribution of $\bar{x}$, construct a relative frequency histogram for the 500 values of $\bar{x}$.

b.   Repeat part a for the sample median. Compare this approximate sampling distribution with the approximate sampling distribution of $\bar{x}$ found in part a.

*5.5.   Consider a population that contains values of $x$ equal to 00, 01, 02, 03, . . . , 96, 97, 98, 99. Assume that these values of $x$ occur with equal probability. Generate 500 samples, each containing $n = 25$ measurements, from this population. Calculate the sample mean $\bar{x}$ and sample variance $s^2$ for each of the 500 samples.

a.   To approximate the sampling distribution of $\bar{x}$, construct a relative frequency histogram for the 500 values of $\bar{x}$.

b.   Repeat part a for the 500 values of $s^2$.

## 5.2
## PROPERTIES OF SAMPLING DISTRIBUTIONS: UNBIASEDNESS AND MINIMUM VARIANCE

Beginning in Chapter 2 we have stressed that valuable information about a sample or population is contained in two descriptive measures: the mean and the standard deviation. Similarly, the means and standard deviations of sampling distributions help us to decide which sample statistic contains the most information about a population parameter.

EXAMPLE 5.3

Suppose two statistics, $A$ and $B$, exist to estimate the same population parameter, $\theta$ (theta). (Note that $\theta$ could be any parameter, $\mu$, $\sigma^2$, $\sigma$, etc.) Suppose the two statistics have sampling distributions as shown in Figure 5.6 (page 184). Based upon these sampling distributions, which statistic is more attractive as an estimator of $\theta$?

Solution

As a first consideration, we would like the sampling distribution to center over the parameter we wish to estimate. One way to characterize this property is in terms of the mean of the sampling distribution. Consequently, we say that a statistic is unbiased if the mean of the sampling distribution is equal to the parameter it is intended to estimate. This situation is shown in Figure 5.6(a) where the mean $\mu_A$ of statistic $A$ is equal to $\theta$. If the mean of a sampling distribution is not equal to the parameter it is intended to estimate, the statistic is said to be biased. The sampling distribution for a biased statistic is shown in Figure 5.6(b). The mean $\mu_B$ of the sampling distribution for statistic $B$ is not equal to $\theta$; in fact, it is shifted to the right of $\theta$.

You can see that biased statistics tend to either overestimate or underestimate a parameter. Consequently, when other properties of statistics tend to be equivalent, we will choose an unbiased statistic to estimate a parameter of interest. *

*Unbiased statistics do not exist for all parameters of interest.

FIGURE 5.6
SAMPLING
DISTRIBUTIONS OF
TWO STATISTICS,
*A* AND *B*

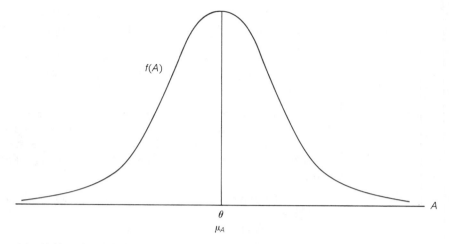

(a)   Unbiased sample statistic for the parameter $\theta$

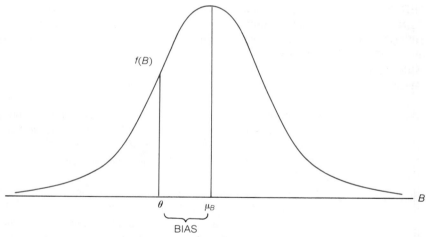

(b)   Biased sample statistic for the parameter $\theta$

---

DEFINITION 5.4

If the sampling distribution of a sample statistic has a mean equal to the population parameter the statistic is intended to estimate, the statistic is said to be an unbiased estimate of the parameter.

If the mean of the sampling distribution is not equal to the parameter, the statistic is said to be a biased estimate of the parameter.

---

The standard deviation of a sampling distribution measures another important property of statistics—the spread of the estimates generated by repeated sampling. Suppose two statistics, A and B, are both unbiased estimators of the population parameter. Since the means of the two sampling distributions are the same, we turn to their standard deviations in order to decide which will provide estimates that fall closer to the unknown population parameter we are estimating. Naturally, we will choose the sample statistic that has the smaller standard deviation. Figure 5.7 depicts sampling distributions for A and B. Note that the standard deviation of the distribution of A is smaller than the standard deviation for B, indicating that over a large number of samples, the values of A cluster more closely around the unknown population parameter than do the values of B.

**FIGURE 5.7
SAMPLING
DISTRIBUTIONS OF
TWO UNBIASED
ESTIMATORS**

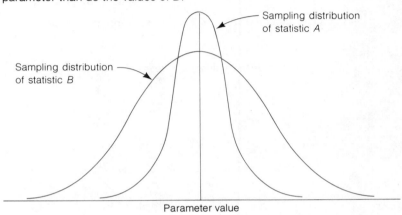

In summary, to make an inference about a population parameter, we will use the sample statistic with a sampling distribution that is unbiased and has a small standard deviation (usually smaller than the standard deviation of other unbiased sample statistics). The derivation of this sample statistic will not concern us, because the "best" statistic for estimating particular parameters is a matter of record. We will simply present an unbiased estimator with its standard deviation for each population parameter we consider. [*Note:* The standard deviation of the sampling distribution of a statistic is also called the standard error of the statistic.]

**EXAMPLE 5.4**

In Example 5.1, we found the sampling distributions of the sample mean $\bar{x}$ and the sample median $m$ for random samples of $n = 3$ measurements from a population defined by the probability distribution:

| $x$ | 0 | 3 | 12 |
|-----|-----|-----|-----|
| $p(x)$ | $\frac{1}{3}$ | $\frac{1}{3}$ | $\frac{1}{3}$ |

The sampling distributions of $\bar{x}$ and $m$ were found to be the following:

**SAMPLING
DISTRIBUTION OF $\bar{x}$**

| $\bar{x}$ | 0 | 1 | 2 | 3 | 4 | 5 | 6 | 8 | 9 | 12 |
|-----|-----|-----|-----|-----|-----|-----|-----|-----|-----|-----|
| $p(\bar{x})$ | $\frac{1}{27}$ | $\frac{3}{27}$ | $\frac{3}{27}$ | $\frac{1}{27}$ | $\frac{3}{27}$ | $\frac{6}{27}$ | $\frac{3}{27}$ | $\frac{3}{27}$ | $\frac{3}{27}$ | $\frac{1}{27}$ |

| $m$ | 0 | 3 | 12 |
|---|---|---|---|
| $p(m)$ | $\frac{7}{27}$ | $\frac{13}{27}$ | $\frac{7}{27}$ |

**a.** Show that $\bar{x}$ is an unbiased estimator of $\mu$ in this situation.

**b.** Show that $m$ is a biased estimator of $\mu$ in this situation.

Solution

**a.** The expected value of a discrete random variable $x$ (See Section 4.3) is defined to be $E(x) = \Sigma x p(x)$ where the summation is over all values of $x$. Then

$$E(x) = \mu = \sum xp(x) = (0)\left(\frac{1}{3}\right) + (3)\left(\frac{1}{3}\right) + (12)\left(\frac{1}{3}\right) = 5$$

The expected value of the discrete random variable $\bar{x}$ is

$$E(\bar{x}) = \sum (\bar{x})p(\bar{x})$$

summed over all values of $\bar{x}$. Or

$$E(\bar{x}) = (0)\left(\frac{1}{27}\right) + (1)\left(\frac{3}{27}\right) + 2\left(\frac{3}{27}\right) + \cdots + (12)\left(\frac{1}{27}\right)$$
$$= 5$$

Since $E(\bar{x}) = \mu$, $\bar{x}$ is an unbiased estimator of $\mu$.

**b.** The expected value of the sample median $m$ is

$$E(m) = \sum mp(m) = (0)\left(\frac{7}{27}\right) + (3)\left(\frac{13}{27}\right) + (12)\left(\frac{7}{27}\right)$$
$$= 4.56$$

Since the expected value of $m$ is not equal to $\mu$ ($\mu = 5$), the sample median $m$ is a biased estimator of $\mu$.

EXAMPLE 5.5

Refer to Example 5.4 and find the standard deviations of the sampling distributions of $\bar{x}$ and $m$. Which statistic would appear to be a better estimator for $\mu$?

Solution

The variance of the sampling distribution of $\bar{x}$ (we will denote it by the symbol $\sigma_{\bar{x}}^2$) is found to be

$$\sigma_{\bar{x}}^2 = E\{[\bar{x} - E(\bar{x})]^2\} = \sum (\bar{x} - \mu)^2 p(\bar{x})$$

where, from Example 5.4,

$$E(\bar{x}) = \mu = 5$$

Then

$$\sigma_{\bar{x}}^2 = (0 - 5)^2\left(\frac{1}{27}\right) + (1 - 5)^2\left(\frac{3}{27}\right) + (2 - 5)^2\left(\frac{3}{27}\right) + \cdots + (12 - 5)^2\left(\frac{1}{27}\right)$$
$$= 8.6667$$

and

$$\sigma_{\bar{x}} = \sqrt{8.6667} = 2.94$$

Similarly, the variance of the sampling distribution of $m$ (we will denote it by $\sigma_m^2$) is

$$\sigma_m^2 = E\{[m - E(m)]^2\}$$

where, from Example 5.4, the expected value of $m$ is $E(m) = 4.56$. Then

$$\sigma_m^2 = E\{[m - E(m)]^2\} = \sum[m - E(m)]^2 p(m)$$

$$= (0 - 4.56)^2\left(\frac{7}{27}\right) + (3 - 4.56)^2\left(\frac{13}{27}\right) + (12 - 4.56)^2\left(\frac{7}{27}\right)$$

$$= 20.9136$$

and

$$\sigma_m = \sqrt{20.9136} = 4.57$$

Which statistic appears to be the better estimator for the population mean $\mu$: the sample mean $\bar{x}$, or the median $m$? To answer this question, we compare the sampling distributions of the two statistics. The sampling distribution of the sample median $m$ is biased (i.e., it is shifted to the left of the mean $\mu$) and its standard deviation $\sigma_m = 4.57$ is much larger than the standard deviation of the sampling distribution of $\bar{x}$, $\sigma_{\bar{x}} = 2.94$. Consequently, the sample mean $\bar{x}$ would be a better estimator of the population mean $\mu$, for the population in question, than would the sample median $m$.

EXERCISES

[*Note:* Starred (*) exercises require the use of a computer.]

**5.6.** Consider the probability distribution:

| $x$ | 0 | 1 | 5 |
|-----|-----|-----|-----|
| $p(x)$ | $\frac{1}{3}$ | $\frac{1}{3}$ | $\frac{1}{3}$ |

a. Find $\mu$ and $\sigma^2$.
b. Find the sampling distribution of the sample mean, $\bar{x}$, for a random sample of $n = 2$ measurements from this distribution.
c. Show that $\bar{x}$ is an unbiased estimator for $\mu$. [*Hint:* Show that $E(\bar{x}) = \Sigma\bar{x}p(\bar{x}) = \mu$.]
d. Find the sampling distribution of the sample variance $s^2$ for a random sample of $n = 2$ measurements from this distribution.
e. Show that $s^2$ is an unbiased estimator for $\sigma^2$.

**5.7.** Consider the following probability distribution:

| $x$ | 2 | 4 | 9 |
|-----|-----|-----|-----|
| $p(x)$ | $\frac{1}{3}$ | $\frac{1}{3}$ | $\frac{1}{3}$ |

a.  Calculate $\mu$ for this distribution.

b.  Find the sampling distribution of the sample mean, $\bar{x}$, for a random sample of $n = 3$ measurements from this distribution and show that $\bar{x}$ is an unbiased estimator of $\mu$.

c.  Find the sampling distribution of the sample median $m$ for a random sample of $n = 3$ measurements from this distribution and show that the median is a biased estimator of $\mu$.

d.  If you wanted to estimate $\mu$ using a sample of 3 measurements from this population, which estimator would you use? Why?

**5.8.**  Consider the following probability distribution:

| $x$ | 0 | 1 | 2 |
|------|-----|-----|-----|
| $p(x)$ | $\frac{1}{3}$ | $\frac{1}{3}$ | $\frac{1}{3}$ |

a.  Find $\mu$.

b.  For a random sample of $n = 3$ observations from this distribution, find the sampling distribution of the sample mean.

c.  Find the sampling distribution of the median of a sample of $n = 3$ observations from this population.

d.  Refer to parts b and c and show that both the mean and median are unbiased estimators of $\mu$ for this population.

e.  Find the variances of the sampling distributions of the sample mean and the sample median.

f.  Which estimator would you use to estimate $\mu$? Why?

*__5.9.__  Generate 500 samples, each containing $n = 25$ measurements, from a population that contains values of $x$ equal to 01, 02, . . . , 48, 49, 50. Assume that these values of $x$ are equally likely. Calculate the sample mean $\bar{x}$ and median $m$ for each sample. Construct relative frequency histograms for the 500 values of $\bar{x}$ and the 500 values of $m$. Use these approximations to the sampling distributions of $\bar{x}$ and $m$ to answer the following questions:

a.  Does it appear that $\bar{x}$ and $m$ are unbiased estimators of the population mean? [*Note:* $\mu = 25.5$]

b.  Which sampling distribution displays greater variation?

---

**5.3**
**THE CENTRAL**
**LIMIT THEOREM**

Estimating the mean useful life of automobiles, the mean number of crimes per month in a large city, and the mean yield per acre of a new soybean hybrid are practical problems with something in common. In each, we are interested in making an inference about the mean $\mu$ of some population. As we mentioned in Chapter 2, the sample mean $\bar{x}$ is, in general, a good estimator of $\mu$. We will now develop pertinent information about the sampling distribution for this useful statistic.

EXAMPLE 5.6

Suppose a population has a uniform probability distribution of the type shown in Figure 5.8 and that a sample of eleven measurements is selected from this population.

FIGURE 5.8
SAMPLED UNIFORM
POPULATION

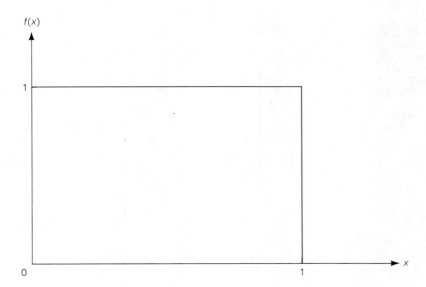

Describe the sampling distribution of the sample mean $\bar{x}$ based on the 1,000 sampling experiments discussed in Example 5.2.

Solution

You will recall that in Example 5.2 we generated 1,000 samples of $n = 11$ measurements each. The relative frequency histogram for the 1,000 sample means is shown in Figure 5.9 (page 190) with a normal probability distribution superimposed. You can see that this normal probability distribution approximates the computer-generated sampling distribution very well.

To describe fully a normal probability distribution, it is necessary to know its mean and standard deviation. Inspection of Figure 5.9 indicates that the mean of the distribution of $\bar{x}$, $\mu_{\bar{x}}$, appears to be very close to .5, the mean of the sampled uniform population. Furthermore, for a mound-shaped distribution such as that shown in Figure 5.9, almost all the measurements should fall within 3 standard deviations of the mean. Since the number of values of $\bar{x}$ is very large (1,000), the range of the observed $\bar{x}$'s divided by 6 (rather than 4) should give a reasonable approximation to the standard deviation of the sample means, $\sigma_{\bar{x}}$. The values of $\bar{x}$ range from about .2 to .8, so we calculate

$$\sigma_{\bar{x}} \approx \frac{\text{Range of } \bar{x}\text{'s}}{6} = \frac{.8 - .2}{6} = .1$$

To summarize our findings based on 1,000 samples, each consisting of eleven measurements from a uniform population, the sampling distribution of $\bar{x}$ appears to be approximately normal, with a mean of about .5 and a standard deviation of about .1.

FIGURE 5.9
RELATIVE FREQUENCY
HISTOGRAM FOR $\bar{x}$
IN 1,000 SAMPLES
OF $n = 11$
MEASUREMENTS, WITH
NORMAL FREQUENCY
FUNCTION
SUPERIMPOSED

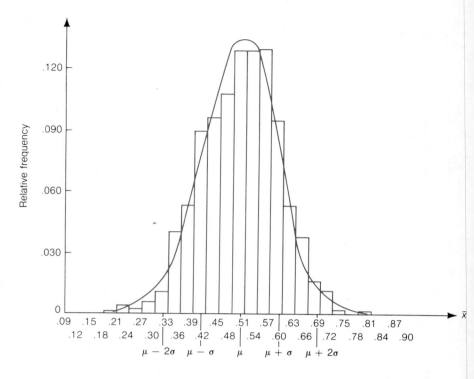

It can be shown in general that the sampling distribution of $\bar{x}$ has the following properties:

---

PROPERTIES OF THE SAMPLING DISTRIBUTION OF $\bar{x}$

1. Mean of sampling distribution = Mean of sampled population

that is, $\mu_{\bar{x}} = E(\bar{x}) = \mu$.

2. Standard deviation of sampling distribution $= \dfrac{\text{Standard deviation of sampled population}}{\text{Square root of sample size}}$

that is, $\sigma_{\bar{x}} = \sigma / \sqrt{n}$.

3. The sampling distribution of $\bar{x}$ is approximately normal for large sample sizes.

---

Does the sampling distribution in Figure 5.9 confirm the properties stated in the box? It can be shown (proof omitted) that the mean and standard deviation of the uniform distribution in Figure 5.8 are $\mu = .5$ and $\sigma = .29$.

You can see that our approximation to $\mu_{\bar{x}}$ in Example 5.6 was precise, since property 1 assures us that the mean is the same as that of the sampled population,

.5. Property 2 tells us how to calculate the standard deviation of the sampling distribution of $\bar{x}$. Substituting $\sigma = .29$, the standard deviation of the sampled uniform distribution, and the sample size $n = 11$ into the formula for $\sigma_{\bar{x}}$, we find

$$\sigma_{\bar{x}} = \frac{\sigma}{\sqrt{n}} = \frac{.29}{\sqrt{11}} = .09$$

Thus, the approximation we obtained in Example 5.6, $\sigma_{\bar{x}} \approx .1$, is very close to the exact value, $\sigma_{\bar{x}} = .09$.

The justification for property 3 is contained in one of the most important theoretical results in statistics, the central limit theorem:

---

CENTRAL LIMIT THEOREM
For large sample sizes, the mean $\bar{x}$ of a sample from a population with mean $\mu$ and standard deviation $\sigma$ possesses a sampling distribution that is approximately normal, regardless of the probability distribution of the sampled population. The larger the sample size, the better will be the normal approximation to the sampling distribution of $\bar{x}$.*

---

In summary, the sampling distribution of $\bar{x}$ will be approximately normal with a mean equal to $\mu$ (i.e., $E(\bar{x}) = \mu$) and a standard deviation equal to $\sigma/\sqrt{n}$, where $n$ is the sample size.

EXAMPLE 5.7

A manufacturer of automobile batteries claims that the distribution of the lengths of life of its best battery has a mean of 54 months, and a standard deviation of 6 months. Suppose a consumer group decides to check the claim by purchasing a sample of fifty of these batteries and subjecting them to tests that determine their lives.

**a.** Assuming the manufacturer's claim is true, describe the sampling distribution of the mean lifetime of a sample of fifty batteries.
**b.** Assuming the manufacturer's claim is true, what is the probability the consumer group's sample has a mean life of 52 or fewer months?

Solution

**a.** Even though we have no information about the shape of the probability distribution of the lives of the batteries, we can use the central limit theorem to deduce that the sampling distribution for a sample mean lifetime of fifty batteries is approximately normally distributed. Furthermore, the mean of this sampling distribution is the same as the mean of the sampled population, which is $\mu = 54$ months, according to the manufacturer's claim. Finally, the standard deviation of the sampling distribution is given by

---

*Also, because of the central limit theorem, the sum of the identically distributed independent random variables, $\Sigma x$, will possess a sampling distribution that will be approximately normal for large samples. This distribution will have a mean equal to $n\mu$ and a variance equal to $n\sigma^2$. Proof of the central limit theorem is beyond the scope of this text. It can be found in many mathematical statistics texts.

$$\sigma_{\bar{x}} = \frac{\sigma}{\sqrt{n}} = \frac{6}{\sqrt{50}} = .85 \text{ month}$$

Note that we used the claimed standard deviation of the sampled population, $\sigma = 6$ months. Thus, if we assume the claim is true, the sampling distribution for the mean life of the fifty batteries sampled is as shown in Figure 5.10.

**FIGURE 5.10**
**SAMPLING**
**DISTRIBUTION OF** $\bar{x}$
**IN EXAMPLE 5.7**
**FOR** $n = 50$

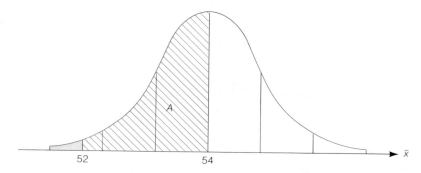

**b.** If the manufacturer's claim is true, the probability that the consumer group observes a mean battery life of 52 or fewer months for their sample of fifty batteries, $P(\bar{x} \leq 52)$, is equivalent to the shaded area in Figure 5.10. Since the sampling distribution is approximately normal, we can find this area by computing the $z$ value:

$$z = \frac{\bar{x} - \mu_{\bar{x}}}{\sigma_{\bar{x}}} = \frac{\bar{x} - \mu}{\sigma_{\bar{x}}} = \frac{52 - 54}{.85} = -2.35$$

where $\mu_{\bar{x}}$, the mean of the sampling distribution of $\bar{x}$, is equal to $\mu$, the mean of the lives of the sampled population, and $\sigma_{\bar{x}}$ is the standard deviation of the sampling distribution of $\bar{x}$. Note that the $z$ is the familiar standardized distance ($z$-score) of Section 2.6 and, since $\bar{x}$ is approximately normally distributed, it will possess the standard normal distribution of Section 4.6.

The area $A$ shown in Figure 5.10 between $\bar{x} = 52$ and $\bar{x} = 54$ (corresponding to $z = -2.35$) is found in Table III of the Appendix to be .4906. Therefore, the area to the left of $\bar{x} = 52$ is

$$P(\bar{x} \leq 52) = .5 - A = .5 - .4906 = .0094$$

Thus, the probability the consumer group will observe a sample mean of 52 or less is only .0094 if the manufacturer's claim is true. If the fifty tested batteries do result in a mean of 52 or fewer months, the consumer group will have strong evidence that the manufacturer's claim is untrue, because such an event is very unlikely to occur if the claim is true. (This is still another application of the rare event approach to statistical inference.)

In addition to providing a very useful approximation for the sampling distribution of a sample mean, the central limit theorem offers an explanation for the fact that many relative frequency distributions of data possess mound-shaped distributions. Many of the macroscopic measurements we take in various areas of research are really means or sums of many microscopic phenomena. For example, a year's growth of a pine seedling is the total of the many individual components that affect the plant's growth. Similarly, the length of time a construction company takes to complete a house might be viewed as the total of the times taken to complete each of the distinct jobs necessary to build the house. The monthly demand for blood at a hospital may be viewed as the total of the individual patients' needs. Whether the observations entering into these sums satisfy the assumptions basic to the central limit theorem is questionable, but it is a fact that many sampling distributions are mound-shaped and possess the appearance of normal distributions. Thus, the central limit thereom offers one explanation for the frequent occurrence of mound-shaped distributions in nature.

EXERCISES

Learning the mechanics
[*Note:* Starred (*) exercises require the use of a computer.]

**5.10.** Suppose that a random sample of $n$ measurements is selected from a population with mean $\mu = 50$ and variance $\sigma^2 = 70$. For each of the following values of $n$, give the mean and standard deviation of the sampling distribution of the sample mean, $\bar{x}$:

a.   $n = 10$     b.   $n = 25$     c.   $n = 100$
d.   $n = 70$     e.   $n = 1,000$     f.   $n = 400$

**5.11.** Suppose that a random sample of $n = 100$ measurements is selected from a population with mean $\mu$ and standard deviation $\sigma$. For each of the following values of $\mu$ and $\sigma$, give the values of $\mu_{\bar{x}}$ and $\sigma_{\bar{x}}$:

a.   $\mu = 10$,   $\sigma = 20$     b.   $\mu = 20$,   $\sigma = 10$
c.   $\mu = 50$,   $\sigma = 300$     d.   $\mu = 30$,   $\sigma = 200$

**5.12.** Consider the following probability distribution:

| $x$ | 1 | 2 | 3 | 10 |
|---|---|---|---|---|
| $p(x)$ | $\frac{1}{4}$ | $\frac{1}{4}$ | $\frac{1}{4}$ | $\frac{1}{4}$ |

a.   Find $\mu$, $\sigma^2$, and $\sigma$.
b.   Find the sampling distribution of $\bar{x}$ for random samples of $n = 2$ measurements from this distribution.
c.   Show that $\mu_{\bar{x}} = \mu$ and $\sigma_{\bar{x}} = \sigma/\sqrt{n} = \sigma/\sqrt{2}$.

**5.13.** Will the sampling distribution of $\bar{x}$ always be approximately normally distributed? Explain.

**5.14.** A random sample of $n = 100$ observations is selected from a population with $\mu = 40$ and $\sigma = 20$. Approximate the following probabilities:

a.   $P(\bar{x} \geq 40)$     b.   $P(37.6 \leq \bar{x} \leq 41.3)$
c.   $P(\bar{x} \leq 34.2)$     d.   $P(\bar{x} \geq 44.1)$

**5.15.** A random sample of $n = 70$ observations is selected from a population with $\mu = 10$ and $\sigma = 3$. Approximate the following probabilities:

    a. $P(\bar{x} \le 11)$             b. $P(\bar{x} \le 9.5)$

    c. $P(9.8 \le \bar{x} \le 10.3)$      d. $P(\bar{x} \ge 9.2)$

**5.16.** A random sample of $n = 800$ observations is selected from a population with $\mu = 100$ and $\sigma = 10$.

    a. What are the largest and smallest values of $\bar{x}$ that you would expect to see?

    b. How far, at the most, would you expect $\bar{x}$ to deviate from $\mu$?

    c. Did you have to know $\mu$ to answer part b? Explain.

***5.17.** Consider a population that contains values of $x$ equal to 00, 01, 02, . . . , 97, 98, 99. Assume that the values of $x$ are equally likely. For each of the following values of $n$, generate 500 random samples and calculate $\bar{x}$ for each sample. For each sample size, construct a relative frequency histogram of the 500 values of $\bar{x}$. What changes occur in the histograms as the value of $n$ increases? What similarities exist? Use $n = 2, n = 5, n = 10, n = 30, n = 50$.

Applying the concepts

**5.18.** The number of violent crimes per day in a particular city possesses a mean equal to 1.3 and a standard deviation equal to 1.7. A random sample of 50 days is observed, and the daily mean number of crimes for this sample, $\bar{x}$, is calculated.

    a. Give the mean and standard deviation of the sampling distribution of $\bar{x}$.

    b. Will the sampling distribution of $\bar{x}$ be approximately normal? Explain.

    c. Find an approximate value of $P(\bar{x} < 1)$.

    d. Find an approximate value of $P(\bar{x} > 1.9)$.

**5.19.** An educational researcher has developed an IQ test that she claims is not biased against black children. It is known that scores of white children on the test have a mean equal to 100 and a standard deviation equal to 15. In order to test the claim of nonbias, a random sample of 200 black children are given the researcher's IQ test.

    a. Assuming the researcher's claim is true, describe the sampling distribution of the mean IQ score for a sample of 200 black children.

    b. Assuming the researcher's claim is true, what is the approximate probability that the sample mean IQ score is less than 97?

    c. Suppose that the sample mean were actually 96.5. How would you interpret this value of $\bar{x}$ in view of the researcher's claim? Explain.

    d. Suppose the sample mean were actually 98.5. How would you interpret this value of $\bar{x}$ in view of the researcher's claim? Explain.

**5.20.** Last year a company initiated a program to compensate its employees for unused sick days, paying each employee a bonus of one-half the usual wage earned for each unused sick day. The question that naturally arises is "Did this policy motivate employees to use fewer allotted sick days?'' *Before* last year, the number of sick days used by employees had a distribution with a mean of 7 days and a standard deviation of 2 days.

    a.  Assuming these parameters did not change last year, find the approximate probability that the sample mean number of sick days used by 100 employees chosen at random was less than or equal to 6.4 last year.

    b.  Suppose the sample mean for the 100 employees was, in fact, 6.4. How would you interpret this result?

**5.21.**  Suppose a sample of $n = 50$ items is drawn from a population of manufactured products and the weight $x$ of each item is recorded. Prior experience has shown that the weight has a probability distribution with $\mu = 6$ ounces and $\sigma = 2.5$ ounces. Then $\bar{x}$, the sample mean, will be approximately normally distributed (because of the central limit theorem).

    a.  Calculate $\mu_{\bar{x}}$ and $\sigma_{\bar{x}}$.

    b.  What is the approximate probability that the manufacturer's sample has a mean weight of between 5.75 and 6.25 ounces?

    c.  What is the approximate probability that the manufacturer's sample has a mean weight of less than 5.5 ounces?

    d.  How would the sampling distribution of $\bar{x}$ change if the sample size $n$ were increased from 50 to, say, 100?

## 5.4
## THE RELATION BETWEEN SAMPLE SIZE AND A SAMPLING DISTRIBUTION

Suppose you draw two random samples, one containing $n = 5$ and the second $n = 10$ observations, from a population with mean $\mu$ and standard deviation $\sigma$. If you compute the mean $\bar{x}$ for each sample and obtain the results shown in the table, which estimate do you think contains more information about $\mu$?

| SAMPLE 1 | SAMPLE 2 |
|---|---|
| $n = 5$ | $n = 10$ |
| $\bar{x} = 12.6$ | $\bar{x} = 13.1$ |

    Intuitively, it would seem that the sample mean based on ten measurements would contain more information than the mean based on five, but to answer the question correctly, we need to compare the sampling distributions of these two statistics.

    From Section 5.3 we know that the expected value of the sample means in repeated sampling is $\mu$, regardless of the sample size. That is, both sample means are unbiased estimators of $\mu$. The main difference in the sampling distributions lies in their standard deviations: the standard deviation of the mean based on $n = 5$ is $\sigma/\sqrt{5}$, while that based on $n = 10$ is $\sigma/\sqrt{10}$. Since the second standard deviation is smaller, we expect the sample means based on ten measurements to cluster more closely around $\mu$ in repeated sampling than those based on five measurements. Thus, our intuitive feeling that $\bar{x}$ for $n = 10$ contains more information about $\mu$ is justified.

    For the statistics you will encounter in this text, the variance of the sampling distribution will be inversely proportional to the sample size.* Or, you can say that

*Note that this is not true of all statistics, but it is true for most.

the standard deviation of the sampling distribution is proportional to $1/\sqrt{n}$. (This relationship can be seen in the standard deviation of a sample mean $\bar{x}$: $\sigma_{\bar{x}} = \sigma/\sqrt{n}$.) So, to reduce the standard deviation of the sampling distribution of a statistic by $\frac{1}{2}$, you will need 4 times as many observations in your sample. Or, to reduce the standard deviation to $\frac{1}{3}$ its original value, you will need 9 times as many observations.

The sampling distributions for the sample mean $\bar{x}$, based on random samplings from a normally distributed population, are shown in Figure 5.11 for $n = 1, 4$, and 16 observations. The curve for $n = 1$ represents the probability distribution for the population. Those for $n = 4$ and $n = 16$ are sampling distributions for $\bar{x}$. Note how the distributions contract (variation decreases) for $n = 4$ and $n = 16$. The standard deviation for $\bar{x}$ based on $n = 16$ measurements is $\frac{1}{2}$ the corresponding standard deviation for the distribution based on $n = 4$ measurements.

**FIGURE 5.11**
**THREE SAMPLING**
**DISTRIBUTIONS FOR $\bar{x}$**

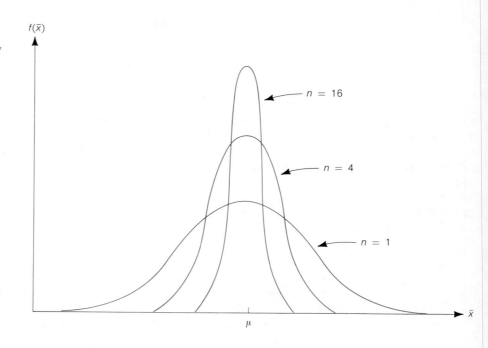

## 5.5
## THE SAMPLING DISTRIBUTION FOR THE DIFFERENCE BETWEEN TWO STATISTICS

Quite often we will wish to compare the proportion of people or objects in one group that possess some special attribute with the proportion in another. For example, we might wish to compare the proportion of Democrats in favor of certain legislation with the proportion of Republicans in favor of the same legislation, or we might wish to compare the proportions of consumers who prefer product 1 to those who prefer product 2. Or, we might wish to compare the means of two populations, say the mean gas mileage for one compact car with the mean mileage of another. All three of these comparisons will utilize a comparison of the corresponding sample statistics. For example, to estimate the difference in the means of two populations we use the difference between the means, $\bar{x}_1$ and $\bar{x}_2$, of independent random samples selected from the two populations. How close will the difference in the sample means, $(\bar{x}_1 - \bar{x}_2)$, lie to the actual difference in the population means? To answer this question, we need to know something about the sampling distribution of the quantity $(\bar{x}_1 - \bar{x}_2)$.

We cannot completely specify the form of the sampling distribution for the difference between two statistics without considering particular cases, but we can say something about the mean, variance, and standard deviation of the sampling distribution.* We will give formulas for these quantities that will always apply and then demonstrate their uses with an example.

Suppose you wish to estimate the difference between two population parameters, $\theta_1$ and $\theta_2$. You have an unbiased statistic for estimating $\theta_1$ (call it A) and another unbiased statistic for estimating $\theta_2$ (call it B). Then, because these estimators are unbiased, it follows that $E(A) = \theta_1$ and $E(B) = \theta_2$ (i.e., the mean of the sampling distribution of A is $\theta_1$ and the mean of the sampling distribution of B is $\theta_2$). Further, assume that the variance of the sampling distribution of A is $\sigma_A^2$ and the variance of B is $\sigma_B^2$. Then it can be shown (the proof is omitted here) that the mean and variance of the sampling distribution of $(A - B)$, assuming A and B are independent, are

$$\mu_{(A-B)} = E(A - B) = \theta_1 - \theta_2$$
$$\sigma_{(A-B)}^2 = \sigma_A^2 + \sigma_B^2$$

We will illustrate the use of these formulas with Example 5.8.[†]

EXAMPLE 5.8

Suppose you have two populations of investment returns that have means $\mu_1$ and $\mu_2$ and variances $\sigma_1^2$ and $\sigma_2^2$, respectively. Independent random samples of $n_1$ and $n_2$ observations are selected from the two populations: $n_1$ from population 1 and $n_2$ from population 2. The sample means $\bar{x}_1$ and $\bar{x}_2$ are computed from the samples. Find the expected value and standard deviation of the sampling distribution for the difference in two sample means, $(\bar{x}_1 - \bar{x}_2)$.

Solution

Since we know from Section 5.3 that $E(\bar{x}_1) = \mu_1$ and $E(\bar{x}_2) = \mu_2$, it follows that the mean of the sampling distribution of $(\bar{x}_1 - \bar{x}_2)$ is

$$E(\bar{x}_1 - \bar{x}_2) = E(\bar{x}_1) - E(\bar{x}_2) = \mu_1 - \mu_2$$

Recall that the standard deviation of the mean $\bar{x}$ of a random sample of n observations is $\sigma/\sqrt{n}$ (where $\sigma$ is the standard deviation of the sampled population). Since the

---

*The theory of statistics provides information on the form of the sampling distribution for the following class of statistics: The sums or differences of any number of normally distributed random variables will have a sampling distribution that is normally distributed. The random variables need not be independent of each other.
[†]Although it is not relevant to our discussion, it also can be shown that the variance of the sum of two independent statistics A and B is $\sigma_{(A+B)}^2 = \sigma_A^2 + \sigma_B^2$.

variance of a random variable is equal to the square of the standard deviation, it follows that the variances of the sample means $\bar{x}_1$ and $\bar{x}_2$ are

$$\sigma_{\bar{x}_1}^2 = \frac{\sigma_1^2}{n_1} \qquad \text{and} \qquad \sigma_{\bar{x}_2}^2 = \frac{\sigma_2^2}{n_2}$$

Then, using the formula for the variance of the difference of two independent statistics,

$$\sigma_{(\bar{x}_1 - \bar{x}_2)}^2 = \sigma_{\bar{x}_1}^2 + \sigma_{\bar{x}_2}^2 = \frac{\sigma_1^2}{n_1} + \frac{\sigma_2^2}{n_2}$$

and the standard deviation of the sampling distribution for $(\bar{x}_1 - \bar{x}_2)$ is

$$\sigma_{(\bar{x}_1 - \bar{x}_2)} = \sqrt{\frac{\sigma_1^2}{n_1} + \frac{\sigma_2^2}{n_2}}$$

It can be shown that the sampling distribution for $(\bar{x}_1 - \bar{x}_2)$ will be approximately normal for large values of $n_1$ and $n_2$. Therefore, for large samples, the sampling distribution for $(\bar{x}_1 - \bar{x}_2)$ will appear as shown in Figure 5.12.

**FIGURE 5.12 SAMPLING DISTRIBUTION FOR $(\bar{x}_1 - \bar{x}_2)$**

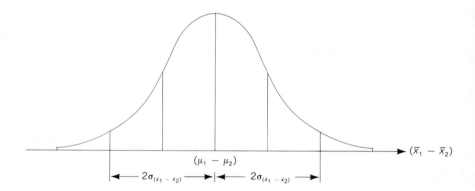

SUMMARY

Many practical problems require that an inference be made about some population parameter (we call it $\theta$). If we want to make this inference on the basis of information in a sample, we will need to compute a sample statistic that contains information about $\theta$. The amount of information a sample statistic contains about $\theta$ is reflected by its sampling distribution, the probability distribution of the sample statistic. In particular, we want a sample statistic that is an unbiased estimator of $\theta$ and has a smaller variance than any other unbiased sample statistic.

When the population parameter of interest is the mean $\mu$, the sample mean provides an unbiased estimator with a standard deviation of $\sigma/\sqrt{n}$. In addition, the central limit theorem assures us that the sampling distribution for the mean of a large sample will be approximately normally distributed, no matter what the shape of the relative frequency distribution of the sampled population.

The amount of information in a sample that is relevant to some population parameter is related to the sample size. For example, the standard deviation of the sampling distribution of the sample mean $\bar{x}$ will be inversely proportional to the square root of the sample size (i.e., $1/\sqrt{n}$).

The sampling distributions for all the many statistics that can be computed from sample data could be discussed in detail, but this would delay discussion of the practical objective of this course—the role of statistical inference in decision-making. Consequently, we will comment further on the sampling distributions of statistics when we use them as estimators or decision-makers in the following chapters.

**SUPPLEMENTARY EXERCISES**

Learning the mechanics

**5.22.** A random sample of $n = 3$ observations is selected from a population that is described by the following probability distribution.

| $x$ | 0 | 2 | 3 |
|------|-----|-----|-----|
| $p(x)$ | ⅓ | ⅓ | ⅓ |

a.  Calculate $\mu$ for this distribution.
b.  List all the possible samples of $n = 3$ measurements from this population, and calculate the sample mean and median for each sample.
c.  Find the sampling distributions of the sample mean and median.
d.  Show that the mean is an unbiased estimator of $\mu$ and that the median is not.

**5.23.** A random sample of $n = 75$ observations is selected from a population with $\mu = 120$ and $\sigma^2 = 410$.

a.  Find $\mu_{\bar{x}}$ and $\sigma_{\bar{x}}$.
b.  What is the shape of the sampling distribution of $\bar{x}$?
c.  Find the approximate value of $P(\bar{x} \leq 118)$.
d.  Find the approximate value of $P(115 \leq \bar{x} \leq 123)$.
e.  Find the approximate value of $P(\bar{x} \leq 124.6)$.
f.  Find the approximate value of $P(\bar{x} \geq 122.7)$.

**5.24.** A random sample of $n = 52$ observations is selected from a population with $\mu = 19.6$ and $\sigma = 2.5$. Approximate each of the following probabilities:

a.  $P(\bar{x} \leq 19.6)$     b.  $P(\bar{x} \leq 19)$
c.  $P(\bar{x} \geq 20.1)$     d.  $P(19.2 \leq \bar{x} \leq 20.6)$

**5.25.** Suppose that two fair dice are tossed. Let $\bar{x}$ be the mean of the two numbers that appear on the upper faces of the dice.

a.  Give the sampling distribution of $\bar{x}$.
b.  Find the approximate value of $P(\bar{x} \leq 5)$.

**5.26.** Refer to Exercise 5.25. Simulate the sampling distribution of $\bar{x}$ by tossing a pair of balanced dice 100 times. Calculate $\bar{x}$ for each toss of the dice, and construct a relative frequency histogram for these 100 values of $\bar{x}$. Compare this graph with the exact sampling distribution for $\bar{x}$ of Exercise 5.25.

**5.27.** The following table contains fifty random samples of random digits, $x = 0, 1, 2, 3, \ldots, 9$, where $p(x) = \frac{1}{10}$. Each sample contains $n = 6$ measurements.

| SAMPLE | SAMPLE | SAMPLE | SAMPLE |
|---|---|---|---|
| 8,1,8,0,6,6 | 7,6,7,0,4,3 | 4,4,5,2,6,6 | 0,8,4,7,6,9 |
| 7,2,1,7,2,9 | 1,0,5,9,9,6 | 2,9,3,7,1,3 | 5,6,9,4,4,2 |
| 7,4,5,7,7,1 | 2,4,4,7,5,6 | 5,1,9,6,9,2 | 4,2,3,7,6,3 |
| 8,3,6,1,8,1 | 4,6,6,5,5,6 | 8,5,1,2,3,4 | 1,2,0,6,3,3 |
| 0,9,8,6,2,9 | 1,5,0,6,6,5 | 2,4,5,3,4,8 | 1,1,9,0,3,2 |
| 0,6,8,8,3,5 | 3,3,0,4,9,6 | 1,5,6,7,8,2 | 7,8,9,2,7,0 |
| 7,9,5,7,7,9 | 9,3,0,7,4,1 | 3,3,8,6,0,1 | 1,1,5,0,5,1 |
| 7,7,6,4,4,7 | 5,3,6,4,2,0 | 3,1,4,4,9,0 | 7,7,8,7,7,6 |
| 1,6,5,6,4,2 | 7,1,5,0,5,8 | 9,7,7,9,8,1 | 4,9,3,7,3,9 |
| 9,8,6,8,6,0 | 4,4,6,2,6,2 | 6,9,2,9,8,7 | 5,5,1,1,4,0 |
| 3,1,6,0,0,9 | 3,1,8,8,2,1 | 6,6,8,9,6,0 | 4,2,5,7,7,9 |
| 0,6,8,5,2,8 | 8,9,0,6,1,7 | 3,3,4,6,7,0 | 8,3,0,6,9,7 |
| 8,2,4,9,4,6 | 1,3,7,3,4,3 | | |

    a.   Use the 300 random digits to construct a relative frequency histogram for the data. This relative frequency distribution should approximate $p(x)$.

    b.   Calculate the mean of the 300 digits. This will give an accurate estimate of $\mu$ (the mean of the population), which is 4.5.

    c.   Calculate $s^2$ for the 300 digits. This should be close to the variance of $x$, $\sigma^2 = 8.25$.

    d.   Suppose you intend to make an inference about the mean $\mu$ using the median of a sample of $n = 6$ measurements. To see how well the sample median will estimate $\mu$, calculate the median $m$ for each of the fifty samples. Construct a relative frequency histogram for the sample medians to see how close they lie to the mean of $\mu = 4.5$. Calculate the mean and standard deviation of the fifty medians.

    e.   Calculate $\bar{x}$ for each of the fifty samples. Construct a relative frequency histogram for the sample means to see how close they lie to the mean of $\mu = 4.5$. Calculate the mean and standard deviation of the fifty means.

    f.   Which estimator, $\bar{x}$ or $m$, seems to be a better estimator of the population mean $\mu$? Explain.

**5.28.** To see the effect of sample size on the standard deviation of the sampling distribution of a statistic, refer to Exercise 5.27 and combine pairs of samples (moving down the columns of the table) to obtain twenty-five samples of $n = 12$ measurements. Calculate the median for each sample.

    a.   Construct a relative frequency histogram for the twenty-five medians. Compare this with the histogram prepared for Exercise 5.27, part d, which is based on samples of $n = 6$ digits.

    b.   Calculate the mean and standard deviation of the twenty-five medians. Compare the standard deviation of this sampling distribution with the standard deviation of the sampling distribution in Exercise 5.27, part d. What relationship would you expect to exist between the two standard deviations?

**5.29.** Refer to Exercise 5.28. Repeat the exercise, but use the means of the samples rather than the medians. Compare the results with those of Exercise 5.27, part e.

### Applying the concepts

**5.30.** Over the last month a large supermarket chain received many consumer complaints about the quantity of chips in 9 ounce bags of a particular brand of potato chips. Suspecting that the complaints were merely the result of the potato chips settling to the bottom of the bags during shipping, but wanting to be able to assure its customers they were getting their money's worth, the chain decided to examine the next shipment of chips received by their largest store. Thirty-five 9 ounce bags were randomly selected from the shipment, their contents weighed, and the sample mean weight computed. The chain's management decided that if the sample mean were less than 8.95 ounces, the shipment would be refused and a complaint registered with the potato chip company. Assume the distribution of weights of the contents of all the potato chip bags in question has a mean of 8.9 ounces and a standard deviation of .13 ounce.

    a.  What is the approximate probability that the supermarket chain's investigation will lead to refusal of the shipment?

    b.  What assumption(s) did you have to make in order to answer part a? Justify the assumptions.

**5.31.** The distribution of the number of barrels of oil produced by a particular oil well each day for the past 3 years has a mean of 400 and a standard deviation of 75.

    a.  Describe the sampling distribution of the mean number of barrels produced per day for samples of 40 production days drawn from the past 3 years.

    b.  What is the approximate probability that the sample mean will be greater than 425?

    c.  What is the approximate probability that the sample mean will be less than 400?

**5.32.** Water availability is of prime importance in the life cycle of most reptiles. To determine the rate of evaporative water loss of a certain species of lizard at a particular desert site, thirty-four such lizards were randomly collected, weighed, and placed under the appropriate experimental conditions. After 24 hours, each lizard was removed, reweighed, and its total water loss was calculated. (Water loss = Initial body weight — Body weight after treatment.) Previous studies have shown that the distribution of water loss for the lizards has a mean of 3.1 grams and a standard deviation of .8 gram.

    a.  Find the approximate probability that the thirty-four lizards have a mean water loss of less than 3.0 grams.

    b.  Between 3.15 and 3.25 grams.

**5.33.** Electric power plants that use water for cooling their condensers sometimes discharge heated water into rivers, lakes, or oceans. It is known that water heated above certain temperatures has a detrimental effect on the plant and animal life in the water. Suppose it is known that the increased temperature of the heated water discharged by a certain power plant on any given day has a distribution with a mean of $5°C$ and a standard deviation of $.5°C$.

    a.  For 50 randomly selected days, what is the approximate probability that the average increase in temperature of the discharged water is greater than $5.0°C$?

    b.  Less than $4.8°C$?

    c.  What assumption(s) must be made for you to answer the questions?

**5.34.** Random number generators* have many uses in statistics. One type is designed to produce a sequence of numbers between 0 and 1. A number $x$ can assume any value in the interval from 0 to 1 with equal probability, and any one value of $x$ is independent of the values of previous numbers that appear in the sequence. Furthermore, the probability distribution of $x$ has a mean $\mu = .5$ and a standard deviation $\sigma = .29$. Let $y$ be the average of $n$ such random numbers.

    a.  Graph the probability distribution for $x$.

    b.  Give the mean and standard deviation of the sampling distribution of $y$.

    c.  What is the approximate form of the sampling distribution of $y$ when $n$ is large?

    d.  Sketch the sampling distribution of $y$ and compare it with your graph from part a.

**5.35.** To determine whether a metal lathe producing machine bearings is properly adjusted, a random sample of twenty-five bearings is collected and the diameter of each is measured. If the standard deviation of the diameters of the machine bearings measured over a long period of time is .001 inch, what is the approximate probability that the mean diameter $\bar{x}$ of the sample of twenty-five bearings will lie within .0001 inch of the population mean diameter of the bearings?

**5.36.** Refer to Exercise 5.35. The mean diameter of the bearings produced by the machine is supposed to be .5 inch. The company decides to use the sample mean (from Exercise 5.35) to decide whether the process is in control, i.e., whether it is producing bearings with a mean diameter of .5 inch. The machine will be considered out of control if the mean of the sample of $n = 25$ diameters is less than .4994 inch or larger than .5006 inch. If the true mean diameter of the bearings produced by the machine is .501 inch, what is the approximate probability that the test will imply that the process is out of control?

**5.37.** Suppose you wish to purchase a case of expensive wine. You plan to open two bottles for immediate use and you will keep the remaining bottles if the two are

---

*Random number generators are used to produce random numbers such as those that appear in Table I of the Appendix.

acceptable. Suppose there are ten bottles in the case, and, unknown to you, the condition of the wine in the bottles is as shown below (1 = good, 0 = bad):

| BOTTLE | 1 | 2 | 3 | 4 | 5 | 6 | 7 | 8 | 9 | 10 |
|---|---|---|---|---|---|---|---|---|---|---|
| CONDITION | 1 | 0 | 0 | 1 | 1 | 1 | 1 | 1 | 0 | 1 |

Since you are interested only in the ten bottles in the case, the collection of ten 0 or 1 responses is the population of interest to you.

    a.  If you randomly sample two bottles from the case, how many different samples (different pairs of bottles) could you select? List them.

    b.  Suppose you are going to accept the case only if both bottles in the sample are good. Identify all samples containing two good bottles. What is the probability that you will accept the case? [*Hint:* See the definition of a random sample, Section 3.7.]

    c.  Let $x$ equal the number of good bottles in the sample of $n = 2$. Construct the sampling distribution of $x$.

**5.38.** The distribution of the number of characters printed per second by a particular kind of line printer at a computer terminal has the following parameters: $\mu = 45$ characters per second, $\sigma = 2$ characters per second.

    a.  Describe the sampling distribution of the mean number of characters printed per second for random samples of 1 minute intervals.

    b.  Find the approximate probability that the sample mean for a random sample of 60 seconds will be between 44.5 and 45.3 characters per second.

    c.  Find the approximate probability that the sample mean will be less than 44 characters per second.

**5.39.** This past year, an elementary school began using a new method to teach arithmetic to first graders. A standardized test, administered at the end of the year, was used to measure the effectiveness of the new method. The distribution of past scores on the standardized test produced a mean of 75 and a standard deviation of 10.

    a.  If the new method is no different from the old method, what is the approximate probability that the mean score $\bar{x}$ of a random sample of thirty-six students will be greater than 79?

    b.  What assumptions must be satisfied to make your answer valid?

**5.40.** As part of a company's quality control program, it is a common practice to monitor the quality characteristics of a product over time. For example, the amount of alkali in soap might be monitored by randomly selecting from the production process and analyzing $n = 5$ test quantities of soap each hour. The mean, $\bar{x}$, of the sample would be plotted against time on a control chart as shown on the top of the next page.

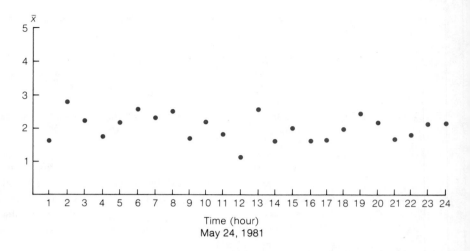

Time (hour)
May 24, 1981

If the process is in control, $\bar{x}$ should assume a distribution with a process mean $\mu$ and standard deviation $\sigma$. The control chart below shows a horizontal line to locate the process mean and two lines, located $3\sigma_{\bar{x}}$ above and below $\mu$, which are called control limits:

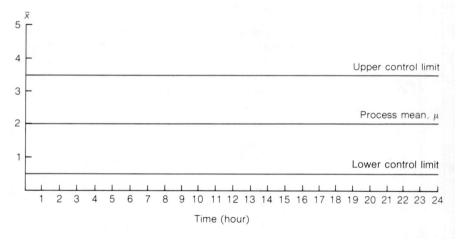

Time (hour)

If $\bar{x}$ falls within the control limits, the process is deemed to be in control. If $\bar{x}$ is outside the limits, the monitor flashes a warning and suggests that something is wrong with the process. Suppose for the soap process that experience has shown $\mu = 2\%$ and $\sigma = 1\%$.

    a.  If $n = 5$, how far away from $\mu$ should you locate the upper and lower control limits?

    b.  If the process is in control, what is the approximate probability that at any fixed point in time, $\bar{x}$ will fall outside the control limits? State any assumptions you must make in reaching a solution.

**5.41.** The distribution of the number of loaves of bread sold per day by a large grocery store over the past 5 years has a mean of 250 and a standard deviation of 45.

    a.   Describe the sampling distribution of the total number of loaves of bread sold per 30 randomly selected shopping days. [*Hint:* See the footnote in Section 5.3 that gives the application of the central limit theorem to the sum of the measurements in a sample.]

    b.   Give the approximate probability that the total number of loaves sold per 30 shopping days is between 7,000 and 8,000.

    c.   Give the approximate probability that the total is greater than 8,100 loaves.

**ON YOUR OWN . . .**

To better understand the central limit theorem and sampling distribution, consider the following experiment: Toss four identical coins and record the number of heads observed. Then repeat this experiment four more times, so that you end up with a total of five observations for the random variable $x$, the number of heads when four coins are tossed.

Now, derive and graph the probability distribution for $x$, assuming the coins are balanced. Note that the mean of this distribution is $\mu = 2$ and the standard deviation is $\sigma = 1$. This probability distribution represents the one from which you are drawing a random sample of five measurements.

Next, calculate the mean $\bar{x}$ of the five measurements, i.e., calculate the mean number of heads you observed in five repetitions of the experiment. Although you have repeated the basic experiment five times, you have only one observed value of $\bar{x}$. To derive the probability distribution or sampling distribution of $\bar{x}$ empirically, you have to repeat the entire process (of tossing four coins five times) many times. Do it 100 times.

The approximate sampling distribution of $\bar{x}$ can be derived theoretically by making use of the central limit theorem. We expect at least an approximate normal probability distribution, with a mean $\mu = 2$ and a standard deviation

$$\sigma_{\bar{x}} = \frac{\sigma}{\sqrt{n}} = \frac{1}{\sqrt{5}} = .45$$

Count the number of your 100 $\bar{x}$'s that fall in each of the following intervals:

| Interval | Interval | Interval | Interval | Interval | Interval |
|:---:|:---:|:---:|:---:|:---:|:---:|
| ←— 1 —→ | ←— 2 —→ | ←— 3 —→ | ←— 4 —→ | ←— 5 —→ | ←— 6 —→ |
| 1.10 | 1.55 | 2 | 2.45 | 2.90 | |
| $\mu - 2\sigma_{\bar{x}}$ | $\mu - \sigma_{\bar{x}}$ | $\mu$ | $\mu + \sigma_{\bar{x}}$ | $\mu + 2\sigma_{\bar{x}}$ | |

Use the normal probability distribution with $\mu = 2$ and $\sigma_{\bar{x}} = .45$ to calculate the expected number of the 100 $\bar{x}$'s in each of the intervals. How closely does the theory describe your experimental results?

**REFERENCES**

Hogg, R. V., & Craig, A. T. *Introduction to mathematical statistics.* 4th ed. New York: Macmillan, 1978. Chapter 4.

Lindgren, B. W. *Statistical theory.* 3d ed. New York: Macmillan, 1976. Chapter 2.

Mendenhall, W., Scheaffer, R. L., & Wackerly, D. *Mathematical statistics with applications.* 2d ed. North Scituate, Mass.: Duxbury, 1980. Chapter 7.

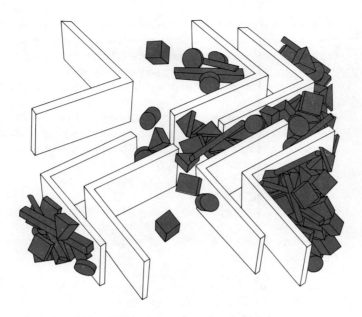

# 6

# ESTIMATION AND TESTS OF HYPOTHESES

**WHERE WE'VE BEEN . . .**
In the preceding chapters we learned that relevant information about populations can be characterized by numerical descriptive measures (called *parameters*) and that decisions about their values are based on sample statistics computed from sample data. Since statistics vary in a random manner from sample to sample, inferences based on sample statistics will be subject to uncertainty. This property is reflected in the sampling (probability) distribution of a statistic.

**WHERE WE'RE GOING . . .**
This chapter begins to put some of the preceding material to practice. That is, we will estimate or make decisions about population means or proportions based on a single sample selected from a population and then use the sampling distribution of a sample statistic to assess the uncertainty associated with an inference.

The estimation of the mean gas mileage for a new car model, the testing of a claim that a certain brand of television tube has a mean life of 5 years, and the estimation of the mean potency of a drug are practical problems with a common element. In each case we are interested in making an inference about the mean of a population. This important problem constitutes one of the primary topics of this chapter.

We will concentrate on two types of inferences about a population parameter: estimation of the parameter and tests of hypotheses, or claims, about the parameter. You will see that different techniques are used for making inferences depending on whether a sample contains a large or small number of measurements. Regardless, our objectives remain the same: We want to make best use of the information in the sample to make an inference and to assess its reliability.

In Sections 6.1 and 6.2 we consider large-sample methods for estimation and tests of hypotheses about population means. The small-sample analogs of these two topics are covered in Section 6.4. We consider large-sample inferences about a binomial population proportion in Section 6.5 and a method for determining the appropriate sample size in Section 6.6.

## 6.1
## LARGE-SAMPLE ESTIMATION OF A POPULATION MEAN

We will illustrate the large-sample method of estimating a population mean with an example. Suppose a large hospital wants to estimate the average length of time patients remain in the hospital. To accomplish this objective, the hospital administrators plan to sample 100 of all previous patients' records and to use the sample mean, $\bar{x}$, of the lengths of stay to estimate the mean stay, $\mu$, of *all* patients' visits. Further, they plan to use the sampling distribution of the sample mean to assess the accuracy of their estimate. How will this be accomplished?

According to the central limit theorem, the sampling distribution of the sample mean is approximately normal for large samples, as shown as Figure 6.1. Let us calculate the interval

$$\bar{x} \pm 2\sigma_{\bar{x}} = \bar{x} \pm \frac{2\sigma}{\sqrt{n}}$$

**FIGURE 6.1
SAMPLING
DISTRIBUTION OF $\bar{x}$**

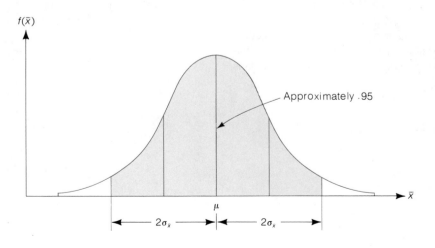

$f(\bar{x})$

Approximately .95

$\mu$

$2\sigma_{\bar{x}}$    $2\sigma_{\bar{x}}$

$\bar{x}$

That is, we will form an interval with endpoints located 2 standard deviations above and below the sample mean. What are the chances (answer before we have drawn a sample) that this interval will enclose $\mu$, the population mean?

To answer this question, refer to Figure 6.1. If the 100 measurements yield a value of $\bar{x}$ that falls between the two lines shown in color, i.e., within 2 standard deviations of $\mu$, then the interval $\bar{x} \pm 2\sigma_{\bar{x}}$ will contain $\mu$; if $\bar{x}$ falls outside these boundaries, the interval $\bar{x} \pm 2\sigma_{\bar{x}}$ will not contain $\mu$. Since the area under the normal curve (the sampling distribution of $\bar{x}$) between these boundaries is about .95 (more precisely, from Table III the area is .9544), we know that the interval $\bar{x} \pm 2\sigma_{\bar{x}}$ will contain $\mu$ with a probability approximately equal to .95.

To illustrate, suppose that the sum and the sum of squared deviations for the sample of 100 lengths of time spent in the hospital are

$$\Sigma x = 465 \text{ days} \qquad \text{and} \qquad \Sigma (x - \bar{x})^2 = 2{,}387$$

Then

$$\bar{x} = \frac{\Sigma x}{n} = \frac{465}{100} = 4.65$$

$$s^2 = \frac{\Sigma (x - \bar{x})^2}{n - 1} = \frac{2{,}387}{99} = 24.11 \qquad \text{and} \qquad s = 4.9$$

To form the interval of 2 standard deviations around $\bar{x}$, we calculate

$$\bar{x} \pm 2\sigma_{\bar{x}} = 4.65 \pm 2\frac{\sigma}{\sqrt{100}}$$

But now we face a problem. You can see that without knowing the standard deviation, $\sigma$, of the original population, i.e., the standard deviation of the lengths of stay of *all* patients, we cannot calculate this interval. However, since we have a large sample ($n = 100$ measurements), we can approximate the interval by using the sample standard deviation, $s$, to approximate $\sigma$. Thus,

$$\bar{x} \pm 2\frac{\sigma}{\sqrt{100}} \approx \bar{x} \pm 2\frac{s}{\sqrt{100}} = 4.65 \pm 2\left(\frac{4.9}{10}\right) = 4.56 \pm .98$$

That is, we estimate the mean length of stay in the hospital for all patients to fall in the interval, 3.67 to 5.63 days.

Can we be sure that $\mu$, the true mean, is in the interval 3.67 to 5.63? We cannot be certain, but we can be reasonably confident that it is. This confidence is derived from the knowledge that if we were to draw repeated random samples of 100 measurements from this population and form an interval of 2 standard deviations around $\bar{x}$ each time, approximately 95% of the intervals would contain $\mu$. We have no way of knowing (without looking at all the patients' records) whether our sample interval is one of the 95% that contain $\mu$ or one of the 5% that do not, but the odds certainly favor its containing $\mu$. Consequently, the interval 3.67 to 5.63 provides an estimate of the mean length of patient stay in the hospital. The formula that tells us how to calculate an interval estimate based on sample data is called an interval estimator.

The probability, .95, that measures the confidence that we can place in the interval estimate, is called a confidence coefficient. The percentage, 95%, is called the confidence level for the interval estimate.

---

DEFINITION 6.1
An interval estimator is a formula that tells us how to use sample data to calculate an interval that estimates a population parameter.

DEFINITION 6.2
The confidence coefficient is the probability that an interval estimator encloses the population parameter if the estimator is used repeatedly a very large number of times.

The confidence level is the confidence coefficient expressed as a percentage.

---

The foregoing is an example of how an interval can be used to estimate a population parameter. This is a common statistical practice, because when we use an interval estimator, we can usually assess the level of confidence we have that the interval actually contains the true value of the parameter. Figure 6.2 shows what happens

**FIGURE 6.2
INTERVAL
ESTIMATORS FOR $\mu$:
TEN SAMPLES**

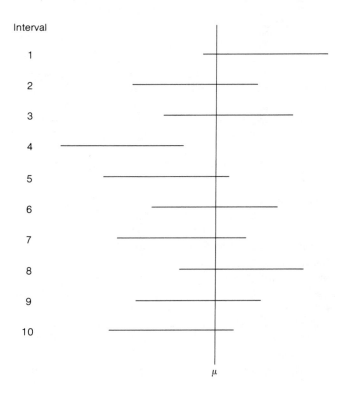

when ten different samples are drawn from a population and a confidence interval for a parameter, say $\mu$, is calculated from each. The true value of $\mu$ is located by the vertical line in the figure. Ten confidence intervals, corresponding to ten samples, are shown as horizontal line segments. Note that the confidence intervals move from sample to sample—sometimes containing $\mu$ and other times missing $\mu$. If our confidence level is 95%, then in the long-run, 95% of our sample confidence intervals will contain $\mu$.

Suppose you wish to choose a confidence coefficient other than .95. Notice in Figure 6.1 that the confidence coefficient .95 is equal to the total area under the sampling distribution, less .05 of the area, which is divided equally between the two tails. Using this idea, we can construct a confidence interval with any desired confidence coefficient by increasing or decreasing the area (call it $\alpha$) assigned to the tails of the sampling distribution (see Figure 6.3). For example, if we place area $\frac{\alpha}{2}$ in each tail and if $z_{\alpha/2}$ is the $z$ value such that the area $\frac{\alpha}{2}$ will lie to its right, then the confidence interval with confidence coefficient $(1 - \alpha)$ is

$$\bar{x} \pm z_{\alpha/2} \sigma_{\bar{x}}$$

FIGURE 6.3
LOCATING $z_{\alpha/2}$ ON
THE STANDARD
NORMAL CURVE

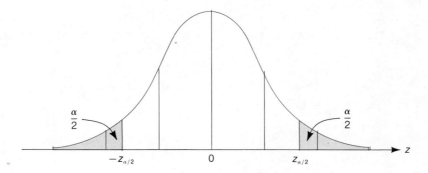

To illustrate, for a confidence coefficient of .90, $(1 - \alpha) = .90$, $\alpha = .10$, $\frac{\alpha}{2} = .05$, and $z_{.05}$ is the $z$ value that locates area .05 in the upper tail of the sampling distribution. Recall that Table III in the Appendix gives the areas between the mean and a specified $z$ value. Since the total area to the right of the mean is .5, $z_{.05}$ will be the $z$ value corresponding to an area of $.5 - .05 = .45$ to the right of the mean (see Figure 6.4, page 212). This $z$ value is $z_{.05} = 1.645$. Confidence coefficients used in practice (in published articles) range from .90 to .99. The most common confidence coefficients with corresponding values of $\alpha$ and $z_{\alpha/2}$ are shown in Table 6.1.

**TABLE 6.1**
**COMMONLY USED**
**VALUES OF $z_{\alpha/2}$**

| CONFIDENCE LEVEL | | | |
|---|---|---|---|
| $100(1 - \alpha)$ | $\alpha$ | $\frac{\alpha}{2}$ | $z_{\alpha/2}$ |
| 90% | .10 | .05 | 1.645 |
| 95% | .05 | .025 | 1.96 |
| 99% | .01 | .005 | 2.58 |

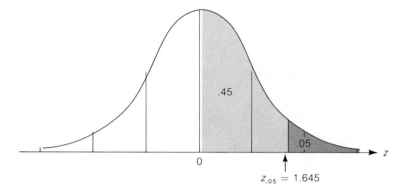

FIGURE 6.4
THE z VALUE ($z_{.05}$)
CORRESPONDING TO
AN AREA EQUAL TO .05
IN THE UPPER TAIL OF
THE z DISTRIBUTION

.45

.05

0

$z_{.05} = 1.645$

z

---

LARGE-SAMPLE $100(1 - \alpha)$PERCENT CONFIDENCE
INTERVAL FOR $\mu$

$$\bar{x} \pm z_{\alpha/2}\sigma_{\bar{x}} = \bar{x} \pm z_{\alpha/2}\frac{\sigma}{\sqrt{n}}$$

where $z_{\alpha/2}$ is the z value with an area $\alpha/2$ to its right (see Figure 6.3) and
$\sigma_{\bar{x}} = \sigma/\sqrt{n}$. The parameter $\sigma$ is the standard deviation of the sampled
population and $n$ is the sample size.

When $n$ is equal to 30 or more, the confidence interval is approximately equal
to

$$\bar{x} \pm z_{\alpha/2}\left(\frac{s}{\sqrt{n}}\right)$$

where $s$ is the sample standard deviation.

---

EXAMPLE 6.1

Unoccupied seats on flights cause the airlines to lose revenue. Suppose a large airline wants to estimate its average number of unoccupied seats per flight over the past
year. To accomplish this, the records of 225 flights are randomly selected from the
files, and the number of unoccupied seats is noted for each of the sampled flights.
The sample mean and standard deviation are

$$\bar{x} = 11.6 \text{ seats} \qquad s = 4.1 \text{ seats}$$

Estimate $\mu$, the mean number of unoccupied seats per flight during the past year,
using a 90% confidence interval.

Solution

The general form of the 90% confidence interval for a population mean is

$$\bar{x} \pm z_{\alpha/2}\sigma_{\bar{x}} = \bar{x} \pm z_{.05}\sigma_{\bar{x}} = \bar{x} \pm 1.645\left(\frac{\sigma}{\sqrt{n}}\right)$$

For the 225 records sampled, we have

$$11.6 \pm 1.645\left(\frac{\sigma}{\sqrt{225}}\right)$$

Since we do not know the value of $\sigma$ (the standard deviation of the number of unoccupied seats per flight for all flights of the year), we use our best approximation, the sample standard deviation, $s$. Then the 90% confidence interval is, approximately,

$$11.6 \pm 1.645\left(\frac{4.1}{\sqrt{225}}\right) = 11.6 \pm .45$$

or, from 11.15 to 12.05. That is, at the 90% confidence level, we estimate the mean number of unoccupied seats per flight to be between 11.15 and 12.05 during the sampled year. We stress that the confidence level refers to the procedure used. If we were to apply this procedure repeatedly to different samples, approximately 90% of the intervals would contain $\mu$.

**CASE STUDY 6.1**
**DANCING TO THE CUSTOMER'S TUNE: THE NEED TO ASSESS CUSTOMER PREFERENCES**

The following quotations have been extracted from the December 13, 1976 issue of *Business Week:*\*

"We're dancing to the tune of the customer as never before," says J. Janvier Wetzel, vice-president for sales promotion at Los Angeles-based Broadway Department Stores. "With population growth down to a trickle compared with its previous level, we're no longer spoiled with instant success every time we open a new store. Traditional department stores are locked in the biggest competitive battle in their history."

The nation's retailers are becoming uncomfortably aware that today's operating environment is vastly different from that of the 1960s. Population growth is slowing, a growing singles market is emerging, family formations are coming at later ages, and more women are embarking on careers. Of the 71 million households in the U.S. today, the dominant consumer buying segment is families headed by persons over 45. But by 1980 this group will have lost its majority status to the 25 to 40 year-old group. Merchants must now reposition their stores to attract these new customers.

To do so retailers are using market research to ferret out new purchasing attitudes and lifestyles and then translating this into customer buying segments. . . . department stores are taking a hard look at some of the basics of their business by . . . spending heavily for far more elaborate market research. Data on demographics, psychographics (measurement of attitudes), and lifestyle are being fed into retailers' computers so they can make marketing decisions based on actual spending patterns and estimate their inventory needs with less risk.

In order to stock their various departments with the type and style of goods that appeal to their potential group of customers, a downtown department store should be interested in estimating the average age of downtown shoppers, not shoppers in general. Suppose a downtown department store questions forty-nine downtown shoppers concerning their age (the offer of a small gift certificate may help convince shoppers to respond to such questions). The sample mean and standard deviation are found to be 40.1 and 8.6, respectively. The store could then estimate the mean age, $\mu$, of all downtown shoppers with a 95% confidence interval as follows:

$$\bar{x} \pm 1.96\left(\frac{s}{\sqrt{n}}\right) = 40.1 \pm 1.96\left(\frac{8.6}{\sqrt{49}}\right) = 40.1 \pm 2.4$$

Thus, the department store should gear its sales to the segment of consumers with average age between 37.7 and 42.5.

EXERCISES

Learning the mechanics

**6.1.** A random sample of sixty observations produced a mean, $\bar{x} = 75$, and a standard deviation, $s = 12$.

    a.   Find a 95% confidence interval for the population mean $\mu$.
    b.   Find a 90% confidence interval for $\mu$.
    c.   Find a 99% confidence interval for $\mu$.

**6.2.** A random sample of $n$ measurements was selected from a population with unknown mean $\mu$ and standard deviation $\sigma$. Calculate a 95% confidence interval for $\mu$ for each of the following situations:

    a.   $n = 50$,  $\bar{x} = 40$,  $s^2 = 30$    b.   $n = 200$,  $\bar{x} = 40$,  $s^2 = 30$
    c.   $n = 100$,  $\bar{x} = 50$,  $s = 20$    d.   $n = 100$,  $\bar{x} = 50$,  $s = 40$

**6.3.** A random sample of fifty observations from a population produced the following summary statistics:

$$\Sigma x = 390 \qquad \Sigma x^2 = 7,212$$

    a.   Find a 99% confidence interval for $\mu$.
    b.   Find a 90% confidence interval for $\mu$.

**6.4.** A random sample of 100 observations from a normally distributed population possesses a mean equal to 13.2 and a standard deviation equal to 3.1.

    a.   Find a 95% confidence interval for $\mu$.
    b.   What is meant when you say that a confidence coefficient is .95?
    c.   Find a 99% confidence interval for $\mu$.
    d.   What happens to the width of a confidence interval as the value of the confidence coefficient is increased while the sample size is held fixed?

**6.5.** Explain what is meant by the statement, "We are 95% confident that an interval estimate contains $\mu$."

**6.6.** Will a large-sample confidence interval be valid if the population from which the sample is taken is not normally distributed? Explain.

**6.7.** The mean and standard deviation of a random sample of $n$ measurements are equal to 13.6 and 4.1, respectively.

    a.   Find a 95% confidence interval for $\mu$ if $n = 100$.
    b.   Find a 95% confidence interval for $\mu$ if $n = 400$.

c. Find the widths of the confidence intervals found in parts a and b. What is the effect on the width of a confidence interval of quadrupling the sample size while holding the confidence coefficient fixed?

## Applying the concepts

**6.8.** A fact long known but little understood is that twins, in their early years, tend to have lower IQ's and pick up language more slowly than nontwins. Recently, psychologists have found that the slower intellectual growth of most twins may be caused by benign parental neglect. Suppose that it is desired to estimate the mean attention time given to twins per week by their parents. A sample of forty-six sets of 2½ year old twin boys is taken, and at the end of 1 week, the attention time given to each pair is recorded. The results are as follows:

$$\bar{x} = 22 \text{ hours} \qquad s = 16 \text{ hours}$$

Using the data, find a 90% confidence interval for the mean attention time given to all twin boys by their parents.

**6.9.** Suppose a large labor union wishes to estimate the mean number of hours per month a union member is absent from work. The union decides to randomly sample 320 of its members and monitor their working time for 1 month. At the end of the month, the total number of hours absent from work is recorded for each employee. If the mean and standard deviation of the sample are $\bar{x} = 9.6$ hours and $s = 6.4$ hours, find a 95% confidence interval for the true mean number of hours absent per month per employee.

**6.10.** Automotive engineers are continually improving their products. Suppose a new type of brake light has been developed by General Motors. As part of a product safety evaluation program, General Motors' engineers wish to estimate the mean driver response time to the new brake light. (Response time is the length of time from the point that the brake is applied until the driver in the following car takes some corrective action.) Fifty drivers are selected at random and the response time (in seconds) for each driver is recorded, yielding the following results: $\bar{x} = .72$, $s^2 = .022$. Estimate the mean driver response time to the new brake light using a 99% confidence interval.

**6.11.** As an aid in the establishment of personnel requirements, the director of a hospital wishes to estimate the mean number of people who are admitted to the emergency room during a 24 hour period. The director randomly selects sixty-four different 24 hour periods and determines the number of admissions for each. For this sample, $\bar{x} = 19.8$ and $s^2 = 25$. Estimate the mean number of admissions per 24 hour period with a 95% confidence interval. Interpret the result.

**6.12.** A sociologist wishes to estimate the average number of television viewing hours per American family per week. A random sample of 400 families yields a mean of 32.6 hours and a standard deviation of 9.9 hours. Estimate the mean viewing time with a 90% confidence interval.

6.2

A LARGE-SAMPLE
TEST OF AN
HYPOTHESIS
ABOUT A
POPULATION
MEAN

Suppose building specifications in a certain city require that the average breaking strength of residential sewer pipe be more than 2,400 pounds per foot of length (i.e., per lineal foot). Each manufacturer who wants to sell pipe in this city must demonstrate that its product meets the specification. Note that we are again interested in making an inference about the mean, $\mu$, of a population. However, in this example we are less interested in estimating the value of $\mu$ than we are in testing an hypothesis about its value. That is, we want to decide whether or not the mean breaking strength of the pipe exceeds 2,400 pounds per lineal foot.

In general, the hypothesis that a researcher wishes to establish—the research hypothesis—is called the alternative hypothesis. To establish this alternative hypothesis, we first define a null hypothesis, a theory that directly opposes the research hypothesis. Then we attempt to gain support for the research (or alternative) hypothesis by producing evidence to show that the null hypothesis is false. This is accomplished using our rare event philosophy of Chapter 2. That is, we attempt to show that the sample is inconsistent with the null hypothesis: if the null hypothesis is true, the sample represents a rare event.

In our example, the sewer pipe manufacturer wishes to show that the mean breaking strength of its pipe exceeds 2,400 pounds per lineal foot, i.e., $\mu > 2,400$. To do this, the researcher will hypothesize that $\mu = 2,400$ and attempt to show that this hypothesis is false. Therefore, the null and research hypotheses are:

Null hypothesis ($H_0$):   $\mu = 2,400$ (i.e., the manufacturer's pipe does not meet specifications)

Research (alternative) hypothesis ($H_a$):   $\mu > 2,400$ (i.e., the manufacturer's pipe does meet specifications)

Next, we need a procedure for using the information in the sample to decide which hypothesis is true. Since we are testing hypotheses about a population mean $\mu$, it is reasonable to use the sample mean, $\bar{x}$, to decide between the two hypotheses. Specifically, we will reject the null hypothesis $H_0$ in favor of the research hypothesis $H_a$ when the sample mean $\bar{x}$ strongly indicates that $\mu$ exceeds 2,400 pounds per lineal foot.

A convenient measure of the distance between $\bar{x}$ and the hypothesized mean value of 2,400 is the $z$-score:

$$z = \frac{\bar{x} - 2,400}{\sigma_{\bar{x}}} = \frac{\bar{x} - 2,400}{\sigma/\sqrt{n}}$$

Note that the $z$-score expresses the distance between $\bar{x}$ and the hypothesized mean $\mu$ in units of standard deviations of $\bar{x}$ (i.e., $\sigma_{\bar{x}}$). How large a $z$-score will be required before you decide to reject the null hypothesis? If you examine Figure 6.5, you will note that the chance of observing $\bar{x}$ more than 1.645 standard deviations above 2,400 is only .05, if in fact the true mean $\mu$ is 2,400. Thus, if the sample mean is more than 1.645 standard deviations above 2,400, either $H_0$ is true and a relatively rare event has occurred (.05 probability) or $H_a$ is true and the population mean exceeds 2,400. Deciding that the research hypothesis is true if in fact it is false is called a Type I decision error. As indicated in Figure 6.5, the risk of making a Type I

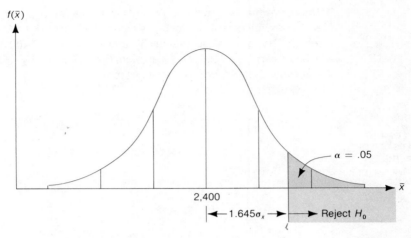

error, that is, deciding in favor of the research hypothesis if in fact the null hypothesis is true, is denoted by the symbol $\alpha$. That is,

$$\alpha = P \text{ (Type I error)}$$
$$= P \text{ (Rejecting the null hypothesis if it is true)}$$

In our example

$$\alpha = P(z > 1.645 \text{ if in fact } \mu = 2,400) = .05$$

We summarize the elements of the test below:

$$H_0: \quad \mu = 2,400 \qquad H_a: \quad \mu > 2,400$$

Test statistic: $\quad z = \dfrac{\bar{x} - 2,400}{\sigma_{\bar{x}}}$

Rejection region: $\quad z > 1.645$ which corresponds to $\alpha = .05$

Note that the rejection region refers to the values of the test statistic for which we will reject the null hypothesis.

To illustrate the use of the test, suppose we tested fifty sections of sewer pipe and found the mean and standard deviation for these fifty measurements to be

$$\bar{x} = 2,460 \text{ pounds per lineal foot} \qquad s = 200 \text{ pounds per lineal foot}$$

As in the case of estimation, we can use $s$ to approximate $\sigma$ when $s$ is calculated from a large set of sample measurements.

The test statistic is

$$z = \frac{\bar{x} - 2,400}{\sigma_{\bar{x}}} = \frac{\bar{x} - 2,400}{\sigma/\sqrt{n}} \approx \frac{\bar{x} - 2,400}{s/\sqrt{n}}$$

Substituting $\bar{x} = 2,460$, $n = 50$, and $s = 200$, we have

$$z \approx \frac{2,460 - 2,400}{200/\sqrt{50}} = \frac{60}{28.28} = 2.12$$

Therefore, the sample mean lies $2.12\sigma_{\bar{x}}$ above the hypothesized value of $\mu$, 2,400, as shown in Figure 6.6. Since this value of $z$ exceeds 1.645, it falls in the rejection region. That is, we reject the null hypothesis that $\mu = 2,400$ and accept the research hypothesis, $\mu > 2,400$. Thus, it appears that the company's pipe has a mean strength that exceeds 2,400 pounds per lineal foot.

**FIGURE 6.6**
**LOCATION OF THE TEST STATISTIC FOR A TEST OF THE HYPOTHESIS**
$H_0$: $\mu = 2,400$

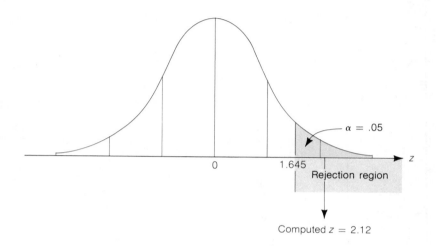

How much faith can be placed in this conclusion? What is the probability that our statistical test could lead us to reject the null hypothesis (and conclude that the company's pipe meets the city's specifications) if in fact the null hypothesis is true? The answer is "$\alpha = .05$." That is, we selected the level of risk, $\alpha$, of making a Type I error when we constructed the test. Thus, the chances are only 1 in 20 that our test could lead us to conclude the manufacturer's pipe satisfies the city's specifications if in fact this conclusion is false.

Now suppose the sample data had not indicated that the sewer pipe met the city's specifications, i.e., what would we have concluded if we had computed a value of $z \le 1.645$? Failure to reject the null hypothesis indicates that the evidence in the sample was not sufficient to support the research hypothesis at the $\alpha = .05$ level of significance. Note that we carefully avoid stating that the null hypothesis is true, for then we would be risking a second type of error—concluding the null hypothesis is true (the pipe fails to meet specifications) if in fact the research hypothesis is true (the pipe does meet specifications). We call this a Type II error. The probability of committing a Type II error is usually denoted by the symbol $\beta$ (beta).

For example, if $\mu$ is in fact equal to 2,475 pounds per lineal foot instead of the hypothesized 2,400 pounds, the sampling distribution of $\bar{x}$ would be shifted to the right (see Figure 6.7). The probability $\beta$ that $\bar{x}$ would fall in the "acceptance" region is the area under this sampling distribution (a normal curve) to the left of the value of $\bar{x}$ that locates the boundary of the rejection region. This boundary (see Figure 6.5) is located $1.645\ \sigma_{\bar{x}}$ to the right of the hypothesized mean, i.e., at

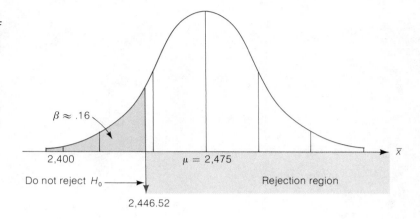

$$2{,}400 + 1.645 \left( \frac{\sigma}{\sqrt{n}} \right) \approx 2{,}400 + 1.645 \left( \frac{s}{\sqrt{n}} \right) = 2{,}400 + 1.645 \left( \frac{200}{\sqrt{50}} \right)$$
$$= 2{,}446.52$$

The $z$-score corresponding to $\bar{x} = 2{,}446.52$ is

$$z = \frac{\bar{x} - \mu}{\sigma_{\bar{x}}} = \frac{2{,}446.52 - 2{,}475}{28.28} = -1.01$$

You can verify (see Table III, Appendix) that the lower tail area under the normal curve to the left of $z = -1.01$ is equal to .1562. Therefore, the probability $\beta$ of failing to reject the null hypothesis if in fact $\mu = 2{,}475$ is approximately equal to .16. The risk of concluding that the pipe does not meet specifications if it is really 75 pounds stronger than the minimum 2,400 pounds per lineal foot is rather large ($\beta = .16$). This risk increases (see Figure 6.8, page 220) if the actual value of $\mu$ is closer to the hypothesized value, $\mu = 2{,}400$. The values of $\beta$, corresponding to the shaded areas in Figure 6.8, are given for $\mu = 2{,}425$, 2,450, and 2,475.

Table 6.2 summarizes the four possible situations that might arise when an hypothesis is tested. The two possible states of nature correspond to the two columns of the table; that is, either $H_0$ is true or $H_a$ is true. The two rows of the table indicate the two possible conclusions that can be reached; either $H_0$ is true or $H_a$ is true. The two kinds of decisions are shown in the body of the table. Either the research hypothesis or the null hypothesis can be accepted. Associated with these two actions are two types of risk: the risk of making a Type I error, measured by $\alpha$, and the risk of making

TABLE 6.2
CONCLUSIONS AND
CONSEQUENCES FOR
A TEST OF
AN HYPOTHESIS

| | | TRUE STATE OF NATURE | |
|---|---|---|---|
| | | $H_0$ true | $H_a$ true |
| CONCLUSION | $H_0$ true | Correct decision | Type II error   (probability $\beta$) |
| | $H_a$ true | Type I error   (probability $\alpha$) | Correct decision |

**FIGURE 6.8**
VALUES OF $\beta$ WHEN
$\mu = 2,425, 2,450,$ AND
2,475 POUNDS PER
LINEAL FOOT

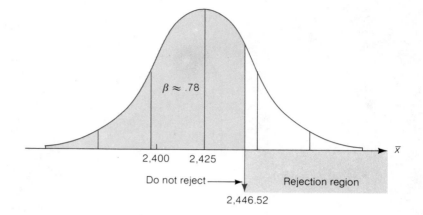

$\beta \approx .78$

2,400    2,425

Do not reject ⟶          Rejection region

2,446.52

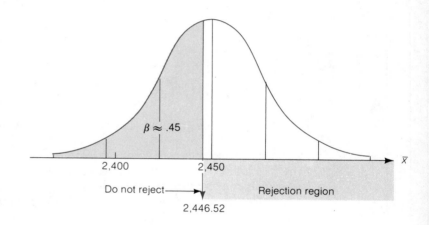

$\beta \approx .45$

2,400    2,450

Do not reject ⟶          Rejection region

2,446.52

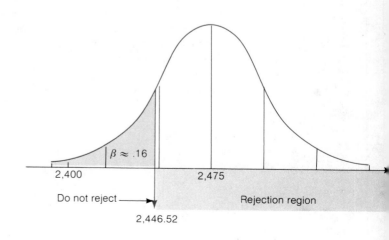

$\beta \approx .16$

2,400          2,475

Do not reject ⟶          Rejection region

2,446.52

a Type II error, measured by $\beta$. Note that a Type I error can be made *only* when the research hypothesis is accepted (which occurs when the null hypothesis is rejected) and a Type II error can be made *only* when the null hypothesis is accepted. Not too surprisingly, the measures of these two types of risk, $\alpha$ and $\beta$, are related.

You can see in Figure 6.9 that $\alpha$ is decreased by moving the rejection region farther out into the tail of the sampling distribution. By doing so, the rejection region becomes smaller and the acceptance region becomes larger. What happens to $\beta$ as the acceptance region becomes larger?

Since $\beta$ is the probability of accepting $H_0$ if in fact some alternative value of the parameter is true, $\beta$ is the probability that the test statistic falls in the acceptance region. And, the larger the acceptance region, the larger will be the value of $\beta$. So, the relationship between $\alpha$ and $\beta$ is what you might expect intuitively: As you decrease one type of risk (say, the risk $\alpha$ of falsely accepting $H_a$), you increase the other (the risk $\beta$ of falsely accepting $H_0$). Fortunately, we can reduce both types of risk by increasing the sample size. The more information you have in the sample, the greater will be the ability of the test statistic to reach the correct decision.

In theory, we could consider the probabilities of the two types of risk, $\alpha$ and $\beta$, the possible losses attached to the Type I and II errors, and choose the rejection region to minimize the expected loss.

In practice, $\beta$ is difficult to calculate for many tests, and it is impossible to specify a meaningful alternative to the null hypothesis for others. So as an introduction to tests of hypotheses, we suggest the following procedure: Select the null hypothesis as the theory opposing the research hypothesis (the one you want to support). Then, if the null hypothesis is true, you will know the probability that the test will lead to an incorrect rejection of $H_0$. It will be $\alpha$ and you can choose this value as large or small as you wish prior to the selection of your sample. If the test statistic does not fall in the rejection region, *do not* accept the null hypothesis unless you know $\beta$. Withhold judgment and seek a larger sample size to lead you closer to a decision. Or, estimate the parameter using a confidence interval. This will give an interval estimate of its true value and give you a measure of the reliability of your inference.

The elements of a test of an hypothesis are summarized in the box on page 222.

**FIGURE 6.9**
**REDUCING $\alpha$ REDUCES THE REJECTION REGION AND ENLARGES THE ACCEPTANCE REGION**

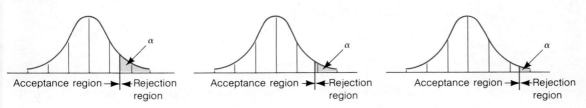

ELEMENTS OF A TEST OF AN HYPOTHESIS

**1.** Null Hypothesis ($H_0$): A theory that is phrased in terms of the values of one or more population parameters. The theory is usually one that we wish to disprove.

**2.** Alternative (Research) Hypothesis ($H_a$): A theory that opposes the null hypothesis and that we wish to establish as true.

**3.** Test Statistic: A sample statistic used to decide whether to reject the null hypothesis.

**4.** Rejection Region: The numerical values of the test statistic for which the null hypothesis will be rejected. The rejection region is chosen so that the probability is $\alpha$ that it will contain the test statistic if the null hypothesis is true (thereby leading to an incorrect conclusion), where $\alpha$ is usually chosen to be small (e.g., .01, .05, or .10).

**5.** Experiment and Calculation of Test Statistic: The sampling experiment is performed and the numerical value of the test statistic is determined.

**6.** Conclusion:

  **a.** If the numerical value of the test statistic falls in the rejection region, we conclude that the alternative hypothesis is true (i.e., reject the null hypothesis), and we know that the test procedure will lead to this conclusion incorrectly only $100\alpha\%$ of the time it is used.

  **b.** If the test statistic does not fall in the rejection region, we reserve judgment about which hypothesis is true. We do not accept the null hypothesis, because we do not (in general) know the probability $\beta$ that our test procedure will lead us to falsely accept $H_0$.

---

EXAMPLE 6.2

A research psychologist plans to administer a test designed to measure self-confidence to a random sample of fifty professional athletes. The psychologist theorizes that professional athletes tend to be more self-confident than others. Since the national norm of the test is known to be 72, the theory may be partially validated if it can be shown that the mean score for all professional athletes, $\mu$, exceeds 72.

Suppose the sample mean and standard deviation of the fifty scores are

$$\bar{x} = 74.1 \qquad s = 13.3$$

Do these data support the research hypothesis of the psychologist? Use $\alpha = .10$.

Solution

The elements of the test are

$$H_0: \ \mu = 72 \qquad H_a: \ \mu > 72$$

Test statistic: $z = \dfrac{\bar{x} - 72}{\sigma_{\bar{x}}} = \dfrac{\bar{x} - 72}{\sigma/\sqrt{n}} \approx \dfrac{\bar{x} - 72}{s/\sqrt{n}}$

Rejection region: $z > 1.28$ (the table value corresponding to $\alpha = .10$)

We now substitute the sample statistics into the test statistic to obtain

$$z \approx \frac{74.1 - 72}{13.3/\sqrt{50}} = 1.12$$

Thus, although the mean score for the sample of athletes exceeds the national norm by more than 2 points, the $z$ value of 1.12 does not fall in the rejection region (see Figure 6.10). Therefore, this sample does not provide sufficient evidence at the $\alpha = .10$ level to support the psychologist's theory.

**FIGURE 6.10**
LOCATION OF THE
TEST STATISTIC FOR
EXAMPLE 6.2

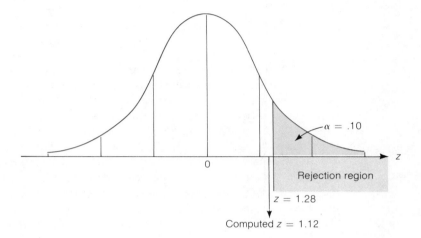

Although the value of the sample mean $\bar{x}$ could be used as a test statistic to test an hypothesis about a population mean $\mu$, we will find it more convenient to use the $z$ statistic. In fact, for the preceding examples, we based the decision either to reject or not reject the null hypothesis on the computed value of $z$. For example, saying that you will reject $H_0$ in Example 6.2 if $\bar{x}$ lies more than $1.28\sigma_{\bar{x}}$ above $\mu = 72$ is the same as saying that you will reject $H_0$ if $z$ is greater than 1.28.

The research hypothesis for Example 6.2, namely that $\mu > 72$, leads to a one-tailed (upper tail) statistical test, because we would reject the null hypothesis only for large values of $z$ (values in the upper tail of the $z$ distribution). See Figure 6.10. Some statistical investigations seek to show that $\mu$ is *either* larger or smaller than some specified value. This type of research (alternative) hypothesis, for example,

$$H_a: \quad \mu > 72 \text{ or } \mu < 72$$

will be supported for large positive or negative values of $z$. Thus, the rejection region will be located in both tails of the $z$ distribution, splitting $\alpha$ between the two tails (see Figure 6.11, page 224). Such a statistical test is said to be two-sided or two-tailed. The value of $z$, denoted by the symbol $z_{\alpha/2}$, that places half of $\alpha$ in the upper tail of the $z$ distribution can be obtained from the table of areas under the normal curve (Table III in the Appendix).

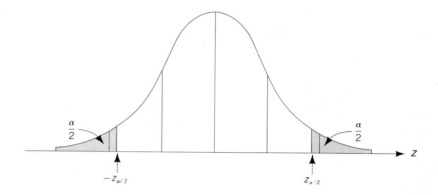

Notice that alternative hypotheses are always expressed as inequalities, i.e., you are always attempting to show that $\mu$ is larger than some value (upper one-tailed test), smaller than some value (lower one-tailed test), or not equal to some value (a two-tailed test). In contrast, the null hypothesis is always expressed as an equality. To illustrate, in Example 6.2, the alternative hypothesis is $H_a$: $\mu > 72$. Although the opposite of this alternative is $\mu \leq 72$, the null hypothesis is given as $H_0$: $\mu = 72$. This is because we need to determine from the sampling distribution of $\bar{x}$ those values that contradict the null hypothesis and that support the alternative hypothesis, $H_a$: $\mu > 72$. Any value of $\bar{x}$ that leads to rejection of $\mu = 72$ in favor of $\mu > 72$ would certainly also lead to rejection of any value of $\mu$ less than 72. For this reason, the null hypothesis is given as the equality $H_0$: $\mu = 72$ rather than the inequality $\mu \leq 72$.

---

**STEPS TO FOLLOW IN SELECTING THE NULL AND ALTERNATIVE HYPOTHESES**

**1.** Specify the hypothesis that you wish to support. Remember this will give a range of possible values for the parameter being tested and will be expressed as an inequality in the alternative hypothesis $H_a$.

*Example:* $H_a$: $\mu > 72$

**2.** Define the opposite of the alternative hypothesis. This will be the set of all possible values of the parameter that are not contained in $H_a$.

*Example:* $\mu \leq 72$

For the null hypothesis, $H_0$, choose the value of the parameter that is nearest in value to those specified in $H_a$.

*Example:* Of the values $\mu \leq 72$, the one nearest in value to those contained in $\mu > 72$ is $\mu = 72$. Thus, the null hypothesis is $H_0$: $\mu = 72$.

---

**EXAMPLE 6.3**

A nutritionist believes that a 12 ounce box of breakfast cereal should contain an average of 1.2 ounces of bran. The nutritionist measures a random sample of sixty boxes of a popular cereal for bran content. Suppose the data yield

$$\bar{x} = 1.170 \text{ ounces of bran} \qquad s = .111 \text{ ounce of bran}$$

Do the data indicate that the mean bran content of all boxes of this brand of cereal differs from 1.2 ounces? Use $\alpha = .05$.

**Solution**

We wish to determine whether $\mu$, the mean amount of bran, *differs* from 1.2 ounces, i.e., we wish to detect $\mu > 1.2$ or $\mu < 1.2$ if either of these situations exists. Therefore, we want to conduct a two-tailed test. The elements of the test are

$$H_0: \quad \mu = 1.2 \qquad H_a: \quad \mu \neq 1.2$$

$$\text{Test statistic: } z = \frac{\bar{x} - 1.2}{\sigma_{\bar{x}}}$$

Rejection region: $z > 1.96$ or $z < -1.96$ (see Figure 6.12)

Note that $z = 1.96$ was chosen for the boundary of the rejection region because $P(z > 1.96) = .025$. This value is obtained from Table III in the Appendix.

We now calculate

$$z = \frac{\bar{x} - 1.2}{\sigma_{\bar{x}}} = \frac{\bar{x} - 1.2}{\sigma/\sqrt{n}} = \frac{1.170 - 1.2}{\sigma/\sqrt{60}}$$

$$\approx \frac{1.170 - 1.2}{s/\sqrt{60}} = \frac{-.030}{.111/\sqrt{60}} = -2.09$$

You can see in Figure 6.12 that the calculated $z$ value, $-2.09$, is in the lower-tail rejection region, and there is evidence to indicate that the mean bran content, $\mu$, differs from 1.2 ounces. It appears that, on average, the cereal boxes contain too little

**FIGURE 6.12**
**TWO-TAILED**
**REJECTION REGION:**
$\alpha = .05$

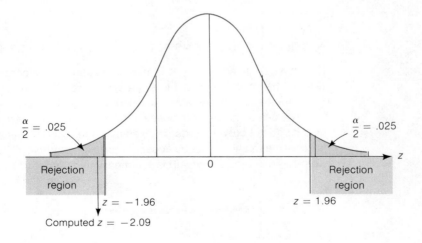

bran (as judged by the nutritionist). How reliable is this conclusion? We know that the test statistic will erroneously reject the null hypothesis only 5% of the time (because $\alpha = .05$). Therefore, we are reasonably confident that this statistical test has led us to a correct conclusion.

---

### LARGE-SAMPLE TEST OF AN HYPOTHESIS ABOUT $\mu$

**One-tailed test**

$H_0: \quad \mu = \mu_0$*

$H_a: \quad \mu < \mu_0$

(or $\quad H_a: \quad \mu > \mu_0$)

Test statistic: $\quad z = \dfrac{\bar{x} - \mu_0}{\sigma_{\bar{x}}}$

Rejection region: $\quad z < -z_\alpha$

(or $\quad z > z_\alpha$ when

$H_a: \quad \mu > \mu_0$)

where $z_\alpha$ is chosen so that

$$P(z > z_\alpha) = \alpha$$

**Two-tailed test**

$H_0: \quad \mu = \mu_0$*

$H_a: \quad \mu \neq \mu_0$

Test statistic: $\quad z = \dfrac{\bar{x} - \mu_0}{\sigma_{\bar{x}}}$

Rejection region: $\quad z < -z_{\alpha/2}$

or $\quad z > z_{\alpha/2}$

where $z_{\alpha/2}$ is chosen so that

$$P(z > z_{\alpha/2}) = \alpha/2$$

---

As we have indicated by the preceding examples, a large-sample statistical test of an hypothesis concerning a population mean can be either one-tailed or two-tailed, depending on the nature of the research (alternative) hypothesis we wish to support. A summary of the test is given in the box. The two possible conclusions resulting from the sample data are given below.

---

### POSSIBLE CONCLUSIONS FOR A TEST OF AN HYPOTHESIS

**1.** If the calculated $z$-score falls in the rejection region, conclude that the research hypothesis is true. If this strategy is used repeatedly, then Type I errors are made approximately $100\alpha\%$ of the time if $H_0$ is true.

**2.** If the calculated $z$-score does not fall in the rejection region, state that the data do not provide evidence to support the research hypothesis. (The null hypothesis should not be accepted unless the probability $\beta$ of making a Type II error is calculated. This is not easy to do for most hypothesis tests.)

---

*Note:* $\mu_0$ is the symbol for the numerical value assigned to $\mu$ under the null hypothesis.

Learning the mechanics

**6.13.** Define each of the following:

    a. Null hypothesis     b. Alternative hypothesis

    c. Test statistic        d. Rejection region

    e. Type I error        f. Type II error

    g. $\alpha$                h. $\beta$

**6.14.** For each of the following rejection regions, sketch the sampling distribution for $z$ and indicate the location of the rejection region:

    a. $z > 1.96$    b. $z > 1.645$    c. $z > 2.58$    d. $z < -1.28$

    e. $z < -1.645$ or $z > 1.645$    f. $z < -2.58$ or $z > 2.58$

**6.15.** If the rejection region is defined as in Exercise 6.14, what is the probability that a Type I error will be made in each case?

**6.16.** A random sample of $n$ observations is taken from a population with unknown mean $\mu$. Give the six elements of a test of hypothesis for each of the following situations:

    a. $H_0$: $\mu = 10$; $H_a$: $\mu \neq 10$; $n = 70$; $\bar{x} = 11.2$; $s = 7.3$; $\alpha = .05$

    b. $H_0$: $\mu = 76$; $H_a$: $\mu > 76$; $n = 50$; $\bar{x} = 80$; $s^2 = 413$; $\alpha = .10$

    c. $H_0$: $\mu = 0.2$; $H_a$: $\mu < 0.2$; $n = 40$; $\bar{x} = .19$; $s = .02$; $\alpha = .01$

**6.17.** A random sample of $n$ observations is selected from a population with unknown mean $\mu$ and variance $\sigma^2$. Give the six elements of a test of hypothesis for each of the following situations:

    a. $H_0$: $\mu = 5,000$; $H_a$: $\mu > 5,000$; $n = 200$; $\bar{x} = 6,000$; $s = 9,000$; $\alpha = .05$

    b. $H_0$: $\mu = 5,000$; $H_a$: $\mu > 5,000$; $n = 200$; $\bar{x} = 6,000$; $s = 5,000$; $\alpha = .01$

    c. $H_0$: $\mu = 12.7$; $H_a$: $\mu \neq 12.7$; $n = 150$; $\bar{x} = 10.3$; $s^2 = 210.6$; $\alpha = .05$

**6.18.** A random sample of $n$ observations is selected from a population with unknown mean $\mu$ and variance $\sigma^2$. Give the six elements of a test of hypothesis for each of the following situations:

    a. $H_0$: $\mu = 0$; $H_a$: $\mu \neq 0$; $n = 175$; $\bar{x} = 6.5$; $s^2 = 496.3$; $\alpha = .01$

    b. $H_0$: $\mu = 65$; $H_a$: $\mu < 65$; $n = 86$; $\bar{x} = 63.8$; $s = 10.3$; $\alpha = .10$

    c. $H_0$: $\mu = 65$; $H_a$: $\mu < 65$; $n = 186$; $\bar{x} = 63.8$; $s = 10.3$; $\alpha = .10$

**6.19.** A random sample of forty-nine observations produced the following sums:

$$\Sigma x = 50.3 \qquad \Sigma x^2 = 68$$

a. Test the null hypothesis that $\mu = 1.18$ against the alternative hypothesis that $\mu < 1.18$. Use $\alpha = .05$.

b. Test the null hypothesis that $\mu = 1.18$ against the alternative hypothesis that $\mu \neq 1.18$. Use $\alpha = .05$.

**6.20.** In a test of hypothesis, who or what determines the size of the rejection region?

**6.21.** If you test an hypothesis and reject the null hypothesis in favor of your research hypothesis, does your test prove that the research hypothesis is correct? Explain.

**6.22.** When do you risk making a Type I error? A Type II error?

Applying the concepts

**6.23.** An automobile manufacturer believes that the mean mileage per gallon of one of its new models exceeds the mean EPA (Environmental Protection Agency) rating of 43 miles per gallon. To gain evidence to support its belief, the manufacturer randomly selected forty of the cars and recorded the miles per gallon for each over a 100 mile course. The mean and standard deviation of the mileages per gallon for the sample of forty cars were $\bar{x} = 43.6$ and $s = 1.3$ miles per gallon.

a. Since the manufacturer wants to show that the mean miles per gallon for the cars exceeds 43, what should you choose for your alternative and null hypotheses?

b. Do the data provide sufficient evidence to support the manufacturer's belief? Use $\alpha = .05$.

**6.24.** Florida's housing market remains strong due to the steady stream of new residents fleeing harsh northern winters. This year, the state association of home builders makes the claim that the mean cost of a new home in Florida is $62,000. However, one realtor believes this figure is too high. In a random sample of thirty new homes sold in Florida this year, the average cost was $58,200 and the standard deviation was $8,000.

a. Identify the alternative and the null hypotheses that are of interest to the realtor.

b. Does the sample information tend to support the realtor's belief? Test using $\alpha = .01$.

**6.25.** A pain reliever currently being used in a hospital is known to bring relief to patients in a mean time of 3.5 minutes. To compare a new pain reliever with the one currently being used, the new drug is administered to a random sample of fifty patients. The mean time to relief for the sample of patients is 2.8 minutes and the standard deviation is 1.14 minutes. Do the data provide sufficient evidence to conclude that the new drug was effective in reducing the mean time until a patient receives relief from pain? Test using $\alpha = .10$.

**6.26.** Small increases in the mean level of bills for monthly long-distance telephone calls produce substantial increases in the profits for telephone companies. A telephone company's records indicate that the amounts paid by private customers per month for long-distance telephone calls have a distribution with mean $17.10 and standard deviation $21.21.

    a.  If a random sample of fifty bills is taken, what is the approximate probability that the sample mean is greater than $20?

    b.  If a random sample of 100 bills is taken, what is the approximate probability that the sample mean is greater than $20?

    c.  Suppose that a random sample of 100 customers' bills during a given month produced a sample mean of $22.10 expended for long-distance calls. Do these data suggest that the mean level of billing per private customer for long-distance calls is in excess of $17.10? Test using $\alpha = .10$.

**6.27.**  The University of Minnesota uses thousands of fluorescent light bulbs each year. The brand of bulb it currently uses has a mean life of 900 hours. A manufacturer claims that its new brand of bulbs, which cost the same as the brand the university currently uses, has a mean life of more than 900 hours. The university has decided to purchase the new brand if, when tested, the test evidence supports the manufacturer's claim at the .05 significance level. Suppose sixty-four bulbs were tested with the following results:

$$\bar{x} = 920 \text{ hours} \qquad s = 80 \text{ hours}$$

Will the University of Minnesota purchase the new brand of fluorescent bulbs?

**6.28.**  A machine is set to produce bolts with a mean length of 1 inch. Bolts that are too long or too short do not meet the customer's specifications and must be rejected. To avoid producing too many rejects, the bolts produced by the machine are sampled from time to time and tested as a check to see whether the machine is still operating properly, i.e., producing bolts with a mean length of 1 inch. Suppose fifty bolts have been sampled, and $\bar{x} = 1.02$ inches and $s = 0.04$ inch. At the $\alpha = .01$ significance level, does the sample evidence indicate that the machine is producing bolts with a mean length not equal to 1 inch; i.e., is the production process out of control?

**6.3
OBSERVED
SIGNIFICANCE
LEVELS:
p VALUES**

According to the statistical test procedure described in Section 6.2, the rejection region and, correspondingly, the value of $\alpha$ are selected prior to conducting the test, and the conclusions are stated in terms of rejecting or not rejecting the null hypothesis. A second method of presenting the results of a statistical test is one that reports the extent to which the test statistic disagrees with the null hypothesis and leaves to the reader the task of deciding whether to reject the null hypothesis. This measure of disagreement is called the observed significance level (or p value) for the test. For example, the value of the test statistic computed for the sample of $n = 50$ sections of sewer pipe was $z = 2.12$. Since the test is one-tailed, i.e., the alternative (research) hypothesis of interest is $H_a$: $\mu > 2,400$, values of the test statistic even more contradictory to $H_0$ than the one observed would be values larger than $z = 2.12$. Therefore, the observed significance level (p value) for this test is

$$p = P\{z \geq 2.12\}$$

or, equivalently, the area under the standard normal curve to the right of $z = 2.12$ (see Figure 6.13).

**FIGURE 6.13**
FINDING THE $p$ VALUE
FOR AN UPPER-TAILED
TEST WHEN $z = 2.12$

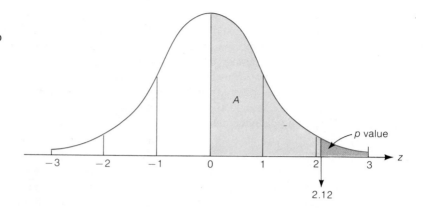

The area $A$ in Figure 6.13 is given in Table III, Appendix, as .4830. Therefore, the upper-tail area corresponding to $z = 2.12$ is

$$p \text{ value} = .5 - .4830 = .0170$$

Consequently, we say that these test results are "very significant", i.e., they disagree rather strongly with the null hypothesis, $H_0$: $\mu = 2,400$, and favor $H_a$: $\mu > 2,400$. The probability of observing a $z$ value as large as 2.12 is only .0170, if in fact the true value of $\mu$ is 2,400.

If you are inclined to select $\alpha = .05$ for this test, then you would reject the null hypothesis because the $p$ value for the test, .0170, is less than .05. In contrast, if you choose $\alpha = .01$, you would not reject the null hypothesis because the $p$ value for the test is larger than .01. Thus, the use of the observed significance level is identical to the test procedure described in the preceding sections except that the choice of $\alpha$ is left to the reader.

---

DEFINITION 6.3
The observed significance level, or $p$ value, for a specific statistical test is the probability (assuming $H_0$ were true) of observing a value of the test statistic that is at least as contradictory to the null hypothesis, and supportive of the alternative hypothesis, as the one computed from the sample data.

---

EXAMPLE 6.4

Find the observed significance level for the test of the mean weight of bran in cereal in Example 6.3.

Solution

Example 6.3 presented a two-tailed test of the hypothesis

$$H_0: \quad \mu = 1.2 \text{ ounces}$$

against the alternative hypothesis,

$H_a$: $\mu \neq 1.2$ ounces

The observed value of the test statistic, Example 6.3, was $z = -2.09$ and any value of $z$ less than $-2.09$ or larger than 2.09 (because this is a two-tailed test) would be even more contradictory to $H_0$. Therefore, the observed significance level for the test is

$p$ value $= P\{z < -2.09 \text{ or } z > 2.09\}$

Consulting Table III in the Appendix, we find

$P\{z > 2.09\} = .5 - .4817 = .0183$

Therefore, the $p$ value for the test is

$2(.0183) = .0366$

These test results would be called "significant" in a statistical sense. Whether the results are "significant" in a practical sense depends upon how much the actual mean weight of bran differs from the desired weight of 1.2 ounces and whether the difference is large enough to be of significance from an economic and nutritional point of view.

When publishing the results of a statistical test of hypothesis in journals, case studies, reports, etc., many researchers make use of $p$ values. Instead of selecting $\alpha$ a priori and then conducting a test as outlined in this chapter, the researcher will compute and report the value of the appropriate test statistic and its associated $p$ value. It is left to the reader of the report to judge the significance of the result, i.e., the reader must determine whether to reject the null hypothesis in favor of the alternative hypothesis, based upon the reported $p$ value. This $p$ value is often referred to as the attained significance level of the test. Usually, the null hypothesis will be rejected if the observed significance level is *less* than the fixed significance level, $\alpha$, chosen by the reader. The inherent advantages of reporting test results in this manner are twofold: (1) readers are permitted to select the maximum value of $\alpha$ that they would be willing to tolerate if they actually carried out a standard test of hypothesis in the manner outlined in this chapter, and (2) a measure of the degree of significance of the result (i.e., the $p$ value) is provided.

---

REPORTING TEST RESULTS AS $p$ VALUES:
HOW TO DECIDE WHETHER TO REJECT $H_0$

**1.** Choose the maximum value of $\alpha$ that you are willing to tolerate.
**2.** If the observed significance level ($p$ value) of the test is less than the maximum value of $\alpha$, then reject the null hypothesis.

---

EXERCISES

Learning the mechanics

**6.29.** Give the observed significance level for each of the following observed values of the $z$ statistic for testing $H_0$: $\mu = 10$ against $H_a$: $\mu > 10$:

a.  $z = 1.72$    b.  $z = 2.49$    c.  $z = 1.35$
d.  $z = 1.4$     e.  $z = 1.18$    f.  $z = 2.95$

**6.30.** Refer to Exercise 6.29. Interpret the observed significance levels that you acquired for parts a and b.

**6.31.** Give the observed significance level for each of the following observed values of the $z$ statistic for testing $H_0$:  $\mu = 16$ against $H_a$:  $\mu \neq 16$:

a.  $z = 1.72$    b.  $z = 2.49$    c.  $z = -1.44$
d.  $z = -2.33$   e.  $z = 1.59$    f.  $z = -1.29$

**6.32.** Refer to Exercise 6.31. Interpret the observed significance levels for parts a and f.

**6.33.** Suppose you were to test $H_0$:  $\mu = 1.4$ against the alternative $H_a$:  $\mu < 1.4$ when $n = 100$, $\bar{x} = 1.1$, and $s = 1.7$. Give the observed significance level for the test and interpret its value.

Applying the concepts

**6.34.** According to advertisements, a strain of soybeans planted on soil prepared with a specified fertilizer treatment has a mean yield of 500 bushels per acre. Fifty farmers who belong to a cooperative plant the soybeans, each using a 40 acre plot, and each records the mean yield per acre. The mean and variance for the sample of fifty farms are $\bar{x} = 485$ and $s^2 = 10{,}045$.

    a.  Do the data provide sufficient evidence to indicate that the mean yield for the soybeans is different than advertised? Give the observed significance level for the test and interpret its value.

    b.  In reaching the conclusion in part a, would you have to qualify your conclusions because of the manner in which the sample was selected? Explain.

**6.35.** Refer to the automobile mileage test in Exercise 6.23. Find the observed significance level for the test and interpret its value.

**6.36.** A new blood pressure drug is advertised to reduce, after one week of medication, a patient's blood pressure an average of 10 units. Blood pressure reductions were recorded for thirty-seven patients after treatment with the drug. The mean and standard deviation for this sample were 8.7 and 6.8, respectively. Do the data appear to contradict the advertising claim? Explain. Find the observed significance level for the test.

**6.37.** Refer to the test of the length of life of the fluorescent light bulbs in Exercise 6.27. Find the observed significance level for this test and interpret its value.

**6.4**

**SMALL-SAMPLE INFERENCES ABOUT A POPULATION MEAN**

Federal legislation requires pharmaceutical companies to perform extensive tests on new drugs before they can be marketed. Initially, a new drug is tested using animals. If the drug is deemed safe after this first phase of testing, the pharmaceutical company is then permitted to begin human testing on a limited basis. During this second phase, inferences must be made about the safety of the drug based upon information in very small samples.

Suppose a pharmaceutical company must demonstrate that a prescribed dose of a certain new drug will result in an average increase in blood pressure of less than 3 points. Assume that only six patients can be used in the initial phase of human testing. The use of a small sample in making an inference about $\mu$ presents two immediate problems when we attempt to use the standard normal $z$ as a test statistic.

PROBLEM 1

The shape of the sampling distribution of the sample mean $\bar{x}$ (and the $z$ statistic) now depends on the shape of the population that is sampled. We can no longer assume that the sampling distribution of $\bar{x}$ is approximately normal, because the central limit theorem only assures normality for samples that are sufficiently large.

Solution to Problem 1

The sampling distribution of $\bar{x}$ (and $z$) will be approximately normal even for relatively small samples if the sampled population is normal.

PROBLEM 2

Although it is still true that $\sigma_{\bar{x}} = \sigma / \sqrt{n}$, the sample standard deviation provides a poor estimate for $\sigma$ when the sample size is small. This affects the sampling distribution of the $z$ statistic when we substitute $s$ for $\sigma$.

Solution to Problem 2

Instead of using the statistic

$$z = \frac{\bar{x} - \mu}{\sigma_{\bar{x}}} = \frac{\bar{x} - \mu}{\sigma / \sqrt{n}}$$

which requires knowledge of, or a good approximation to, $\sigma$ (in order that the sampling distribution of $z$ be normally distributed), we use the statistic

$$t = \frac{\bar{x} - \mu}{s / \sqrt{n}}$$

(which replaces the population standard deviation $\sigma$ by the sample standard deviation $s$) and determine its exact sampling distribution.

The distribution of the $t$ statistic in repeated sampling was discovered by W. S. Gosset, a scientist in the Guinness brewery, who published his discovery in 1908 under the pen name of Student. The main result of Gosset's work is that if we are sampling from a normal distribution, the $t$ statistic will have a sampling distribution very much like that of the $z$ statistic: mound-shaped, symmetric, with mean zero. The primary difference between the sampling distributions of $t$ and $z$ is that the $t$ distribution is more variable than the $z$, which follows intuitively when you realize that $t$ contains two random quantities ($\bar{x}$ and $s$), while $z$ contains only one ($\bar{x}$).

The actual amount of variability in the sampling distribution of $t$ depends on the sample size $n$. A convenient way of expressing this dependence is to say that the $t$ statistic has $(n - 1)$ degrees of freedom. Recall that the quantity $(n - 1)$ is the divisor that appears in the formula for $s^2$. This number plays a key role in the sampling distribution of $s^2$ and will appear in discussions of other statistics in later chapters. Particularly, the smaller the number of degrees of freedom associated with the $t$ statistic, the more variable will be its sampling distribution.

FIGURE 6.14
STANDARD NORMAL (z)
DISTRIBUTION AND
t DISTRIBUTION
WITH 4 df

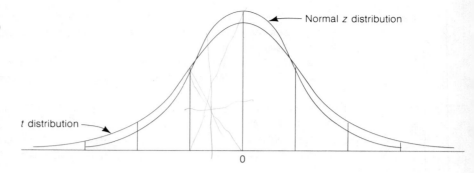

In Figure 6.14 we show both the sampling distribution of z and the sampling distribution of a t statistic with 4 degrees of freedom (df). You can see that the increased variability of the t statistic means that the t value, $t_\alpha$, that locates an area $\alpha$ in the upper tail of the t distribution will be larger than the corresponding value $z_\alpha$. For any given value of $\alpha$, $t_\alpha$ will increase as the number of degrees of freedom (df) decreases. Values of t that will be used in forming small-sample confidence intervals for $\mu$ and rejection regions for small-sample tests of hypotheses about $\mu$ are given in Table IV of the Appendix. A partial reproduction of this table is shown in Figure 6.15.

**FIGURE 6.15**
**REPRODUCTION OF**
**PART OF TABLE IV**
**IN THE APPENDIX**

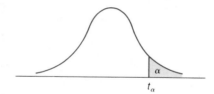

| DEGREES OF FREEDOM | $t_{.100}$ | $t_{.050}$ | $t_{.025}$ | $t_{.010}$ | $t_{.005}$ |
|---|---|---|---|---|---|
| 1 | 3.078 | 6.314 | 12.706 | 31.821 | 63.657 |
| 2 | 1.886 | 2.920 | 4.303 | 6.965 | 9.925 |
| 3 | 1.638 | 2.353 | 3.182 | 4.541 | 5.841 |
| 4 | 1.533 | 2.132 | 2.776 | 3.747 | 4.604 |
| 5 | 1.476 | 2.015 | 2.571 | 3.365 | 4.032 |
| 6 | 1.440 | 1.943 | 2.447 | 3.143 | 3.707 |
| 7 | 1.415 | 1.895 | 2.365 | 2.998 | 3.499 |
| 8 | 1.397 | 1.860 | 2.306 | 2.896 | 3.355 |
| 9 | 1.383 | 1.833 | 2.262 | 2.821 | 3.250 |
| 10 | 1.372 | 1.812 | 2.228 | 2.764 | 3.169 |
| 11 | 1.363 | 1.796 | 2.201 | 2.718 | 3.106 |
| 12 | 1.356 | 1.782 | 2.179 | 2.681 | 3.055 |
| 13 | 1.350 | 1.771 | 2.160 | 2.650 | 3.012 |
| 14 | 1.345 | 1.761 | 2.145 | 2.624 | 2.977 |
| 15 | 1.341 | 1.753 | 2.131 | 2.602 | 2.947 |

Note that $t_\alpha$ values are listed for degrees of freedom from 1 to 29, where $\alpha$ refers to the tail area under the $t$ distribution to the right of $t_\alpha$. For example, if we want the $t$ value with an area of .025 to its right and 4 df, we look in the table under the column $t_{.025}$ for the entry in the row corresponding to 4 df. This entry is $t_{.025} = 2.776$, as shown in Figure 6.16. The corresponding standard normal $z$-score is $z_{.025} = 1.96$.

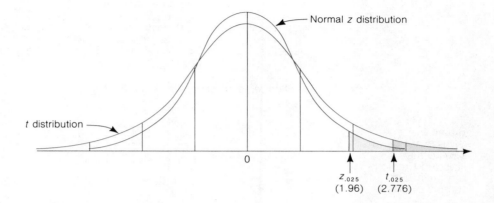

Note that the last row of Table IV, where df = infinity, contains the standard normal $z$ values. This follows from the fact that as the sample size $n$ grows very large, $s$ becomes closer to $\sigma$, and thus $t$ becomes closer in distribution to $z$. In fact, when df = 29, there is little difference between corresponding tabulated values of $z$ and $t$. Thus, we choose the arbitrary cutoff of $n = 30$ (df = 29) to distinguish between the large- and small-sample inferential techniques.

Returning to the example of testing a new drug, suppose that the six test patients have blood pressure increases of 1.7, 3.0, 0.8, 3.4, 2.7, and 2.1 points. We calculate

$$\bar{x} = \frac{\Sigma x}{n} = \frac{13.7}{6} = 2.28$$

$$s^2 = \frac{\Sigma (x - \bar{x})^2}{n - 1} = \frac{\Sigma x^2 - \frac{(\Sigma x)^2}{n}}{n - 1} = \frac{35.79 - \frac{(13.7)^2}{6}}{5} = .9017$$

$$s = \sqrt{s^2} = .950$$

We can now use these results to determine whether there is evidence that the new drug satisfies the requirement that the resulting increase in blood pressure averages less than 3 points. This can be accomplished by testing the null hypothesis that the true mean increase, $\mu$, is equal to 3 against the alternative hypothesis that $\mu$ is less than 3. The elements of the test are

Null hypothesis $H_0$: $\mu = 3$

Alternative hypothesis $H_a$: $\mu < 3$

For the small sample we use the $t$ statistic.

Test statistic: $t = \dfrac{\bar{x} - \mu_0}{s/\sqrt{n}} = \dfrac{\bar{x} - 3}{s/\sqrt{n}}$

Assumption: The relative frequency distribution of the population of blood pressure increases associated with patients taking the drug is approximately normal.

If we want to test at the $\alpha = .05$ level, the rejection region will be

Rejection region: $t < -t_{.05} = -2.015$ where df (degrees of freedom) is equal to $n - 1 = 5$

The rejection region is shown in Figure 6.17.

FIGURE 6.17
A $t$ DISTRIBUTION WITH
5 df AND THE
REJECTION REGION
FOR THE DRUG
TESTING EXAMPLE

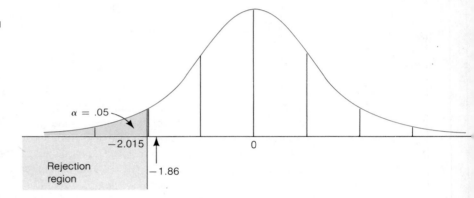

We now calculate

$$t = \frac{\bar{x} - 3}{s/\sqrt{n}} = \frac{2.28 - 3}{.95/\sqrt{6}} = -1.86$$

Since the value $t = -1.86$ that is calculated from the sample is not less than the tabulated value $-2.015$, we cannot conclude that the mean increase in blood pressure resulting from taking the drug is less than 3 points. The pharmaceutical company may decide to perform further tests on animals or to modify the composition of the drug before further testing.

It is interesting to note that the calculated $t$ value, $-1.86$, *is less than* the .05 level $z$ value, $-1.645$. The implication is that if we had *incorrectly* used a $z$ statistic for this test, we would have rejected the null hypothesis at the .05 level, concluding that the mean increase in blood pressure is less than 3. The important point is that the statistical procedure to be used must always be closely scrutinized and all the assumptions understood. Many statistical "lies" are the result of misapplications of otherwise valid procedures.

SMALL-SAMPLE TEST OF AN HYPOTHESIS ABOUT $\mu$

| One-tailed test | Two-tailed test |
|---|---|
| $H_0$: $\mu = \mu_0$ | $H_0$: $\mu = \mu_0$ |
| $H_a$: $\mu < \mu_0$ | $H_a$: $\mu \neq \mu_0$ |
| (or $H_a$: $\mu > \mu_0$) | |

Test statistic:  $t = \dfrac{\bar{x} - \mu_0}{s/\sqrt{n}}$          Test statistic:  $t = \dfrac{\bar{x} - \mu_0}{s/\sqrt{n}}$

Rejection region:  $t < -t_\alpha$          Rejection region:  $t < -t_{\alpha/2}$

(or $t > t_\alpha$          or $t > t_{\alpha/2}$

when $H_a$: $\mu > \mu_0$)

where $t_\alpha$ and $t_{\alpha/2}$ are based on $(n - 1)$ degrees of freedom

Assumption:   A random sample is selected from a population with a relative frequency distribution that is approximately normal.

EXAMPLE 6.5

A major car manufacturer wants to test a new engine to see whether it meets new air pollution standards. The mean emission, $\mu$, of all engines of this type must be less than 20 parts per million of carbon. Ten engines are manufactured for testing purposes, and the mean and standard deviation of the emissions for this sample of engines are determined to be

$$\bar{x} = 17.1 \text{ parts per million} \qquad s = 3.0 \text{ parts per million}$$

Do the data supply sufficient evidence to allow the manufacturer to conclude that this type of engine meets the pollution standard? Assume that the manufacturer is willing to risk a Type I error with probability $\alpha = .01$.

Solution

The manufacturer wants to support the research hypothesis that the mean emission level, $\mu$, for all engines of this type is less than 20 parts per million. The elements of this small-sample one-tailed test are

$$H_0: \ \mu = 20 \qquad H_a: \ \mu < 20$$

Test statistic:  $t = \dfrac{\bar{x} - 20}{s/\sqrt{n}}$

Assumption:   The relative frequency distribution of the population of emission levels for all engines of this type is approximately normal.

Rejection region:   For $\alpha = .01$ and df $= n - 1 = 9$, the one-tailed rejection region (see Figure 6.18) is $t < -t_{.01} = 2.821$.

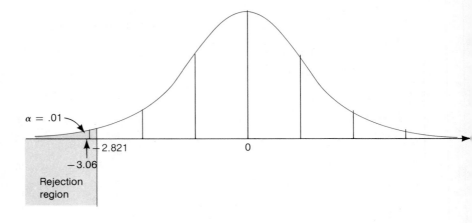

$\alpha = .01$

$-2.821$

$-3.06$

0

Rejection region

We now calculate the test statistic:

$$t = \frac{\bar{x} - 20}{s/\sqrt{n}} = \frac{17.1 - 20}{3.0/\sqrt{10}} = -3.06$$

Since the calculated $t$ falls in the rejection region (see Figure 6.18), the manufacturer concludes that $\mu < 20$ parts per million and the new engine type meets the pollution standard. Are you satisfied with the reliability associated with this inference? The probability is only $\alpha = .01$ that the test would support the research hypothesis if in fact it were false.

EXAMPLE 6.6

Find the observed significance level for the test described in Example 6.5.

Solution

The test of Example 6.5 was a lower-tailed test: $H_0$: $\mu = 20$ versus $H_a$: $\mu < 20$. Since the value of $t$ computed from the sample data was $t = -3.06$, the observed significance level (or $p$ value) for the test is equal to the probability that $t$ would assume a value less than or equal to $-3.06$, if in fact $H_0$ were true. This is equal to the area in the

FIGURE 6.19
THE OBSERVED
SIGNIFICANCE LEVEL
FOR THE TEST,
EXAMPLE 6.5

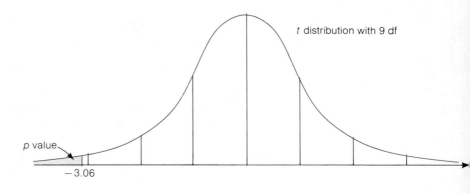

$t$ distribution with 9 df

$p$ value

$-3.06$

lower tail of the $t$ distribution (shaded in Figure 6.19). To find this area, i.e., the $p$ value for the test, we consult the $t$ table (Table IV in the Appendix). Unlike the table of areas under the normal curve, Table IV gives only the $t$ values corresponding to the areas .100, .050, .025, .010, and .005. Therefore, we can only approximate the $p$ value for the test. Since the observed $t$ value was based on 9 degrees of freedom, we use the df $= 9$ row in Table IV and move across the row until we reach the $t$ values that are closest to the observed $t = -3.06$. [*Note:* We ignore the minus sign.] The $t$ values corresponding to $p$ values of .010 and .005 are 2.821 and 3.250, respectively. Since the observed $t$ value falls between $t_{.010}$ and $t_{.005}$, the $p$ value for the test lies between .005 and .010. We could interpolate to more accurately locate the $p$ value for the test, but it is easier and adequate for our purposes to choose the larger area as the $p$ value and report it as .010. Thus, we would reject the null hypothesis, $H_0$: $\mu = 20$ parts per million, for any value of $\alpha$ larger than .01.

We may also use the $t$ distribution to form a small-sample confidence interval for a population mean $\mu$, if the population is approximately normally distributed. Recall that the large-sample confidence interval for $\mu$ is

$$\bar{x} \pm z_{\alpha/2}\sigma_{\bar{x}} = \bar{x} \pm z_{\alpha/2}\left(\frac{\sigma}{\sqrt{n}}\right)$$

where $100(1 - \alpha)$ percent is the desired confidence level. To form the small-sample confidence interval, replace $\sigma$ by $s$ and $z_{\alpha/2}$ by $t_{\alpha/2}$, where the number of degrees of freedom for the tabulated $t$ value is $(n - 1)$.

---

SMALL-SAMPLE CONFIDENCE INTERVAL FOR $\mu$

$$\bar{x} \pm t_{\alpha/2}\left(\frac{s}{\sqrt{n}}\right)$$

where $t_{\alpha/2}$ is based on $(n - 1)$ degrees of freedom

Assumption: A random sample is selected from a population with a relative frequency distribution that is approximately normal.

---

EXAMPLE 6.7

When food prices began their rapid increase in the early 1970's, some of the major television networks began periodically to purchase a grocery basket full of food at supermarkets around the country. They always bought the same items at each store so they could compare food prices. Suppose you want to estimate the mean price for a grocery basket in a specific geographical region of the country. You purchase the specified items at a random sample of twenty supermarkets in the region. The mean and standard deviation of the costs at the twenty supermarkets are

$$\bar{x} = \$26.84 \qquad s = \$2.63$$

Form a 95% confidence interval for the mean cost, $\mu$, of a grocery basket for this region.

Solution

If we assume that the distribution of costs for the grocery basket at all supermarkets in the region is approximately normal, we can use the $t$ statistic to form the confidence interval. For a confidence level of 95%, we need the tabulated value of $t$ with df $= n - 1 = 19$:

$$t_{\alpha/2} = t_{.025} = 2.093$$

Then the confidence interval is

$$\bar{x} \pm t_{.025}\left(\frac{s}{\sqrt{n}}\right) = 26.84 \pm 2.093\left(\frac{2.63}{\sqrt{20}}\right)$$

$$= 26.84 \pm 1.23 = (25.61, 28.07)$$

Thus, we are reasonably confident that the interval from \$25.61 to \$28.07 contains the true mean cost, $\mu$, of the grocery basket. This is because, if we were to employ our interval estimator on repeated occasions, 95% of the intervals constructed would contain $\mu$.

We have emphasized throughout this section that the assumption of a normally distributed population is necessary for making small-sample inferences about $\mu$ when using the $t$ statistic. While many phenomena do have approximately normal distributions, it is also true that many random phenomena have distributions that are not normal or even mound-shaped. Empirical evidence acquired over the years has shown that the $t$ distribution is rather insensitive to moderate departures from normality. That is, the use of the $t$ statistic when sampling from mound-shaped populations generally produces credible results; however, for cases in which the distribution is distinctly nonnormal, either take a larger sample or use a nonparametric statistical method. These nonparametric methods are described in the references.

EXERCISES*

Learning the mechanics

**6.38.** A random sample of $n$ observations is selected from a normal population to test the null hypothesis that $\mu = 10$. Specify the rejection region for each of the following combinations of $H_a$, $\alpha$, and $n$:

a. $H_a$: $\mu \neq 10$; $\alpha = .05$; $n = 20$
b. $H_a$: $\mu > 10$; $\alpha = .01$; $n = 25$
c. $H_a$: $\mu > 10$; $\alpha = .10$; $n = 10$
d. $H_a$: $\mu < 10$; $\alpha = .01$; $n = 13$

*Although we will not always ask you to list the assumptions necessary for valid implementation of the procedures used in these exercises, you should develop the habit of listing the assumptions in each application of a statistical procedure.

e. $H_a$:  $\mu \neq 10$;  $\alpha = .10$;  $n = 17$
f. $H_a$:  $\mu < 10$;  $\alpha = .05$;  $n = 8$

**6.39.** A random sample of $n$ observations is selected from a normal population with mean $\mu$. For each of the following combinations of sample size and confidence level, specify the value of $t$ needed to form a confidence interval for $\mu$.

   a.  $n = 15$,  confidence level of 90%
   b.  $n = 6$,  confidence level of 99%
   c.  $n = 16$,  confidence level of 99%
   d.  $n = 28$,  confidence level of 95%
   e.  $n = 28$,  confidence level of 90%

**6.40.** A random sample of five measurements from a normally distributed population yielded $\bar{x} = 4.8$ and $s = 1.5$.

   a.  Test the null hypothesis that the mean of the population is 6 against the alternative hypothesis, $\mu < 6$. Use $\alpha = .10$.
   b.  Test the null hypothesis that the mean of the population is 6 against the alternative hypothesis, $\mu \neq 6$. Use $\alpha = .10$.
   c.  Form a 95% confidence interval for $\mu$.
   d.  Form a 90% confidence interval for $\mu$.

**6.41.** A random sample of six measurements from a normally distributed population produced the following observations: 4, 6, 2, 7, 6, 5.

   a.  Test the null hypothesis that $\mu = 2.4$ against the alternative hypothesis that $\mu > 2.4$. Use $\alpha = .01$.
   b.  Test the null hypothesis that $\mu = 2.4$ against the alternative hypothesis that $\mu \neq 2.4$. Use $\alpha = .01$.
   c.  Test the null hypothesis that $\mu = 2.4$ against the alternative hypothesis that $\mu \neq 2.4$. Use $\alpha = .05$.
   d.  Find a 95% confidence interval for $\mu$.
   e.  Find a 99% confidence interval for $\mu$.

Applying the concepts

**6.42.** Pulse rate is an important measure of the fitness of a person's cardiovascular system. The mean pulse rate for American adult males is approximately 72 heart beats per minute. A random sample of twenty-one American adult males who jog at least 15 miles per week had a mean pulse rate of 52.6 beats per minute and a standard deviation of 3.22 beats per minute.

   a.  Find a 95% confidence interval for the mean pulse rate of all American adult males who jog at least 15 miles per week.
   b.  Interpret the interval found in part a.
   c.  What assumptions are required for the validity of the confidence interval?

**6.43.** A consumer protection group is concerned that a catsup manufacturer is filling its 20 ounce family size containers with less than 20 ounces of catsup. The group purchases ten family size bottles of this catsup, weighs the contents of each, and finds that the mean weight is equal to 19.86 ounces and the standard deviation is equal to .22 ounce.

   a.   Do the data provide sufficient evidence for the consumer group to conclude that the mean fill per family size bottle is less than 20 ounces? Test using $\alpha = .05$.

   b.   If the test in part a were conducted on a periodic basis by the company's quality control department, is the consumer group more concerned about making a Type I error or a Type II error? The probability of making this type of error is called the consumer's risk.

   c.   The catsup company is also interested in the mean amount of catsup per bottle. It does not wish to overfill them. For the test conducted in part a, which type of error is more serious from the company's point of view — a Type I error or a Type II error? The probability of making this type of error is called the producer's risk.

**6.44.** Refer to Exercise 6.43. Find a 90% confidence interval for the mean number of ounces of catsup being dispensed.

**6.45.** The application of adrenalin is the prevailing treatment to reduce eye pressure in glaucoma patients. Theoretically, a new synthetic drug will cause the same mean drop in pressure (5.5 units) without the side effects caused by adrenalin. The new drug is given to five glaucoma patients and the reductions in eye pressure for the patients are 4.0, 3.8, 5.7, 5.3, and 4.6 units.

   a.   Look at the data. Based on your intuition, do you think that the mean reduction in pressure for the new drug differs from the mean reduction produced by adrenalin?

   b.   Now use a statistical test to answer the question in part a. Do the data provide sufficient evidence to indicate that the mean reduction in eye pressure due to the new drug is different from that produced by adrenalin? Test using $\alpha = .05$.

   c.   Give the approximate observed significance level for the test and interpret its value.

**6.46.** One of the most feared predators in the ocean is the great white shark. Although it is known that the white shark grows to a mean length of 21 feet, a marine biologist believes that the great white sharks off the Bermuda coast grow much longer due to unusual feeding habits. To test this claim, a number of full-grown great white sharks are captured off the Bermuda coast, measured, and then set free. However, because the capture of sharks is difficult, costly, and very dangerous, only three are sampled. Their lengths are 24, 20, and 22 feet.

   a.   Do the data provide sufficient evidence to support the marine biologist's claim? Use $\alpha = .10$.

   b.   Give the approximate observed significance level for the test in part a, and interpret its value.

c. What assumptions must be made in order to carry out the test?

d. Do you think these assumptions are likely to be satisfied in this particular sampling situation?

**6.47.** One way of determining whether red pine trees are growing properly is to measure the diameter of the main stem of the tree at the age of 4 years. The main stem growth for a random sample of seventeen 4 year old red pine seedlings produced a mean and standard deviation equal to 11.3 and 3.1 inches, respectively.

a. Find a 99% confidence interval for the mean main stem growth of a population of 4 year old red pine trees. Interpret the interval.

b. What assumptions are necessary in order for your confidence interval to be valid?

**6.48.** A cigarette manufacturer advertises that its new low-tar cigarette "contains on average no more than 4 milligrams of tar." You have been asked to test the claim using the following sample information: $n = 25$, $\bar{x} = 4.10$ milligrams, $s = .14$ milligram.

a. Does the sample information disagree with the manufacturer's claim? Test using $\alpha = .01$.

b. Give the approximate observed significance level for the test in part a, and interpret its value.

c. What assumptions are needed to ensure the validity of the testing procedure?

**6.49.** Refer to Exercise 6.48. Find a 98% confidence interval for the mean amount of tar in the manufacturer's new low-tar brand. Interpret your result.

**6.50.** A company purchases large quantities of naphtha in 50 gallon drums. Because the purchases are on-going, small shortages in the drums can represent a sizable loss to the company. The weights of the drums vary slightly from drum to drum, so the weight of the naphtha is determined by removing it from the drums and measuring it. Suppose the company samples the contents of twenty drums, measures the naphtha in each, and calculates $\bar{x} = 49.70$ gallons and $s = .32$ gallon. Do the sample statistics provide sufficient evidence to indicate that the mean fill per 50 gallon drum is less than 50 gallons? Use $\alpha = .10$.

**6.51.** Refer to Exercise 6.50. Find a 90% confidence interval for the mean number of gallons of naphtha per drum.

**6.52.** An important problem facing strawberry growers is the control of nematodes. These organisms compete with the plants for nutrients in the soil, thereby reducing yield. For this reason, fumigation is normally a part of field preparation. In the past, the fumigants used yielded an average of 8 pounds of marketable fruit for a certain standard sized plot. Recently, a new fumigant has been developed. It is applied to six standard plots of strawberries, and the yield of marketable fruit (in pounds) for each plot is 9, 9, 13, 9, 10, 8.

a. Do the data indicate a significant increase in average yield at the .05 level of significance?

b. What assumptions are necessary for the procedure used to be valid?

**6.53.** A psychologist was interested in knowing whether male heroin addicts' assessments of self-worth differ from those of the general male population. On a test designed to measure assessment of self-worth, the mean score for males from the general population is 48.6. A random sample of twenty-five scores achieved by heroin addicts yielded a mean of 44.1 and a standard deviation of 6.2.

a. Do the data indicate a difference in assessment of self-worth between male heroin addicts and the general male population? Test using $\alpha = .01$.
b. Give the approximate observed significance level for the test and interpret its value.

**6.54.** Four fossils of humerus bones were unearthed at an archeological site in East Africa. The length-to-width ratios of the bones were 6, 9, 10, 10. Humerus bones from the same species of animal tend to have approximately the same length-to-width ratio. It is a known fact that species A has a mean ratio of 8.5. It can be assumed that the four unearthed bones are all from an unknown species and all are from the same species of animal.

a. If an archeologist believes the bones to be from an animal of species A, do the data provide sufficient evidence to contradict the archeologist's theory? Use $\alpha = .10$.
b. State the assumptions necessary to make your test valid.

**6.55.** Suppose you want to estimate the mean percentage of gain in per share value for growth-type mutual funds over a specific 2 year period. Ten mutual funds are randomly selected from the population of all the commonly listed funds. The percentage gain figures are shown below (negative gains indicate losses):

| 12.1 | −3.7 | 7.6 | 6.8 | −2.3 |
|------|------|-----|-----|------|
| 4.6 | 8.4 | 18.1 | 9.2 | 3.0 |

Find a 99% confidence interval for the mean percentage of gain in per share value for the population of funds.

**6.56.** A random sample of twelve market analysts gave the following forecasts for the price change (in dollars) over a 6 month period of a particular steel stock:

| 0 | 3 | 12 | −4 | 10 | 7 |
|---|---|----|----|----|---|
| 6 | 5 | 5 | −1 | 4 | 3 |

Suppose the mean of the population of forecasts for all market analysts is an accurate measure of the actual gain that the stock will experience. Estimate the mean forecast gain using a 95% confidence interval.   [*Note:* Although it is likely that the population of forecasts will not be normal, assume the distribution will be adequately approximated by a normal distribution.]

**6.57.** A problem that occurs with certain types of mining is that some byproducts tend to be mildly radioactive and these products sometimes get into our freshwater supply. The EPA has issued regulations concerning a limit on the amount of radioactivity in supplies of drinking water. Particularly, the maximum level for naturally occurring radiation is 5 picocuries per liter of water. A random sample of twenty-four water specimens from a city's water supply produced the sample statistics $\bar{x} = 4.61$ picocuries per liter and $s = .87$ picocurie per liter.

    a.   Do these data provide sufficient evidence to indicate that the mean level of radiation is safe (below the maximum level set by the EPA)? Test using $\alpha = .01$.

    b.   Why should you want to use a small value of $\alpha$ for the test in part a?

## 6.5 LARGE-SAMPLE INFERENCES ABOUT A BINOMIAL POPULATION PROPORTION

In recent years the number of public opinion polls has grown at an astounding rate. Almost daily, the news media report the results of some poll. Pollsters regularly determine the percentage of people in favor of the President's energy program, the fraction of voters in favor of a certain candidate, the fraction of customers who favor a particular brand of wine, and the proportion of people who smoke cigarettes. In each case, we are interested in estimating the percentage (or proportion) of some group with a particular characteristic. In this section we will consider methods for making inferences about population proportions.

**EXAMPLE 6.8**

The mid-1970's may well be remembered for the political unrest at the national level. Since the days of Watergate, public opinion polls have been conducted to estimate the fraction of Americans who trust the President. Suppose that 1,000 people are randomly chosen and 637 answer that they trust the President. How would you estimate the true fraction of *all* American people who trust the President?

**Solution**

What we have really asked is how would you estimate the probability $p$ of success in a binomial experiment, where $p$ is the probability a person chosen trusts the President. One logical method of estimating $p$ for the population is to use the proportion of successes in the sample. That is, we can estimate $p$ by calculating

$$\hat{p} \text{ (read ''p hat'')} = \frac{\text{Number of people sampled who trust the President}}{\text{Number of people sampled}}$$

Thus, in this case,

$$\hat{p} = \frac{637}{1,000} = .637$$

To determine the reliability of the estimator $\hat{p}$, we need to know its sampling distribution. That is, if we were to draw samples of 1,000 people over and over again, each time calculating a new estimate $\hat{p}$, what would the frequency distribution of all the $\hat{p}$'s be? The answer lies in viewing $\hat{p}$ as the average or mean number of successes per trial over the $n$ trials. If each success is assigned a value equal to 1 and a failure is assigned a value of 0, then the sum of all $n$ sample observations is $x$, the total number of successes, and $\hat{p} = x/n$ is the average or mean number of successes per trial in the $n$ trials. The central limit theorem tells us that the relative frequency distribution of the sample mean for any population is approximately normal for large samples. Therefore, the sampling distribution of $\hat{p}$ has the characteristics listed in the box on page 246 (see Figure 6.20).

**FIGURE 6.20**
**SAMPLING**
**DISTRIBUTION OF $\hat{p}$**

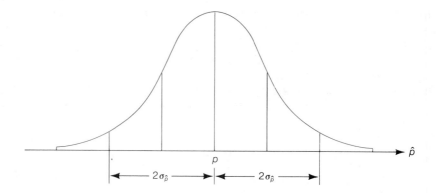

The fact that the sampling distribution of $\hat{p}$ is approximately normal for large samples allows us to form confidence intervals and test hypotheses about $p$ in a manner which is completely analogous to that used for large-sample inferences about $\mu$.

Thus, if 637 of 1,000 Americans say they trust the President, a 95% confidence interval for the proportion of *all* Americans who trust the President is

$$\hat{p} \pm z_{\alpha/2}\sigma_{\hat{p}} = .637 \pm 1.96\sqrt{pq/1{,}000}$$

where $q = 1 - p$. Just as we needed an approximation for $\sigma$ in calculating a large-sample confidence interval for $\mu$, we now need an approximation for $p$. As Table 6.3

**TABLE 6.3**

**VALUES OF $pq$ FOR SEVERAL DIFFERENT $p$ VALUES**

| $p$ | $pq$ | $\sqrt{pq}$ |
| --- | --- | --- |
| .5 | .25 | .50 |
| .6 or .4 | .24 | .49 |
| .7 or .3 | .21 | .46 |
| .8 or .2 | .16 | .40 |
| .9 or .1 | .09 | .30 |

shows, the approximation for $p$ does not have to be especially accurate, because the value of $\sqrt{pq}$ needed for the confidence interval is relatively insensitive to changes in $p$. Therefore, we can use $\hat{p}$ to approximate $p$. Keeping in mind that $\hat{q} = 1 - \hat{p}$, we substitute these values into the formula for the confidence interval:

$$\hat{p} \pm 1.96\sqrt{pq/1{,}000} \approx \hat{p} \pm 1.96\sqrt{\hat{p}\hat{q}/1{,}000}$$
$$= .637 \pm 1.96\sqrt{(.637)(.363)/1{,}000}$$
$$= .637 \pm .030$$
$$= (.607, .667)$$

Then we can be 95% confident that the interval from 60.7% to 66.7% contains the true percentage of *all* Americans who trust the President. That is, in repeated construction of confidence intervals, approximately 95% of all samples would produce confidence intervals that enclose $p$.

The general form of a large-sample confidence interval for $p$ is shown in the box.

---

LARGE-SAMPLE CONFIDENCE INTERVAL FOR $p$

$$\hat{p} \pm z_{\alpha/2}\sigma_{\hat{p}} \approx \hat{p} \pm z_{\alpha/2}\sqrt{\hat{p}\hat{q}/n} \qquad \text{where} \quad \hat{q} = 1 - \hat{p}$$

---

Tests of hypotheses concerning $p$ are also analogous to those for population means (large samples).

---

LARGE-SAMPLE TEST OF AN HYPOTHESIS ABOUT $p$

One-tailed test

$H_0$: $p = p_0$ ($p_0$ = hypothesized $p$ value)

$H_a$: $p < p_0$
(or $H_a$: $p > p_0$)

Test statistic: $z = \dfrac{\hat{p} - p_0}{\sigma_{\hat{p}}}$

where, according to $H_0$, $\sigma_{\hat{p}} = \sqrt{p_0 q_0/n}$ and $q_0 = 1 - p_0$

Rejection region: $z < -z_\alpha$
(or $z > z_\alpha$
when $H_a$: $p > p_0$)

Two-tailed test

$H_0$: $p = p_0$

$H_a$: $p \neq p_0$

Test statistic: $z = \dfrac{\hat{p} - p_0}{\sigma_{\hat{p}}}$

Rejection region: $z < -z_{\alpha/2}$
or $z > z_{\alpha/2}$

---

EXAMPLE 6.9     The reputations (and hence, sales) of many businesses can be severely damaged by shipments of manufactured items that contain an unusually large percentage of defectives. For example, a manufacturer of flashbulbs for cameras may want to be reasonably certain that less than 5% of its bulbs are defective. Suppose 300 bulbs are randomly selected from a very large shipment, each is tested, and ten defective bulbs are found. Does this provide sufficient evidence for the manufacturer to conclude that the fraction defective in the entire shipment is less than .05? Use $\alpha = .01$.

Solution     The objective of the sampling is to determine whether there is sufficient evidence to indicate that the fraction defective, $p$, is less than .05. Consequently, we will test the null hypothesis that $p = .05$ against the alternative hypothesis that $p < .05$. The elements of the test are

$$H_0: \quad p = .05 \qquad H_a: \quad p < .05$$

Test statistic: $\quad z = \dfrac{\hat{p} - .05}{\sigma_{\hat{p}}}$

Rejection region: $\quad z < -z_{.01} = -2.33 \quad$ (see Figure 6.21)

We now calculate the test statistic

$$z = \frac{\hat{p} - .05}{\sigma_{\hat{p}}} = \frac{(10/300) - .05}{\sqrt{p_0 q_0 / n}} = \frac{.033 - .05}{\sqrt{p_0 q_0 / 300}}$$

Notice that we use $p_0$ to calculate $\sigma_{\hat{p}}$ because, in contrast to calculating $\sigma_{\hat{p}}$ for a confidence interval, the test statistic is computed on the assumption that the null hypothesis is true, i.e., $p = p_0$. Therefore, substituting the values for $\hat{p}$ and $p_0$ into the $z$ statistic, we obtain

FIGURE 6.21
REJECTION REGION
FOR EXAMPLE 6.9

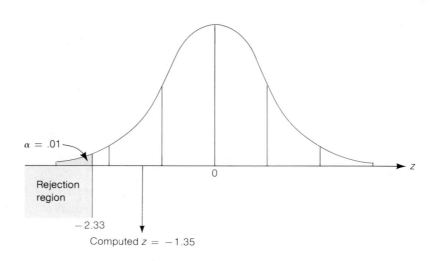

$\alpha = .01$

Rejection
region

$-2.33$

Computed $z = -1.35$

0

$z$

$$z \approx \frac{-.017}{\sqrt{(.05)(.95)/300}} = \frac{-.017}{.0126} = -1.35$$

As shown in Figure 6.21, the calculated $z$ value does not fall in the rejection region. Therefore, based on this test, there is insufficient evidence to indicate that the shipment contains fewer than 5% defective bulbs.

**EXAMPLE 6.10**

In Example 6.9, we found that we did not have sufficient evidence, at the $\alpha = .01$ level of significance, to indicate that the fraction defective $p$ of flashbulbs was less than $p = .05$. How strong was the weight of evidence favoring the alternative hypothesis ($H_a$: $p < .05$)? Find the observed significance level for the test.

**Solution**

The computed value of the test statistic $z$ was $z = -1.35$. Therefore, for this lower-tailed test, the observed significance level is

$$\text{Observed significance level} = P\{z \le -1.35\}$$

This lower-tail area is shown in Figure 6.22. The area between $z = 0$ and $z = 1.35$ is given in Table III in the Appendix as .4115. Therefore, the observed significance level is $.5 - .4115 = .0885$. Note that this probability is quite small. Although we did not reject $H_0$: $p = .05$ at $\alpha = .01$, the probability of observing a $z$ value as small or smaller than $-1.35$ is only .0885 if in fact $H_0$ is true. Therefore, we would reject $H_0$ if we choose $\alpha = .10$ (since the $p$ value is less than .10), and we would not reject $H_0$ (the conclusion of Example 6.9) if we choose $\alpha = .05$.

**FIGURE 6.22**
**THE OBSERVED**
**SIGNIFICANCE LEVEL**
**FOR EXAMPLE 6.10**

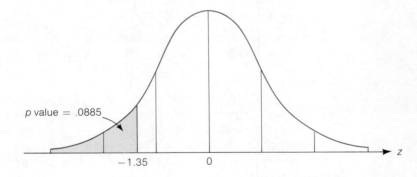

The confidence interval and test of an hypothesis for $p$ in the previous examples are based on the assumption that the sample size $n$ is large enough so that $\hat{p}$ will have an approximately normal sampling distribution (according to the central limit theorem). As a rule of thumb, this condition will be satisfied if the interval $\hat{p} \pm 3\sigma_{\hat{p}}$ does not contain 0 or 1.

Small-sample estimators and test procedures are also available for $p$. These are omitted from our discussion, because most surveys use samples that are large enough to employ the large-sample estimators and tests presented in this section.

Learning the mechanics

**6.58.** A random sample of sixty-four observations is selected from a binomial population with unknown proportion of successes $p$. The computed value of $\hat{p}$ is equal to .45.

    a. Construct a 95% confidence interval for $p$.
    b. Construct a 90% confidence interval for $p$.
    c. Test $H_0$: $p = .5$ against $H_a$: $p \neq .5$. Use $\alpha = .05$.
    d. Test $H_0$: $p = .5$ against $H_a$: $p < .5$. Use $\alpha = .05$.

**6.59.** A random sample of 100 observations is selected from a binomial population with unknown probability of success $p$. The computed value of $\hat{p}$ is equal to .9.

    a. Test $H_0$: $p = .8$ against $H_a$: $p > .8$. Use $\alpha = .01$.
    b. Test $H_0$: $p = .8$ against $H_a$: $p > .8$. use $\alpha = .10$.
    c. Test $H_0$: $p = .85$ against $H_a$: $p \neq .85$. Use $\alpha = .05$.
    d. Form a 95% confidence interval for $p$.
    e. Form a 99% confidence interval for $p$.

**6.60.** Suppose that a random sample of 100 observations from a binomial population gave a value of $\hat{p} = .63$ and that you wish to test the null hypothesis that the population parameter $p$ is equal to .70 against the alternative hypothesis, $p < .70$.

    a. Noting that $\hat{p} = .63$, what does your intuition tell you? Does the value of $\hat{p}$ appear to contradict the null hypothesis?
    b. Use the large-sample $z$ test to test $H_0$: $p = .70$ against the alternative hypothesis, $H_a$: $p < .70$. Use $\alpha = .05$. How do the test results compare with your intuitive decision from part a?

**6.61.** What conditions must be satisfied in order that a large-sample confidence interval for $p$, or large-sample test of hypothesis for $p$, be valid?

Applying the concepts

**6.62.** In July 1977, the United States Congress was considering three bills that would make it illegal for an employer to force an employee to retire at age 65. A government survey was undertaken in order to estimate $p$, the percentage of workers retiring at age 65 who would have preferred to stay on the job. In a sample of 300 elderly Americans, it was found that twenty-one would have preferred to stay on the job. Use this sample to find a 95% confidence interval for $p$. Interpret the interval.

**6.63.** The United States Commission on Crime wishes to estimate the fraction of crimes related to firearms in an area with one of the highest crime rates in the country. The commission randomly selects 600 files of recently committed crimes in the area and finds 380 in which a firearm was reportedly used. Find a 99% confidence interval for $p$, the true fraction of crimes in the area in which some type of firearm was reportedly used.

**6.64.** After losing five games of backgammon in a row, you notice that one die you have been using is flawed. You suspect that the flaw is causing the die to show a 1 or 2 more frequently than would be expected for a fair die. Design a test to confirm (or allay) your suspicion. [*Hint:* Identify the parameter of interest, the null and alternative hypotheses, the sampling method, the test statistic, the rejection region, and explain how you would complete the test of the null hypothesis once the data are collected.]

**6.65.** A recent report stated that only 20% of all college graduates find work in the field of their undergraduate major. A sample of 400 graduates from across the country found 100 working in the field of their undergraduate major.

    a.  Does the sample provide evidence at the $\alpha = .05$ level of significance to indicate the percentage given in the report was too low?

    b.  Find the observed significance level for the test and interpret its value.

**6.66.** A method currently used by doctors to screen women for possible breast cancer fails to detect cancer in 15% of the women who actually have the disease. A new method has been developed which researchers hope will be able to detect cancer more accurately. A random sample of seventy women known to have breast cancer were screened using the new method. Of these, the new method failed to detect cancer in six.

    a.  Do the data provide sufficient evidence to indicate that the new screening method is better than the one currently in use? Test using $\alpha = .05$.

    b.  Find the observed significance level for the test and interpret its value.

**6.67.** A tire manufacturer interested in estimating the proportion of defective automobile tires it produces tested a sample of 490 tires and found twenty-seven to be defective. Find a 95% confidence interval for $p$, the true fraction of defective tires produced by the firm. Interpret this interval.

**6.68.** Interested in how well their new computer billing operation is working, a company statistician samples 400 bills that are ready for mailing and checks them for errors. Twenty-four are found to contain at least one error. Find a 90% confidence interval for $p$, the true proportion of bills that contain errors.

**6.69.** A producer of frozen orange juice claims that 20% or more of all orange juice drinkers prefer its product. To test the validity of this claim, a competitor samples 200 orange juice drinkers and finds that only thirty-three prefer the producer's brand. Does the sample evidence refute the producer's claim? Test at the $\alpha = .10$ level of significance.

**6.70.** Last year a local television station determined that 70% of the people who watch news at 11:00 PM watch its station. The station's management believes that the current audience share may have changed. In an attempt to determine whether the audience share had in fact changed, the station questioned a random sample of eighty local viewers and found that sixty watched its news show. Does the sample evidence support the management's belief? Test at the $\alpha = .10$ level of significance.

**6.71.** Shoplifting is a constant problem for retailers. In past years, a large department store found that one out of twelve people entering its store engaged in shoplifting. To reduce the incidence of shoplifting, the store recently hired more security guards and introduced other methods of surveillance. After instituting this new security program, the store randomly selected 750 shoppers and observed their behavior. Forty were found to be shoplifting. Is there sufficient evidence to conclude that the incidence of shoplifting is lower since the new security measures were introduced? Test using $\alpha = .01$.

**6.72.** The jobless figures compiled by the United States Bureau of the Census showed that in early 1981 16.4% of the labor force of a large industrial city was unemployed. However, some critics believe that these figures underestimate actual unemployment since so-called "discouraged workers," who have given up hope of finding a job, are not counted by the bureau. Suppose that in a random sample of 1,000 members of this city's labor force, 186 are found to be unemployed (some discouraged workers are included). Estimate the unemployment rate for this city when discouraged workers are included. Use a 99% confidence interval.

## 6.6 DETERMINING THE SAMPLE SIZE

When you are planning an experiment with the intention of estimating a population parameter, say a mean $\mu$ or a binomial population proportion $p$, one of the first concerns is sample size. How can the appropriate sample size be determined? To answer this question you must decide how reliable you wish your estimate to be and then you will need to know how many measurements to include in your sample. Consider Example 6.11.

EXAMPLE 6.11

In the hospital example in Section 6.1, we estimated the mean length of time, $\mu$, that patients remain in a hospital. A sample of 100 patients' records produced an estimate, $\bar{x}$, that was within .98 day of $\mu$ with probability equal to .95. Suppose we wish to estimate the true mean to within .5 day with a probability equal to .95. How large a sample would be required?

Solution

For the sample size $n = 100$, we found an approximate 95% confidence interval to be

$$\bar{x} \pm 2\sigma_{\bar{x}} \approx 4.65 \pm .98$$

If we now want our estimator $\bar{x}$ to be within .5 day of $\mu$, we must have

$$2\sigma_{\bar{x}} = .5$$

or

$$\frac{2\sigma}{\sqrt{n}} = .5$$

The necessary sample size is found by solving the above equation for $n$.

To solve for $n$, we need an approximation for $\sigma$, which in this case comes from the original pilot sample of $n = 100$ records, $s = 4.9$ (see Section 6.1). Thus, we have

$$\frac{2\sigma}{\sqrt{n}} \approx \frac{2s}{\sqrt{n}} = .5$$

or

$$\frac{2(4.9)}{\sqrt{n}} = .5$$

$$\sqrt{n} = \frac{2(4.9)}{.5} = 19.6$$

$$n = (19.6)^2$$

$$= 384.16 \approx 384$$

The company will have to sample approximately 384 patients' records in order to estimate the mean length of stay, $\mu$, to within .5 day with probability equal to .95.

A similar argument follows if we want to determine the sample size necessary for estimating a binomial population proportion to within a given bound $B$ with a specified confidence level. The general equations for determining the sample size to estimate both $\mu$ and $p$ are given in the box.

---

SAMPLE SIZE DETERMINATION WITH $100(1 - \alpha)$ PERCENT CONFIDENCE DESIRED

For estimating $\mu$ to within a bound $B$ with probability $(1 - \alpha)$, solve the following equation for $n$:

$$z'_{\alpha/2} \left( \frac{\sigma}{\sqrt{n}} \right) = B$$

The solution is

$$n = \frac{z_{\alpha/2}^2 \sigma^2}{B^2}$$

The value of $\sigma$ substituted into these expressions is obtained from an estimate $s$ of $\sigma$ obtained from a prior sample (pilot study, etc.) or from an educated guess as to the value of the range $R$ of the observations in the population. The value of $\sigma$ is taken to be approximately $R/4$.

- - - - - - - - - - - - - - - - - - - - - - - - - - - - - - - - - - - - - - - - - - - - - -

For estimating $p$ to within a bound $B$ with probability $(1 - \alpha)$, solve the following equation for $n$:

$$z_{\alpha/2} \sqrt{pq/n} = B$$

The solution is

$$n = \frac{z_{\alpha/2}^2 pq}{B^2}$$

The value of $p$ substituted into these expressions is obtained from an estimate based on a prior sample, obtained from an educated guess, or, most conservatively, chosen equal to .5.

---

EXAMPLE 6.12    Refer to Example 6.9 in which a flashbulb manufacturer was making an inference about the fraction defective in a shipment. Suppose the manufacturer wants to estimate the true fraction $p$ to within .01 (i.e., $B = .01$) with a confidence coefficient equal to .90. How large a sample would be needed? (Assume the true $p$ value is near .05.)

Solution    Since we want the error of estimation to be less than $B = .01$ with probability .90, we must have $\alpha = 1 - .90 = .10$. Then, $z_{\alpha/2} = z_{.05} = 1.645$. Substituting these values into the formula for $n$, we get

$$n = \frac{z_{\alpha/2}^2 pq}{B^2} = \frac{(1.645)^2(.05)(.95)}{(.01)^2}$$

$$= 1{,}285.4 \approx 1{,}285$$

Thus, a fairly large sample—about 1,285 bulbs—must be tested if the manufacturer wants to be 90% certain that the estimate of the fraction defective will fall within .01 of the true value of $p$.

Note in Example 6.12 that the manufacturer had prior information concerning the approximate value of the fraction defective, $p$. Knowing that $p$ would tend to be small, the manufacturer assumed a value of $p$ equal to .05 in order to obtain an approximate sample size necessary to estimate its true value. In many sampling situations, you will have no information on the approximate value of $p$. Then you may wish to substitute $p = .5$ into the formula for the sample size. The value of $n$ you obtain will be the largest sample size needed to achieve the specified bound $B$ on the error of estimation, regardless of the true value of $p$.

EXERCISES    Learning the mechanics
6.73.    Find the sample size needed to estimate $\mu$ for each of the following situations:

   a.   Bound is 3;   $\sigma = 40$;   95% confidence is desired
   b.   Bound is 1;   $\sigma = 40$;   95% confidence is desired
   c.   Bound is 3;   $\sigma = 80$;   99% confidence is desired
   d.   Bound is 6;   $\sigma = 80$;   90% confidence is desired

6.74.    Find the sample size needed to estimate $p$ for each of the following situations:

   a.   Bound is .01;   $p$ is near .9;   99% confidence is desired
   b.   Bound is .05;   $p$ is near .9;   99% confidence is desired
   c.   Bound is .02;   $p$ is near .6;   95% confidence is desired
   d.   Bound is .03;   $p$ is near .4;   90% confidence is desired
   e.   Bound is .04;   no prior estimate of $p$ is available;   95% confidence is desired

**6.75.** If you wish to estimate a population mean correct to within .1 with probability .95 and you know, from prior sampling, that $\sigma^2$ is approximately equal to 1.5, how many observations would have to be included in your sample?

**6.76.** Find the approximate sample size necessary to estimate a binomial proportion $p$ correct to within .02 with probability equal to .90.

    a. Assume that you know $p$ is near .7.

    b. Suppose that you have no knowledge of the value of $p$ but you wish to be certain that your sample is large enough to achieve the specified accuracy for the estimate.

**6.77.** Suppose you wish to estimate a population mean correct to within .15 with probability equal to .90. You do not know $\sigma^2$, but you know that the observations will range in value between 39 and 42. Find the approximate sample size that will produce the desired accuracy of the estimate. You wish to be conservative to ensure that the sample size will be ample to achieve the desired accuracy of the estimate. [*Hint:* Using your knowledge of data variation from Section 2.5, assume that the range of the observations will equal $4\sigma$.]

**6.78.** Assume that you wish to estimate a population mean using a 95% confidence interval and that you know from prior information that $\sigma^2 \approx 1$.

    a. To see the effect of the sample size on the width of the confidence interval calculate the width of the confidence interval for $n = 16, 25, 49, 100, 400$.

    b. Plot the widths as a function of sample size $n$ on graph paper. Connect the points by a smooth curve and note how the width decreases as $n$ increases.

## Applying the concepts

**6.79.** A psychologist hired by the National Football League wishes to estimate $\mu$, the mean number of days spent away from home by professional football players during a year. Past records indicate that the distribution of the number of days spent away from home has a standard deviation of 10 days.

    a. How many professional football players should be included in the sample if the researcher wishes to be 90% confident that the estimate is within 2 days of the true value of $\mu$?

    b. To reduce the sample size required, would the psychologist have to increase or decrease the desired confidence?

**6.80.** Before a bill to increase federal price supports for farmers comes before the United States Congress, a Congressman would like to know how nonfarmers feel about the issue. Approximately how many nonfarmers should the Congressman survey in order to estimate the true proportion favoring this bill to within .05 with probability equal to .99?

**6.81.** The owner of a large turkey farm knows from previous years that the largest profit is made by selling turkeys when their average weight is 15 pounds. Approximately how many turkeys must be sampled in order to estimate their true mean weight to within .5 pound with probability equal to .95? Assume prior knowledge indicates that the standard deviation of the weights of the turkeys is approximately 2 pounds.

**6.82.** To estimate the mean age (in months) at which an American Indian child learns to walk, how many Indian children must be sampled if the researcher desires an estimate to be within 1 month of the true mean with 99% confidence? Assume the researcher knows only that the age of walking of these children ranges from 8 to 26 months.

**6.83.** If you want to estimate the proportion of operating automobiles that are equipped with air pollution devices, approximately how large a sample would be required to estimate $p$ to within .02 with probability equal to .95?

**6.84.** Suppose a department store wants to estimate $\mu$, the average age of the customers in its contemporary apparel department, correct to within 2 years with probability equal to .95. Approximately how large a sample would be required? [*Note:* The management does not know $\sigma$, but guesses that the age of its customers ranges from 15 to 45. If you take this range to equal $4\sigma$, you will have a conservative approximation to $\sigma$ that can be used to calculate $n$.]

**6.85.** The EPA standards on the amount of suspended solids that can be discharged into rivers and streams is a maximum of 60 milligrams per liter daily, with a maximum monthly average of 30 milligrams per liter. Suppose you want to test a randomly selected sample of $n$ water specimens in order to estimate the mean daily rate of pollution produced by a mining operation. If you want your estimate correct to within 1 milligram with probability equal to .90, how many water specimens would you have to include in your sample? Assume prior knowledge indicates that pollution readings in water samples taken during a day have a standard deviation equal to 5 milligrams.

**6.86.** Suppose you are a retailer and you want to estimate the proportion of your customers who are shoplifters. You decide to select a random sample of shoppers and check closely to determine whether they steal any mechandise while in the store. Suppose that experience suggests the percentage of shoplifters is near 5%. How many customers should you include in your sample if you want to estimate the proportion of shoplifters in your store correct to within .02 with probability equal to .90?

SUMMARY

The objective of statistics is to make inferences about a population based on information in a sample. In this chapter, we have presented several methods for accomplishing this objective.

The inference-making techniques we discussed are estimation and hypothesis testing. Estimation of a population parameter is accomplished by using an interval estimate with a probability of coverage (confidence coefficient) that is fixed by the experimenter at a high level (usually .90, .95, or .99). On the other hand, when a specific research (alternative) hypothesis about a parameter is tested, the probability $\alpha$ of falsely rejecting the null hypothesis and accepting the research hypothesis is chosen to be small. Thus, we try to control the chance of error in both these inference-making procedures.

One of the most important parameters about which inferences are made is the population mean $\mu$. The sample mean $\bar{x}$ is used for making the inference, but the method depends on the sample size. When the sample size is large (we have specified $n > 30$ as large), the standard normal $z$ statistic is used. The $t$ statistic is employed when $\sigma$ is unknown and a small sample is drawn from a normally (or approximately normally) distributed population.

Another important parameter in practical applications is the binomial population proportion $p$. This probability of success is estimated by the sample fraction of successes, $\hat{p}$, and the $z$ statistic is again used to form confidence intervals or to test an hypothesis.

To conclude this chapter, we showed that the sample size necessary for estimating a population mean $\mu$ or a binomial population proportion $p$ can be determined by specifying the confidence level and the desired bound on the error of the estimate.

[*Note: List the assumptions necessary for the valid implementation of the statistical procedures you use in solving all these exercises.*]

Learning the mechanics

**6.87.** A random sample of twenty observations selected from a normal population produced $\bar{x} = 72.6$ and $s^2 = 19.4$.

a. Form a 90% confidence interval for the population mean.
b. Test $H_0$: $\mu = 80$ against $H_a$: $\mu < 80$. Use $\alpha = .05$.
c. Test $H_0$: $\mu = 80$ against $H_a$: $\mu \neq 80$. Use $\alpha = .01$.
d. Form a 99% confidence interval for $\mu$.
e. How large a sample would be required to estimate $\mu$ to within 1 unit with 95% confidence?

**6.88.** A random sample of $n = 200$ observations from a binomial population yields $\hat{p} = .29$.

a. Test $H_0$: $p = .35$ against $H_a$: $p < .35$. Use $\alpha = .05$.
b. Test $H_0$: $p = .35$ against $H_a$: $p \neq .35$. Use $\alpha = .05$.
c. Form a 95% confidence interval for $p$.
d. Form a 99% confidence interval for $p$.
e. How large a sample would be required to estimate $p$ to within .05 with 99% confidence?

**6.89.** A random sample of 175 measurements possessed a mean $\bar{x} = 8.2$ and a standard deviation $s = .79$.

a. Form a 95% confidence interval for $\mu$.
b. Test $H_0$: $\mu = 8.3$ against $H_a$: $\mu \neq 8.3$. Use $\alpha = .05$.
c. Test $H_0$: $\mu = 8.4$ against $H_a$: $\mu \neq 8.4$. Use $\alpha = .05$.

Applying the concepts

**6.90.** Failure to meet payments on student loans guaranteed by the United States government has been a major problem for both banks and the government. Approximately 50% of all student loans guaranteed by the government are in default. A random sample of 350 loans to college students in one region of the United States indicates that 147 loans are in default.

> a. Do the data indicate that the proportion of student loans in default in this area of the country differs from the proportion of all student loans in the United States that are in default? Use $\alpha = .01$.
> b. Find the observed significance level for the test and interpret its value.

**6.91.** In order to be effective, the mean length of life of a particular mechanical component used in a space craft must be larger than 1,100 hours. Due to the prohibitive cost of the components, only three components can be tested under simulated space conditions. The lifetimes (hours) of the components were recorded and the following statistics were computed: $\bar{x} = 1,173.6$, $s = 36.3$. Do the data provide sufficient evidence to conclude that the component will be effective? Use $\alpha = .01$.

**6.92.** The mean score on a Peace Corps application test, based on many tests conducted over a long period of time, is 80. Ten prospective applicants have taken a course designed to improve their scores on the test. The scores of the ten applicants who completed the course had a mean equal to 86.1 and a standard deviation equal to 12.4.

> a. Do the data provide sufficient evidence to conclude that students taking the course will have a higher mean score than those who do not? Test using $\alpha = .10$.
> b. Find the approximate observed significance level for the test and interpret its value.
> c. What assumptions must be made in order for the procedure that you used in part a to be valid?

**6.93.** A sporting goods manufacturer who produces both white and yellow tennis balls claims that more than 75% of all tennis balls sold are yellow. A marketability study of the purchases of white and yellow tennis balls at a number of stores showed that of 470 cans sold, 410 were yellow and 60 were white.

> a. Is there sufficient evidence to support the manufacturer's claim? Test using $\alpha = .01$.
> b. Find the observed significance level for the test and interpret its value.

**6.94.** During past harvests, a farmer has averaged 68.2 bushels of corn per acre. A new fertilizer has been placed on the market, and after using the new fertilizer the farmer notes the yield of corn for four randomly selected fields of equal size. The mean yield is 72.4 bushels per acre and the standard deviation is 2.2 bushels.

> a. If these data truly represent a random sample of corn yields that the farmer might expect (now and in the future) when using the new fertilizer, do they

suggest that the mean yield of corn per acre has changed from past years? Test using $\alpha = .05$.

b.   Note that the four yield measurements were selected from within the same year. Are these measurements a random sample selected from the population of interest to the farmer? If not, what information do the data provide the farmer?

**6.95.**   The EPA sets a limit of 5 parts per million on PCB (a dangerous substance) in water. A major manufacturing firm producing PCB for electrical insulation discharges small amounts from the plant. The company management, attempting to control the PCB in its discharge, has given instructions to halt production if the mean amount of PCB in the effluent exceeds 3 parts per million. A random sample of fifty water specimens produced the following statistics: $\bar{x} = 3.1$ parts per million, $s = .5$ part per million.

a.   Do these statistics provide sufficient evidence to halt the production process? Use $\alpha = .01$.

b.   If you were the plant manager, would you want to use a large or a small value for $\alpha$ for the test in part a?

**6.96.**   If the rejection of the null hypothesis of a particular test would cause your firm to go out of business, would you want $\alpha$ to be small or large? Explain.

**6.97.**   A company is interested in estimating $\mu$, the mean number of days of sick leave taken by all its employees. The firm's statistician selects at random 100 personnel files and notes the number of sick days taken by each employee. The following sample statistics are computed:

$$\bar{x} = 12.2 \text{ days} \qquad s = 10 \text{ days}$$

a.   Estimate $\mu$ using a 90% confidence interval.

b.   How many personnel files would the statistician have to select in order to estimate $\mu$ to within 2 days with 99% confidence?

**6.98.**   A large mail-order company has placed an order for 5,000 electric can openers with a supplier on condition that no more than 2% of the can openers will be defective. To check the shipment, the company tests a random sample of 400 of the can openers and finds eleven are defective. Does this provide sufficient evidence to indicate that the proportion of defective can openers in the shipment exceeds 2%? Test using $\alpha = .05$.

**6.99.**   Refer to Exercise 6.98. Suppose the company wants to estimate the proportion, $p$, of defective can openers in the shipment correct to within .04 with probability equal to .95. Approximately how large a sample would be required?

**6.100.**   In checking the reliability of a bank's records, auditing firms sometimes ask a sample of the bank's customers to confirm the accuracy of their savings account balances as reported by the bank. Suppose an auditing firm is interested in estimating $p$, the proportion of a bank's savings accounts on whose balances the bank and the customer disagree. Of 200 savings account customers questioned by the auditors, fifteen said their balance disagreed with that reported by the bank.

    a.   Estimate the actual proportion of the bank's savings accounts on whose balances the bank and customer disagree using a 95% confidence interval.

    b.   The bank claims that the true fraction of accounts on which there is disagreement is at most .05. You, as an auditor, doubt this claim. Test the bank's claim at the .10 significance level.

**6.101.**    Refer to Exercise 6.100. How many savings account customers should the auditors question if they want to estimate $p$ to within .02 with probability equal to .95?

**6.102.**    The Chamber of Commerce of a small seaside resort would like to know the mean number of hours of labor required daily to clear litter from its public beach on weekends. A random sample of the labor expended on each of fifteen randomly selected Sunday mornings produced a mean of 3.6 hours and a standard deviation of 0.6 hour. Estimate the mean amount of labor required per Sunday morning to clear the beach of litter. Use a 95% confidence interval.

**6.103.**    The mean grade-point average (GPA) at a certain university was 3.20 in 1974. To show that grade inflation has been reversed, a dean sets out to show that the mean GPA is now lower than 3.20. A random sample of 100 students yields a mean GPA of 3.05 and a standard deviation equal to .90. Do the data provide sufficient evidence to indicate that grade inflation has been reversed? (Test using $\alpha = .10$.)

**6.104.**    The strength of a pesticide dosage is often measured by the proportion of pests that dosage will kill. To determine this proportion for a particular dosage of rat poison, 250 rats are fed the dosage of poison and 215 die. Use a 90% confidence interval to estimate the true proportion of rats that will succumb to the dosage. Interpret your result.

**6.105.**    A meteorologist wishes to estimate the mean amount of snowfall per year in Spokane, Washington. A random sample of the recorded snowfalls for 20 years produces a sample mean equal to 54 inches and a standard deviation of 9.59 inches.

    a.   Estimate the true mean amount of snowfall in Spokane using a 99% confidence interval.

    b.   If you were purchasing snow-removal equipment for a city, what numerical descriptive measure of the distribution of depth of snowfall would be of most interest to you? Would it be the mean?

**6.106.**    A university dean is interested in determining the proportion of students who receive some sort of financial aid. Rather than examine the records for all students, the dean randomly selects 200 students and finds that 118 of them are receiving financial aid. Use a 95% confidence interval to estimate the true proportion of students who receive aid.

**6.107.**    Before approval is given for the use of a new insecticide, the United States Department of Agriculture (USDA) requires that several tests be performed to see how the substance will affect wildlife. In particular, the USDA would like to know the proportion of starlings that will die after being exposed to the insecticide. A random sample of eighty starlings were caught and fed their regular food, which had been treated with the substance. After 10 days, ten starlings had died. Use a 99% confidence interval to estimate the true proportion of starlings that will be killed by the substance.

**6.108.**   Many people think that a national lobby's successful fight against gun control legislation is reflecting the will of a minority of Americans. A random sample of 4,000 citizens yielded 2,250 who are in favor of gun control legislation. Use a 99% confidence interval to estimate the true proportion of Americans who favor gun control legislation. Interpret the result.

**6.109.**   To help consumers assess the risks they are taking, the Food and Drug Administration (FDA) publishes the amount of nicotine found in all commercial brands of cigarettes. A new cigarette has recently been marketed. The FDA tests on this cigarette gave a mean nicotine content of 26.4 milligrams and standard deviation of 2.0 milligrams for a sample of $n = 9$ cigarettes. Estimate the true mean nicotine content per cigarette for the brand using a 95% confidence interval. Interpret the results.

**6.110.**   A health researcher wished to estimate the mean number of cavities per child for children under the age of twelve who live in a specified environment. The number of cavities per child for a random sample of thirty-five children under the age of twelve had a mean of 2 and a standard deviation of 1.7. Construct a 90% confidence interval for the mean number of cavities per child under the age of twelve who lives in the sampled environment.

**ON YOUR OWN . . .**

Choose a population pertinent to your major area of interest that has an unknown mean (or, if the population is binomial, that has an unknown probability of success). For example, a marketing major may be interested in the proportion of consumers who prefer a particular product. A sociology major may be interested in estimating the proportion of people in a certain socioeconomic group or the mean income of people living in a certain part of a city. A political scientist may wish to estimate the proportion of an electorate in favor of a certain candidate, a certain amendment, or a certain presidential policy. A person interested in medicine might want to find the average length of time patients stay in the hospital or the average number of people treated daily in the emergency room. We could continue with examples, but the point should be clear—choose something of interest to you.

Define the parameter you want to estimate and conduct a *pilot study* to obtain an initial estimate of the parameter of interest, and more importantly, an estimate of the variability associated with the estimator. A pilot study is a small experiment (perhaps twenty to thirty observations) used to gain some information about the population of interest. The purpose is to help plan more elaborate future experiments. Based upon the results of your pilot study, determine the sample size necessary to estimate the parameter to within a reasonable bound (of your choice) with a 95% confidence interval.

**REFERENCES**

Environmental Protection Agency. *Environment Midwest,* Sept–Oct. 1976, Region V.

Freedman, D., Pisani, R., & Purves, R. *Statistics.* New York: W. W. Norton, 1978. Chapter 26.

Mendenhall, W. *Introduction to probability and statistics.* 6th ed. Boston: Duxbury, 1983. Chapters 8, 9, and 10.

Snedecor, G. W., & Cochran, W. G. *Statistical methods.* 7th ed. Ames, Iowa: Iowa State University Press, 1980. Chapters 4 and 5.

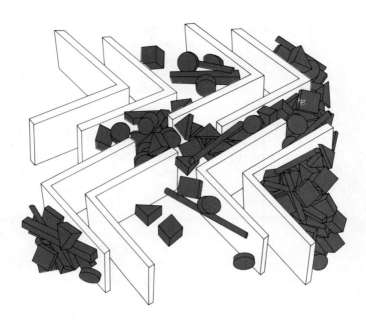

# 7

# COMPARING TWO OR MORE POPULATION MEANS

## WHERE WE'VE BEEN . . .

Two methods for making statistical inferences, estimation and tests of hypotheses, were presented in Chapter 6. Confidence intervals and tests of hypotheses based on single samples were used to make inferences about the nature of sampled populations. Particularly, we gave confidence intervals and tests of hypotheses concerning a population mean $\mu$ and a binomial proportion $p$, and we learned how to select the sample size necessary to obtain a specified amount of information concerning a parameter.

## WHERE WE'RE GOING . . .

Now that we have learned to make inferences about a single population, we will learn how to compare two populations. Such problems often arise in practice. We may wish to compare the mean gas mileages for two models of automobiles, the mean retirement ages of workers in the public and private sectors, or the mean reaction times of men and women to a visual stimulus. We may also wish to compare two population proportions, say the proportions of subjects in a psychological experiment that respond to two different types of stimuli. How to decide whether differences exist and how to estimate the differences between population means will be the subject of this chapter. We will learn how to compare two or more population proportions in Chapter 8.

Many experiments involve a comparison of two or more population means. For example, a sales manager for a steel company may want to estimate the difference in mean sales per customer between two different salespeople. A consumer group may want to test whether two major brands of food freezers differ in the mean amount of electricity they use. We will consider techniques for solving these two-sample inference problems in this chapter.

## 7.1
## LARGE-SAMPLE INFERENCES ABOUT THE DIFFERENCE BETWEEN TWO POPULATION MEANS: INDEPENDENT SAMPLING

Many of the same procedures that are used to estimate and test hypotheses about a single parameter can be modified to make inferences about two parameters. Both the $z$ and $t$ statistics may be adapted to make inferences about the difference between two population means.

In this section we develop the large-sample $z$ statistic for comparing two population means. The $t$ statistic for making small-sample inferences about the difference between two population means is introduced in Section 7.2.

We will use Example 7.1 to introduce the procedures for making large-sample inferences about the difference between two population means.

EXAMPLE 7.1

A dietitian has developed a diet that is low in fats, carbohydrates, and cholesterol. Although the diet was initially intended to be used by people with heart disease, the dietitian wishes to examine the effect this diet has on the weights of obese people. Two random samples of 100 obese people each are selected, and one group of 100 is placed on the low-fat diet. The other 100 are placed on a diet that contains approximately the same quantity of food, but is not as low in fats, carbohydrates, and cholesterol. For each person, the amount of weight lost (or gained) in a 3 week period is recorded. Based on the data given in the table, form a 95% confidence interval for the difference between the population mean weight losses for the two diets.

|  | LOW-FAT DIET | OTHER DIET |
|---|---|---|
| SAMPLE SIZE | 100 | 100 |
| SAMPLE MEAN WEIGHT LOSS | 9.3 pounds | 3.7 pounds |
| SAMPLE VARIANCE | 22.4 | 16.3 |

Solution

Recall that the general form of a large-sample confidence interval for a single mean, $\mu$, is $\bar{x} \pm z_{\alpha/2}\sigma_{\bar{x}}$. That is, we add and subtract $z_{\alpha/2}$ standard deviations of the sample estimate, $\bar{x}$, to the value of the estimate. We will employ a similar procedure to form the confidence interval for the difference between two population means.

Let $\mu_1$ represent the mean of the conceptual population of weight losses for all obese people who could be placed on the low-fat diet. Let $\mu_2$ be similarly defined for the other diet. We wish to form a confidence interval for $(\mu_1 - \mu_2)$. An intuitively appealing estimator for $(\mu_1 - \mu_2)$ is the difference between the sample means, $(\bar{x}_1 - \bar{x}_2)$. Thus, we will form the confidence interval of interest by

7 COMPARING TWO OR MORE POPULATION MEANS

$$(\bar{x}_1 - \bar{x}_2) \pm z_{\alpha/2}\sigma_{(\bar{x}_1 - \bar{x}_2)}$$

In Section 5.5, we noted that

$$\sigma_{(\bar{x}_1 - \bar{x}_2)} = \sqrt{\frac{\sigma_1^2}{n_1} + \frac{\sigma_2^2}{n_2}} \approx \sqrt{\frac{s_1^2}{n_1} + \frac{s_2^2}{n_2}}$$

Using the sample data and noting that $\alpha = .05$, $z_{.025} = 1.96$, we find that the 95% confidence interval is, approximately,

$$(9.3 - 3.7) \pm 1.96 \sqrt{\frac{22.4}{100} + \frac{16.3}{100}} = 5.6 \pm (1.96)(.62)$$

$$= 5.6 \pm 1.22$$

or, (4.38, 6.82). Using this estimation procedure over and over again for different samples, we know that approximately 95% of the confidence intervals formed in the above manner will enclose the difference in population means, $(\mu_1 - \mu_2)$. Therefore, we are reasonably confident that the mean weight loss for the low-fat diet is between 4.38 and 6.82 pounds more than the mean weight loss for the other diet. Based on this information, the dietitian better understands the potential of the low-fat diet as a weight-reducing diet.

The justification for the procedure used in Example 7.1 to estimate $(\mu_1 - \mu_2)$ relies on the properties of the sampling distribution of $(\bar{x}_1 - \bar{x}_2)$. These properties are summarized in the box, and the performance of the estimator in repeated sampling is pictured in Figure 7.1 on page 266.

---

PROPERTIES OF THE SAMPLING DISTRIBUTION OF $(\bar{x}_1 - \bar{x}_2)$

1.  The sampling distribution of $(\bar{x}_1 - \bar{x}_2)$ is approximately normal for large samples.
2.  The mean of the sampling distribution of $(\bar{x}_1 - \bar{x}_2)$ is $(\mu_1 - \mu_2)$.
3.  If the two samples are independent, the standard deviation of the sampling distribution is

$$\sigma_{(\bar{x}_1 - \bar{x}_2)} = \sqrt{\frac{\sigma_1^2}{n_1} + \frac{\sigma_2^2}{n_2}}$$

where $\sigma_1^2$ and $\sigma_2^2$ are the variances of the two populations being sampled, and $n_1$ and $n_2$ are the respective sample sizes.

---

In Example 7.1, the similarity in the procedures for forming a large-sample confidence interval for one population mean and a large-sample confidence interval for the difference between two population means was noted. When testing hypotheses, the procedures are again very similar. The general large-sample procedures for forming confidence intervals and testing hypotheses about $(\mu_1 - \mu_2)$ are summarized in the boxes on page 266.

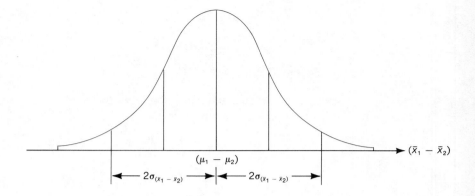

FIGURE 7.1
SAMPLING
DISTRIBUTION OF
$(\bar{x}_1 - \bar{x}_2)$

$(\mu_1 - \mu_2)$

$(\bar{x}_1 - \bar{x}_2)$

$2\sigma_{(\bar{x}_1 - \bar{x}_2)}$          $2\sigma_{(\bar{x}_1 - \bar{x}_2)}$

---

## LARGE-SAMPLE CONFIDENCE INTERVAL FOR $(\mu_1 - \mu_2)$

$$(\bar{x}_1 - \bar{x}_2) \pm z_{\alpha/2}\sigma_{(\bar{x}_1 - \bar{x}_2)} = (\bar{x}_1 - \bar{x}_2) \pm z_{\alpha/2} \sqrt{\frac{\sigma_1^2}{n_1} + \frac{\sigma_2^2}{n_2}}$$

Assumptions:   The two samples are randomly selected in an independent
manner from the two populations. The sample sizes, $n_1$
and $n_2$, are large enough so that $\bar{x}_1$ and $\bar{x}_2$ each have
approximately normal sampling distributions and so
that $s_1^2$ and $s_2^2$ provide good approximations to $\sigma_1^2$ and $\sigma_2^2$.
This will be true if $n_1 \geq 30$ and $n_2 \geq 30$.

---

## LARGE-SAMPLE TEST OF AN HYPOTHESIS FOR $(\mu_1 - \mu_2)$

One-tailed test                     Two-tailed test

$H_0$: $(\mu_1 - \mu_2) = D_0$             $H_0$: $(\mu_1 - \mu_2) = D_0$

$H_a$: $(\mu_1 - \mu_2) < D_0$             $H_a$: $(\mu_1 - \mu_2) \neq D_0$

[or   $H_a$: $(\mu_1 - \mu_2) > D_0$]

where   $D_0 =$ Hypothesized difference between the means (this is
often zero)

Test statistic:   $z = \dfrac{(\bar{x}_1 - \bar{x}_2) - D_0}{\sigma_{(\bar{x}_1 - \bar{x}_2)}}$     Test statistic:   $z = \dfrac{(\bar{x}_1 - \bar{x}_2) - D_0}{\sigma_{(\bar{x}_1 - \bar{x}_2)}}$

where   $\sigma_{(\bar{x}_1 - \bar{x}_2)} = \sqrt{\dfrac{\sigma_1^2}{n_1} + \dfrac{\sigma_2^2}{n_2}}$

Rejection region:   $z < -z_\alpha$             Rejection region:   $z < -z_{\alpha/2}$

[or   $z > z_\alpha$ when                     or   $z > z_{\alpha/2}$

$H_a$: $(\mu_1 - \mu_2) > D_0$]

Assumptions:   Same as for the large-sample confidence interval above.

EXAMPLE 7.2

The management of a restaurant wants to determine whether a new advertising campaign has increased its mean daily income (net). The incomes for each of 50 business days prior to the campaign's beginning are recorded. After conducting the advertising campaign and allowing a 20 day period for the advertising to take effect, the restaurant management records the income for 30 business days. These two samples will allow the management to make an inference about the effect of the advertising campaign on the restaurant's daily income. A summary of the results of the two samples is shown below:

| BEFORE CAMPAIGN | AFTER CAMPAIGN |
|---|---|
| $n_1 = 50$ | $n_2 = 30$ |
| $\bar{x}_1 = \$1,255$ | $\bar{x}_2 = \$1,330$ |
| $s_1 = \$215$ | $s_2 = \$238$ |

Do these samples provide sufficient evidence for the management to conclude that the mean income has been increased by the advertising campaign? Test using $\alpha = .05$.

Solution

We can best answer this question by performing a test of an hypothesis. Defining $\mu_1$ as the mean daily income before the campaign and $\mu_2$ as the mean daily income after the campaign, we will attempt to support the research (alternative) hypothesis that $\mu_2 > \mu_1$ [i.e., that $(\mu_1 - \mu_2) < 0$]. Thus, we will test the null hypothesis, $(\mu_1 - \mu_2) = 0$, rejecting this hypothesis if $(\bar{x}_1 - \bar{x}_2)$ equals a large negative value. The elements of the test are as follows:

$$H_0: \quad (\mu_1 - \mu_2) = 0 \qquad (\text{i.e., } D_0 = 0)$$
$$H_a: \quad (\mu_1 - \mu_2) < 0 \qquad (\text{i.e., } \mu_1 < \mu_2)$$

Test statistic: $\quad z = \dfrac{(\bar{x}_1 - \bar{x}_2) - D_0}{\sigma_{(\bar{x}_1 - \bar{x}_2)}} = \dfrac{(\bar{x}_1 - \bar{x}_2) - 0}{\sigma_{(\bar{x}_1 - \bar{x}_2)}}$

Rejection region: $z < -z_\alpha = -1.645$ (see Figure 7.2)

FIGURE 7.2
REJECTION REGION
FOR ADVERTISING
CAMPAIGN EXAMPLE

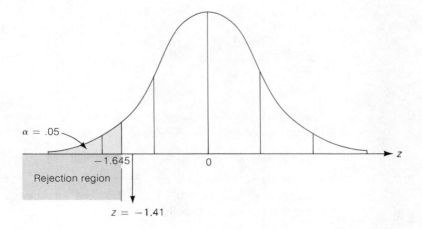

$\alpha = .05$

$-1.645$

$0$

$z$

Rejection region

$z = -1.41$

We now calculate

$$z = \frac{(\bar{x}_1 - \bar{x}_2) - 0}{\sigma_{(\bar{x}_1 - \bar{x}_2)}} = \frac{(1{,}255 - 1{,}330)}{\sqrt{\dfrac{\sigma_1^2}{n_1} + \dfrac{\sigma_2^2}{n_2}}}$$

$$\approx \frac{-75}{\sqrt{\dfrac{s_1^2}{n_1} + \dfrac{s_2^2}{n_2}}} = \frac{-75}{\sqrt{\dfrac{(215)^2}{50} + \dfrac{(238)^2}{30}}} = \frac{-75}{53.03} = -1.41$$

As you can see in Figure 7.2, the calculated $z$ value does not fall in the rejection region. The samples do not provide sufficient evidence, at the $\alpha = .05$ significance level, for the restaurant management to conclude that the advertising campaign has increased the mean daily income.

EXAMPLE 7.3    Find the observed significance level for the test, Example 7.2.

Solution    The alternative hypothesis in Example 7.2, $H_a$: $\mu_1 - \mu_2 < 0$, required a lower one-tailed test. Since the value of the test statistic calculated from the sample data was $z = -1.41$, the observed significance level ($p$ value) for the test is the probability of observing a value of $z$ at least as contradictory to the null hypothesis as $z = -1.41$ if in fact $H_0$ is true, i.e.,

$p$ value $= P\{z \le -1.41\}$

This probability is equal to the shaded area shown in Figure 7.3.

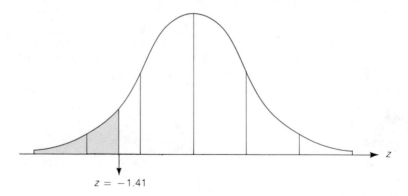

$z = -1.41$

The tabulated area corresponding to $z = 1.41$ (Table III of the Appendix) is .4207. Therefore, the observed significance level for the test is

$p$ value $= .5 - .4207 = .0793$

Therefore, we would reject the null hypothesis (and conclude that $\mu_1 - \mu_2 < 0$) only for values of $\alpha$ larger than .0793. Consistent with the conclusion of Example 7.2, we cannot reject the null hypothesis for $\alpha = .05$.

EXAMPLE 7.4    Find a 95% confidence interval for the difference in mean daily incomes before and after the advertising campaign of Example 7.2 and discuss the implications of the confidence interval.

Solution    The 95% confidence interval for $(\mu_1 - \mu_2)$ is

$$(\bar{x}_1 - \bar{x}_2) \pm z_{\alpha/2} \sqrt{\frac{\sigma_1^2}{n_1} + \frac{\sigma_2^2}{n_2}}$$

Once again, we will substitute $s_1^2$ and $s_2^2$ for $\sigma_1^2$ and $\sigma_2^2$, because these quantities will provide good approximations to $\sigma_1^2$ and $\sigma_2^2$ for samples as large as $n_1 = 50$ and $n_2 = 30$. Then, the 95% confidence interval for $(\mu_1 - \mu_2)$ is

$$(1{,}255 - 1{,}330) \pm 1.96 \sqrt{\frac{(215)^2}{50} + \frac{(238)^2}{30}} = -75 \pm 103.94$$

Thus, we estimate the difference in mean daily income to fall in the interval $-\$178.94$ to $\$28.94$. In other words, we estimate that $\mu_2$, the mean daily income *after* the advertising campaign, could be larger than $\mu_1$, the mean daily income *before* the campaign, by as much as $\$178.94$ per day or it could be less than $\mu_1$ by $\$28.94$ per day.

Now what should the restaurant management do? You can see that the sample sizes collected in the experiment were not large enough to detect a difference in $(\mu_1 - \mu_2)$. To be able to detect a difference (if in fact a difference exists), the management will have to repeat the experiment and increase the sample sizes. This will reduce the width of the confidence interval for $(\mu_1 - \mu_2)$. The restaurant management's best estimate of $(\mu_1 - \mu_2)$ is the point estimate $(\bar{x}_1 - \bar{x}_2) = -\$75$. Thus, the management must decide whether the cost of conducting the advertising campaign is overshadowed by a possible gain in mean daily income estimated at $\$75$ (but which might be as large as $\$178.94$ or could be as low as $-\$28.94$). Based on this analysis, the management will decide whether to continue the experiment or reject the new advertising program as a poor investment.

EXERCISES    Learning the mechanics
**7.1.**    Select two independent random samples: forty observations from population 1 and fifty from population 2. The sample means and variances are shown in the table.

| SAMPLE 1 | SAMPLE 2 |
|---|---|
| $n_1 = 40$ | $n_2 = 50$ |
| $\bar{x}_1 = 3.7$ | $\bar{x}_2 = 4.2$ |
| $s_1^2 = .27$ | $s_2^2 = .33$ |

a.    Form a 95% confidence interval for $(\mu_1 - \mu_2)$, the difference between the means of population 1 and population 2.
b.    Test the null hypothesis $H_0$: $(\mu_1 - \mu_2) = 0$ against the alternative hypothesis $H_a$: $(\mu_1 - \mu_2) \neq 0$. Use $\alpha = .05$.
c.    Find the observed significance level for the test and interpret its value.

**7.2.** Two independent random samples were selected from populations with means $\mu_1$ and $\mu_2$, respectively. The sample sizes, means, and standard deviations are shown in the table.

| SAMPLE 1 | SAMPLE 2 |
|---|---|
| $n_1 = 75$ | $n_2 = 75$ |
| $\bar{x}_1 = 20.3$ | $\bar{x}_2 = 24.4$ |
| $s_1 = 6.1$ | $s_2 = 8.9$ |

a. Form a 90% confidence interval for $(\mu_1 - \mu_2)$.
b. Form a 99% confidence interval for $(\mu_1 - \mu_2)$.
c. Test the null hypothesis $H_0$: $(\mu_1 - \mu_2) = 0$ against the alternative hypothesis $H_a$: $(\mu_1 - \mu_2) < 0$. Use $\alpha = .01$.
d. Find the observed significance level for the test and interpret its value.

**7.3.** Two independent random samples produced the following results:

| SAMPLE 1 | SAMPLE 2 |
|---|---|
| $n_1 = 110$ | $n_2 = 160$ |
| $\bar{x}_1 = 6.8$ | $\bar{x}_2 = 4.7$ |
| $s_1^2 = 93.2$ | $s_2^2 = 117.6$ |

a. Test $H_0$: $(\mu_1 - \mu_2) = 0$ against $H_a$: $(\mu_1 - \mu_2) > 0$. Use $\alpha = .10$.
b. Test $H_0$: $(\mu_1 - \mu_2) = 0$ against $H_a$: $(\mu_1 - \mu_2) \neq 0$. Use $\alpha = .05$.
c. Form a 95% confidence interval for $(\mu_1 - \mu_2)$.

**7.4.** Are the hypothesis test and confidence interval procedures given in this section valid if the sampled populations are not normally distributed? Explain.

Applying the concepts
**7.5.** An experiment has been conducted at a university to compare the mean number of study hours expended per week by student athletes with the mean number of hours expended by nonathletes. A random sample of 55 athletes produced a mean equal to 20.6 hours studied per week and a standard deviation equal to 5.3 hours. A second random sample of 200 nonathletes produced a mean equal to 23.5 hours per week and a standard deviation equal to 4.1 hours.

a. Describe the two populations involved in the comparison.
b. Do the samples provide sufficient evidence to conclude that there is a difference in the mean number of hours of study per week between athletes and nonathletes? Test using $\alpha = .01$.
c. Construct a 99% confidence interval for $(\mu_1 - \mu_2)$.
d. Would a 95% confidence interval for $(\mu_1 - \mu_2)$ be narrower or wider than the one you found in part c? Why?

**7.6.** A new type of band has been developed by a dental laboratory for children who have to wear braces. The new bands are designed to be more comfortable, look better, and hopefully provide more rapid progress in realigning teeth. An experiment was conducted to compare the mean wearing time necessary to correct a specific type of misalignment between the old braces and the new bands. One hundred children were randomly assigned, fifty to each group. A summary of the data is shown in the table.

|   | OLD BRACES | NEW BANDS |
|---|------------|-----------|
| $\bar{x}$ | 410 days | 380 days |
| $s$ | 45 days | 60 days |

a. Is there sufficient evidence to conclude that the new bands do not have to be worn as long as the old braces? Use $\alpha = .01$.

b. Find a 95% confidence interval for the difference in mean wearing times for the two types of braces. Interpret the interval.

**7.7.** Suppose it is desired to compare two physical education training programs for preadolescent girls. A total of eighty girls are randomly selected, with forty assigned to each program. After three 6 week periods on the program, each girl is given a fitness test that yields a score between 0 and 100. The means and variances of the scores for the two groups are shown in the table. Calculate a 99% confidence interval for the true difference in mean fitness scores for girls trained using these two programs.

|   | $n$ | $\bar{x}$ | $s^2$ |
|---|-----|-----------|-------|
| PROGRAM 1 | 40 | 78.7 | 201.6 |
| PROGRAM 2 | 40 | 75.3 | 259.2 |

**7.8.** A distributor of soft drink vending machines knows from experience that the mean number of drinks a machine will sell per day varies according to the location of the machine. At a boys' club, two machines are placed in what the distributor believes to be two different optimum locations. The machines are observed for 30 days, and the number of drinks sold per day for each machine is recorded. The means and standard deviations of the numbers of drinks sold per day at the two locations are given in the table. Based on the data, can the distributor conclude that either location is better than the other? Test at the $\alpha = .05$ level of significance.

|   | MACHINE AT LOCATION 1 | MACHINE AT LOCATION 2 |
|---|----------------------|----------------------|
| $\bar{x}$ | 32.5 | 28.5 |
| $s$ | 6.0 | 5.5 |

**7.9.** An experiment was conducted to compare the yield of two varieties of tomatoes, A and B. Forty plants of each variety were randomly selected and planted within the same field. The yields, recorded in kilograms of tomatoes produced for each plant, possessed means of 10.5 kilograms per plant for variety A and 9.3 kilograms per plant for variety B. The variances for samples A and B were 2.1 and 2.8, respectively. Do the data provide sufficient evidence to conclude there is a difference between the mean weights of tomatoes produced per plant for the two varieties? Test using $\alpha = .05$.

**7.10.** It is often said that economic status is related to the commission of crimes. To test this theory, a sociologist selected a random sample of seventy people (who had no record of criminal conviction) from the census records of a certain city and recorded their annual incomes. Similarly, a random sample of sixty people, each of whom had committed their first crime, was selected from court records and the annual income (prior to arrest) was recorded for each. The means and variances of the annual incomes (in thousands of dollars) for the people in the two groups are shown in the table. Do the data provide sufficient evidence to indicate that the mean income of criminals, prior to committing their first offense, is lower than that for the noncriminal public? Test using $\alpha = .05$.

|  | $\bar{x}$ | $s^2$ |
|---|---|---|
| CRIMINALS | 13.3 | 24.2 |
| NONCRIMINALS | 15.4 | 42.6 |

**7.11.** A large supermarket chain is interested in determining whether a difference exists between the mean shelf-life (in days) of brand S bread and brand H bread. Random samples of fifty freshly baked loaves of each brand were tested, with the following results:

| BRAND S | BRAND H |
|---|---|
| $\bar{x}_1 = 4.1$ | $\bar{x}_2 = 5.2$ |
| $s_1 = 1.2$ | $s_2 = 1.4$ |

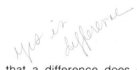

a. Is there sufficient evidence to conclude that a difference does exist between the mean shelf-lives of brand S and brand H bread? Test at the $\alpha = .05$ level.

b. Find the observed significance level for the test and interpret its value.

c. Let $\mu_1$ and $\mu_2$ represent the mean shelf-lives for brands S and H, respectively. Construct a 90% confidence interval for $(\mu_1 - \mu_2)$. Give an interpretation of your confidence interval.

**7.12.** Two manufacturers of corrugated fiberboard each claim that the strength of their product tests on the average at more than 360 pounds per square inch. As a result of consumer complaints, a consumer products testing firm believes that firm A's product is stronger than firm B's. To test its belief, 100 fiberboards were chosen randomly from firm A's inventory and 100 were chosen from firm B's inventory. The following are the results of tests run on the samples:

| A | B |
|---|---|
| $\bar{x}_1 = 365$ | $\bar{x}_2 = 352$ |
| $s_1 = 23$ | $s_2 = 41$ |

a. Does the sample information support the consumer products testing firm's belief? Test at the .05 significance level.

b. What assumptions did you make in conducting the test in part a? Do you think such assumptions could comfortably be made in practice? Why or why not?

c. Find the observed significance level for the test and interpret its value.

**7.13.** Give a practical example where each of the following hypotheses would be appropriate.

a. $H_0$: $(\mu_1 - \mu_2) = 0$ and $H_a$: $(\mu_1 - \mu_2) > 0$
b. $H_0$: $(\mu_1 - \mu_2) = 0$ and $H_a$: $(\mu_1 - \mu_2) < 0$
c. $H_0$: $(\mu_1 - \mu_2) = 0$ and $H_a$: $(\mu_1 - \mu_2) \neq 0$

## 7.2 SMALL-SAMPLE INFERENCES ABOUT THE DIFFERENCE BETWEEN TWO POPULATION MEANS: INDEPENDENT SAMPLING

Suppose a television network wanted to determine whether major sports events or first-run movies attract more viewers in the prime-time hours. It selected twenty-eight prime-time evenings; of these, thirteen had programs devoted to major sports events and the remaining fifteen had first-run movies. The number of viewers (estimated by a television viewer rating firm) was recorded for each program. If $\mu_1$ is the mean number of sports viewers per evening of sports programming and $\mu_2$ is the mean number of movie viewers per evening, we want to detect a difference between $\mu_1$ and $\mu_2$—if such a difference exists. Therefore, we want to test the null hypothesis

$$H_0: \quad (\mu_1 - \mu_2) = 0$$

against the alternative hypothesis

$$H_a: \quad (\mu_1 - \mu_2) \neq 0 \quad \text{(i.e., either } \mu_1 > \mu_2 \text{ or } \mu_2 > \mu_1\text{)}$$

Since the sample sizes are small, estimates of $\sigma_1^2$ and $\sigma_2^2$ will be unreliable and the $z$ test statistic will be inappropriate for the test. But, as in the case of a single mean (Section 6.4), we can construct a Student's $t$ statistic. This statistic (formula to be given subsequently) has the familiar $t$ distribution described in Chapter 6. To use the $t$ statistic, both sampled populations must be approximately normally distributed with equal population variances, and the random samples must be selected independently of each other. The normality and equal variances assumptions would imply relative frequency distributions for the populations that would appear as shown in Figure 7.4.

**FIGURE 7.4**
ASSUMPTIONS FOR THE TWO-SAMPLE $t$: (1) NORMAL POPULATIONS, (2) EQUAL VARIANCES

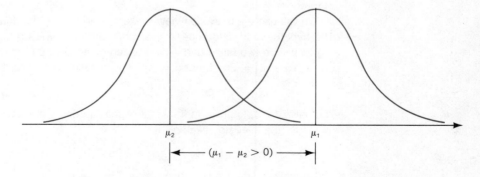

Will these assumptions be satisfied for the television viewing problem? We think that both assumptions will be adequately satisfied in this particular sampling situation. Since we assume the two populations have equal variances ($\sigma_1^2 = \sigma_2^2 = \sigma^2$), it is reasonable to use the information contained in both samples to construct a pooled sample estimator of $\sigma^2$ for use in the $t$ statistic. Thus, if $s_1^2$ and $s_2^2$ are the two sample variances (both estimating the variance $\sigma^2$ common to both populations), the pooled estimator of $\sigma^2$, denoted as $s_p^2$, is

$$s_p^2 = \frac{(n_1 - 1)s_1^2 + (n_2 - 1)s_2^2}{(n_1 - 1) + (n_2 - 1)}$$

$$= \frac{(n_1 - 1)s_1^2 + (n_2 - 1)s_2^2}{n_1 + n_2 - 2}$$

or

$$s_p^2 = \frac{\overbrace{\Sigma (x_1 - \bar{x}_1)^2}^{\substack{\text{From} \\ \text{sample 1}}} + \overbrace{\Sigma (x_2 - \bar{x}_2)^2}^{\substack{\text{From} \\ \text{sample 2}}}}{n_1 + n_2 - 2}$$

where $x_1$ represents a measurement from sample 1 and $x_2$ represents a measurement from sample 2. Recall that the term *degrees of freedom* was defined in Section 6.4 as 1 less than the sample size for each sample, i.e., $(n_1 - 1)$ for sample 1 and $(n_2 - 1)$ for sample 2. Since we are pooling the information on $\sigma^2$ obtained from both samples, the degrees of freedom associated with the pooled variance $s_p^2$ is equal to the sum of the degrees of freedom for the two samples, namely, the denominator of $s_p^2$, i.e., $(n_1 - 1) + (n_2 - 1) = n_1 + n_2 - 2$.

To obtain the small-sample test statistic for testing $H_0$: $(\mu_1 - \mu_2) = D_0$, substitute the pooled estimate of $\sigma^2$ into the formula for the two-sample $z$ statistic (Section 7.1) to obtain

$$t = \frac{(\bar{x}_1 - \bar{x}_2) - D_0}{\sqrt{s_p^2 \left( \frac{1}{n_1} + \frac{1}{n_2} \right)}}$$

It can be shown that this statistic has a $t$ distribution with $(n_1 + n_2 - 2)$ degrees of freedom.

We will use the television viewer example to outline the final steps for this $t$ test: The hypothesized difference in mean number of viewers is $D_0 = 0$. The rejection region will be two-tailed and will be based on a $t$ distribution with $(n_1 + n_2 - 2)$ or $(13 + 15 - 2) = 26$ df. For $\alpha = .05$, the rejection region for the test would be

$$t < -t_{\alpha/2} \qquad \text{or} \qquad t > t_{\alpha/2}$$

The value for $t_{.025}$ given in Table IV of the Appendix is 2.056. Thus, the rejection region for the television example is

$$t < -2.056 \qquad \text{or} \qquad t > 2.056$$

This rejection region is shown in Figure 7.5.

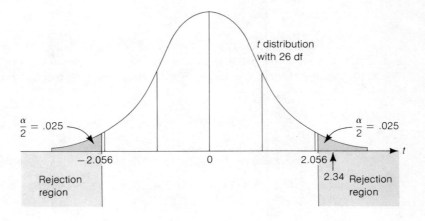

## EXAMPLE 7.5

Suppose the television network's samples produce the results shown below:

| SPORTS | MOVIE |
|---|---|
| $n_1 = 13$ | $n_2 = 15$ |
| $\bar{x}_1 = 6.8$ million | $\bar{x}_2 = 5.3$ million |
| $s_1 = 1.8$ million | $s_2 = 1.6$ million |

Do the data provide sufficient evidence to indicate a difference between the mean numbers of viewers for major sports events and first run movies shown in prime time? Test using $\alpha = .05$.

## Solution

We first calculate

$$s_p^2 = \frac{(n_1 - 1)s_1^2 + (n_2 - 1)s_2^2}{n_1 + n_2 - 2} = \frac{(13 - 1)(1.8)^2 + (15 - 1)(1.6)^2}{13 + 15 - 2}$$

$$= \frac{74.72}{26} = 2.87$$

Then

$$t = \frac{(\bar{x}_1 - \bar{x}_2) - D_0}{\sqrt{s_p^2 \left(\frac{1}{n_1} + \frac{1}{n_2}\right)}} = \frac{(6.8 - 5.3) - 0}{\sqrt{2.87 \left(\frac{1}{13} + \frac{1}{15}\right)}}$$

$$= \frac{1.5}{.64} = 2.34$$

Since the observed value of $t$ ($t = 2.34$) falls in the rejection region (see Figure 7.5), the samples provide sufficient evidence to indicate that the mean numbers of viewers differ for major sports events and first-run movies shown in prime time. Or, we say that the test results are statistically significant at the $\alpha = .05$ level of significance. Because the rejection was in the positive or upper tail of the $t$ distribution, it appears that the mean number of viewers for sports events exceeds that for movies.

EXAMPLE 7.6    Find the approximate observed significance level for the test in Example 7.5.

Solution    The observed significance level ($p$ value) for the test is the probability of observing a value of the test statistic that is *at least* as contradictory to $H_0$: $\mu_1 = \mu_2$ as the value observed, if in fact $H_0$ is true. Thus, since the test was two-sided, we have

$$p \text{ value} = P\{t < -2.34 \quad \text{or} \quad t > 2.34\}$$

$$= 2P\{t > 2.34\} \quad \text{(because the } t \text{ distribution is symmetric about its mean, 0)}$$

Turning to Table IV, Appendix, for 26 degrees of freedom, you can see that

$$P\{t < -2.479 \quad \text{or} \quad t > 2.479\} = 2P\{t > 2.479\} = 2(.01) = .02$$

and

$$P\{t < -2.056 \quad \text{or} \quad t > 2.056\} = 2P\{t > 2.056\} = 2(.025) = .05$$

Since the observed value of $t$ lies between 2.479 and 2.056, the $p$ value for the test lies between .02 and .05. We have agreed to choose the larger of these as the approximate observed significance level, and would thus give the approximate $p$ value for the test as .05. A reader of this reported $p$ value would reject the null hypothesis (and conclude that a difference exists between the mean numbers of sports and movie viewers) for values of $\alpha$ larger than or equal to .05.

The same $t$ statistic can also be used to construct confidence intervals for the difference between population means. Both the confidence interval and the test of hypothesis procedures are summarized in the boxes.

---

SMALL-SAMPLE CONFIDENCE INTERVAL FOR $(\mu_1 - \mu_2)$
(INDEPENDENT SAMPLES)

$$(\bar{x}_1 - \bar{x}_2) \pm t_{\alpha/2} \sqrt{s_p^2 \left( \frac{1}{n_1} + \frac{1}{n_2} \right)}$$

where

$$s_p^2 = \frac{(n_1 - 1)s_1^2 + (n_2 - 1)s_2^2}{n_1 + n_2 - 2}$$

and $t_{\alpha/2}$ is based on $(n_1 + n_2 - 2)$ degrees of freedom.

Assumptions:   1.   Both sampled populations have relative frequency distributions that are approximately normal.
2.   The population variances are equal.
3.   The samples are randomly and independently selected from the populations.

---

SMALL-SAMPLE TEST OF AN HYPOTHESIS FOR $(\mu_1 - \mu_2)$
(INDEPENDENT SAMPLES)

| One-tailed test | Two-tailed test |
|---|---|
| $H_0$: $(\mu_1 - \mu_2) = D_0$ | $H_0$: $(\mu_1 - \mu_2) = D_0$ |
| $H_a$: $(\mu_1 - \mu_2) < D_0$ | $H_a$: $(\mu_1 - \mu_2) \neq D_0$ |
| [or $H_a$: $(\mu_1 - \mu_2) > D_0$] | |

One-tailed test

Test statistic:

$$t = \frac{(\bar{x}_1 - \bar{x}_2) - D_0}{\sqrt{s_p^2\left(\dfrac{1}{n_1} + \dfrac{1}{n_2}\right)}}$$

Rejection region:

$t < -t_\alpha$

[or $t > t_\alpha$ when

$H_a$: $(\mu_1 - \mu_2) > D_0$]

Two-tailed test

Test statistic:

$$t = \frac{(\bar{x}_1 - \bar{x}_2) - D_0}{\sqrt{s_p^2\left(\dfrac{1}{n_1} + \dfrac{1}{n_2}\right)}}$$

Rejection region:

$t < -t_{\alpha/2}$

or $t > t_{\alpha/2}$

where $t_\alpha$ and $t_{\alpha/2}$ are based on $(n_1 + n_2 - 2)$ degrees of freedom.

Assumptions: Same as for the small-sample confidence interval for $(\mu_1 - \mu_2)$ in the previous box (page 276).

EXAMPLE 7.7

Suppose you wish to compare a new method of teaching reading to "slow learners" to the current standard method. You decide to base this comparison upon the results of a reading test given at the end of a learning period of 6 months. Of a random sample of twenty slow learners, eight are taught by the new method and twelve are taught by the standard method. All twenty children are taught by qualified instructors under similar conditions for a 6 month period. The results of the reading test at the end of this period are summarized in the table. Estimate the true mean difference $(\mu_1 - \mu_2)$ between the test scores for the new method and the standard method. Use a 90% confidence interval, and interpret the interval. What assumptions must be made in order that the estimate be valid?

| NEW METHOD | STANDARD METHOD |
|---|---|
| $n_1 = 8$ | $n_2 = 12$ |
| $\bar{x}_1 = 76.9$ | $\bar{x}_2 = 72.7$ |
| $s_1 = 4.85$ | $s_2 = 6.35$ |

Solution

The objective of this experiment is to obtain a 90% confidence interval for $(\mu_1 - \mu_2)$. To use the small-sample confidence interval for $(\mu_1 - \mu_2)$, the following assumptions must be satisfied:

**1.** We assume that the populations of test scores are normally distributed for both the new and the standard methods of instruction. Since a test score can be viewed as a sum of the results on the various components of the test, the central limit theorem lends credence to this assumption.

**2.** The variance of the test scores is assumed to be the same for the two populations. Under the circumstances, we might expect the variation in test scores to be approximately the same for both methods.

**3.** The samples are randomly and independently selected from the two populations. We have randomly chosen twenty different slow learners for the two samples in such a way that the test score for one child is not dependent upon the test score for any other child. Therefore, this assumption would probably be valid.

The first step in constructing the confidence interval is to calculate the pooled estimate of variance

$$s_p^2 = \frac{(n_1 - 1)s_1^2 + (n_2 - 1)s_2^2}{n_1 + n_2 - 2}$$

$$= \frac{(8 - 1)(4.85)^2 + (12 - 1)(6.35)^2}{8 + 12 - 2}$$

$$= 33.7892$$

where $s_p^2$ is based on $(n_1 + n_2 - 2) = (8 + 12 - 2) = 18$ degrees of freedom. Then the 90% confidence interval for $(\mu_1 - \mu_2)$, the difference between mean test scores for the two methods, is

$$(\bar{x}_1 - \bar{x}_2) \pm t_{\alpha/2} \sqrt{s_p^2 \left(\frac{1}{n_1} + \frac{1}{n_2}\right)} = (76.9 - 72.7) \pm t_{.05} \sqrt{33.7892 \left(\frac{1}{8} + \frac{1}{12}\right)}$$

$$= 4.20 \pm 1.734(2.653)$$

$$= 4.20 \pm 4.60$$

or $(-.40, 8.80)$. This means that with a confidence coefficient equal to .90, we estimate the difference in mean test scores between using the new method of teaching and the standard method to fall in the interval $-.40$ to $8.80$. In other words, we estimate the mean test score for the new method to be anywhere from .40 less than to 8.80 more than the mean test score for the standard method. Although the sample means seem to suggest that the new method is associated with a higher mean test score, there is insufficient evidence to indicate that $(\mu_1 - \mu_2)$ differs from 0 [because the interval includes 0 as a possible value for $(\mu_1 - \mu_2)$]. To show a difference in mean test scores (if it exists), you could increase the sample size and thereby narrow the width of the confidence interval for $(\mu_1 - \mu_2)$. An alternative to this is to design the experiment differently. This will be discussed in the next section.

The two-sample $t$ statistic is a powerful tool for comparing population means when the assumptions are satisfied. It has also been shown to retain its usefulness when the sampled populations are only approximately normally distributed. And, when the sample sizes are equal, the assumption of equal population variances can be relaxed. That is, when $n_1 = n_2$, $\sigma_1^2$ and $\sigma_2^2$ can be quite different and the test statistic will still possess, approximately, a Student's $t$ distribution. When the experimental situation does not satisfy the assumptions, you can select larger samples from the populations or you can use other available statistical tests (nonparametric statistical tests, which are described in the references).

EXERCISES

Learning the mechanics

7.14. Independent random samples from normal populations produced these results:

| SAMPLE 1 | SAMPLE 2 |
|----------|----------|
| 2.1 | 3.4 |
| 3.6 | 2.8 |
| 1.4 | 4.1 |
| 3.0 | 3.9 |
| 2.9 | |

a. Calculate the pooled estimate of $\sigma^2$.

b. Do the data provide sufficient evidence to indicate that $\mu_2 > \mu_1$? Test using $\alpha = .10$.

c. Find a 90% confidence interval for $(\mu_1 - \mu_2)$.

7.15. Independent random samples selected from two normal populations produced the sample means and standard deviations shown in the table.

| SAMPLE 1 | SAMPLE 2 |
|----------|----------|
| $n_1 = 15$ | $n_2 = 10$ |
| $\bar{x}_1 = 7.8$ | $\bar{x}_2 = 10.3$ |
| $s_1 = 3.5$ | $s_2 = 5.4$ |

a. Test $H_0$: $(\mu_1 - \mu_2) = 0$ against $H_a$: $(\mu_1 - \mu_2) \neq 0$. Use $\alpha = .05$.

b. Form a 95% confidence interval for $(\mu_1 - \mu_2)$.

7.16. To use the $t$ statistic to test for a difference between the means of two populations, what assumptions must be made about the two populations? About the two samples?

7.17. In the $t$ tests of this section, $\sigma_1^2$ and $\sigma_2^2$ are assumed to be equal. Thus, we say $\sigma_1^2 = \sigma_2^2 = \sigma^2$. Why is a pooled estimator of $\sigma^2$ used instead of either $s_1^2$ or $s_2^2$?

Applying the concepts

**7.18.** An industrial plant wants to determine which of two types of fuel—gas or electric—will produce more useful energy at the lower cost. One measure of economical energy production, called the *plant investment per delivered quad,* is calculated by taking the amount of money (in dollars) invested in the particular utility by the plant, and dividing by the delivered amount of energy (in quadrillion British thermal units). The smaller this ratio, the less an industrial plant pays for its delivered energy.

Random samples of eleven plants using electrical utilities and sixteen plants using gas utilities were taken, and the plant investment/quad was calculated for each. The data produced the results shown in the table.

|  | ELECTRIC | GAS |
|---|---|---|
| SAMPLE SIZE | 11 | 16 |
| MEAN INVESTMENT/QUAD (BILLIONS) | $44.5 | $34.5 |
| VARIANCE | 76.4 | 63.8 |

a. Do these data provide sufficient evidence at the $\alpha = .05$ level of significance to indicate a difference in the average investment/quad between the plants using gas and those using electrical utilities?

b. Find a 90% confidence interval for $(\mu_1 - \mu_2)$. Give a practical interpretation of this interval.

**7.19.** Some statistics students complain that pocket calculators give other students an unfair advantage during statistics examinations. To check this contention, forty-five students were randomly assigned to two groups, twenty-three to use calculators and twenty-two to perform calculations by hand. The students then took a statistics examination that required a modest amount of arithmetic. The means and variances of the test scores for the two groups are shown in the table. Do the data provide sufficient evidence to indicate that students taking *this particular examination* obtain higher scores when using a calculator? Test using $\alpha = .10$.

|  | $n$ | $\bar{x}$ | $s^2$ |
|---|---|---|---|
| CALCULATORS | 23 | 80.7 | 49.5 |
| NO CALCULATORS | 22 | 78.9 | 60.4 |

**7.20.** An experiment is conducted to investigate the effect of a drug on the time to complete a task. Twenty people are divided at random into two groups, ten in each group. One group of ten people is given a placebo, while the second experimental group is administered a drug thought to increase the ability to complete the task quickly. For the control group, the times required to complete the task had a mean of 14.8 minutes and a variance of 3.9; for the experimental group, the average was 12.3 minutes and the variance was 4.3.

a. Test the null hypothesis that the drug has no effect in reducing the mean length of time to complete the task against the alternative hypothesis that the mean time is less for those subjects who receive the drug. Conduct the test at the $\alpha = .10$ level of significance.

b. Find the approximate observed significance level for the test and interpret its value.

**7.21.** To compare two methods of teaching reading, randomly selected groups of elementary school children were assigned to each of the two teaching methods for a 6 month period. The criterion for measuring achievement was a reading comprehension test. The results are shown in the table. Do the data provide sufficient evidence to indicate a difference in mean scores on the comprehension test for the two teaching methods? Test using $\alpha = .05$.

| | NUMBER OF CHILDREN PER GROUP | $\bar{x}$ | $s^2$ |
|---|---|---|---|
| METHOD 1 | 11 | 64 | 52 |
| METHOD 2 | 14 | 69 | 71 |

*small sample*

**7.22.** A manufacturing company is interested in determining whether there is a significant difference between the average number of units produced per day by two different machine operators. A random sample of ten daily outputs was selected for each operator from the outputs over the past year. The data on number of items produced per day are shown below:

| OPERATOR 1 | OPERATOR 2 |
|---|---|
| $n_1 = 10$ | $n_2 = 10$ |
| $\bar{x}_1 = 35$ | $\bar{x}_2 = 31$ |
| $s_1^2 = 17.2$ | $s_2^2 = 19.1$ |

a. Do the samples provide sufficient evidence at the .05 significance level to conclude that a difference does exist between the mean daily outputs of the machine operators?

b. Find a 90% confidence interval for $(\mu_1 - \mu_2)$. Explain clearly the meaning of your confidence interval.

**7.23.** Suppose you are the personnel manager for a company and you suspect a difference in the mean length of work time lost due to sickness for two types of employees: those who work at night versus those who work during the day. Particularly, you suspect that the mean time lost for the night shift exceeds the mean for the day shift. To check your theory, you randomly sample the records for ten employees for each shift category and record the number of days lost due to sickness within the past year. The data are shown in the table at the top of page 282.

| NIGHT SHIFT, 1 | | DAY SHIFT, 2 | |
| --- | --- | --- | --- |
| 21 | 2 | 13 | 18 |
| 10 | 19 | 5 | 17 |
| 14 | 6 | 16 | 3 |
| 33 | 4 | 0 | 24 |
| 7 | 12 | 7 | 1 |
| $\bar{x}_1 = 12.8$ | | $\bar{x}_2 = 10.4$ | |
| $\Sigma x_1^2 = 2,436$ | | $\Sigma x_2^2 = 1,698$ | |

a. Calculate $s_1^2$ and $s_2^2$.

b. Show that the pooled estimate of the common population standard deviation, $\sigma$, is 8.86. Look at the range of the observations within each of the two samples. Does it appear that the estimate, 8.86, is a reasonable value for $\sigma$?

c. If $\mu_1$ and $\mu_2$ represent the mean number of days per year lost due to sickness for the night and day shifts, respectively, test the null hypothesis $H_0$: $\mu_1 = \mu_2$ against the alternative $H_a$: $\mu_1 > \mu_2$. Use $\alpha = .05$. Do the data provide sufficient evidence to indicate that $\mu_1 > \mu_2$?

d. What assumptions must be satisfied so that the $t$ test from part c is valid?

**7.24.** Suppose your plant purifies its liquid waste and discharges the water into a local river. An EPA inspector has collected water specimens of the discharge of your plant and also water specimens in the river upstream from your plant. Each water specimen is divided into five parts, the bacteria count is read on each, and the mean count for each specimen is reported. The average bacteria count readings for each of six specimens are reported below for the two locations:

| PLANT DISCHARGE | UPSTREAM |
| --- | --- |
| 30.1 | 29.7 |
| 36.2 | 30.3 |
| 33.4 | 26.4 |
| 28.2 | 27.3 |
| 29.8 | 31.7 |
| 34.9 | 32.3 |

a. Why might the bacteria count readings shown above tend to be approximately normally distributed?

b. Do the data provide sufficient evidence to indicate that the mean of the bacteria count for the discharge exceeds the mean of the count upstream? Use $\alpha = .05$.

c. Find the approximate observed significance level for the test and interpret its value.

7.3
INFERENCES
ABOUT THE
DIFFERENCE
BETWEEN TWO
POPULATION
MEANS: PAIRED
DIFFERENCE
EXPERIMENTS

In Example 7.7 we compared two methods of teaching reading to slow learners by means of a 90% confidence interval. Suppose it is possible to measure the slow learners' "reading IQ's" *before* they are subjected to a teaching method. Eight pairs of slow learners with similar reading IQ's are found and one member of each pair is randomly assigned to the standard teaching method while the other is assigned to the new method. Do the data in Table 7.1 support the hypothesis that the population mean reading test score for slow learners taught by the new method is greater than the mean reading test score for those taught by the standard method?

TABLE 7.1
READING TEST
SCORES FOR
EIGHT PAIRS OF
SLOW LEARNERS

| PAIR | NEW METHOD | STANDARD METHOD |
|------|-----------|-----------------|
| 1 | 77 | 72 |
| 2 | 74 | 68 |
| 3 | 82 | 76 |
| 4 | 73 | 68 |
| 5 | 87 | 84 |
| 6 | 69 | 68 |
| 7 | 66 | 61 |
| 8 | 80 | 76 |
| | $\bar{x}_1 = 76.0$ | $\bar{x}_2 = 71.625$ |
| | $s_1^2 = 48.0$ | $s_2^2 = 49.1$ |

We want to test

$$H_0: \ (\mu_1 - \mu_2) = 0 \qquad H_a: \ (\mu_1 - \mu_2) > 0$$

If we use the two-sample $t$ statistic (Section 7.2), we first calculate

$$s_p^2 = \frac{(n_1 - 1)s_1^2 + (n_2 - 1)s_2^2}{n_1 + n_2 - 2}$$

$$= \frac{(8 - 1)(48.0) + (8 - 1)(49.1)}{8 + 8 - 2}$$

$$= 48.55$$

and then the test statistic

$$t = \frac{(\bar{x}_1 - \bar{x}_2) - 0}{\sqrt{s_p^2\left(\frac{1}{n_1} + \frac{1}{n_2}\right)}} = \frac{76.0 - 71.625}{\sqrt{48.55\left(\frac{1}{8} + \frac{1}{8}\right)}} = \frac{4.375}{3.485} = 1.26$$

This small $t$ value will not lead to rejection of $H_0$ when compared to the $t$ distribution with $n_1 + n_2 - 2 = 14$ df, even if $\alpha$ is chosen as large as .10 ($t_{.10} = 1.345$). Thus, based on *this* analysis we might conclude that there is insufficient evidence to infer a difference in the mean test scores for the two methods.

However, if you carefully examine the data in Table 7.1, you will find this result difficult to accept. The test score of the new method is larger than the corresponding test score for the standard method for every one of the eight pairs of slow learners. This, in itself, seems to provide strong evidence to indicate that $\mu_1$ exceeds $\mu_2$. Why, then, did the $t$ test fail to detect this difference? The answer is: The two-sample $t$ is not a valid procedure to use with this set of data.

The two-sample $t$ is inappropriate because the assumption of independent samples is invalid. We have randomly chosen pairs of test scores, and thus, once we have chosen the sample for the new method, we have *not* independently chosen the sample for the standard method. The dependence between observations within pairs can be seen by examining the pairs of test scores, which tend to rise and fall together as we go from pair to pair. This pattern provides strong visual evidence of a violation of the assumption of independence required for the two-sample $t$ test of Section 7.2. In this particular situation, you will note the large variation within samples (reflected by the large value of $s_p^2$) in comparison to the relatively small difference between the sample means. Because $s_p^2$ was so large, the $t$ test of Section 7.2 was unable to detect a difference between $\mu_1$ and $\mu_2$.

We now consider a valid method of analyzing the data of Table 7.1. In Table 7.2 we add the column of differences between the test scores of the pairs of slow learners.

**TABLE 7.2**

| PAIR | NEW METHOD | STANDARD METHOD | DIFFERENCE (NEW METHOD − STANDARD METHOD) |
|------|-----------|-----------------|-------------------------------------------|
| 1 | 77 | 72 | 5 |
| 2 | 74 | 68 | 6 |
| 3 | 82 | 76 | 6 |
| 4 | 73 | 68 | 5 |
| 5 | 87 | 84 | 3 |
| 6 | 69 | 68 | 1 |
| 7 | 66 | 61 | 5 |
| 8 | 80 | 76 | 4 |

$$\bar{x}_D = 4.375$$
$$s_D = 1.69$$

We can regard these differences in test scores as a random sample of differences for all pairs (matched on reading IQ) of slow learners, past and present. Then we can use this sample to make inferences about the mean of the population of differences, $\mu_D$, which is equal to the difference $(\mu_1 - \mu_2)$, i.e., the mean of the population (and sample) of differences equals the difference between the population (and sample) means. Thus, our test becomes

$$H_0: \quad \mu_D = 0 \qquad (\text{i.e.,} \quad \mu_1 - \mu_2 = 0)$$
$$H_a: \quad \mu_D > 0 \qquad (\text{i.e.,} \quad \mu_1 - \mu_2 > 0)$$

The test statistic is a one-sample $t$, since we are now analyzing a single sample of differences.

$$\text{Test statistic:} \quad t = \frac{\bar{x}_D - 0}{s_D / \sqrt{n_D}}$$

where

$\bar{x}_D$ = Sample mean difference

$s_D$ = Sample standard deviation of differences

$n_D$ = Number of differences = Number of pairs

Assumptions: The population of differences in test scores is approximately normally distributed. The sample differences are randomly selected from the population of differences.

[*Note:* We do not need to assume that $\sigma_1^2 = \sigma_2^2$.]

Rejection region: At significance level $\alpha = .05$, we will reject $H_0$ if $t > t_{.05}$, where $t_{.05}$ is based on $(n_D - 1)$ degrees of freedom.

Referring to Table IV in the Appendix, we find the $t$ value corresponding to $\alpha = .05$ and $n_D - 1 = 8 - 1 = 7$ df to be $t_{.05} = 1.895$. Then we will reject the null hypothesis if $t > 1.895$. Note that the number of degrees of freedom has decreased from $n_1 + n_2 - 2 = 14$ to 7 by using the paired difference experiment rather than the two independent random samples design. Now calculate

$$t = \frac{\bar{x}_D - 0}{s_D / \sqrt{n_D}} = \frac{4.375}{1.69 / \sqrt{8}} = 7.32$$

Because this value of $t$ falls in the rejection region, we conclude that the mean test score for slow learners taught by the new method exceeds the mean score for those taught by the standard method. We have 95% confidence in this conclusion (since $\alpha = .05$).

This kind of experiment, in which observations are paired and the differences are analyzed, is called a paired difference experiment. In many cases, a paired difference experiment can provide more information about the difference between population means than an independent samples experiment. The idea is to compare population means by comparing the differences between pairs of experimental units (objects, people, etc.) that were very similar prior to the experiment. The differencing removes sources of variation that tend to inflate $\sigma^2$. For example, when two children are taught to read by two different methods, the observed difference in achievement may be due to a difference in the effectiveness of the two teaching methods *or* it may be due to differences in the initial reading levels and IQ's of the two children (random error). To reduce the effect of differences in the children on the observed differences in reading achievement, the two methods of reading are imposed on two children who are more likely to possess similar intellectual potentials, namely children with nearly equal IQ's. The effect of this pairing is to remove the larger source of variation that would be present if children with different abilities were randomly assigned to the two samples. Making comparisons within groups of similar experimental units is called blocking, and the paired difference experiment is a simple example of a randomized block experiment. In our example, pairs of children with matching IQ scores represent the blocks.

Some other examples for which the paired difference experiment might be appropriate are the following:

**1.** Suppose you want to estimate the difference $(\mu_1 - \mu_2)$ in mean price per gallon between two major brands of premium gasoline. If you choose two independent random samples of stations for each brand, the variability in price due to geographical location may be large. To eliminate this source of variability you could choose pairs of stations of similar size, one station for each brand, in close geographical proximity and use the sample of differences between the prices of the brands to make an inference about $(\mu_1 - \mu_2)$.

**2.** Suppose a college placement center wants to estimate the difference $(\mu_1 - \mu_2)$ in mean starting salaries for men and women graduates who seek jobs through the center. If it independently samples men and women, the starting salaries may vary due to their different college majors and differences in grade-point averages. To eliminate these sources of variability, the placement center could match male and female jobseekers according to their majors and grade-point averages. Then the differences between the starting salaries of each pair in the sample could be used to make an inference about $(\mu_1 - \mu_2)$.

**3.** Suppose you wish to estimate the difference $(\mu_1 - \mu_2)$ in mean absorption rate into the bloodstream for two drugs that relieve pain. If you independently sample people, the absorption rates might vary due to factors such as age, weight, sex, blood pressure, etc. In fact, there are many possible sources of nuisance variability, and pairing individuals who are similar in all the possible sources would be quite difficult. However, it may be possible to obtain two measurements *on the same person.* First, we administer one of the two drugs and record the time until absorption. After a sufficient amount of time, the other drug is administered and a second measurement on absorption time is obtained. The differences between the measurements for each person in the sample could then be used to estimate $(\mu_1 - \mu_2)$. This procedure would be advisable only if the amount of time allotted between drugs is sufficient to guarantee little or no carryover effect. Otherwise, it would be better to use different people matched as closely as possible on the factors thought to be most important.

The one- and two-tailed hypothesis testing procedures and the method of forming confidence intervals for the difference between two means using a paired difference experiment are summarized in the boxes.

---

PAIRED DIFFERENCE CONFIDENCE INTERVAL

$$\bar{x}_D \pm t_{\alpha/2}(s_D / \sqrt{n_D})$$

where $t_{\alpha/2}$ is based on $(n_D - 1)$ degrees of freedom.

Assumptions:  1.  The relative frequency distribution of the population of differences is normal.

                2.  The sample differences are randomly selected from the population of differences.

---

## PAIRED DIFFERENCE TEST OF AN HYPOTHESIS

**One-tailed test**

$H_0$: $(\mu_1 - \mu_2) = D_0$,
     i.e., $(\mu_D = D_0)$

$H_a$: $(\mu_1 - \mu_2) < D_0$,
     i.e., $(\mu_D < D_0)$

     [or   $H_a$: $(\mu_1 - \mu_2) > D_0$,
     i.e., $(\mu_D > D_0)$]

Test statistic:   $t = \dfrac{\bar{x}_D - D_0}{s_D / \sqrt{n_D}}$

Rejection region:
   $t < -t_\alpha$
   [or   $t > t_\alpha$
   when   $H_a$: $(\mu_1 - \mu_2) > D_0$]

**Two-tailed test**

$H_0$: $(\mu_1 - \mu_2) = D_0$,
     i.e., $(\mu_D = D_0)$

$H_a$: $(\mu_1 - \mu_2) \neq D_0$,
     i.e., $(\mu_D \neq D_0)$

Test statistic:   $t = \dfrac{\bar{x}_D - D_0}{s_D / \sqrt{n_D}}$

Rejection region:
   $t < -t_{\alpha/2}$
   or   $t > t_{\alpha/2}$

where $t_\alpha$ and $t_{\alpha/2}$ are based on $(n_D - 1)$ degrees of freedom

Assumptions:   1.   The relative frequency distribution of the population of differences is normal.
                2.   The differences are randomly selected from the population of differences.

**EXAMPLE 7.8**

A paired difference experiment is conducted to compare the starting salaries of male and female college graduates who find jobs. Pairs are formed by choosing a male and a female with the same major and similar grade-point averages. Suppose a random sample of ten pairs is formed in this manner and the starting annual salary of each person is recorded. The results are shown in Table 7.3. Test to see whether there is evidence that the mean starting salary, $\mu_1$, for males exceeds the mean starting salary, $\mu_2$, for females. Use $\alpha = .05$.

**TABLE 7.3**

| PAIR | MALE | FEMALE | DIFFERENCE (MALE − FEMALE) |
|------|------|--------|----------------------------|
| 1 | $14,300 | $13,800 | $ 500 |
| 2 | 16,500 | 16,600 | −100 |
| 3 | 15,400 | 14,800 | 600 |
| 4 | 13,500 | 13,500 | 0 |
| 5 | 18,500 | 17,600 | 900 |
| 6 | 12,800 | 13,000 | −200 |
| 7 | 14,500 | 14,200 | 300 |
| 8 | 16,200 | 15,100 | 1,100 |
| 9 | 13,400 | 13,200 | 200 |
| 10 | 14,200 | 13,500 | 700 |

Solution

The elements of the paired difference test are

$$H_0: \quad \mu_D = 0 \qquad (\mu_1 - \mu_2 = 0)$$
$$H_a: \quad \mu_D > 0 \qquad (\mu_1 - \mu_2 > 0)$$

Note that we propose a one-sided research hypothesis, since we are interested in determining whether the data indicate that $\mu_1$ exceeds $\mu_2$, i.e., male mean starting salary exceeds female mean starting salary.

Test statistic: $\quad t = \dfrac{\bar{x}_D - 0}{s_D / \sqrt{n_D}}$

Assumptions: 1. The relative frequency distribution for the population of differences is normal.
2. The sample differences are randomly selected from the population.

Since the test is upper-tailed, we will reject $H_0$ if $t > t_\alpha$, where $t_{.05} = 1.833$ is based upon $(n_D - 1) = 9$ degrees of freedom. The rejection region is shown in Figure 7.6.

FIGURE 7.6
REJECTION REGION
FOR EXAMPLE 7.8

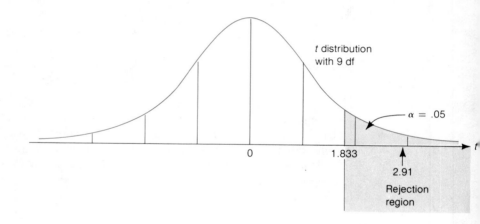

We now calculate

$$\Sigma\, x_D = 500 + (-100) + \cdots + 700 = 4{,}000$$

and

$$\Sigma\, x_D^2 = 3{,}300{,}000$$

Then,

$$\bar{x}_D = \frac{\Sigma\, x_D}{10} = \frac{4{,}000}{10} = 400$$

$$s_D^2 = \frac{\Sigma (x_D - \bar{x}_D)^2}{n_D - 1} = \frac{\Sigma x_D^2 - (\Sigma x_D)^2/10}{9}$$

$$= \frac{3{,}300{,}000 - (4{,}000)^2/10}{9} = 188{,}888.89$$

$$s_D = \sqrt{s_D^2} = 434.61$$

Substituting these values into the formula for the test statistic, we find that

$$t = \frac{\bar{x}_D - 0}{s_D/\sqrt{n_D}} = \frac{400}{434.61/\sqrt{10}} = \frac{400}{137.44} = 2.91$$

As you can see in Figure 7.6, the calculated $t$ falls in the rejection region. Thus, we conclude at the $\alpha = .05$ level of significance that the mean starting salary for males exceeds the mean starting salary for females.

One measure of the amount of information about $(\mu_1 - \mu_2)$ gained by using a paired difference experiment rather than an independent samples experiment in Example 7.8 is the relative widths of the confidence intervals obtained by the two methods. A 95% confidence interval for $(\mu_1 - \mu_2)$ using the paired difference experiment is

$$\bar{x}_D \pm t_{\alpha/2}\frac{s_D}{\sqrt{n_D}} = 400 \pm t_{.025}\frac{434.61}{\sqrt{10}}$$

$$= 400 \pm 2.262\frac{434.61}{\sqrt{10}}$$

$$= 400 \pm 310.88$$

$$\approx 400 \pm 311 = (\$89, \$711)$$

If we analyzed the same data as though it were an independent samples experiment,* we would first calculate the following quantities:

| MALES | FEMALES |
|---|---|
| $n_1 = 10$ | $n_2 = 10$ |
| $\bar{x}_1 = \$14{,}930$ | $\bar{x}_2 = \$14{,}530$ |
| $s_1^2 = 3{,}009{,}000$ | $s_2^2 = 2{,}331{,}222.22$ |

Then

$$s_p^2 = \frac{(n_1 - 1)s_1^2 + (n_2 - 1)s_2^2}{n_1 + n_2 - 2} = \frac{9(3{,}009{,}000) + 9(2{,}331{,}222.22)}{10 + 10 - 2}$$

$$= 2{,}670{,}111.11$$

$$s_p = \sqrt{s_p^2} = 1{,}634.05$$

*This is done only to provide a measure of the increase in the amount of information obtained by a paired design in comparison to an unpaired design. Actually, if an experiment is designed using pairing, an unpaired analysis would be invalid because the assumption of independent samples would not be satisfied.

The 95% confidence interval is

$$(\bar{x}_1 - \bar{x}_2) \pm t_{.025} \sqrt{s_p^2 \left(\frac{1}{n_1} + \frac{1}{n_2}\right)} = 400 \pm (2.101) \sqrt{2{,}670{,}111.11 \left(\frac{1}{10} + \frac{1}{10}\right)}$$

$$= 400 \pm 1{,}535.35$$
$$\approx 400 \pm 1{,}535$$
$$= (-\$1{,}135, \$1{,}935)$$

The confidence interval for the independent sampling experiment is about five times wider than for the corresponding paired difference confidence interval. Blocking out the variability due to differences in majors and grade-point averages significantly increases the information about the difference in male and female mean starting salaries by providing a much more accurate (smaller confidence interval for the same confidence coefficient) estimate of $(\mu_1 - \mu_2)$.

You may wonder whether conducting a paired difference experiment is always superior to an independent samples experiment. The answer is: Most of the time, but not always. We sacrifice half of the degrees of freedom in the $t$ statistic when a paired difference design is used instead of an independent samples design. This is a loss of information, and unless this loss is more than compensated for by the reduction in variability obtained by blocking (pairing), the paired difference experiment will result in a net loss of information about $(\mu_1 - \mu_2)$. Thus, we should be convinced that the pairing will significantly reduce variability before performing the paired difference experiment. Most of the time this will happen.

*One final note:* The pairing of the observations is determined *before* the experiment is performed (that is, by the *design* of the experiment). A paired difference experiment is *never* obtained by pairing the sample observations after the measurements have been acquired.

CASE STUDY 7.1
MATCHED PAIRING
IN STUDYING THE
MENTALLY RETARDED

The statistical implications underlying matching procedures have frequently been overlooked in educational research with the mentally retarded population. It is, therefore, the purpose of this paper to point out some of the advantages and disadvantages of different matching procedures.

Stainback and Stainback (1973) describe a number of experimental situations in which blocking (matching) subjects before experimentation might reduce variability and thereby increase the amount of information obtained. One of the matching procedures discussed by the authors can be summarized as follows: Suppose it is desired to form two groups of mentally retarded subjects to compare two methods of educational therapy. Subjects could be randomly selected from an existing (large) group of subjects and ordered (from lowest to highest) on the scores of an appropriate matching variable. Several variables suggested by the authors are "pre-test measures of the experimental criterion, measures of learning rate (mental age and Intelligence Quotients), chronological age, personal characteristics (sex, race), environmental conditions (socioeconomic level), or combinations of two or more variables." The two highest-ranking subjects on the matching variable(s) would then form pair 1, the next two pair 2, etc., and one member from each pair would be randomly assigned to each therapy group. This experiment is a practical example of a paired difference experiment, and if the matching variable(s) are correctly chosen, the responses of subjects will be more homogeneous within pairs than between pairs. The authors conclude:

> It is important that when comparing two groups on a criterion variable the two groups be as equal as possible on all relevant factors excepting only the independent variables. This consideration deserves particular emphasis in the area of mental retardation since the researchers are constantly dealing with diverse groups. It should be restated, therefore, that researchers in the area of mental retardation should become acutely aware of advantages and disadvantages of matching procedures and their alternatives.

EXERCISES

Learning the mechanics

**7.25.** The data for a random sample of six paired observations are shown in the table.

| PAIR | SAMPLE FROM POPULATION 1 Observation 1 | SAMPLE FROM POPULATION 2 Observation 2 |
|------|------|------|
| 1 | 6 | 3 |
| 2 | 2 | 0 |
| 3 | 5 | 7 |
| 4 | 10 | 6 |
| 5 | 8 | 5 |
| 6 | 4 | 2 |

a. Calculate the difference between each pair of observations by subtracting observation 2 from observation 1. Use the differences to calculate $\bar{x}_D$ and $s_D^2$.

b. If $\mu_1$ and $\mu_2$ are the means of populations 1 and 2, respectively, express $\mu_D$ in terms of $\mu_1$ and $\mu_2$.

c. Form a 95% confidence interval for $\mu_D$.

d. Test the null hypothesis $H_0$: $\mu_D = 0$ against the alternative hypothesis $H_a$: $\mu_D \neq 0$. Use $\alpha = .05$.

**7.26.** The data for a random sample of ten paired observations are shown in the table at the top of page 292.

| PAIR | SAMPLE FROM POPULATION 1 | SAMPLE FROM POPULATION 2 |
|------|--------------------------|--------------------------|
| 1 | 50 | 55 |
| 2 | 55 | 61 |
| 3 | 47 | 50 |
| 4 | 63 | 68 |
| 5 | 51 | 54 |
| 6 | 48 | 54 |
| 7 | 50 | 53 |
| 8 | 47 | 52 |
| 9 | 65 | 65 |
| 10 | 50 | 56 |

a. Do the data provide sufficient evidence to indicate that $\mu_2$, the mean of population 2, is larger than $\mu_1$? (Test $H_0$: $\mu_D = 0$ against $H_a$: $\mu_D < 0$, where $\mu_D = \mu_1 - \mu_2$.) Test using $\alpha = .05$.

b. Form a 90% confidence interval for $\mu_D$.

Applying the concepts

7.27. A new weight-reducing technique, consisting of a liquid protein diet, is currently undergoing tests by the Food and Drug Administration (FDA) before its introduction into the market. A typical test performed by the FDA is the following: The weights of a random sample of five people are recorded before they are introduced to the liquid protein diet. The five individuals are then instructed to follow the liquid protein diet for 3 weeks. At the end of this period, their weights (in pounds) are again recorded. The results are listed in the table. Let $\mu_1$ be the true mean weight of individuals before starting the diet and let $\mu_2$ be the true mean weight of individuals after 3 weeks on the diet. Construct a 95% confidence interval for the difference between the true mean weights before and after the diet is used. What assumptions are necessary to ensure the validity of the procedure you used?

| PERSON | WEIGHT BEFORE DIET | WEIGHT AFTER DIET |
|--------|--------------------|--------------------|
| 1 | 150 | 143 |
| 2 | 195 | 190 |
| 3 | 188 | 185 |
| 4 | 197 | 191 |
| 5 | 204 | 200 |

7.28. A company is interested in hiring a new secretary. Several candidates are interviewed and the choice is narrowed to two possibilities. The final choice will be based on typing ability. Six letters were randomly selected from the company's files and each candidate is required to type each of the six letters. The number of words typed per minute is recorded for each candidate—letter combination. These data are shown in the table at the top of page 293.

a. Do the data provide sufficient evidence to indicate a difference between the mean numbers of words typed per minute by the two candidates? Test using $\alpha = .10$.

| LETTER | CANDIDATE | |
|--------|-----------|-----|
|        | 1 | 2 |
| 1 | 62 | 59 |
| 2 | 60 | 60 |
| 3 | 65 | 61 |
| 4 | 58 | 57 |
| 5 | 59 | 55 |
| 6 | 64 | 60 |

b.  Find the approximate observed significance level for the test and interpret its value.

c.  Find a 90% confidence interval for the difference between the mean typing rates of the two candidates.

**7.29.**  An assertiveness training course has just been added to the services offered by a college counseling center. To measure its effectiveness, ten students are given a test at the beginning of the course and again at the end. A high score on the test implies high assertiveness. The test scores are shown in the table. Do the data provide sufficient evidence to conclude that people are more assertive after taking the course? Use $\alpha = .025$. What assumptions did you make?

| STUDENT | BEFORE | AFTER |
|---------|--------|-------|
| 1 | 50 | 65 |
| 2 | 62 | 68 |
| 3 | 51 | 52 |
| 4 | 41 | 43 |
| 5 | 63 | 60 |
| 6 | 56 | 70 |
| 7 | 49 | 48 |
| 8 | 67 | 69 |
| 9 | 42 | 53 |
| 10 | 57 | 61 |

**7.30.**  A farmer was interested in determining which of two soil fumigants, A or B, is more effective in controlling the number of parasites in a particular agricultural crop. To compare the fumigants, four small fields were divided into equal areas: fumigant A was applied to one part and fumigant B to the other. Crop samples of equal size were taken from each of the eight plots and the numbers of parasites per square foot were counted. The data are shown in the table. Do the data provide sufficient evidence to indicate a difference in the mean level of parasites for the two fumigants?

| FIELD | A | B |
|-------|-----|---|
| 1 | 15 | 9 |
| 2 | 5 | 3 |
| 3 | 8 | 6 |
| 4 | 8 | 4 |

**7.31.** In the past, many bodily functions were thought to be beyond conscious control. However, recent experimentation suggests that it may be possible for a person to control certain body functions if that person is trained in a program of *biofeedback* exercises. An experiment is conducted to show that blood pressure levels can be consciously reduced in people trained in this program. The blood pressure measurements (in millimeters of mercury) listed in the table represent readings before and after the biofeedback training of six subjects.

| SUBJECT | BEFORE | AFTER |
|---------|--------|-------|
| 1 | 136.9 | 130.2 |
| 2 | 201.4 | 180.7 |
| 3 | 166.8 | 149.6 |
| 4 | 150.0 | 153.2 |
| 5 | 173.2 | 162.6 |
| 6 | 169.3 | 160.1 |

a. Is there sufficient evidence to conclude that the mean blood pressure after people are trained in this program is less than the mean blood pressure before training? Use $\alpha = .05$.

b. Find the approximate observed significance level for the test and interpret its value.

**7.32.** A manufacturer of automobile shock absorbers was interested in comparing the durability of its shocks with that of the biggest competitor. To make the comparison, one of the manufacturer's and one of the competitor's shocks were randomly selected and installed on the rear wheels of each of six cars. After the cars had been driven 20,000 miles, the strength of each test shock was measured, coded, and recorded. The following are the results of the examination:

| CAR NUMBER | MANUFACTURER'S | COMPETITOR'S |
|------------|----------------|--------------|
| 1 | 8.8 | 8.4 |
| 2 | 10.5 | 10.1 |
| 3 | 12.5 | 12.0 |
| 4 | 9.7 | 9.3 |
| 5 | 9.6 | 9.0 |
| 6 | 13.2 | 13.0 |

a. Do the data present sufficient evidence to conclude there is a difference in the mean strength of the two types of shocks after 20,000 miles of use? Let $\alpha = .05$.

b. What assumptions are necessary in order to apply a paired difference analysis to the data?

c. Construct a 95% confidence interval for $(\mu_1 - \mu_2)$. Interpret the meaning of your confidence interval.

**7.33.** Suppose the data in Exercise 7.32 are based on independent random samples.

a. Do the data provide sufficient evidence to indicate a difference between the mean strengths for the two types of shocks? Let $\alpha = .05$.

b. Construct a 95% confidence interval for $(\mu_1 - \mu_2)$. Interpret your result.

c. Compare the confidence intervals you obtained in Exercise 7.32 and part b of this exercise. Which is larger? To what do you attribute the difference in size? Assuming in each case that the appropriate assumptions are satisfied, which interval provides you with more information about $(\mu_1 - \mu_2)$? Explain.

d. Are the results of an unpaired analysis valid when the data have been collected from a paired experiment?

7.34. A company has five plants that each produce the same product. The number of units produced at one of the plants on a randomly selected day was recorded. This was done for all five plants. After the physical arrangement of the assembly line at each plant was modified, the sampling procedure was repeated. The sampling results are contained in the following table:

| PLANT NUMBER | BEFORE | AFTER |
|---|---|---|
| 1 | 90 | 93 |
| 2 | 94 | 96 |
| 3 | 91 | 92 |
| 4 | 85 | 88 |
| 5 | 88 | 90 |

a. Are the two samples independent? Explain.

b. Do the data present sufficient evidence to conclude that there is a difference in the mean daily output of the company's plants before and after the change in their assembly line structure?

c. Construct a 90% confidence interval for $(\mu_1 - \mu_2)$. Interpret the interval.

7.35. The *National Survey for Professional, Administrative, Technical, and Clerical Pay* is conducted annually to determine whether federal pay scales are commensurate with private sector salaries. The government and private workers in the study are matched as closely as possible before the salaries are compared. Suppose that the following represent annual salaries for twelve pairs of individuals in the sample matched on job level and experience:

| PAIR | PRIVATE | GOVERNMENT |
|---|---|---|
| 1 | $16,250 | $15,280 |
| 2 | 28,990 | 27,170 |
| 3 | 18,850 | 19,240 |
| 4 | 49,900 | 38,810 |
| 5 | 27,040 | 27,950 |
| 6 | 24,960 | 23,920 |
| 7 | 20,540 | 18,860 |
| 8 | 22,750 | 23,270 |
| 9 | 30,260 | 27,830 |
| 10 | 54,730 | 56,160 |
| 11 | 21,840 | 19,790 |
| 12 | 18,820 | 18,460 |

a. Use these data to place a 99% confidence interval on the difference between the mean salaries of the private and government sectors.

b. What assumptions are necessary for the validity of the procedure you used in part a?

## 7.4 DETERMINING THE SAMPLE SIZE

You can find the appropriate sample size to estimate the difference between a pair of population means with a specified degree of reliability by using the method described in Section 6.6. That is, to estimate the difference between a pair of means correct to within $B$ units with probability $(1 - \alpha)$, let $z_{\alpha/2}$ standard deviations of the sampling distribution of the estimator equal $B$. Then solve for the sample size. To do this, you have to solve the problem for a specific ratio between $n_1$ and $n_2$. Most often, you will want to have equal sample sizes, i.e., $n_1 = n_2 = n$. We will illustrate the procedure with two examples.

### EXAMPLE 7.9

New fertilizer compounds are often advertised with the promise of increased yields. Suppose we want to compare the mean yield $\mu_1$ of wheat when a new fertilizer is used to the mean yield $\mu_2$ with a fertilizer in common use. The estimate of the difference in mean yield per acre is to be correct to within .25 bushel with a confidence coefficient of .95. If the sample sizes are to be equal, find $n_1 = n_2 = n$, the number of 1 acre plots of wheat assigned to each fertilizer.

### Solution

To solve the problem, you need to know something about the variation in the bushels of yield per acre. Suppose that from past records you know the yields of wheat possess a range of approximately 10 bushels per acre. You could then approximate $\sigma_1 = \sigma_2 = \sigma$ by letting the range equal $4\sigma$. Thus,

$$4\sigma \approx 10 \text{ bushels}$$

$$\sigma \approx 2.5 \text{ bushels}$$

The next step is to solve the equation

$$z_{\alpha/2}\sigma_{(\bar{x}_1 - \bar{x}_2)} = B$$

or

$$z_{\alpha/2}\sqrt{\frac{\sigma_1^2}{n_1} + \frac{\sigma_2^2}{n_2}} = B$$

for $n$, where $n = n_1 = n_2$. Since we want the estimate to lie within $B = .25$ of $(\mu_1 - \mu_2)$ with confidence coefficient equal to .95, $z_{\alpha/2} = z_{.025} = 1.96$. Then, letting $\sigma_1 = \sigma_2 = 2.5$ and solving for $n$, we have

$$1.96\sqrt{\frac{(2.5)^2}{n} + \frac{(2.5)^2}{n}} = .25$$

$$1.96\sqrt{\frac{2(2.5)^2}{n}} = .25$$

$$n = 768.32 \approx 768$$

Consequently, you will have to sample 768 acres of wheat for each fertilizer to estimate the difference in mean yield per acre to within .25 bushel. Since this would necessitate extensive and costly experimentation, you might decide to allow a larger bound (say, $B = .50$ or $B = 1$) in order to reduce the sample size, or you might decrease the confidence coefficient. The point is, we can obtain an idea of the experimental effort necessary to achieve a specified precision in our final estimate by determining the approximate sample size *before* the experiment is begun.

EXAMPLE 7.10    A laboratory manager wishes to compare the difference in the mean readings of two instruments, A and B, designed to measure the potency (in parts per million) of an antibiotic. To conduct the experiment, the manager plans to select $n_D$ specimens of the antibiotic from a vat and to measure each specimen with both instruments. The difference $(\mu_A - \mu_B)$ will be estimated based on the $n_D$ paired differences $(x_A - x_B)$ obtained in the experiment. If preliminary measurements suggest that the differences will range between plus or minus 10 parts per million, how many differences will be needed to estimate $(\mu_A - \mu_B)$ correct to within 1 part per million with confidence coefficient equal to .99?

Solution    The estimator for $(\mu_A - \mu_B)$, based on a paired difference experiment, is $\bar{x}_D = (\bar{x}_A - \bar{x}_B)$ and

$$\sigma_{\bar{x}_D} = \frac{\sigma_D}{\sqrt{n}}$$

Thus, the number $n_D$ of pairs of measurements needed to estimate $(\mu_A - \mu_B)$ to within 1 part per million can be obtained by solving for $n_D$ in the equation

$$z_{\alpha/2} \frac{\sigma_D}{\sqrt{n_D}} = B$$

where $z_{.005} = 2.58$ and $B = 1$. To solve this equation for $n_D$, we need to have an approximate value for $\sigma_D$.

We are given the information that the differences are expected to range from $-10$ to 10 parts per million. Letting the range equal $4\sigma_D$, we find

$$\text{Range} = 20 \approx 4\sigma_D$$

$$\sigma_D \approx 5$$

Substituting $\sigma_D = 5$, $B = 1$, and $z_{.005} = 2.58$ into the equation and solving for $n_D$, we obtain

$$2.58 \frac{5}{\sqrt{n_D}} = 1$$

$$n_D = [(2.58)(5)]^2$$

$$= 166.41$$

Therefore, it will require approximately $n_D = 166$ pairs of measurements to estimate $(\mu_A - \mu_B)$ correct to within 1 part per million using the paired difference experiment.

Learning the mechanics

**7.36.** Find the appropriate values of $n_1$ and $n_2$ (assume $n_1 = n_2$) needed to estimate $(\mu_1 - \mu_2)$ to within:

    a. A bound on the error of estimation equal to 3 with 95% confidence. From prior experience it is known that $\sigma_1 \approx 12$ and $\sigma_2 \approx 15$.

    b. A bound on the error of estimation equal to 5 with 99% confidence. The range of each population is 40.

    c. A bound on the error of estimation equal to .4 with 90% confidence. Assume that $\sigma_1^2 \approx 3.4$ and $\sigma_2^2 \approx 5.3$.

**7.37.** One reason high school seniors are encouraged to attend college is that the job opportunities are much better for those with college degrees than for those without. A high school counselor wants to estimate the difference in mean income per day between high school graduates who have a college education and those who have not gone on to college. Suppose it is decided to compare the daily incomes using pairs of 30 year old graduates who have been matched according to their high school curricula and grade-point averages. If the range in daily income within pairs is approximately $200 per day, how many pairs of graduates should be sampled in order to estimate the true difference between mean daily incomes correct to within $10 per day with probability .9?

**7.38.** Suppose you are interested in the growth rate of dividends. Consider investing $1,000 in a stock and suppose you want to estimate the dividend rate on your $1,000 investment at the end of 5 years. Particularly, you want to compare two types of stocks, electrical utilities and oil companies. To conduct your study, you plan to randomly select $n$ oil stocks and $n$ electrical utility stocks. For each stock, you will check the records, calculate the number of shares of stock you could have purchased 5 years ago for $1,000, and then calculate the dividend rate (in percent) that the stock would be paying today on your $1,000 investment. Suppose you think the dividend rates will vary over a range of roughly 25%. How large should $n$ be if you want to estimate the difference between the mean rates of dividend return correct to within 3% with probability equal to .95?

**7.5
COMPARING
MORE THAN TWO
POPULATION
MEANS:
AN ANALYSIS OF
VARIANCE
(OPTIONAL)**

More than two population means may be compared by using an analysis of variance (ANOVA). The logic of an analysis of variance can be seen by examining Table 7.4. Suppose that you drew independent random samples of five observations from each of three normally distributed populations and that the data appeared as shown in part A of Table 7.4. Do the data provide sufficient evidence to indicate that the population means, $\mu_1$, $\mu_2$, $\mu_3$, differ? We think you will agree that there is sufficient evidence to indicate differences among the population means. This is because the differences among the sample means are large in comparison to the variation of the $x$ values within the samples (which is zero for this illustration). In contrast, suppose the sample data appeared as shown in part B of Table 7.4. Do you think there is sufficient evidence to indicate differences among the population means, $\mu_1$, $\mu_2$ and $\mu_3$? We think your answer will be "no." Although the sample means in part B are identical to those shown

**TABLE 7.4**
**A VISUAL COMPARISON OF SAMPLE MEANS**

| A. | SAMPLE | | | B. | SAMPLE | | |
|---|---|---|---|---|---|---|---|
| | 1 | 2 | 3 | | 1 | 2 | 3 |
| | 5 | 2 | 7 | | 15 | −16 | 7 |
| | 5 | 2 | 7 | | −18 | 9 | 25 |
| | 5 | 2 | 7 | | 23 | 19 | −16 |
| | 5 | 2 | 7 | | 12 | −10 | −12 |
| | 5 | 2 | 7 | | −7 | 8 | 31 |
| | $\bar{x}_1 = 5$ | $\bar{x}_2 = 2$ | $\bar{x}_3 = 7$ | | $\bar{x}_1 = 5$ | $\bar{x}_2 = 2$ | $\bar{x}_3 = 7$ |

in part A of Table 7.4, the differences among the sample means are relatively small in comparison to the very large amount of variation in the $x$ values within the respective samples. Thus, the logic of an analysis of variance is clear. We will conclude that there is sufficient evidence of differences among population means if the variation of the sample means is large in relation to the variation of the $x$ values within their respective samples.

The analysis of variance test discussed here is based on the assumption that independent random samples of $n_1, n_2, \ldots, n_k$ observations are selected from $k$ normally distributed populations that possess a common variance $\sigma^2$. Thus, the sampling situation and the assumptions are identical to those given in Section 7.2 for the comparison of two population means. A summary of the notation is shown in Table 7.5.

**TABLE 7.5**
**SUMMARY NOTATION FOR AN ANALYSIS OF VARIANCE: INDEPENDENT SAMPLING**

| | POPULATIONS (TREATMENTS) | | | |
|---|---|---|---|---|
| | 1 | 2 | 3 . . . k | |
| MEAN | $\mu_1$ | $\mu_2$ | $\mu_3 \cdots \mu_k$ | |
| VARIANCE | $\sigma^2$ | $\sigma^2$ | $\sigma^2 \ldots \sigma^2$ | |
| | INDEPENDENT RANDOM SAMPLES | | | |
| | 1 | 2 | 3 . . . k | |
| SAMPLE SIZE | $n_1$ | $n_2$ | $n_3 \ldots n_k$ | |
| SAMPLE TOTALS | $T_1$ | $T_2$ | $T_3 \ldots T_k$ | |
| SAMPLE MEANS | $\bar{x}_1$ | $\bar{x}_2$ | $\bar{x}_3 \ldots \bar{x}_k$ | |

Total number of measurements $= n = n_1 + n_2 + n_3 + \cdots + n_k$

Sum of all $n$ measurements $= \Sigma x$

Mean of all $n$ measurements $= \bar{\bar{x}}$

Sum of squares of all $n$ measurements $= \Sigma x^2$

To test the null hypothesis

$H_0$:  $\mu_1 = \mu_2 = \cdots = \mu_k$

i.e., that all the population means are equal, against the alternative hypothesis

$H_a$:  At least two of the population means differ

we calculate a measure of the variation within treatment means, called mean square for treatments,* where

$$MST = \frac{SST}{k - 1}$$

The quantity $k$ is equal to the number of population means and SST, called the sum of squares for treatments, is

$$SST = \sum_{i=1}^{k} [n_i (\bar{x}_i - \bar{\bar{x}})^2]$$

The larger the deviations of the individual sample means about the overall mean, $\bar{\bar{x}}$, the larger will be the value of SST and, consequently, MST.

The variation within samples is the mean square for error,

$$MSE = \frac{SSE}{n - k}$$

where SSE is the sum of squared errors and is calculated as

$$SSE = \Sigma (x_i - \bar{x}_1)^2 + \Sigma (x_i - \bar{x}_2)^2 + \cdots + \Sigma (x_i - \bar{x}_k)^2$$

This is the sum of the sum of squares of deviations of the sample $x$ values about their respective sample means. Therefore,

$$MSE = s_p^2 = \frac{\Sigma (x_1 - \bar{x}_1)^2 + \Sigma (x_2 - \bar{x}_2)^2 + \cdots + \Sigma (x_k - \bar{x}_k)^2}{n - k}$$

$$= \frac{(n_1 - 1)s_1^2 + (n_2 - 1)s_2^2 + \cdots + (n_k - 1)s_k^2}{n - k}$$

is a pooled estimator of $\sigma^2$, which is simply an extension of the formula given in Section 7.2.[†]

The test statistic used to test $H_0: \mu_1 = \mu_2 = \cdots = \mu_k$ compares the variation among sample means (measured by MST) with the variation of the $x$ values within samples (measured by MSE). This test statistic,

$$F = \frac{MST}{MSE}$$

will be large when MST is large relative to the value of MSE. Thus, the rejection

---

*The term *treatment* originates from methods where the populations that are being compared represent experimental units subjected to different types of treatments. For example, the populations might correspond to the yields of a variety of wheat subjected to different types (treatments) of fertilizer.

[†]For $k > 2$, MSE is an extension of the pooled estimator of $\sigma^2$ discussed in Section 7.2. For the two-sample case,

$$s_p^2 = \frac{\Sigma (x_1 - \bar{x}_1)^2 + \Sigma (x_2 - \bar{x}_2)^2}{n_1 + n_2 - 2}$$

The numerator of $s_p^2$ is SSE.

region for the test will always consist of *large* values of $F$, i.e., values of $F$ larger than some value, say $F_\alpha$.

In practice, the quantities SST and SSE can be calculated more easily using the shortcut computational formulas shown in the box. In Example 7.11, we will use these formulas to calculate an $F$ statistic. The analysis of variance $F$ test will be illustrated in Example 7.12.

---

FORMULAS FOR THE ANALYSIS OF VARIANCE CALCULATIONS: INDEPENDENT RANDOM SAMPLING

$$CM = \text{Correction for mean}$$

$$= \frac{(\text{Total of all observations})^2}{\text{Total number of observations}}$$

$$= \frac{(\Sigma x)^2}{n} = \frac{(T_1 + \cdots + T_k)^2}{n}$$

$$SS(\text{Total}) = \text{Total sum of squares}$$

$$= (\text{Sum of squares of all observations}) - CM$$

$$= \Sigma x^2 - CM$$

$$SST = \text{Sum of squares for treatments}$$

$$= \left( \begin{array}{c} \text{Sum of squares of treatment totals with} \\ \text{each square divided by the number of} \\ \text{observations for that treatment} \end{array} \right) - CM$$

$$= \frac{T_1^2}{n_1} + \frac{T_2^2}{n_2} + \cdots + \frac{T_k^2}{n_k} - CM$$

$$SSE = \text{Sum of squares for error}$$

$$= SS(\text{Total}) - SST$$

$$MST = \text{Mean square for treatments} = \frac{SST}{k - 1}$$

$$MSE = \text{Mean square for error} = \frac{SSE}{n - k}$$

$$F = \text{Test statistic} = \frac{MST}{MSE}$$

where $k$ is the total number of treatments and $n$ is the total number of observations.

---

EXAMPLE 7.11

An animal nutritionist wished to compare the mean weight gain of pigs fed on three different diets. The data for this experiment are given in Table 7.6 (page 302). Perform the calculations required to obtain the $F$ statistic.

| | DIET A | DIET B | DIET C |
|---|---|---|---|
| TABLE 7.6 | 30.1 | 35.5 | 27.8 |
| DIET: | 31.6 | 31.3 | 26.1 |
| WEIGHT GAINS | 25.4 | 32.7 | 27.3 |
| (IN KILOGRAMS) | 33.3 | 30.0 | 29.2 |
| | 29.0 | 31.5 | 30.5 |
| | 30.9 | 33.6 | 28.4 |
| | 28.5 | 34.1 | 29.4 |
| | 29.3 | 32.9 | 23.6 |
| | 31.2 | 33.2 | 31.3 |
| | 32.7 | 27.7 | 27.7 |
| Totals | 302.0 | 322.5 | 281.3 |

Solution

From the table, the totals for the three samples are $T_1 = 302.0$, $T_2 = 322.5$, and $T_3 = 281.3$.

$$\Sigma x = T_1 + T_2 + T_3 = 905.8$$
$$\Sigma x^2 = (30.1)^2 + (31.6)^2 + \cdots + (27.7)^2$$
$$= 27{,}570.18$$

Then, following the order of calculations listed in the box, we find

$$CM = \frac{(\Sigma x)^2}{n} = \frac{(905.8)^2}{30} = 27{,}349.121$$

$$SS(Total) = \Sigma x^2 - CM$$
$$= 27{,}570.18 - 27{,}349.121$$
$$= 221.059$$

$$SST = \frac{T_1^2}{n_1} + \frac{T_2^2}{n_2} + \frac{T_3^2}{n_3} - CM$$
$$= \frac{(302.0)^2}{10} + \frac{(322.5)^2}{10} + \frac{(281.3)^2}{10} - 27{,}349.121$$
$$= 27{,}433.994 - 27{,}349.121$$
$$= 84.873$$

$$SSE = SS(Total) - SST$$
$$= 221.059 - 84.873$$
$$= 136.186$$

$$MST = \frac{SST}{k - 1} = \frac{84.873}{2} = 42.437$$

$$MSE = \frac{SSE}{n - k} = \frac{136.186}{27} = 5.044$$

Finally, the value of the $F$ statistic is

7 COMPARING TWO OR MORE POPULATION MEANS

$$F = \frac{MST}{MSE} = \frac{42.437}{5.044} = 8.41$$

**EXAMPLE 7.12**

Do the data in Example 7.11 provide sufficient evidence to indicate a difference in mean weight gains for pigs fed on diets A, B, and C?

**Solution**

We wish to test the null hypothesis

$$H_0: \quad \mu_1 = \mu_2 = \mu_3$$

i.e., that the mean gains in weight for the three diets are equal, against the alternative hypothesis

$$H_a: \quad \text{At least two of the three means differ}$$

The test statistic, $F = MST/MSE$, will have a sampling distribution that will depend upon the number of degrees of freedom associated with MST and MSE, the quantities that appear in the numerator and the denominator, respectively, of the $F$ statistic. These will be $\nu_1 = k - 1$ numerator df and $\nu_2 = n - k$ denominator df.

The rejection region for the test will be large values of $F$, i.e., values that exceed $F_\alpha$, where $F_\alpha$ locates an area $\alpha$ in the upper tail of the $F$ distribution. The values of $F_\alpha$ corresponding to $\alpha = .10, .05, .025,$ and $.01$ are given in Appendix Tables V, VI, VII, and VIII, respectively. Table VI, for $\alpha = .05$, is partially reproduced in Figure 7.7.

**FIGURE 7.7**
**REPRODUCTION OF PART OF TABLE VI IN THE APPENDIX: $\alpha = .05$**

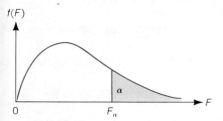

| $\nu_2$ \ $\nu_1$ | 1 | 2 | 3 | 4 | 5 | 6 | 7 | 8 | 9 |
|---|---|---|---|---|---|---|---|---|---|
| 25 | 4.24 | 3.39 | 2.99 | 2.76 | 2.60 | 2.49 | 2.40 | 2.34 | 2.28 |
| 26 | 4.23 | 3.37 | 2.98 | 2.74 | 2.59 | 2.47 | 2.39 | 2.32 | 2.27 |
| 27 | 4.21 | 3.35 | 2.96 | 2.73 | 2.57 | 2.46 | 2.37 | 2.31 | 2.25 |
| 28 | 4.20 | 3.34 | 2.95 | 2.71 | 2.56 | 2.45 | 2.36 | 2.29 | 2.24 |
| 29 | 4.18 | 3.33 | 2.93 | 2.70 | 2.55 | 2.43 | 2.35 | 2.28 | 2.22 |
| 30 | 4.17 | 3.32 | 2.92 | 2.69 | 2.53 | 2.42 | 2.33 | 2.27 | 2.21 |
| 40 | 4.08 | 3.23 | 2.84 | 2.61 | 2.45 | 2.34 | 2.25 | 2.18 | 2.12 |
| 60 | 4.00 | 3.15 | 2.76 | 2.53 | 2.37 | 2.25 | 2.17 | 2.10 | 2.04 |
| 120 | 3.92 | 3.07 | 2.68 | 2.45 | 2.29 | 2.17 | 2.09 | 2.02 | 1.96 |
| $\infty$ | 3.84 | 3.00 | 2.60 | 2.37 | 2.21 | 2.10 | 2.01 | 1.94 | 1.88 |

NUMERATOR DEGREES OF FREEDOM

DENOMINATOR DEGREES OF FREEDOM

The values of $\nu_1$ are shown across the top of a table, and the values of $\nu_2$ are given in the column at the left of the table. For $\nu_1 = k - 1 = 3 - 1 = 2$ and $\nu_2 = n - k = 30 - 3 = 27$, Table VI gives $F_{.05} = 3.35$. Therefore, the rejection region for the test is $F \geq 3.35$ (see Figure 7.8).

Since the computed value of $F$ found in Example 7.11, $F = 8.41$, exceeds the tabulated value, $F_{.05} = 3.35$, it falls in the rejection region. Consequently, there is sufficient evidence to indicate a difference among the mean weight gains for the three diets.

A confidence interval for a single population mean or the difference between a pair of means is constructed in the same way that we constructed a confidence interval for a single population mean in Section 6.4, or for a pair of population means in Section 7.2:

---

CONFIDENCE INTERVALS FOR MEANS

Single population mean (say, population $i$): $\quad \bar{x}_i \pm t_{\alpha/2} \dfrac{s_p}{\sqrt{n_i}}$

Difference between two population means (say, populations $i$ and $j$):

$$(\bar{x}_i - \bar{x}_j) \pm t_{\alpha/2}\, s_p \sqrt{\frac{1}{n_i} + \frac{1}{n_j}}$$

where $s_p = \sqrt{\text{MSE}}$ and $t_{\alpha/2}$ is the tabulated value of $t$ (Table IV in the Appendix) that locates $\alpha/2$ in the upper tail of the $t$ distribution and has $(n - k)$ degrees of freedom (the degrees of freedom associated with error in the analysis of variance).

---

EXAMPLE 7.13

Find a 95% confidence interval for the difference $(\mu_A - \mu_B)$ in Example 7.11.

Solution

The 95% confidence interval for the difference in the mean gains in weight for diets A and B is

$$(\bar{x}_A - \bar{x}_B) \pm t_{.025} s_p \sqrt{\frac{1}{n_A} + \frac{1}{n_B}}$$

where

$$s_p = \sqrt{\text{MSE}} = \sqrt{5.044} = 2.246$$

based on $n - k = 27$ df, we find $t_{.025} = 2.052$ in Table IV, and $n_A = n_B = 10$. Substituting into the formula for the confidence interval, we obtain

$$(30.20 - 32.25) \pm (2.052)(2.246)\sqrt{\frac{1}{10} + \frac{1}{10}} = -2.05 \pm 2.06$$

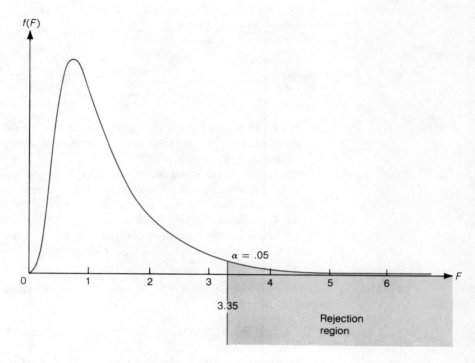

Therefore, our interval estimate of $(\mu_A - \mu_B)$ is

$-4.11$ to $0.01$

Note that this interval includes zero as a possible value. Consequently, there would be insufficient evidence to reject the null hypothesis, $H_0$: $(\mu_A - \mu_B) = 0$.

EXERCISES

Learning the mechanics

**7.39.** Independent random samples were selected from three normally distributed populations with common (but unknown) variance, $\sigma^2$. The data are shown in the table.

| SAMPLE 1 | SAMPLE 2 | SAMPLE 3 |
|----------|----------|----------|
| 3.1 | 5.4 | 1.1 |
| 4.3 | 3.6 | .2 |
| 1.2 | 4.0 | 3.0 |
|  | 2.9 |  |

a. Compute the appropriate sums of squares and mean squares and fill in the appropriate entries in the analysis of variance (ANOVA) table shown at the top of the next page.

**ANOVA TABLE**

| SOURCE | df | SS | MS | F |
|--------|----|----|----|----|
| Treatments | | | | |
| Error | | | | |
| Total | | | | |

b.   Test the hypothesis that the population means are equal (i.e., $\mu_1 = \mu_2 = \mu_3$) against the alternative hypothesis that at least one mean is different from the other two. Test using $\alpha = .05$.

c.   Find a 90% confidence interval for $(\mu_2 - \mu_3)$. Interpret the interval.

d.   What would happen to the width of the confidence interval in part c if you quadrupled the number of observations in the two samples?

e.   Find a 95% confidence interval for $\mu_2$.

f.   Approximately how many observations would be required if you wished to be able to estimate a population mean correct to within .4 with probability equal to .95?

**7.40.**   A partially completed ANOVA table for an independent random sampling design is shown below:

| SOURCE | df | SS | MS | F |
|--------|----|------|----|----|
| Treatments | 4 | 24.7 | | |
| Error | | | | |
| Total | 34 | 62.4 | | |

a.   Complete the ANOVA table.

b.   How many treatments are involved in the experiment?

c.   Do the data provide sufficient evidence to indicate a difference among the population means? Test using $\alpha = .10$.

d.   Suppose that $\bar{x}_1 = 3.7$ and $\bar{x}_2 = 4.1$. Do the data provide sufficient evidence to indicate a difference between $\mu_1$ and $\mu_2$? Assume that there are seven observations for each treatment. Test using $\alpha = .10$.

e.   Refer to part d. Find a 90% confidence interval for $(\mu_1 - \mu_2)$.

f.   Refer to part d. Find a 90% confidence interval for $\mu_1$.

Applying the concepts

**7.41.**   Studies conducted at the University of Melbourne (Australia) indicate that there may be a difference between the pain thresholds of blonds and brunettes (*Family Weekly, Gainesville Sun,* Gainesville, Florida, February 5, 1978). Men and women of various ages were divided into four categories according to hair color: light blond, dark blond, light brunette, and dark brunette. The purpose of the experiment was to determine whether hair color is related to the amount of pain produced by common types of mishaps and assorted types of trauma. Each person in the experiment was given a pain threshold score based upon his or her performance in a pain sensitivity test (the higher the score, the higher the person's pain tolerance). The results are given in the table.

| | HAIR COLOR | | |
|---|---|---|---|
| Light blond | Dark blond | Light brunette | Dark brunette |
| 62 | 63 | 42 | 32 |
| 60 | 57 | 50 | 39 |
| 71 | 52 | 41 | 51 |
| 55 | 41 | 37 | 30 |
| 48 | 43 | | 35 |

a. Is there evidence that mean pain thresholds differ among people possessing the four types of hair color? Use $\alpha = .01$.

b. Estimate the difference in mean pain threshold scores between light blonds and dark brunettes using a 95% confidence interval. Interpret the interval.

c. Estimate the difference in mean pain threshold scores between light brunettes and dark brunettes using a 95% confidence interval.

7.42. Some varieties of nematodes (roundworms that live in the soil and that are frequently so small they are invisible to the naked eye) feed upon the roots of lawn grasses and crops such as strawberries and tomatoes. This pest, which is particularly troublesome in warm climates, can be treated by the application of nematicides. However, because of the size of the worms, it is very difficult to measure the effectiveness of these pesticides directly. To compare four nematicides, the yields of equal-sized plots of one variety of tomatoes were collected. The data (yields in pounds per plot) are shown in the table.

| | NEMATICIDE | | |
|---|---|---|---|
| 1 | 2 | 3 | 4 |
| 18.6 | 18.7 | 19.4 | 19.0 |
| 18.2 | 19.3 | 19.9 | 18.5 |
| 17.6 | 18.9 | 19.7 | 18.6 |
| | | 19.1 | |

a. Do the data provide sufficient evidence to indicate a difference among the mean yields of tomatoes per plot for the four nematicides? Test using $\alpha = .05$.

b. Find the approximate observed significance level for the test in part a, and interpret its value.

c. Estimate the difference in mean yield between nematicide 2 and nematicide 3. Use a 90% confidence interval.

7.43. A research psychologist wishes to investigate the difference in maze test scores for a strain of laboratory mice trained under different laboratory conditions. The experiment is conducted using eighteen randomly selected mice of this strain, with six receiving no training at all (control group), six trained under condition 1, and six trained under condition 2. Then each of the mice is given a test score between 0 and 100, depending on its performance in a test maze. The experiment produced the following results:

| CONTROL | CONDITION 1 | CONDITION 2 |
|---------|-------------|-------------|
| 58 | 73 | 53 |
| 32 | 70 | 74 |
| 59 | 68 | 72 |
| 64 | 71 | 62 |
| 55 | 60 | 58 |
| 49 | 62 | 61 |

a. Is there sufficient evidence to indicate a difference among mean maze test scores for mice trained under the three different laboratory conditions? Use $\alpha = .05$.

b. Estimate the mean maze test score for the mice in the control group using a 90% confidence interval.

c. Estimate the difference in mean maze test scores between mice trained under condition 1 and those trained under condition 2 using a 90% confidence interval.

**7.44.** A company that employs a large number of salespeople is interested in learning which of the salespeople sell the most: those strictly on commission, those with a fixed salary, or those with a reduced fixed salary plus a commission. The previous month's records for a sample of salespeople are inspected and the amount of sales (in dollars) is recorded for each, as shown in the table.

| COMMISSION | FIXED SALARY | COMMISSION PLUS SALARY |
|------------|--------------|------------------------|
| $425 | $420 | $430 |
| 507 | 448 | 492 |
| 450 | 437 | 470 |
| 483 | 432 | 501 |
| 466 | 444 | |
| 492 | | |

a. Do the data provide sufficient evidence to indicate a difference in the mean sales for the three types of compensation? Use $\alpha = .05$.

b. Use a 90% confidence interval to estimate the mean sales for salespeople who receive a commission plus salary.

c. Use a 90% confidence interval to estimate the difference in mean sales between salespeople on commission plus salary versus those on fixed salary.

**7.45.** In Section 7.2 we compared two population means using a Student's $t$ test. Reanalyze the data of Exercise 7.23 using an analysis of variance. Then:

a. Calculate the value of the $F$ statistic.

b. Show that the $F$ value computed in part a is equal to $t^2$, where $t$ is the value of the test statistic calculated in Exercise 7.23. (This demonstrates a theoretical fact: An $F$ statistic with $\nu_1 = 1$ numerator degrees of freedom and $\nu_2$ denominator degrees of freedom is equal to $t^2$, where $t$ possesses $\nu_2$ degrees of freedom.)

c. Use the $F$ test to test $H_0$: $\mu_1 = \mu_2$ against $H_a$: $\mu_1 \neq \mu_2$. Use $\alpha = .05$.

d. Examine the alternative hypotheses used for the $t$ test in Exercise 7.23 and for the $F$ test in part c. How do they differ? In what sense is the $F$ test less versatile than the $t$ test for comparing two population means?

## SUMMARY

We have presented various techniques for using the information in samples to make inferences about the difference between two or more population means. As you would expect, we are able to make reliable inferences with fewer assumptions about the sampled populations when the sample sizes are large. When we cannot take large samples from the populations, the two-sample $t$ statistic permits us to use the limited sample information to make inferences about the difference between means when the assumptions of normality and equal population variances are at least approximately true. The paired difference experiment offers the possibility of increasing the information about $(\mu_1 - \mu_2)$ by pairing similar observational units to control variability. In designing a paired difference experiment, we expect that the reduction in variability will more than compensate the loss in degrees of freedom.

More than two population means may be compared (optional Section 7.5) using an analysis of variance and an $F$ test. The $F$ test is based on the assumption that the samples have been randomly and independently selected from normally distributed populations that possess a common variance, $\sigma^2$. Thus, the sampling procedure and the assumptions are the same as those for a comparison of two population means (Section 7.2).

## SUPPLEMENTARY EXERCISES

[*Note: Starred (\*) exercises refer to the optional section.*]

Learning the mechanics

**7.46.** Independent random samples were selected from two normally distributed populations with means $\mu_1$ and $\mu_2$, respectively. The sample sizes, means, and variances are shown in the table.

| SAMPLE 1 | SAMPLE 2 |
|---|---|
| $n_1 = 12$ | $n_2 = 14$ |
| $\bar{x}_1 = 17.8$ | $\bar{x}_2 = 15.3$ |
| $s_1^2 = 74.2$ | $s_2^2 = 60.5$ |

a. Test the null hypothesis $H_0$: $(\mu_1 - \mu_2) = 0$ against the alternative hypothesis $H_a$: $(\mu_1 - \mu_2) > 0$. Use $\alpha = .05$.

b. Form a 99% confidence interval for $(\mu_1 - \mu_2)$.

c. How large must $n_1$ and $n_2$ be if you wish to estimate $(\mu_1 - \mu_2)$ to within 2 units with 99% confidence? Assume that $n_1 = n_2$.

**7.47.** A random sample of five pairs of observations were selected, one of each pair from a population with mean $\mu_1$, the other from a population with mean $\mu_2$. The data are shown in the table.

| PAIR | VALUE FROM POPULATION 1 | VALUE FROM POPULATION 2 |
|------|------------------------|------------------------|
| 1 | 28 | 22 |
| 2 | 31 | 27 |
| 3 | 24 | 20 |
| 4 | 30 | 27 |
| 5 | 22 | 20 |

a. Test the null hypothesis $H_0$: $\mu_D = 0$ against $H_a$: $\mu_D \neq 0$, where $\mu_D = \mu_1 - \mu_2$. Use $\alpha = .05$.
b. Form a 95% confidence interval for $\mu_D$.
c. When are the procedures you used in parts a and b valid?

**7.48.** List the assumptions necessary for each of the following inferential technique

a. Large-sample inferences about the difference $(\mu_1 - \mu_2)$ between population means using a two-sample $z$ statistic
b. Small-sample inferences about $(\mu_1 - \mu_2)$ using an independent sample design and a two-sample $t$ statistic
c. Small-sample inferences about $(\mu_1 - \mu_2)$ using a paired difference design and a single-sample $t$ statistic to analyze the differences
*d. Inferences about more than two population means using an analysis of variance.

*__7.49.__ Independent random samples were selected from four normally distributed populations with common (but unknown) variance $\sigma^2$. The data are shown in the table

| SAMPLE 1 | SAMPLE 2 | SAMPLE 3 | SAMPLE 4 |
|----------|----------|----------|----------|
| 8 | 6 | 9 | 12 |
| 10 | 9 | 10 | 13 |
| 9 | 8 | 8 | 10 |
| 10 | 8 | 11 | 11 |
| 11 | 7 | 12 | 11 |

a. Do the data provide sufficient evidence to indicate a difference among population means? Test using $\alpha = .05$.
b. Form a 90% confidence interval for $(\mu_1 - \mu_2)$.
c. Form a 95% confidence interval for $\mu_4$.

Applying the concepts
**7.50.** To compare the rate of return an investor can expect on tax-free municipal bonds with the rate of return on taxable bonds, an investment advisory firm randomly

samples ten bonds of each type and computes the annual rate of return over the past 3 years for each bond. The rate of return is then adjusted for taxes, assuming the investor is in a 30% tax bracket. The means and standard deviations for the adjusted returns are as follows:

| TAX-FREE BONDS | TAXABLE BONDS |
|---|---|
| $\bar{x}_1 = 9.8\%$ | $\bar{x}_2 = 9.3\%$ |
| $s_1 = 1.1\%$ | $s_2 = 1.0\%$ |

a. Test to see whether there is a difference in the mean rates of return between tax-free and taxable bonds for investors in the 30% tax bracket. Use $\alpha = .05$.

b. What assumptions were necessary for the validity of the procedure you used in part a?

**7.51.** Management training programs are often instituted in order to teach supervisory skills and thereby increase productivity. Suppose a company psychologist administers a set of examinations to each of ten supervisors before such a training program begins and then administers similar examinations at the end of the program. The examinations are designed to measure supervisory skills, with higher scores indicating increased skill. The results of the tests are shown in the table. Test to see whether the data indicate that the training program is effective. Use $\alpha = .10$.

| SUPERVISOR | BEFORE TRAINING PROGRAM | AFTER TRAINING PROGRAM |
|---|---|---|
| 1 | 63 | 78 |
| 2 | 93 | 92 |
| 3 | 84 | 91 |
| 4 | 72 | 80 |
| 5 | 65 | 69 |
| 6 | 72 | 85 |
| 7 | 91 | 99 |
| 8 | 84 | 82 |
| 9 | 71 | 81 |
| 10 | 80 | 87 |

**7.52.** Lack of motivation is a problem of many students in inner-city schools. To cope with this problem, an experiment was conducted to determine whether motivation could be improved by allowing students greater choice in the structures of their curricula. Two schools with similar student populations were chosen and fifty students were randomly selected from each to participate in the experiment. School A permitted its fifty students to choose only the courses they wanted to take. School B permitted its students to choose their courses and also to choose when and from which instructors to take the courses. The measure of student motivation was the number of times each student was absent from or late for a class during a 20 day period. The means and variances for the two samples are shown in the table. Do the data provide sufficient evidence to indicate that students from school B were late or absent less frequently than those from school A? (Use $\alpha = .1$.)

| SCHOOL A | SCHOOL B |
|---|---|
| $\bar{x}_A = 20.5$ | $\bar{x}_B = 19.6$ |
| $s_A^2 = 26.2$ | $s_B^2 = 24.1$ |

**7.53.** Two banks, bank 1 and bank 2, each independently sampled forty and fifty of their business accounts, respectively, and determined the number of the bank's services (loans, checking, savings, investment counseling, etc.) each sampled business was using. Both banks offer the same services. A summary of the data supplied by the samples is given in the table. Do the samples yield sufficient evidence to conclude that the average number of services used by bank 1's business customers is significantly greater (at the $\alpha = .10$ level) than the average number of services used by bank 2's business customers?

| BANK 1 | BANK 2 |
|---|---|
| $\bar{x}_1 = 2.2$ | $\bar{x}_2 = 1.8$ |
| $s_1 = 1.15$ | $s_2 = 1.10$ |

**7.54.** Find a 99% confidence interval for $(\mu_1 - \mu_2)$ in Exercise 7.53. Does the interval include zero? Interpret the confidence interval.

**7.55.** The interocular pressure of glaucoma patients is often reduced by treatment with adrenaline. To compare a new synthetic drug with adrenaline, seven glaucoma patients were treated with both drugs, one eye with adrenaline and one with the synthetic drug. The reduction in pressure in each eye was then recorded, as shown in the table. Do the data provide sufficient evidence to indicate a difference in the mean reductions in eye pressure for the two drugs? Test using $\alpha = .10$.

| PATIENT | ADRENALINE | SYNTHETIC |
|---|---|---|
| 1 | 3.5 | 3.2 |
| 2 | 2.6 | 2.8 |
| 3 | 3.0 | 3.1 |
| 4 | 1.9 | 2.4 |
| 5 | 2.9 | 2.9 |
| 6 | 2.4 | 2.2 |
| 7 | 2.0 | 2.2 |

**7.56.** Some power plants are located near rivers or oceans so that the available water can be used for cooling the condensers. As part of an environmental impact study, suppose a power company wants to estimate the difference in mean water temperature between the discharge of its plant and the off-shore waters. How many sample measurements must be taken at each site in order to estimate the true difference between means to within .2°C with 95% confidence? Assume the range in readings will be about 4°C at each site and the same number of readings will be taken at each site.

**7.57.** The use of preservatives by food processors has become a controversial issue. Suppose two preservatives are extensively tested and determined safe for use in meats. A processor wants to compare the preservatives for their effects on retarding spoilage. Suppose fifteen cuts of fresh meat are treated with preservative A and fifteen with B, and the number of hours until spoilage begins is recorded for each of the thirty cuts of meat. The results are summarized below:

| PRESERVATIVE A | PRESERVATIVE B |
|---|---|
| $\bar{x}_1 = 106.4$ hours | $\bar{x}_2 = 96.5$ hours |
| $s_1 = 10.3$ hours | $s_2 = 13.4$ hours |

    a. Is there evidence of a difference in mean time until spoilage begins between the two preservatives at the $\alpha = .05$ level?
    b. Can you recommend an experimental design that the processor could have used to reduce the variability in the data?

**7.58.** A physiologist wished to study the effect of birth-control pills on exercise capacity. Five female subjects who had never taken the pill had their maximal oxygen uptake measured (in milliliters per kilogram of their body weight) during a treadmill session. The five subjects then took the pill for a specified length of time and their uptakes were measured again, as given in the table. Do the data provide sufficient evidence to indicate that the mean maximal oxygen uptake after taking birth-control pills is less than the mean uptake before taking the pill? Use $\alpha = .01$.

| SUBJECT | MAXIMAL OXYGEN UPTAKE | |
|---|---|---|
| | Before | After |
| 1 | 35.0 | 29.5 |
| 2 | 36.5 | 33.5 |
| 3 | 36.0 | 32.0 |
| 4 | 39.0 | 36.5 |
| 5 | 37.5 | 35.0 |

**7.59.** The following experiment was conducted to compare two coatings designed to improve the durability of the soles of jogging shoes. A ⅛ inch thick layer of coating 1 was applied to one of a pair of shoes and a layer of equal thickness of coating 2 was applied to the other shoe. Ten joggers were given pairs of shoes treated in this manner and were instructed to record the number of miles covered in each shoe before the

$\frac{1}{8}$ inch coating was worn through in any one place. The results are given in the table.

| JOGGER | COATING 1 | COATING 2 |
|--------|-----------|-----------|
| 1 | 892 | 985 |
| 2 | 904 | 953 |
| 3 | 775 | 775 |
| 4 | 435 | 510 |
| 5 | 946 | 895 |
| 6 | 853 | 875 |
| 7 | 780 | 895 |
| 8 | 695 | 725 |
| 9 | 825 | 858 |
| 10 | 750 | 812 |

a. Do the data provide sufficient evidence to indicate a difference between the mean numbers of miles of wear that a runner might expect from the two coatings? Test using $\alpha = .05$.

b. Use a 95% confidence interval to estimate the true difference between the mean numbers of miles of wear for the two sole coatings. Interpret this interval.

c. Why is the design used for this experiment preferable to independent random sampling?

**7.60.** An experiment is conducted to determine whether there is a difference among the mean increases in growth produced by five inoculins of growth hormones for plants. The experimental material consists of twenty cuttings of a shrub (all of equal weight), with four cuttings randomly assigned to each of the five different inoculins. The results of the experiment are given in the table; all measurements represent an increase in weight (in grams).

| | | INOCULIN | | |
|---|---|---|---|---|
| A | B | C | D | E |
| 15 | 21 | 22 | 10 | 6 |
| 18 | 13 | 19 | 14 | 11 |
| 9 | 20 | 24 | 21 | 15 |
| 16 | 17 | 21 | 13 | 8 |

a. Based on this information, can we conclude that there is a difference among the mean increases in weight for the five inoculins of growth hormone? Test at the $\alpha = .05$ level of significance.

b. What assumptions must the data satisfy to make the test in part a valid?

c. Find the approximate observed significance level for the test in part a, and interpret its value.

**7.61.** An experiment was conducted to compare corn yield per acre for four different fertilizer applications. Sixteen plots of equal size were prepared for planting and four each were assigned to receive the fertilizer treatments. Unfortunately, the crops in two of the plots were damaged during the growing season and were removed from the experiment. The yields of corn in bushels per plot are shown in the table.

| FERTILIZER APPLICATION (POUNDS PER PLOT) | | | |
|---|---|---|---|
| 5 | 10 | 15 | 20 |
| 57 | 62 | 65 | 43 |
| 51 | 59 | 72 | 55 |
| 45 | 75 | 67 | 49 |
| | 69 | 78 | |

a. Do the data suggest a difference in mean number of bushels of corn produced among the four fertilizer treatments?

b. Use a 95% confidence interval to estimate the difference between the mean number of bushels produced by plots that received 5 and 20 pounds of fertilizer.

c. Use a 95% confidence interval to estimate the mean number of bushels produced by plots that received 15 pounds of fertilizer.

**7.62.** A fast-food chain expects mean gross sales of $800,000 per year per franchise. A random sample of the chain's stores was selected in Miami, Los Angeles, and Chicago, and the results are given in the table.

GROSS SALES (IN UNITS OF $100,000)

| MIAMI | LOS ANGELES | CHICAGO |
|---|---|---|
| 8.7 | 8.7 | 7.8 |
| 7.4 | 8.0 | 7.6 |
| 7.9 | 9.0 | 6.9 |
| 8.0 | 8.3 | 5.7 |
| 8.5 | 9.0 | |
| 7.9 | | |

a. Is there evidence of a difference in mean gross sales among these three cities? Use $\alpha = .01$.

b. Form a 90% confidence interval for the mean gross sales of the Miami stores in this fast-food chain. Interpret this interval.

**7.63.** In hopes of attracting more riders, a city transit company plans to have express bus service from a suburban terminal to the downtown business district. These buses will travel along a major city street where there are numerous traffic lights that will affect travel time. The city decides to perform a study of the effect of four different plans (a special bus lane, traffic signal progression, etc.) on the travel times for the buses. Travel times (in minutes) are measured for several weekdays during a morning rush-hour trip while each plan is in effect. The results are recorded in the table at the top of the next page.

|   | PLAN | | |
|---|---|---|---|
| 1 | 2 | 3 | 4 |
| 27 | 25 | 34 | 30 |
| 25 | 28 | 29 | 33 |
| 29 | 30 | 32 | 31 |
| 26 | 27 | 31 | |
|    | 24 | 36 | |

a.   Is there evidence of a difference in the mean travel times for the four plans? Use $\alpha = .01$.

b.   Form a 95% confidence interval for the difference between the mean travel times for plan 1 (express lane) and plan 3 (a control—no special travel arrangements).

c.   State the assumptions upon which the inferences in parts a and b are based.

## ON YOUR OWN . . .

We have now discussed two methods of collecting data to compare two population means. In many experimental situations a decision must be made either to collect two independent samples or to conduct a paired difference experiment. The importance of this decision cannot be overemphasized, since the amount of information obtained and the cost of the experiment are both directly related to the method of experimentation that is chosen.

Choose two populations (pertinent to your major area) that have unknown means and for which you could both collect two independent samples and collect paired observations. Before conducting the experiment, state which method of sampling you think will provide more information (and why). Then, to compare the two methods, first perform the independent sampling procedure by collecting ten observations from each population (a total of twenty measurements), and then perform the paired difference experiment by collecting ten pairs of observations.

Construct two 95% confidence intervals, one for each experiment you conducted. Which method provided the shorter confidence interval and thus more information on this performance of the experiment? Does this agree with your preliminary expectations?

## REFERENCES

Freedman, D., Pisani, R., & Purves, R. *Statistics.* New York: W. W. Norton and Co., 1978. Chapter 27.

Mendenhall, W. *Introduction to probability and statistics.* 6th ed. Boston: Duxbury, 1983. Chapters 8, 9, and 10.

Snedecor, G. W., & Cochran, W. *Statistical methods.* 7th ed. Ames, Iowa: Iowa University Press, 1980.

Stainback, S., & Stainback, W. C. "Matched procedures and research in mental retardation." *Training School Bulletin,* May 1973, *70,* 33–37.

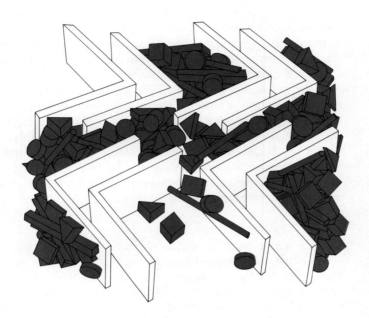

# 8

# COMPARING TWO OR MORE
# POPULATION PROPORTIONS

**WHERE WE'VE BEEN . . .**

Chapter 7 presented methods for comparing two population means using $z$ and $t$ statistics and for comparing two or more population means using an analysis of variance.

**WHERE WE'RE GOING . . .**

In this chapter, we will consider a problem of comparable importance—comparing two or more population proportions. The need to compare population proportions arises because many business and social experiments involve questioning people and classifying their responses. We will learn how to determine whether the proportion of consumers favoring product A differs from the proportion favoring product B, and we will learn how to estimate the difference with a confidence interval.

Many experiments are conducted in the biological, physical, and social sciences to compare two or more population proportions. Those conducted in business and the social sciences to sample the opinions of people are called **sample surveys**. For example, a state government might wish to estimate the difference in the proportions of people in two regions of the state who would qualify for a new welfare program. Or, after an innovative process change, an engineer might wish to determine whether the proportion of defective items produced by a manufacturing process was less than the proportion of defectives produced before the change. This chapter will show you how to test hypotheses about the difference between two population proportions based on independent random sampling. We will also show how to find a confidence interval for the difference. Then, in optional Section 8.3 we will compare more than two population proportions, and in optional Section 8.4 we will present a related problem.

## 8.1 INFERENCES ABOUT THE DIFFERENCE BETWEEN POPULATION PROPORTIONS: INDEPENDENT BINOMIAL EXPERIMENTS

Suppose a presidential candidate wants to compare the preference of registered voters in the northeastern United States (NE) to those in the southeastern United States (SE). Such a comparison would help determine where to concentrate campaign efforts. The candidate hires a professional pollster to randomly choose 1,000 registered voters in the northeast and 1,000 registered voters in the southeast and interview each to learn her or his voting preference. The objective is to use this sample information to make an inference about the difference $(p_1 - p_2)$ between the proportion $p_1$ of *all* registered voters in the northeast and the proportion $p_2$ of *all* registered voters in the southeast who plan to vote for the presidential candidate.

The two samples represent independent binomial experiments (see Section 4.4 for the characteristics of binomial experiments), with the binomial random variables being the numbers $x_1$ and $x_2$ of the 1,000 sampled voters in each area who indicate they will vote for the candidate. The results are summarized as follows:

| NE | SE |
|---|---|
| $n_1 = 1,000$ | $n_2 = 1,000$ |
| $x_1 = 546$ | $x_2 = 475$ |

We can now calculate the sample proportions $\hat{p}_1$ and $\hat{p}_2$ of the voters in favor of the candidate in the northeast and southeast, respectively:

$$\hat{p}_1 = \frac{x_1}{n_1} = \frac{546}{1,000} = .546 \qquad \hat{p}_2 = \frac{x_2}{n_2} = \frac{475}{1,000} = .475$$

The difference between the sample proportions $(\hat{p}_1 - \hat{p}_2)$ makes an intuitively appealing estimator of the difference between the population parameters $(p_1 - p_2)$. For our example, the estimate is

$$(\hat{p}_1 - \hat{p}_2) = .546 - .475 = .071$$

To judge the reliability of the estimator $(\hat{p}_1 - \hat{p}_2)$, we must observe its performance in repeated sampling from the two populations. That is, we need to know the sampling distribution of $(\hat{p}_1 - \hat{p}_2)$. The properties of the sampling distribution are given in the box. Remember that $\hat{p}_1$ and $\hat{p}_2$ can be viewed as means of the number of successes per trial in the respective samples, so the central limit theorem will apply when the sample sizes are large.

Since the distribution of $(\hat{p}_1 - \hat{p}_2)$ in repeated sampling is approximately normal, we can use the $z$ statistic to derive confidence intervals for $(p_1 - p_2)$ or to test an hypothesis about $(p_1 - p_2)$.

For the voter example, a 95% confidence interval for the difference $(p_1 - p_2)$ is

$$(\hat{p}_1 - \hat{p}_2) \pm 1.96 \sigma_{(\hat{p}_1 - \hat{p}_2)} \qquad \text{or} \qquad (\hat{p}_1 - \hat{p}_2) \pm 1.96 \sqrt{\frac{p_1 q_1}{n_1} + \frac{p_2 q_2}{n_2}}$$

The quantities $p_1 q_1$ and $p_2 q_2$ must be estimated in order to complete the calculation of the standard deviation, $\sigma_{(\hat{p}_1 - \hat{p}_2)}$, and hence the calculation of the confidence interval. In Section 6.5 we showed that the value of $pq$ is relatively insensitive to the value chosen to approximate $p$. Therefore, $\hat{p}_1 \hat{q}_1$ and $\hat{p}_2 \hat{q}_2$ will provide satisfactory estimates to approximate $p_1 q_1$ and $p_2 q_2$, respectively. Then

$$\sqrt{\frac{p_1 q_1}{n_1} + \frac{p_2 q_2}{n_2}} \approx \sqrt{\frac{\hat{p}_1 \hat{q}_1}{n_1} + \frac{\hat{p}_2 \hat{q}_2}{n_2}}$$

and we will approximate the 95% confidence interval by

$$(\hat{p}_1 - \hat{p}_2) \pm 1.96 \sqrt{\frac{\hat{p}_1 \hat{q}_1}{n_1} + \frac{\hat{p}_2 \hat{q}_2}{n_2}}$$

Substituting the sample quantities yields

$$(.546 - .475) \pm 1.96 \sqrt{\frac{(.546)(.454)}{1,000} + \frac{(.475)(.525)}{1,000}}$$

or $.071 \pm .044$. Thus, we estimate the difference $(p_1 - p_2)$ to fall in the interval .027 to .115. We infer that there are between 2.7% and 11.5% more registered

---

*The mean and variance of the sampling distribution of $(\hat{p}_1 - \hat{p}_2)$ can be derived using the formulas given in Section 5.5.

voters in the northeast than in the southeast who plan to vote for the presidential candidate. It seems that the candidate should direct a greater campaign effort in the southeast as compared to the northeast. The confidence coefficient associated with our interval estimate is .95.

The general form of a confidence interval for the difference $(p_1 - p_2)$ between binomial proportions is given in the box.

---

LARGE-SAMPLE $100(1 - \alpha)$PERCENT CONFIDENCE INTERVAL FOR $(p_1 - p_2)$

$$(\hat{p}_1 - \hat{p}_2) \pm z_{\alpha/2}\sigma_{(\hat{p}_1 - \hat{p}_2)} = (\hat{p}_1 - \hat{p}_2) \pm z_{\alpha/2}\sqrt{\frac{p_1 q_1}{n_1} + \frac{p_2 q_2}{n_2}}$$

$$\approx (\hat{p}_1 - \hat{p}_2) \pm z_{\alpha/2}\sqrt{\frac{\hat{p}_1 \hat{q}_1}{n_1} + \frac{\hat{p}_2 \hat{q}_2}{n_2}}$$

---

The $z$ statistic,

$$z = \frac{(\hat{p}_1 - \hat{p}_2) - (p_1 - p_2)}{\sigma_{(\hat{p}_1 - \hat{p}_2)}}$$

is used to test the null hypothesis that $(p_1 - p_2)$ equals some specified difference, say $D_0$. For the special case where $D_0 = 0$, i.e., where we want to test the null hypothesis $H_0$: $(p_1 - p_2) = 0$ (or, equivalently, $H_0$: $p_1 = p_2$), the best estimate of $p_1 = p_2 = p$ is obtained by dividing the total number of successes $(x_1 + x_2)$ for the two samples by the total number of observations $(n_1 + n_2)$, i.e.,

$$\hat{p} = \frac{x_1 + x_2}{n_1 + n_2}$$

Then the best estimate of $\sigma_{(\hat{p}_1 - \hat{p}_2)}$ is

$$\sigma_{(\hat{p}_1 - \hat{p}_2)} = \sqrt{\frac{p_1 q_1}{n_1} + \frac{p_2 q_2}{n_2}} \approx \sqrt{\frac{\hat{p}\hat{q}}{n_1} + \frac{\hat{p}\hat{q}}{n_2}} = \sqrt{\hat{p}\hat{q}\left(\frac{1}{n_1} + \frac{1}{n_2}\right)}$$

The test is summarized in the box on page 321.

EXAMPLE 8.1    In the past decade there have been intensive antismoking campaigns sponsored by both federal and private agencies. Suppose that the American Cancer Society randomly sampled 1,500 adults in 1979 and then sampled 2,000 adults in 1981 to determine whether there was evidence that the percentage of smokers had decreased. The results of the two sample surveys are shown in the table, where $x_1$ and $x_2$

| 1979 | 1981 |
|------|------|
| $n_1 = 1,500$ | $n_2 = 2,000$ |
| $x_1 = 576$ | $x_2 = 652$ |

LARGE-SAMPLE TEST OF AN HYPOTHESIS ABOUT $(p_1 - p_2)$

One-tailed test

$H_0$: $(p_1 - p_2) = D_0$

$H_a$: $(p_1 - p_2) < D_0$

[or $H_a$: $(p_1 - p_2) > D_0$]

where $D_0 = $ Hypothesized value of $(p_1 - p_2)$

Test statistic: $z = \dfrac{(\hat{p}_1 - \hat{p}_2) - D_0}{\sigma_{(\hat{p}_1 - \hat{p}_2)}}$

Rejection region: $z < -z_\alpha$

[or $z > z_\alpha$ when

$H_a$: $(p_1 - p_2) > D_0$]

Two-tailed test

$H_0$: $(p_1 - p_2) = D_0$

$H_a$: $(p_1 - p_2) \neq D_0$

Test statistic: $z = \dfrac{(\hat{p}_1 - \hat{p}_2) - D_0}{\sigma_{(\hat{p}_1 - \hat{p}_2)}}$

Rejection region: $z < -z_{\alpha/2}$

or $z > z_{\alpha/2}$

Note: $\sigma_{(\hat{p}_1 - \hat{p}_2)} = \sqrt{\dfrac{p_1 q_1}{n_1} + \dfrac{p_2 q_2}{n_2}}$

To calculate $\sigma_{(\hat{p}_1 - \hat{p}_2)}$, approximate $p_1$ and $p_2$ using $\hat{p}_1$ and $\hat{p}_2$ except for the special case when $D_0 = 0$. Then use

$$\hat{p}_1 = \hat{p}_2 = \hat{p} = \frac{x_1 + x_2}{n_1 + n_2}$$

For this special case,

$$\sigma_{(\hat{p}_1 - \hat{p}_2)} \approx \sqrt{\hat{p}\hat{q}\left(\frac{1}{n_1} + \frac{1}{n_2}\right)}$$

represent the numbers of smokers in the 1979 and 1981 samples, respectively. Do these data indicate that the fraction of smokers decreased over this 2 year period? Use $\alpha = .05$.

Solution

If we define $p_1$ and $p_2$ as the true proportions of adult smokers in 1979 and 1981, the elements of our test are

$H_0$: $(p_1 - p_2) = 0$     $H_a$: $(p_1 - p_2) > 0$

(The test is one-tailed since we are interested only in determining whether the proportion of smokers *decreased*.)

Test statistic: $z = \dfrac{(\hat{p}_1 - \hat{p}_2) - 0}{\sigma_{(\hat{p}_1 - \hat{p}_2)}}$

Rejection region: $\alpha = .05$

$z > z_\alpha = z_{.05} = 1.645$   (See Figure 8.1.)

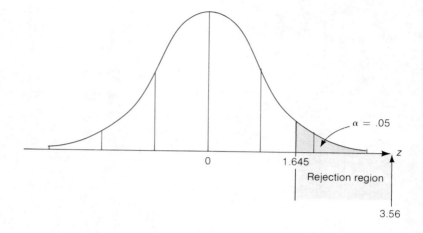

FIGURE 8.1
REJECTION REGION
FOR EXAMPLE 8.1

We now calculate the sample proportions of smokers

$$\hat{p}_1 = \frac{576}{1,500} = .384$$

$$\hat{p}_2 = \frac{652}{2,000} = .326$$

Then

$$z = \frac{(\hat{p}_1 - \hat{p}_2) - 0}{\sigma_{(\hat{p}_1 - \hat{p}_2)}} \approx \frac{(\hat{p}_1 - \hat{p}_2)}{\sqrt{\hat{p}\hat{q}\left(\frac{1}{n_1} + \frac{1}{n_2}\right)}}$$

where

$$\hat{p} = \frac{x_1 + x_2}{n_1 + n_2} = \frac{576 + 652}{1,500 + 2,000} = .351$$

Thus,

$$z = \frac{.384 - .326}{\sqrt{(.351)(.649)\left(\frac{1}{1,500} + \frac{1}{2,000}\right)}}$$

$$= \frac{.058}{.0164} = 3.56$$

There is sufficient evidence at the $\alpha = .05$ level to conclude that the proportion of all adults who smoke has decreased over the 1979–1981 period. We could place a confidence interval on $(p_1 - p_2)$ if we were interested in estimating the extent of the decrease.

## Learning the mechanics

**8.1.** Construct a 95% confidence interval for $(p_1 - p_2)$ in each of the following situations:

    a. $n_1 = 100$, $\hat{p}_1 = .72$; $n_2 = 100$, $\hat{p}_2 = .61$
    b. $n_1 = 130$, $\hat{p}_1 = .16$; $n_2 = 210$, $\hat{p}_2 = .25$
    c. $n_1 = 70$, $\hat{p}_1 = .53$; $n_2 = 60$, $\hat{p}_2 = .48$

**8.2.** Independent random samples, each containing 500 observations, were selected from two binomial populations. The samples from populations 1 and 2 produced 200 and 250 successes, respectively.

    a. Test the null hypothesis $H_0$: $(p_1 - p_2) = 0$ against the alternative hypothesis $H_a$: $(p_1 - p_2) \neq 0$. Use $\alpha = .05$.
    b. Test $H_0$: $(p_1 - p_2) = 0$ against $H_a$: $(p_1 - p_2) \neq 0$. Use $\alpha = .01$.
    c. Test $H_0$: $(p_1 - p_2) = 0$ against $H_a$: $(p_1 - p_2) < 0$. Use $\alpha = .01$.
    d. Form a 90% confidence interval for $(p_1 - p_2)$.

**8.3.** The quantities $\hat{p}_1$ and $\hat{p}_2$ have been defined to be $x_1/n_1$ and $x_2/n_2$, respectively. What assumptions do we make about $x_1$ and $x_2$?

**8.4.** What are the characteristics of a binomial experiment?

**8.5.** Explain why the central limit theorem is important in finding an approximate distribution for $(\hat{p}_1 - \hat{p}_2)$. See Section 5.5.

**8.6.** Sketch the sampling distribution of $(\hat{p}_1 - \hat{p}_2)$ based on independent random samples of $n_1 = 100$ and $n_2 = 200$ observations from two binomial populations with success probabilities $p_1 = .1$ and $p_2 = .5$, respectively.

**8.7.** Explain how knowing the sampling distribution of $(\hat{p}_1 - \hat{p}_2)$ can help us measure the reliability of the estimator $(\hat{p}_1 - \hat{p}_2)$.

## Applying the concepts

**8.8.** Two surgical procedures are widely used to treat a certain type of cancer. To compare the success rates of the two procedures, random samples of the two types of surgical patients were obtained and the numbers of patients who showed no recurrence of the disease after a 1 year period were recorded. The data are shown in the table. Do the data provide sufficient evidence to indicate a difference in the success rates of the two procedures?

| | $n$ | NUMBER OF SUCCESSES |
|---|---|---|
| PROCEDURE A | 100 | 78 |
| PROCEDURE B | 100 | 87 |

**8.9.** A new insect spray, type A, is to be compared with a spray, type B, that is currently in use. Two rooms of equal size are sprayed with the same amount of spray, one room with A, the other with B. Two hundred insects are released into each room and after 1 hour, the numbers of dead insects are counted. The results are given in the table at the top of page 324.

|  | SPRAY A | SPRAY B |
|---|---|---|
| NUMBER OF INSECTS | 200 | 200 |
| NUMBER OF DEAD INSECTS | 120 | 80 |

a. Do the data provide sufficient evidence to indicate that spray A is more effective than spray B in controlling the insects? Test using $\alpha = .05$.

b. Find a 90% confidence interval for $(p_1 - p_2)$, the difference in the rates of kill for the two sprays. Interpret this interval.

**8.10.** When purchasing merchandise from stores, people are often given incorrect change. Is it more likely that an error will be reported if the change received is too little rather than too much? To investigate this problem, a psychologist arranged for the observation of sales made to 100 people. Fifty randomly selected purchasers received $1.00 too much and fifty received $1.00 too little. Twenty-three people who received too much change reported the error, in comparison with forty-one in the group who received too little change. Estimate the true difference in the proportions of people who report the two types of errors using a 99% confidence interval.

**8.11.** To market a new cigarette, a tobacco company decides to use two different advertising agencies, one operating in the east and one in the west. After the cigarette has been on the market for 6 months, random samples of smokers are taken from each of the two regions and questioned concerning their cigarette preference. The numbers in the samples favoring the new brand are shown in the table.

|  | SAMPLE SIZE | NUMBER PREFERRING NEW BRAND |
|---|---|---|
| EAST | 500 | 12 |
| WEST | 450 | 15 |

a. Do the data provide sufficient evidence to indicate a difference in the proportions preferring the new brand between the two regions? Test using $\alpha = .05$.

b. Find the observed significance level for the test and interpret its value.

**8.12.** Moving companies (home movers, etc.) are required by the government to publish a Carrier Performance Report each year. One of the descriptive statistics they must include in this report is the percentage of shipments on which a $50 or greater claim for loss or damage was filed in the previous year. Suppose company A and company B each decide to estimate this figure by sampling their records, and they obtain the following data:

|  | COMPANY A | COMPANY B |
|---|---|---|
| SHIPMENTS DELIVERED | 9,542 | 6,631 |
| NUMBER OF SHIPMENTS ON WHICH A CLAIM OF $50 OR GREATER WAS FILED | 1,653 | 501 |

a. Estimate the true proportion of shipments on which a claim of $50 or greater was made against company A. Use $1 - \alpha = .95$.

b. Repeat part a for company B.

c. Use a 95% confidence interval to estimate the difference between the true proportions of claims made against company A and company B.

**8.13.** Refer to Exercise 8.12. Test the null hypothesis that no difference exists between the true proportions of claims made against company A and company B against the alternative hypothesis that a difference does exist. Use $\alpha = .05$. Do your test results indicate that one carrier is superior to the other? If so, which one? Explain how you arrived at your conclusion.

**8.14.** Suppose a firm switches its table salt container from a cylinder (expensive) to a rectangular box (inexpensive). The firm samples 1,000 households nationwide, both before and after the switch, to estimate the percentage of households that purchase its brand of salt. The following results are obtained:

|  | BEFORE | AFTER |
|---|---|---|
| SAMPLE SIZE | 1,000 | 1,000 |
| NUMBER OF HOUSEHOLDS USING FIRM'S BRAND | 475 | 305 |

a. Estimate the difference between the true proportions of households using the firm's salt before and after the packaging switch. Use a 90% confidence interval.

b. Interpret the confidence interval of part a. Express the reliability of the interval.

**8.15.** Refer to Exercise 8.14. The firm's vice president in charge of sales claims the switch to the box has seriously hurt the market share. Does the sample evidence support this claim at the .05 significance level?

**8.16.** The Reserve Mining Company of Minnesota commissioned a team of physicians to study the breathing patterns of its miners who were exposed to taconite dust. The physicians compared the breathing of 307 miners who had been employed in Reserve's Babbit, Minnesota, mine for more than 20 years with thirty-five Duluth area men with no history of exposure to taconite dust. The physicians concluded that "there is no significant difference in respiratory symptoms or breathing ability between the group of men who have worked in the taconite industry for more than 20 years and a group of men of similar smoking habits but without exposure to taconite dust." Using the statistical procedures you have learned in this chapter, design an hypothesis test (give $H_0$, $H_a$, test statistics, etc.) that would have been appropriate for use in the physicians' study. [Source:  Associated Press, *Minneapolis Tribune,* February 20, 1977.]

**8.17.** Refer to Exercise 8.16. Suppose the physicians determined that sixty-one of the 307 miners had breathing irregularities, and that five of the thirty-five Duluth men had breathing irregularities.

a. Test to see whether these data indicate that a higher proportion of breathing irregularities exists among those who have been exposed to taconite dust than among those who have not been exposed.
b. Find the observed significance level for the test and interpret its value.

## 8.2
## DETERMINING THE SAMPLE SIZE

The sample sizes $n_1$ and $n_2$ required to compare two population proportions can be found in a manner similar to the method described in Section 7.4 for comparing two population means. We will assume equal sample sizes, i.e., $n_1 = n_2 = n$, and then choose $n$ so that $(\hat{p}_1 - \hat{p}_2)$ will differ from $(p_1 - p_2)$ by no more than a bound $B$ with a specified probability. We will illustrate the procedure with an example.

EXAMPLE 8.2

A production supervisor suspects a difference exists between the proportions $p_1$ and $p_2$ of defective items produced by two different machines. Experience has shown that the proportion defective for each of the two machines is in the neighborhood of .03. If the supervisor wants to estimate the difference in the proportions correct to within .005 with probability .95, how many items must be randomly sampled from the production of each machine? (Assume that you want $n_1 = n_2 = n$.)

Solution

For the specified level of reliability, $z_{\alpha/2} = z_{.025} = 1.96$. Then, letting $p_1 = p_2 = .03$ and $n_1 = n_2 = n$, we find the required sample size per machine by solving the following equation for $n$:

$$z_{\alpha/2}\, \sigma_{(\hat{p}_1 - \hat{p}_2)} = B$$

or

$$z_{\alpha/2}\sqrt{\frac{p_1 q_1}{n_1} + \frac{p_2 q_2}{n_2}} = B$$

$$1.96\sqrt{\frac{(.03)(.97)}{n} + \frac{(.03)(.97)}{n}} = .005$$

$$1.96\sqrt{\frac{2(.03)(.97)}{n}} = .005$$

$$n = 8,943.2$$

You can see that this may be a tedious sampling procedure. If the supervisor insists on estimating $(p_1 - p_2)$ correct to within .005 with probability equal to .95, approximately 9,000 items will have to be inspected for each machine.

From the calculations in Example 8.2 you can see that $\sigma_{(\hat{p}_1 - \hat{p}_2)}$ (and hence the solution, $n_1 = n_2 = n$) depends on the actual (but unknown) values of $p_1$ and $p_2$. In fact, the required sample size $n_1 = n_2 = n$ is largest when $p_1 = p_2 = \frac{1}{2}$. Therefore, if you have no prior information on the approximate values of $p_1$ and $p_2$, use $p_1 = p_2 = $

$\frac{1}{2}$ in the formula for $\sigma_{(\hat{p}_1 - \hat{p}_2)}$. If $p_1$ and $p_2$ are in fact close to $\frac{1}{2}$, then the values of $n_1$ and $n_2$ that you have calculated will be correct. If $p_1$ and $p_2$ differ substantially from $\frac{1}{2}$, then your solutions for $n_1$ and $n_2$ will be larger than needed. Consequently, using $p_1 = p_2 = \frac{1}{2}$ when solving for $n_1$ and $n_2$ is a conservative procedure because the sample sizes $n_1$ and $n_2$ will be at *least* as large as (and probably larger than) needed.

EXERCISES

### Learning the mechanics

**8.18.** Assuming that $n_1 = n_2$, find the appropriate sample sizes needed to estimate $(p_1 - p_2)$ for each of the following situations:

    a.   Bound = .01 with 99% confidence. Assume that $p_1 \approx .3$ and $p_2 \approx .6$.

    b.   Bound = .05 with 90% confidence. Assume there is no prior information available to obtain approximate values of $p_1$ and $p_2$.

    c.   Bound = .05 with 90% confidence. Assume that $p_1 \approx .1$ and $p_2 \approx .2$.

### Applying the concepts

**8.19.** As new car prices continue to escalate, buyers are tending to take loans for longer than 36 months; 48 months is becoming a popular alternative. Suppose you plan to survey potential buyers in your sales region to estimate the proportion of buyers in the over 40 age group who favor 48 month automobile loans and the proportion in the 40 and under age group who favor the 48 month automobile loans. If you intend to select random samples of the same size from each of these two groups:

    a.  Approximately how many potential automobile buyers should be included in your samples to estimate the difference between the proportions correct to within .05 with probability equal to .95?

    b.  Suppose you want to obtain individual estimates for the proportions in the two age groups. Will the sample size found in part a be large enough to provide estimates of each proportion correct to within .05 with probability equal to .95?

**8.20.** Rat damage creates a large financial loss in the production of sugar cane. One aspect of the problem that has been investigated by the United States Department of Agriculture concerns the optimal place to locate rat poison. To be most effective in reducing rat damage, should the poison be located in the middle of the field or on the outer perimeter? One way to answer this question is to determine where the greater amount of damage occurs. If damage is measured by the proportion of cane stalks that have been damaged by rats, how many stalks from each section of the field should be sampled in order to estimate the true difference between proportions of stalks damaged in the two sections to within .02 with probability .95?

**8.21.** A television manufacturer wants to compare with a competitor the proportions of its best sets that need repair within 1 year. If it is desired to estimate the difference between proportions to within .05 with 90% confidence, and if the manufacturer plans to sample twice as many buyers ($n_1$) of its sets as buyers ($n_2$) of the competitor's sets, how many buyers of each brand must be sampled? Assume the proportion of sets that need repair will be about .2 for both brands.

## 8.3
## COMPARING TWO OR MORE POPULATION PROPORTIONS: A CHI SQUARE TEST (OPTIONAL)

Two or more, say $k$, binomial populations may be compared by using a method known as a contingency* table analysis. To illustrate, suppose that a manufacturer wishes to compare the fractions of defective fuses produced by three different production lines. One thousand fuses are randomly sampled from each production line and the numbers of defectives, $x_1$, $x_2$, and $x_3$ are recorded. The data appear in Table 8.1.

**TABLE 8.1**
NUMBERS OF DEFECTIVES FOR SAMPLES OF 1,000 FUSES SELECTED FROM EACH OF THREE PRODUCTION LINES

|  | PRODUCTION LINE | | | TOTALS. |
|---|---|---|---|---|
|  | 1 | 2 | 3 |  |
| NUMBER OF DEFECTIVES IN SAMPLE | 49 | 24 | 31 | 104 |
| NUMBER OF NONDEFECTIVES IN SAMPLE | 951 | 976 | 969 | 2,896 |
| TOTALS | 1,000 | 1,000 | 1,000 | 3,000 |

A contingency table analysis compares the observed numbers in the cells of Table 8.1 with the numbers that would be expected if the null hypothesis were true, i.e., if $p_1 = p_2 = p_3 = p$. The expected number of defectives for each cell in row 1 is the product of the sample size (column total) and the common fraction defective, $p$. Since $p$ is unknown, we estimate its value as

$$\hat{p} = \frac{\text{Total number of defectives}}{\text{Total number of fuses}}$$

$$= \frac{(\text{Row 1 total})}{n} = \frac{r_1}{n}$$

We then estimate the expected number for a cell in row 1, column $j$, as

$$\hat{E}(n_{1j}) = \frac{(\text{Row 1 total})(\text{Column } j \text{ total})}{n} = \frac{r_1 c_j}{n}$$

Similarly, the estimated expected numbers of nondefectives in row 2 are the products of $\hat{q} = r_2/n$ and the column totals, or

$$\hat{E}(n_{2j}) = \frac{(\text{Row 2 total})(\text{Column } j \text{ total})}{n} = \frac{r_2 c_j}{n}$$

If you examine the formulas for the estimated expected values of the cell counts, you can see that the estimated expected count for a cell in row $i$ and column $j$ is equal to the product of its row and column totals divided by the total $n$ of all counts. The box at the top of the next page summarizes this information.

---

*The reason for the use of the adjective *contingency* will become apparent when you read optional Section 8.4.

---

ESTIMATED EXPECTED CELL COUNTS

The estimated expected cell count in row $i$ and column $j$ is equal to

$$\hat{E}(n_{ij}) = \frac{r_i c_j}{n}$$

where

$r_i$ = Total for row $i$
$c_j$ = Total for column $j$
$n$ = Total number of trials

---

The estimated expected cell count for the cell in row 1, column 1, Table 8.1 is

$$\hat{E}(n_{11}) = \frac{r_1 c_1}{n} = \frac{(104)(1,000)}{3,000} = 34.667$$

Similarly,

$$\hat{E}(n_{21}) = \frac{r_2 c_1}{n} = \frac{(2,896)(1,000)}{3,000} = 965.333$$

The estimated expected cell counts for all the data of Table 8.1 are shown in parentheses, along with the observed counts, in Table 8.2.

**TABLE 8.2**
**OBSERVED AND ESTIMATED EXPECTED (IN PARENTHESES) CELL COUNTS FOR THE DATA, TABLE 8.1**

| | PRODUCTION LINE | | | TOTALS |
| | 1 | 2 | 3 | |
| --- | --- | --- | --- | --- |
| NUMBER OF DEFECTIVES | 49 (34.667) | 24 (34.667) | 31 (34.667) | 104 |
| NUMBER OF NONDEFECTIVES | 951 (965.333) | 976 (965.333) | 969 (965.333) | 2,896 |
| TOTALS | 1,000 | 1,000 | 1,000 | 3,000 |

Do the observed cell counts in Table 8.2 differ sufficiently from the estimated expected cell counts to indicate that the null hypothesis

$$H_0: \quad p_1 = p_2 = p_3$$

is false and that the population fractions of defectives differ in value for at least two of the production lines? To answer this question, we must calculate the following statistic:

$$X^2 = \sum_{j=1}^{k} \sum_{i=1}^{2} \frac{[n_{ij} - \hat{E}(n_{ij})]^2}{\hat{E}(n_{ij})}$$

When $n$ is large enough so that all the estimated expected cell counts are at least equal to 5, then $X^2$ will possess a sampling distribution known as a **chi square distribution**. The shape of the chi square distribution will depend upon the number of degrees of freedom (df) associated with the contingency table. For this application, the number of degrees of freedom will always be calculated as indicated in the box:

---

NUMBER OF DEGREES OF FREEDOM FOR CHI SQUARE

$\quad$ df $= (r - 1)(c - 1)$

where

$\quad r =$ Number of rows in the contingency table
$\quad c =$ Number of columns

[For the contingency table discussed in this section, $r = 2$, $c = 3$, and df $= (2 - 1)(3 - 1) = 2$.]

---

$\quad$ The rejection region for the chi square test will always correspond to large values of $X^2$, say $X^2 > \chi_\alpha^2$, where $\chi_\alpha^2$ locates an area $\alpha$ in the upper tail of the chi square distribution. (See Figure 8.2.) These values of $\chi_\alpha^2$ are given in Table IX in the Appendix. We will illustrate the use of this table and the chi square test with the following example.

**FIGURE 8.2**
**THE REJECTION REGION FOR A CONTINGENCY TABLE CHI SQUARE TEST**

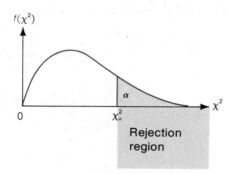

EXAMPLE 8.3

Do the data given in Table 8.2 provide sufficient evidence to indicate a difference in the fractions of defectives produced by the three production lines?

Solution

We wish to test

$\quad H_0: \quad p_1 = p_2 = p_3$

against the alternative hypothesis

$\quad H_a: \quad$ At least two of the fractions of defectives differ

The value of $X^2$ for the data is

$$X^2 = \sum_{j=1}^{3} \sum_{i=1}^{2} \frac{[n_{ij} - \hat{E}(n_{ij})]^2}{\hat{E}(n_{ij})} = \frac{(49 - 34.667)^2}{34.667} + \frac{(24 - 34.667)^2}{34.667}$$

$$+ \frac{(31 - 34.667)^2}{34.667} + \frac{(951 - 965.333)^2}{965.333}$$

$$+ \frac{(976 - 965.333)^2}{965.333} + \frac{(969 - 965.333)^2}{965.333}$$

$$= 9.940675$$

The values of $\chi_\alpha^2$ that locate the rejection region for the chi square test are given in Table IX in the Appendix, a portion of which is shown in Figure 8.3. For $(r - 1)(c - 1) = (2 - 1)(3 - 1) = 2$ df and $\alpha = .05$, we will reject $H_0$: $p_1 = p_2 = p_3$ if $X^2 > \chi_{.05}^2 = 5.99147$ (see Figure 8.2). Since the computed value of $X^2$ is greater than this value, there is sufficient evidence to indicate differences in the fractions of defectives produced by the three production lines.

**FIGURE 8.3**
**REPRODUCTION OF**
**PART OF TABLE IX**
**IN THE APPENDIX**

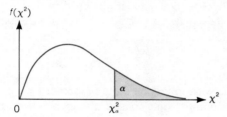

| DEGREES OF FREEDOM | $\chi_{.100}^2$ | $\chi_{.050}^2$ | $\chi_{.025}^2$ | $\chi_{.010}^2$ | $\chi_{.005}^2$ |
|---|---|---|---|---|---|
| 1 | 2.70554 | 3.84146 | 5.02389 | 6.63490 | 7.87944 |
| 2 | 4.60517 | 5.99147 | 7.37776 | 9.21034 | 10.5966 |
| 3 | 6.25139 | 7.81473 | 9.34840 | 11.3449 | 12.8381 |
| 4 | 7.77944 | 9.48773 | 11.1433 | 13.2767 | 14.8602 |
| 5 | 9.23635 | 11.0705 | 12.8325 | 15.0863 | 16.7496 |
| 6 | 10.6446 | 12.5916 | 14.4494 | 16.8119 | 18.5476 |
| 7 | 12.0170 | 14.0671 | 16.0128 | 18.4753 | 20.2777 |

The chi square contingency table analysis is most useful for comparing $k = 3$ or more binomial population proportions. If $k = 2$, it can be shown (proof omitted) that the chi square statistic is equal to the square of the $z$ statistic discussed in Section 8.1. Therefore, although the chi square test can be used to test $H_0$: $(p_1 - p_2) = 0$, it always results in a two-tailed test. If you wish to conduct a one-tailed test for comparing binomial population proportions, i.e., $H_0$: $(p_1 - p_2) = 0$ against $H_a$: $(p_1 - p_2) < 0$ or $H_a$: $(p_1 - p_2) > 0$, you should use the $z$ test of Section 8.1.

CHI SQUARE TEST FOR COMPARING $k$ BINOMIAL POPULATION
PROPORTIONS, $p_1, p_2, \ldots, p_k$

$H_0$:  $p_1 = p_2 = \cdots = p_k$
$H_a$:  At least two of the proportions, $p_1, p_2, \ldots, p_k$, differ

Test statistic:  $X^2 = \sum \dfrac{[n_{ij} - \hat{E}(n_{ij})]^2}{\hat{E}(n_{ij})}$

where

$\hat{E}(n_{ij}) = \dfrac{r_i c_j}{n}$

Rejection region:  $X^2 > \chi^2_\alpha$, where $\chi^2_\alpha$ has $(r-1)(c-1)$ df.
Assumptions:   1.  The samples have been randomly and independently selected from the respective populations.
              2.  The total number $n$ of observations is large enough so that
all estimated expected cell counts are equal to or larger than 5.

EXERCISES

Learning the mechanics
**8.22.** Suppose you wish to compare the proportions $p_1$, $p_2$, $p_3$, and $p_4$ of four binomial populations, and the data, based on independent random samples, are as
shown in the table.

|  | SAMPLE | | | |
|---|---|---|---|---|
|  | 1 | 2 | 3 | 4 |
| SAMPLE SIZE | 100 | 200 | 100 | 400 |
| NUMBER OF SUCCESSES | 30 | 45 | 25 | 150 |

a.  Do the data provide sufficient evidence to indicate differences among $p_1$,
$p_2$, $p_3$, and $p_4$? Test using $\alpha = .05$
b.  Form a 90% confidence interval for $(p_1 - p_2)$.  [*Hint:* Use the method of
Section 8.1.]
c.  Form a 90% confidence interval for $p_1$.  [*Hint:* Use the method of
Section 6.5.]

Applying the concepts
**8.23.** A study was conducted to determine whether a relationship exists between
obesity in children and obesity in their parents. Random samples of fifty obese
and fifty nonobese children were obtained. Then, for each child, it was determined
whether one or both parents were obese. A summary of the data is shown in the
table at the top of the next page. Do the data provide sufficient evidence to indicate
that child obesity is dependent upon parental obesity?

| | | CHILD | | TOTALS |
|---|---|---|---|---|
| | | Obese | Nonobese | |
| PARENT | Obese | 34 | 29 | 63 |
| | Nonobese | 16 | 21 | 37 |
| TOTALS | | 50 | 50 | 100 |

8.24. An experiment was conducted to compare two methods for operating a group family medical practice. Four hundred patients were randomly assigned to two groups: one group received the conventional direct contact with physicians, while the other group made first contact with a nurse–practitioner and were then referred to a physician if a physician's services were deemed necessary. At the conclusion of the experiment the quality of each person's medical care was rated as satisfactory or unsatisfactory by an impartial medical observer in consultation with the patient. The results of the experiment are shown in the table. Do the data present sufficient evidence to indicate a difference in the proportions of satisfactory ratings for the two methods of patient care? Test using $\alpha = .05$.

| | CONVENTIONAL | NURSE–PRACTITIONER | TOTALS |
|---|---|---|---|
| SATISFACTORY | 148 | 161 | 309 |
| UNSATISFACTORY | 52 | 39 | 91 |
| TOTALS | 200 | 200 | 400 |

8.25. A private hospital corporation conducted a survey at four of its hospitals to compare job satisfaction among the nurses. Independent random samples of 200 nurses were selected from each of the four hospitals. The number of dissatisfied nurses in each sample is given in the table. Do the data provide sufficient evidence to indicate differences among the proportions of dissatisfied nurses at the three hospitals? Test using $\alpha = .05$.

| | HOSPITAL | | | |
|---|---|---|---|---|
| | 1 | 2 | 3 | 4 |
| SAMPLE SIZE | 200 | 200 | 200 | 200 |
| NUMBER OF DISSATISFIED NURSES | 23 | 16 | 31 | 34 |

8.4
CONTINGENCY
TABLE ANALYSIS
(OPTIONAL)

In Section 8.3 we used a contingency table analysis and a chi square test to detect differences among the proportions of two or more binomial populations. Contingency tables may also be used to detect a dependency between two methods of classifying data. Suppose an automobile manufacturer wishes to determine whether a relationship exists between the size of newly purchased automobiles and their manufacturers. One thousand recent buyers of American-made cars are randomly sampled, and each purchase is classified with respect to the size and manufacturer of the purchased automobile. The data are summarized in the two-way contingency table shown in Table 8.3.

TABLE 8.3
CONTINGENCY TABLE
FOR AUTOMOBILE SIZE
EXAMPLE

| | MANUFACTURER | | | | TOTALS |
| | A | B | C | D | |
|---|---|---|---|---|---|
| SMALL | 157 | 65 | 181 | 10 | 413 |
| INTERMEDIATE | 126 | 82 | 142 | 46 | 396 |
| LARGE | 58 | 45 | 60 | 28 | 191 |
| TOTALS | 341 | 192 | 383 | 84 | 1,000 |

Note that each of the $n = 1000$ buyers included in the manufacturer's survey is classified according to (1) the manufacturer and (2) the size of the newly purchased car. Each buyer's response falls in one and only one cell of the contingency table. If the buyers are randomly and independently selected from a large population of buyers, this sampling procedure is known as a multinomial experiment.

If the size of a purchased car is independent of the car's manufacturer, we would expect the proportions of small, intermediate, and large cars to be the same for all manufacturers. On the other hand, if the proportions vary from one manufacturer to another, we would say that the choice of car size depends upon the manufacturer, i.e., that car size and car manufacturer are dependent classification variables.

The general form of a two-way contingency table containing $r$ rows and $c$ columns, called an $r \times c$ contingency table, is shown in Table 8.4. Note that the observed count in the cell located in row $i$ and column $j$ is denoted by $n_{ij}$, the $i$th row total is $r_i$, the $j$th column total is $c_j$, and the total sample size is $n$.

TABLE 8.4
GENERAL $r \times c$
CONTINGENCY TABLE

| | | COLUMN | | | | ROW TOTALS |
| | | 1 | 2 | . . . | c | |
|---|---|---|---|---|---|---|
| ROW | 1 | $n_{11}$ | $n_{12}$ | . . . | $n_{1c}$ | $r_1$ |
| | 2 | $n_{21}$ | $n_{22}$ | . . . | $n_{2c}$ | $r_2$ |
| | . | . | . | | . | . |
| | r | $n_{r1}$ | $n_{r2}$ | . . . | $n_{rc}$ | $r_r$ |
| COLUMN TOTALS | | $c_1$ | $c_2$ | . . . | $c_c$ | $n$ |

A test for dependence (or contingency) between two classification variables is conducted in exactly the same manner as the chi square test described in Section 8.3. The only difference is that the contingency table may contain more than two rows. The test is summarized in the box. We will illustrate its use with an example.

GENERAL FORM OF A CONTINGENCY TABLE ANALYSIS:
A TEST FOR INDEPENDENCE

$H_0$:   The two classifications are independent

$H_a$:   The two classifications are dependent

Test statistic:   $X^2 = \sum \dfrac{[n_{ij} - \hat{E}(n_{ij})]^2}{\hat{E}(n_{ij})}$

where

$$\hat{E}(n_{ij}) = \frac{r_i \hat{c}_j}{n}$$

Rejection region:   $X^2 > \chi_\alpha^2$, where $\chi_\alpha^2$ has $(r - 1)(c - 1)$ df.

Assumptions:   1.   The $n$ observed counts are a random sample from the population of interest. We may then consider this to be a multinomial experiment with $r \times c$ possible outcomes.
2.   For the $\chi^2$ approximation to be valid, we require that the estimated expected counts exceed 5 in all cells.

EXAMPLE 8.4

Do the data given in Table 8.3 provide sufficient evidence to indicate a dependence between purchased car size and car manufacturer? Test using $\alpha = .05$.

Solution

Using the data in Table 8.3, we find

$$\hat{E}(n_{11}) = \frac{r_1 c_1}{n} = \frac{(413)(341)}{1,000} = 140.833$$

$$\hat{E}(n_{12}) = \frac{r_1 c_2}{n} = \frac{(413)(192)}{1,000} = 79.296$$

$$\vdots \qquad \vdots \qquad \vdots \qquad \vdots$$

$$\hat{E}(n_{34}) = \frac{r_3 c_4}{n} = \frac{(191)(84)}{1,000} = 16.044$$

The observed data and the estimated expected values (in parentheses) are shown in Table 8.5.

**TABLE 8.5
OBSERVED AND
ESTIMATED EXPECTED
(IN PARENTHESES)
COUNTS**

|  | MANUFACTURER | | | |
|---|---|---|---|---|
|  | A | B | C | D |
| SMALL | 157 (140.833) | 65 (79.296) | 181 (158.179) | 10 (34.692) |
| INTERMEDIATE | 126 (135.036) | 82 (76.032) | 142 (151.668) | 46 (33.264) |
| LARGE | 58 (65.131) | 45 (36.672) | 60 (73.153) | 28 (16.044) |

We now use the $X^2$ statistic to compare the observed and (estimated) expected counts in each cell of the contingency table:

$$X^2 = \frac{[n_{11} - \hat{E}(n_{11})]^2}{\hat{E}(n_{11})} + \frac{[n_{12} - \hat{E}(n_{12})]^2}{\hat{E}(n_{12})} + \cdots + \frac{[n_{34} - \hat{E}(n_{34})]^2}{\hat{E}(n_{34})}$$

$$= \sum \frac{[n_{ij} - \hat{E}(n_{ij})]^2}{\hat{E}(n_{ij})}$$

Substituting the data of Table 8.5 into this expression,

$$X^2 = \frac{(157 - 140.833)^2}{140.833} + \frac{(65 - 79.296)^2}{79.296} + \cdots + \frac{(28 - 16.044)^2}{16.044} = 45.81$$

Large values of $X^2$ imply that the observed and expected counts do not closely agree, and therefore imply that the hypothesis of independence is false. To determine how large $X^2$ must be before it is too large to be attributed to chance, we make use of the fact that the sampling distribution of $X^2$ is approximately a $\chi^2$ probability distribution when the classifications are independent. The appropriate number of degrees of freedom for the approximating chi square distribution will be $(r - 1)(c - 1)$, where $r$ is the number of rows and $c$ is the number of columns in the table. For this example, the degrees of freedom for $\chi^2$ is $(r - 1)(c - 1) = (3 - 1)(4 - 1) = 6$. Then, for $\alpha = .05$, we reject the hypothesis of independence when

$$X^2 > \chi^2_{.05} = 12.5916$$

Since the computed $X^2 = 45.81$ exceeds the value 12.5916, we conclude that the size and manufacturer of a car selected by a purchaser are dependent events.

EXERCISES

Learning the mechanics

**8.26.** a. Test the null hypothesis that the rows and columns for the $4 \times 3$ contingency table shown below are independent. Test using $\alpha = .01$.

|  |  | COLUMN | |  |
|---|---|---|---|---|
|  |  | 1 | 2 | 3 |
|  | 1 | 20 | 30 | 50 |
| ROW | 2 | 40 | 20 | 40 |
|  | 3 | 100 | 50 | 50 |
|  | 4 | 40 | 0 | 60 |

b. Form a 90% confidence interval for $p_{11}$, the probability of observing a response in the first row and first column of the table.

Applying the concepts

**8.27.** One criterion used to evaluate employees in the assembly section of a large factory is the number of defective pieces per 1,000 parts produced. The quality control department wants to find out whether there is a relationship between years of experience and defect rate. Since the job is rather repetitive, after the initial training period, any improvement due to a learning effect might be offset by a decrease in the motivation of a worker. A defect rate is calculated for each worker for a yearly evaluation. The results for 100 workers are given in the table. Is there evidence of a relationship between defect rate and years of experience? Use $\alpha = .05$.

| | | YEARS OF EXPERIENCE (AFTER TRAINING PERIOD) | | |
|---|---|---|---|---|
| | | 1 | 2–5 | 6–10 |
| DEFECT RATE | High | 6 | 9 | 9 |
| | Average | 9 | 19 | 23 |
| | Low | 7 | 8 | 10 |

**8.28.** A team of market researchers conducted a study involving 200 inhabitants of the United States to determine what people fear the most. The sex of each person polled was noted, and then each was asked which of the following was his or her greatest fear: speaking before a group, heights, bugs/insects, financial problems, sickness/death, and other. The results of the poll are given in the table. Do the data provide sufficient information to indicate a relationship between sex and greatest fear? Test at the $\alpha = .05$ level.

| | | GREATEST FEAR | | | | | |
|---|---|---|---|---|---|---|---|
| | | Speaking before a group | Heights | Bugs/insects | Financial problems | Sickness/ death | Other |
| SEX | Male | 21 | 10 | 7 | 23 | 15 | 21 |
| | Female | 16 | 22 | 15 | 9 | 18 | 23 |

**8.29.** Along with the technological age comes the problem of workers being replaced by machines. A labor management organization wants to study the problem of workers displaced by automation within three industries. Case reports for 100 workers whose loss of job is directly attributable to technological advances are selected within each industry. For each worker selected it is determined whether he or she was given another job within the same company, found a job with another company in the same industry, found a job in a new industry, or has been unemployed for longer than 6 months. The results are given in the table.

CURRENT STATUS OF AUTOMATION-DISPLACED WORKERS

| | | SAME COMPANY | NEW COMPANY (Same industry) | NEW INDUSTRY | UNEMPLOYED |
|---|---|---|---|---|---|
| INDUSTRY | A | 62 | 11 | 20 | 7 |
| | B | 45 | 8 | 38 | 9 |
| | C | 68 | 19 | 8 | 5 |

a. Does the plight of automation-displaced workers depend on the industry? Use $\alpha = .01$.
b. Estimate the difference between the proportions of displaced workers who find work in another industry for industries A and C. Use a 95% confidence interval.

This chapter showed you how to compare two population proportions using methodology that is very similar to the methodology used to compare two population means (Chapter 7). When the samples have been randomly and independently selected from two binomial populations and when the sample sizes are large, the difference $(\hat{p}_1 - \hat{p}_2)$ between the sample proportions will be approximately normally distributed (the central limit theorem, Section 6.5). This enables us to form a large sample standard normal $z$ statistic similar to the one used to compare two population means. This statistic is then used as a test statistic for testing hypotheses about the difference between two population proportions and to construct a confidence interval for $(p_1 - p_2)$. The width of this confidence interval can be controlled by adjusting the sample sizes $n_1$ and $n_2$.

Optional Section 8.3 discussed how to compare more than two population proportions using a chi square contingency table analysis. This methodology was applied in optional Section 8.4 to test for a dependence between two methods of classification.

Learning the mechanics

**8.30.** Independent random samples were selected from two binomial populations. The sizes and numbers of observed successes for each sample are shown below:

| SAMPLE 1 | SAMPLE 2 |
|----------|----------|
| $n_1 = 200$ | $n_2 = 200$ |
| $x_1 = 110$ | $x_2 = 130$ |

a. Test the null hypothesis $H_0$: $(p_1 - p_2) = 0$ against the alternative hypothesis $H_a$: $(p_1 - p_2) < 0$. Use $\alpha = .10$.
b. Form a 95% confidence interval for $(p_1 - p_2)$.
c. What sample sizes would be required if we wish to estimate $(p_1 - p_2)$ to within .01 with 95% confidence? Assume that $n_1 = n_2$.

**8.31.** A random sample of 250 observations was classified according to the row and column categories shown in the table. Do the data provide sufficient evidence to conclude that the rows and columns are dependent? Test using $\alpha = .05$.

|  |  | COLUMN | | |
|--|--|--------|--|--|
|  |  | 1 | 2 | 3 |
|  | 1 | 20 | 20 | 10 |
| ROW | 2 | 10 | 20 | 70 |
|  | 3 | 20 | 50 | 30 |

Applying the concepts

**8.32.** A careful auditing is essential to all businesses, large and small. Suppose a firm wants to compare the performances of two auditors it employs. One measure of auditing performance is error rate, so the firm decides to sample 200 pages at random from the work of each auditor and carefully examine each page for errors. Suppose the number of pages on which at least one error is found is seventeen for auditor A and twenty-five for auditor B. Test to see whether these data indicate a difference in the true error rates for the two auditors. Use $\alpha = .01$.

**8.33.** An economist wants to investigate the difference in unemployment rates between an urban industrial community and a university community in the same state. She interviews 525 potential members of the work force in the industrial community and 375 in the university community. Of these, forty-seven and twenty-two, respectively, are unemployed. Use a 95% confidence interval to estimate the difference in unemployment rates in the two communities. Interpret this interval.

**8.34.** A large shipment of produce contains Valencia and navel oranges. To determine whether there is a difference in the proportions of nonmarketable fruit between the two varieties, random samples of 850 Valencia oranges and 1,500 navel oranges were independently selected and the number of nonmarketable oranges of each type were counted. It was found that thirty Valencia and ninety navel oranges from these samples were nonmarketable. Do these data provide sufficient evidence to indicate a difference between the proportions of nonmarketable Valencia and navel oranges? Test at the $\alpha = .05$ level of significance.

**8.35.** A political candidate conducted a sample survey to determine whether a television advertising campaign was worthwhile. Both before and after the advertising campaign random samples were taken from the candidate's constituency and each person was asked his or her voter preference. The results of the survey are shown in the table. Estimate the difference in the proportion of voters who favor the candidate before and after the campaign using a 95% confidence interval.

|  | SAMPLE SIZE | NUMBER WHO PREFER THE CANDIDATE |
|---|---|---|
| BEFORE ADVERTISING CAMPAIGN | 200 | 85 |
| AFTER ADVERTISING CAMPAIGN | 300 | 139 |

**8.36.** Refer to Exercise 8.35. What size samples need to be taken in order to estimate the difference between proportions to within .04 with probability .95? Assume that $n_1 = n_2$.

**8.37.** Smoke detectors are highly recommended safety devices for early fire detection in homes and businesses. It is extremely important that the devices are not defective. Suppose that 100 brand A smoke detectors are tested and twelve fail to emit a warning signal. Subjected to the same test, fifteen out of ninety brand B detectors fail to operate. Form a 90% confidence interval to estimate the difference between the fractions of defective smoke detectors produced by the two companies. Interpret this confidence interval.

**8.38.** A football fan decides to place bets according to the predictions of one of two newspaper columnists. To decide which columnist to use, both columnists' predictions are randomly sampled over the preceding weeks and the number of correct predictions for each is determined. The results are shown in the table (page 340). Do

the data provide sufficient evidence to indicate that one of the columnists is better a
picking winners than the other?

| COLUMNIST | TOTAL NUMBER OF PREDICTIONS | NUMBER OF CORRECT PREDICTIONS |
|-----------|---------------------------|------------------------------|
| 1 | 60 | 48 |
| 2 | 50 | 42 |

**8.39.** A city commission wished to analyze resident reaction to the purchase of land
for a new city park. Random samples of 500 residents were selected from within each
of four distinctly different sections of the city. The numbers of residents who favor
the land purchase and those who oppose it are given in the table.

| | SECTION | | | |
|---|---|---|---|---|
| | 1 | 2 | 3 | 4 |
| NUMBER IN FAVOR | 373 | 269 | 327 | 394 |
| NUMBER OPPOSED | 127 | 231 | 173 | 106 |
| TOTAL NUMBER IN SAMPLE | 500 | 500 | 500 | 500 |

a. Do the data provide sufficient evidence to indicate differences in the proportions of people, in the four sections of the city, who favor the land purchase? Test using $\alpha = .05$.
b. Find a 95% confidence interval for the proportion of people in section 4
who favor the purchase.
c. Find a 95% confidence interval for the difference in the proportions of
people in sections 1 and 2 who favor purchase of the land.

**8.40.** Despite a good winning percentage, a certain major league baseball team has
not drawn as many fans as one would expect. In hopes of finding ways to increase attendance, the management plans to interview fans who come to the games to
find out why they come. One thing that the management might want to know is
whether there are differences in support for the team among various age groups.
Suppose the information in the table was collected during interviews with fans selected at random. Can you conclude that there is a relationship between age and
number of games attended per year? Use $\alpha = .05$.

| | | NUMBER OF GAMES ATTENDED PER YEAR | | |
|---|---|---|---|---|
| | | 1 or 2 | 3–5 | Over 5 |
| | Under 20 | 78 | 107 | 17 |
| AGE OF FAN | 21–30 | 147 | 87 | 13 |
| | 31–40 | 129 | 86 | 19 |
| | 41–55 | 55 | 103 | 40 |
| | Over 55 | 23 | 74 | 22 |

**8.41.** Several life insurance firms have policies geared to college students. To get more information about this group, a major insurance firm interviewed college students to find out the type of life insurance preferred, if any. The following table was produced after surveying 1,600 students:

| | PREFERENCES | | |
|---|---|---|---|
| | Term insurance | Whole-life insurance | No preference |
| FEMALES | 116 | 27 | 676 |
| MALES | 215 | 33 | 533 |

a. Is there evidence that the life insurance preference of students depends on sex?

b. Estimate the proportion of female students who have no preference in life insurance when the null hypothesis is true.

**ON YOUR OWN . . .**

As you have seen from the examples and exercises in this chapter, the need to compare the proportions $p_1$ and $p_2$ from two binomial populations is a very common problem. Identify a problem of this type, based on either opinion surveys or on laboratory experiments.

Clearly define the two binomial populations of interest to you. Choose the bound $B$ on the error of estimation that you are willing to tolerate and determine the sample sizes necessary to achieve this bound. Then conduct the survey, observe the sample values of $x_1$ and $x_2$, and find a confidence interval for the difference between $p_1$ and $p_2$.

**REFERENCES**

Agresti, A., & Agresti, B. F. *Statistical methods for the social sciences.* San Francisco: Dellen, 1979.

Cochran, W. G. "The $\chi^2$ Test of Goodness of Fit." *Annals of Mathematical Statistics,* 1952, *23,* 315–345.

Freedman, D., Pisani, R., & Purves, R. *Statistics.* New York: W. W. Norton and Co., 1978.

Mendenhall, W. *Introduction to probability and statistics.* 6th ed. Boston: Duxbury, 1983.

Snedecor, G. W., & Cochran, W. *Statistical methods.* 7th ed. Ames, Iowa: Iowa University Press, 1980.

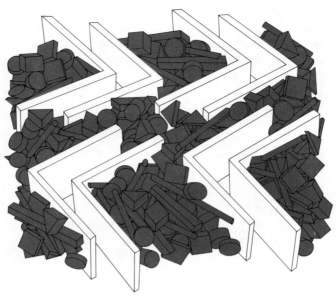

# 9

# SIMPLE LINEAR REGRESSION

## WHERE WE'VE BEEN . . .

As we have previously learned, a population is generated by observations on a single random variable and is described by its mean, variance, the proportion of observations possessing some specific characteristic and other parameters. We learned how to estimate and test hypotheses about these parameters based on a random sample of observations from the population and then extended these methods to allow for a comparison of parameters (means, proportions, etc.) from two or more populations.

## WHERE WE'RE GOING . . .

For many sampling situations, we have much more information available on a random variable (and the population it generates) than that contained in a single random sample. For example, if we wanted to predict the rainfall at a given location on a given day, we could select a single random sample of $n$ daily rainfalls, use the methods of Chapter 6 to estimate the mean daily rainfall $\mu$, and then use this quantity to predict the rainfall for any given day. But this method fails to utilize scientific information that is available to any forecaster. We know the daily rainfall is related to barometric pressure, cloud cover, etc., and we also know it is more likely that the daily rainfall will be greater if the barometric pressure is low and it is a cloudy day than if the pressure is high and the day is sunny. By measuring barometric pressure and cloud cover at the same time that we sample the daily rainfall, we hope to establish the relationship between these variables and to utilize them for prediction.

This chapter is devoted to the most elementary situation—relating two variables.

Much practical research is devoted to the topic of modeling—that is, trying to describe how variables are related. For example, an econometrician might be interested in modeling the relationship between the Gross National Product (GNP) and the current rate of unemployment. A behavioral psychologist might want to model the relationship between a child's motor activity and the concentration of a certain enzyme in the bloodstream. Or, a sociologist might be interested in relating the rate of juvenile delinquency in a neighborhood to the percentage of children in broken homes in that neighborhood.

One method of modeling the relationship between variables is called regression analysis. In this chapter, we will discuss a simple linear (straight-line) model for the relationship between two variables, the least squares method of fitting regression models using sample data, how to make inferences about the model, and how to use the model for prediction.

## 9.1 PROBABILISTIC MODELS

An important consideration when taking any drug is how it may affect one's perception or general awareness. Suppose you want to model the length of time it takes to respond to a stimulus (a measure of awareness) as a function of the percentage of a certain drug in the bloodstream. The first question to be answered is this: "Do you think an exact relationship exists between these two variables?" That is, do you think it is possible to state the exact length of time it takes an individual (subject) to respond if the amount of the drug in the bloodstream is known? We think you will agree with us that this is not possible for several reasons. The reaction time depends on many variables other than the percentage of the drug in the bloodstream; for example, the time of day, the amount of sleep the subject had the night before, the subject's visual acuity, the subject's general reaction time without the drug, and the subject's age would all probably affect reaction time. However, even if many variables are included in a model, it is still unlikely that we would be able to predict *exactly* the subject's reaction time. There will almost certainly be some variation in response times due strictly to random phenomena that cannot be modeled or explained.

If we were to construct a model that hypothesized an exact relationship between variables, it would be called a deterministic model. For example, if we believe that $y$, reaction time (in seconds), will be exactly one and one-half times $x$, the amount of drug in the blood, we write

$$y = 1.5x$$

This represents a deterministic relationship between the variables $y$ and $x$. It implies that $y$ can always be determined exactly when the value of $x$ is known. There is no allowance for error in this prediction.

If, on the other hand, we believe there will be unexplained variation in reaction times—perhaps caused by important but unincluded variables or by random phenomena—we will discard the deterministic model and use a model that accounts for this random error. This probabilistic model includes both a deterministic component and a random error component. For example, if we hypothesize that the response time $y$ is related to the percentage of drug $x$ by

$$y = 1.5x + \text{Random error}$$

we are hypothesizing a probabilistic relationship between $y$ and $x$. Note that the deterministic component of this probabilistic model is $1.5x$.

Figure 9.1(a) shows the possible responses for five different values of $x$, the percentage of drug in the blood, when the model is deterministic. All the responses must fall exactly on the line because a deterministic model leaves no room for error.

**FIGURE 9.1**
POSSIBLE
REACTION TIMES, $y$,
FOR FIVE DIFFERENT
DRUG PERCENTAGES, $x$

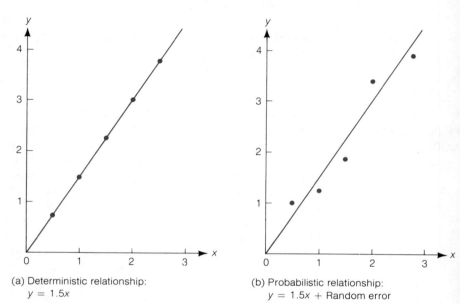

(a) Deterministic relationship:
  $y = 1.5x$

(b) Probabilistic relationship:
  $y = 1.5x + \text{Random error}$

Figure 9.1(b) shows a possible set of responses for the same values of $x$ when we are using a probabilistic model. Note that the deterministic part of the model (the straight line itself) is the same. Now, however, the inclusion of a random error component allows the response times to vary from this line. Since we know that the response time does vary randomly for a given value of $x$, the probabilistic model provides a more realistic model for $y$ than does the deterministic model.

---

GENERAL FORM OF PROBABILISTIC MODELS

  $y = \text{Deterministic component} + \text{Random error}$

where $y$ is the variable to be predicted. We will always assume that the mean value of the random error equals 0. This is equivalent to assuming that the mean value of $y$, $E(y)$, equals the deterministic component of the model, i.e.,

  $E(y) = \text{Deterministic component}$

---

We begin with the simplest of probabilistic models—the straight-line model—which derives its name from the fact that the deterministic portion of the model graphs as a straight line. Fitting this model to a set of data is an example of regression analysis or regression modeling. The elements of the straight-line model are summarized in the box.

---

**A FIRST-ORDER (STRAIGHT-LINE) PROBABILISTIC MODEL**

$$y = \beta_0 + \beta_1 x + \varepsilon$$

where

$y$ = Dependent variable (variable to be modeled)

$x$ = Independent* variable (variable used as a predictor of $y$)

$\varepsilon$ (epsilon) = Random error component

$\beta_0$ (beta zero) = $y$-intercept of the line, i.e., point at which the line intercepts or cuts through the $y$-axis (see Figure 9.2)

$\beta_1$ (beta one) = Slope of the line, i.e., amount of increase (or decrease) in the deterministic component of $y$ for every 1 unit increase in $x$. You can see (Figure 9.2) that $E(y)$ increases by the amount $\beta_1$ as $x$ increases from 2 to 3.

---

**FIGURE 9.2 THE STRAIGHT-LINE MODEL**

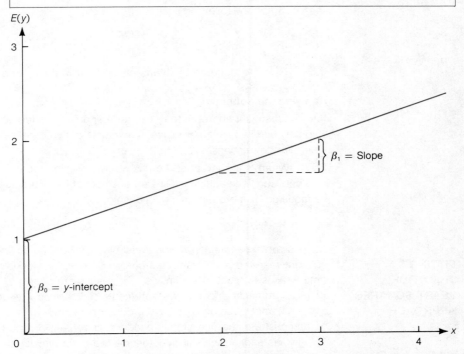

*The word *independent* should not be interpreted in a probabilistic sense, as defined in Chapter 3. The phrase *independent variable* is used in regression analysis to refer to a predictor variable for the response $y$.

Note that we use Greek symbols, $\beta_0$ and $\beta_1$, to represent the $y$-intercept and slope of the model. They are population parameters with numerical values that will be known only if we have access to the entire population of $(x, y)$ measurements.

It is helpful to think of regression modeling as a five step procedure:

**Step 1.** Hypothesize the deterministic component of the probabilistic model.

**Step 2.** Use sample data to estimate unknown parameters in the model.

**Step 3.** Specify the probability distribution of the random error term, and estimate any unknown parameters of this distribution.

**Step 4.** Statistically check the usefulness of the model.

**Step 5.** When satisfied that the model is useful, use it for prediction, estimation, etc.

In this chapter we will skip step 1 (which is difficult) and deal only with the straight-line model.

EXERCISES

Learning the mechanics

**9.1.** Graph the line that passes through the following points:

a. (0, 2) and (2, 6)  b. (0, 4) and (2, 6)
c. (0, 2) and (2, 0)  d. (0, −4) and (3, −7)

**9.2.** Give the slope and $y$-intercept for each of the lines graphed in Exercise 9.1.

**9.3.** Give the equations of the lines graphed in Exercise 9.1.

**9.4** Graph the following lines:

a. $y = 3 + 2x$  b. $y = 1 + x$  c. $y = -2 + 3x$
d. $y = 5x$  e. $y = 4 - 2x$

**9.5.** Give the slope and $y$-intercept for each of the lines defined in Exercise 9.4.

Applying the concepts

**9.6.** When is it appropriate to use a deterministic model to describe the relationship between two variables? Give an example of two variables that you think have a deterministic relationship.

**9.7.** When is it appropriate to use a probabilistic model to describe the relationship between two variables? Give an example of two variables that you think should be modeled by a probabilistic relationship.

9.2
FITTING
THE MODEL:
LEAST SQUARES
APPROACH

Suppose an experiment involving five subjects is conducted to determine the relationship between the percentage of a certain drug in the bloodstream and the length of time it takes to react to a stimulus. The results are shown in Table 9.1 (the number of measurements and the measurements themselves are unrealistically simple to avoid arithmetic confusion in this introductory example). This set of data will be used to demonstrate the five step procedure of regression modeling given in Section 9.1. In this section we will hypothesize the deterministic component of the model and estimate its unknown parameters (steps 1 and 2). Discussion of the model assumptions and the random error component (step 3) are the subjects of Sections 9.3 and 9.4, while Sections 9.5–9.7 assess the utility of the model (step 4). Finally, using the model for prediction and estimation (step 5) is the subject of Section 9.8.

**Step 1.** Hypothesize the deterministic component of the probabilistic model. As stated before, we will consider only straight-line models in this chapter, and thus the complete model to relate mean response time $E(y)$ to drug percentage $x$ is given by

$$E(y) = \beta_0 + \beta_1 x$$

**Step 2.** Use sample data to estimate unknown parameters in the model. This step is the subject of this section—namely, how can we best use the information in the sample of five observations in Table 9.1 to estimate the unknown $y$-intercept $\beta_0$ and slope $\beta_1$?

**TABLE 9.1**
**REACTION TIME VERSUS DRUG PERCENTAGE**

| SUBJECT | AMOUNT OF DRUG $x$, % | REACTION TIME $y$, seconds |
|---------|----------------------|---------------------------|
| 1 | 1 | 1 |
| 2 | 2 | 1 |
| 3 | 3 | 2 |
| 4 | 4 | 2 |
| 5 | 5 | 4 |

**FIGURE 9.3**
SCATTERGRAM FOR DATA IN TABLE 9.1

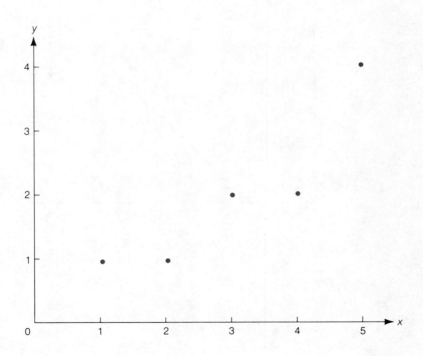

To determine whether a linear relationship between $y$ and $x$ is plausible, it is helpful to plot the sample data. Such a plot, called a scattergram, locates each of the five data points on a graph, as shown in Figure 9.3 (page 347). Note that the scattergram suggests a general tendency for $y$ to increase as $x$ increases. If you place a ruler on the scattergram, you will see that a line may be drawn through three of the five points, as shown in Figure 9.4. To obtain the equation of this visually fitted line, note that the line intersects the $y$-axis at $y = -1$, so the $y$-intercept is $-1$. Also, $y$ increases exactly 1 unit for every 1 unit increase in $x$, indicating that the slope is $+1$. Therefore, the equation is

$$\tilde{y} = -1 + 1(x) = -1 + x$$

where $\tilde{y}$ is used to denote the predicted $y$ from the visual model.

One way to decide quantitatively how well a straight line fits a set of data is to note the extent to which the data points deviate from the line. For example, to evaluate the model in Figure 9.4, we calculate the magnitude of the deviations, i.e., the differ-

**FIGURE 9.4**
**VISUAL**
**STRAIGHT-LINE FIT**
**TO THE DATA**

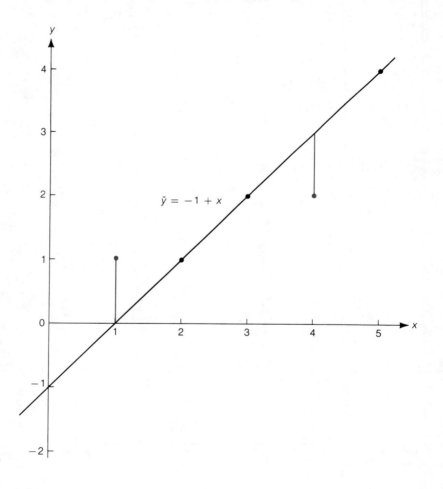

ences between the observed and the predicted values of $y$. These deviations, or errors, are the vertical distances between observed and predicted values (see Figure 9.4). The observed and predicted values of $y$, their differences, and their squared differences are shown in Table 9.2. Note that the sum of errors equals 0 and the sum of squares of the errors (SSE), which gives greater emphasis to large deviations of the points from the line, is equal to 2.

**TABLE 9.2**
**COMPARING OBSERVED AND PREDICTED VALUES FOR THE VISUAL MODEL**

| $x$ | $y$ | $\hat{y} = -1 + x$ | $(y - \hat{y})$ | $(y - \hat{y})^2$ |
|---|---|---|---|---|
| 1 | 1 | 0 | $(1 - 0) = \phantom{-}1$ | 1 |
| 2 | 1 | 1 | $(1 - 1) = \phantom{-}0$ | 0 |
| 3 | 2 | 2 | $(2 - 2) = \phantom{-}0$ | 0 |
| 4 | 2 | 3 | $(2 - 3) = -1$ | 1 |
| 5 | 4 | 4 | $(4 - 4) = \phantom{-}0$ | 0 |
| | | | Sum of errors $= \phantom{-}0$ | Sum of squared errors (SSE) $= 2$ |

You can see by shifting the ruler around the graph that it is possible to find many lines for which the sum of the errors is equal to 0, but it can be shown that there is one (and only one) line for which the SSE is a minimum. This line is called the least squares line, the regression line, or least squares prediction equation.

To find the least squares line for a set of data, assume that we have a sample of $n$ data points which can be identified by corresponding values of $x$ and $y$, say $(x_1, y_1), (x_2, y_2), \ldots, (x_n, y_n)$. For example, the $n = 5$ data points shown in Table 9.2 are (1, 1), (2, 1), (3, 2), (4, 2), and (5, 4). The straight-line model for the response $y$ in terms of $x$ is

$$y = \beta_0 + \beta_1 x + \varepsilon$$

The line that gives the mean (or expected) value of $y$ for a given value of $x$ is

$$E(y) = \beta_0 + \beta_1 x$$

and the fitted line, which we hope to find, is represented as

$$\hat{y} = \hat{\beta}_0 + \hat{\beta}_1 x$$

The "hats" can be read as "estimator of." Thus, $\hat{y}$ is an estimator of the mean value of $y$, $E(y)$, and a predictor of some future value of $y$; and $\hat{\beta}_0$ and $\hat{\beta}_1$ are estimators of $\beta_0$ and $\beta_1$, respectively.

For a given data point, say the point $(x_i, y_i)$, the observed value of $y$ is $y_i$ and the predicted value of $y$ would be obtained by substituting $x_i$ into the prediction equation:

$$\hat{y}_i = \hat{\beta}_0 + \hat{\beta}_1 x_i$$

And the deviation of the $i$th value of $y$ from its predicted value is

$$(y_i - \hat{y}_i) = [y_i - (\hat{\beta}_0 + \hat{\beta}_1 x_i)]$$

Then the sum of squares of the deviations of the $y$ values about their predicted values for all of the $n$ data points is

$$\text{SSE} = \sum [y_i - (\hat{\beta}_0 + \hat{\beta}_1 x_i)]^2$$

The quantities $\hat{\beta}_0$ and $\hat{\beta}_1$ that make the SSE a minimum are called the least squares estimates of the population parameters $\beta_0$ and $\beta_1$, and the prediction equation $\hat{y} = \hat{\beta}_0 + \hat{\beta}_1 x$ is called the least squares line.

---

**DEFINITION 9.1**
The least squares line is one that has a smaller SSE than any other straight-line model.

---

The values of $\hat{\beta}_0$ and $\hat{\beta}_1$ that minimize the SSE are (proof omitted) given by the formulas in the box.[*]

---

**FORMULAS FOR THE LEAST SQUARES ESTIMATES**

Slope:   $\hat{\beta}_1 = \dfrac{\text{SS}_{xy}}{\text{SS}_{xx}}$

$y$-intercept:   $\hat{\beta}_0 = \bar{y} - \hat{\beta}_1 \bar{x}$

where

$$\text{SS}_{xy} = \sum x_i y_i - \frac{\left(\sum x_i\right)\left(\sum y_i\right)}{n}$$

$$\text{SS}_{xx} = \sum x_i^2 - \frac{\left(\sum x_i\right)^2}{n}$$

$n$ = Sample size

---

Preliminary computations for finding the least squares line for the drug reaction time example are contained in Table 9.3. We can now calculate

$$\text{SS}_{xy} = \sum x_i y_i - \frac{\left(\sum x_i\right)\left(\sum y_i\right)}{5} = 37 - \frac{(15)(10)}{5}$$

$$= 37 - 30 = 7$$

[*]Students who are familiar with the calculus should note that the values of $\beta_0$ and $\beta_1$ that minimize SSE $= \Sigma (y_i - \hat{y}_i)^2$ are obtained by setting the two partial derivatives $\partial \text{SSE}/\partial \beta_0$ and $\partial \text{SSE}/\partial \beta_1$ equal to zero. The solutions to these two equations yield the formulas shown in the box. Furthermore, we denote the *sample* solutions to the equations by $\hat{\beta}_0$ and $\hat{\beta}_1$, where the ^(hat) denotes that these are sample estimates of the true population intercept $\beta_0$ and slope $\beta_1$.

| | $x_i$ | $y_i$ | $x_i^2$ | $x_i y_i$ |
|---|---|---|---|---|
| | 1 | 1 | 1 | 1 |
| | 2 | 1 | 4 | 2 |
| | 3 | 2 | 9 | 6 |
| | 4 | 2 | 16 | 8 |
| | 5 | 4 | 25 | 20 |
| Totals | $\Sigma x_i = 15$ | $\Sigma y_i = 10$ | $\Sigma x_i^2 = 55$ | $\Sigma x_i y_i = 37$ |

**TABLE 9.3 PRELIMINARY COMPUTATIONS FOR THE DRUG REACTION TIME EXAMPLE**

$$SS_{xx} = \sum x_i^2 - \frac{\left(\sum x_i\right)^2}{5} = 55 - \frac{(15)^2}{5}$$

$$= 55 - 45 = 10$$

Then, the slope of the least squares line is

$$\hat{\beta}_1 = \frac{SS_{xy}}{SS_{xx}} = \frac{7}{10} = .7$$

and the $y$-intercept is

$$\hat{\beta}_0 = \bar{y} - \hat{\beta}_1 \bar{x} = \frac{\sum y_i}{5} - \hat{\beta}_1 \frac{\left(\sum x_i\right)}{5} = \frac{10}{5} - (.7)\frac{(15)}{5}$$

$$= 2 - (.7)(3) = 2 - 2.1 = -.1$$

The least squares line is thus

$$\hat{y} = \hat{\beta}_0 + \hat{\beta}_1 x = -.1 + .7x$$

The graph of this line is shown in Figure 9.5 on page 352.

The predicted value of $y$ for a given value of $x$ can be obtained by substituting into the formula for the least squares line. Thus, when $x = 2$, we predict $y$ to be

$$\hat{y} = -.1 + .7x = -.1 + .7(2) = 1.3$$

We will show how to find a prediction interval for $y$ in Section 9.8.

The observed and predicted values of $y$, the deviations of the $y$ values about their predicted values, and the squares of these deviations are shown in Table 9.4. Note

**TABLE 9.4 COMPARING OBSERVED AND PREDICTED VALUES FOR THE LEAST SQUARES PREDICTION EQUATION**

| $x$ | $y$ | $\hat{y} = -.1 + .7x$ | $(y - \hat{y})$ | | $(y - \hat{y})^2$ |
|---|---|---|---|---|---|
| 1 | 1 | .6 | $(1 - .6) =$ | .4 | 0.16 |
| 2 | 1 | 1.3 | $(1 - 1.3) = -.3$ | | 0.09 |
| 3 | 2 | 2.0 | $(2 - 2.0) =$ | 0 | 0.00 |
| 4 | 2 | 2.7 | $(2 - 2.7) = -.7$ | | 0.49 |
| 5 | 4 | 3.4 | $(4 - 3.4) =$ | .6 | 0.36 |
| | | | Sum of errors $=$ | 0 | SSE $= 1.10$ |

FIGURE 9.5
THE LINE
$\hat{y} = -.1 + .7x$
FIT TO THE DATA

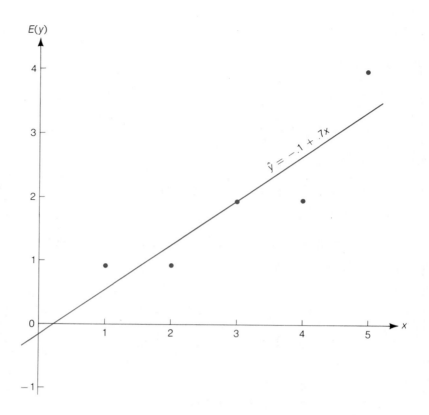

that the sum of squares of the deviations, SSE, is 1.10, and (as we would expect) this is less than the SSE = 2.0 obtained in Table 9.2 for the visually fitted line.

To summarize, we have defined the best-fitting straight line to be the one that satisfies the least squares criterion; that is, the sum of the squared errors will be smaller than for any other straight-line model. This line is called the least squares line, and its equation is called the least squares prediction equation.

EXERCISES

Learning the mechanics

**9.8.** Consider the four data points shown below:

| $x$ | 0 | 1 | 2 | 3 |
|---|---|---|---|---|
| $y$ | 3 | 5 | 7 | 9 |

a. Plot the points and visually fit a straight line through them.
b. Find the slope and the $y$-intercept of the line drawn in part a.
c. Fit a line to the data using the method of least squares, i.e., find the least squares estimates of $\beta_0$ and $\beta_1$, and give the equation of the least squares line.

d. Compare the answers to parts b and c. Are corresponding estimates of $\beta_0$ and $\beta_1$ equal? Should they be?

**9.9.** Consider the six data points shown below:

| $x$ | 1 | 2 | 3 | 4 | 5 | 6 |
|---|---|---|---|---|---|---|
| $y$ | 1 | 2 | 2 | 3 | 5 | 5 |

a. Plot the points and visually fit a straight line through them.
b. Find the slope and the $y$-intercept of the line drawn in part a.
c. Fit a line to the data using the method of least squares, i.e., find the least squares estimates of $\beta_0$ and $\beta_1$.
d. Graph the least squares line and compare it to the line drawn in part a. Comment.

**9.10.** Use the method of least squares to fit a straight line to the following five data points:

| $x$ | −2 | −1 | 0 | 1 | 2 |
|---|---|---|---|---|---|
| $y$ | 4 | 3 | 3 | 1 | −1 |

a. What are the least squares estimates of $\beta_0$ and $\beta_1$?
b. Plot the data points and graph the least squares line. Does the line pass through the data points?
c. Predict the value of $y$ for $x = 1$.
d. Predict the value of $y$ for $x = -1.5$.

Applying the concepts

**9.11.** In recent years, physicians have used the so-called "dividing reflex" to reduce abnormally rapid heartbeats in humans by briefly submerging the patient's face in cold water. The reflex, triggered by cold water temperatures, is an involuntary neural response that shuts off circulation to the skin, muscles, and internal organs, and diverts extra oxygen-carrying blood to the heart, lungs, and brain. A research physician conducted an experiment to investigate the effects of various cold water temperatures on the pulse rate of small children. The data for seven 6 year old children are shown in the table.

| CHILD | TEMPERATURE OF WATER $x$, °F | DECREASE IN PULSE RATE $y$, beats/minute |
|---|---|---|
| 1 | 68 | 2 |
| 2 | 65 | 5 |
| 3 | 70 | 1 |
| 4 | 62 | 10 |
| 5 | 60 | 9 |
| 6 | 55 | 13 |
| 7 | 58 | 10 |

a.   Find the least squares line for the data.

b.   Construct a scattergram for the data; then graph the least squares line as a check on your calculations.

c.   If the water temperature is 60°F, predict the drop in pulse rate for a 6 year old child. [*Note:*   A measure of the reliability of these predictions will be discussed in Section 9.8.]

**9.12.**   A car dealer is interested in modeling the relationship between the number of cars sold by the firm each week and the number of salespeople who work on the showroom floor. The dealer believes the relationship between the two variables can best be described by a straight line. The sample data shown in the table were supplied by the car dealer.

| WEEK OF | NUMBER OF CARS SOLD $y$ | NUMBER OF SALESPEOPLE ON DUTY $x$ |
|---|---|---|
| January 30 | 20 | 6 |
| June 3 | 18 | 6 |
| March 2 | 10 | 4 |
| October 26 | 6 | 2 |
| February 7 | 11 | 3 |

a.   Construct a scattergram for the data.

b.   Assuming the relationship between the variables is best described by a straight line, use the method of least squares to estimate the $y$-intercept and slope of the line.

c.   Plot the least squares line on your scattergram.

d.   According to your least squares line, approximately how many cars should the dealer expect to sell in a week if five salespeople are kept on the showroom floor each day? [*Note:*   A measure of the reliability of these predictions will be discussed in Section 9.8.]

**9.13.**   To investigate the relationship between yield of potatoes, $y$, and level of fertilizer application, $x$, an experimenter divided a field into eight plots of equal size and applied differing amounts of fertilizer to each. The yield of potatoes (in pounds) and the fertilizer application (in pounds) were recorded for each plot. The data are shown below:

| $x$ | 1 | 1.5 | 2 | 2.5 | 3 | 3.5 | 4 | 4.5 |
|---|---|---|---|---|---|---|---|---|
| $y$ | 25 | 31 | 27 | 28 | 36 | 35 | 32 | 34 |

a.   Construct a scattergram for the data.

b.   Find the least squares estimates for $\beta_0$ and $\beta_1$.

c.   According to your least squares line, approximately how many pounds of potatoes would you expect from a plot to which 3.75 pounds of fertilizer had been applied? [*Note:*   A measure of the reliability of these predictions will be discussed in Section 9.8.]

**9.14.** Find the least squares line to describe the relationship between a company's sales and the total sales for a particular industry. The data are shown in the table. Plot the data points and graph the least squares line as a check.

| YEAR | COMPANY SALES $y$, millions of dollars | INDUSTRY SALES $x$, millions of dollars |
|------|------|------|
| 1975 | 1.4 | 15 |
| 1976 | 1.3 | 14 |
| 1977 | 1.6 | 15 |
| 1978 | 1.5 | 16 |
| 1979 | 1.6 | 18 |
| 1980 | 1.8 | 19 |

**9.15.** An appliance company is interested in relating the sales rate of 17 inch color television sets to the price per set. To do this, the company randomly selected 15 weeks in the past year and recorded the number of sets sold per week and the price of the set at the time it was sold. The data are shown in the table.

| WEEK | NUMBER OF 17 INCH COLOR TELEVISION SETS SOLD PER WEEK $y$ | PRICE $x$, dollars |
|------|------|------|
| 1 | 55 | 350 |
| 2 | 54 | 360 |
| 3 | 25 | 385 |
| 4 | 18 | 400 |
| 5 | 51 | 370 |
| 6 | 20 | 390 |
| 7 | 45 | 375 |
| 8 | 19 | 390 |
| 9 | 20 | 400 |
| 10 | 45 | 340 |
| 11 | 50 | 350 |
| 12 | 35 | 335 |
| 13 | 30 | 330 |
| 14 | 30 | 325 |
| 15 | 53 | 365 |

a. Find the least squares line relating $y$ to $x$.
b. Plot the data and graph the least squares line as a check on your calculations.

**9.3
MODEL
ASSUMPTIONS**

In Section 9.2 we assumed that the probabilistic model relating drug reaction time $y$ to the percentage of drug $x$ in the bloodstream is

$$y = \beta_0 + \beta_1 x + \varepsilon$$

and recall that the least squares estimate of the deterministic component of the model, $\beta_0 + \beta_1 x$, is

$$\hat{y} = \hat{\beta}_0 + \hat{\beta}_1 x = -.1 + .7x$$

Now we turn our attention to the random component $\varepsilon$ of the probabilistic model and its relation to the errors of estimating $\beta_0$ and $\beta_1$. In particular, we will see how the probability distribution of $\varepsilon$ determines how well the model describes the true relationship between the dependent variable $y$ and the independent variable $x$.

Step 3 in a regression analysis requires us to specify the probability distribution of the random error $\varepsilon$. We will make four basic assumptions about the general form of this probability distribution.

**Assumption 1.** The mean of the probability distribution of $\varepsilon$ is zero. That is, the average of the errors over an infinitely long series of experiments is zero for each setting of the independent variable $x$. This assumption implies that the mean value of $y$, $E(y)$, for a given value of $x$ is $E(y) = \beta_0 + \beta_1 x$.

**Assumption 2.** The variance of the probability distribution of $\varepsilon$ is constant for all settings of the independent variable $x$. For our straight-line model, this assumption means that the variance of $\varepsilon$ is equal to a constant, say $\sigma^2$, for all values of $x$.

**Assumption 3.** The probability distribution of $\varepsilon$ is normal.

**Assumption 4.** The errors associated with any two different observations are independent. That is, the error associated with one value of $y$ has no effect on the errors associated with other $y$ values.

The implications of the first three assumptions can be seen in Figure 9.6, which shows distributions of errors for three particular values of $x$, namely $x_1$, $x_2$, and $x_3$. Note that the relative frequency distributions of the errors are normal, with a mean of zero and a constant variance $\sigma^2$ (all the distributions shown have the same amount of spread or variability). The straight line shown in Figure 9.6 is the mean value of $y$ for a given value of $x$. We will denote this mean value as $E(y)$.

**FIGURE 9.6**
**THE PROBABILITY**
**DISTRIBUTION OF** $\varepsilon$

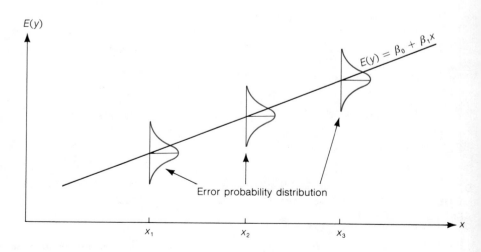

Then, the line of means is given by the equation

$$E(y) = \beta_0 + \beta_1 x$$

Various techniques exist for checking the validity of these assumptions, and there are remedies to be applied when they appear to be invalid. These topics are beyond the scope of this text, but they are discussed in some of the references listed at the end of this chapter. Fortunately, the assumptions need not hold exactly in order for least squares estimators and test statistics (to be described subsequently) to possess the measures of reliability that we would expect from a regression analysis. The assumptions will be satisfied adequately for many applications encountered in practice.

**9.4**
**AN ESTIMATOR**
**OF $\sigma^2$**

It seems reasonable to assume that the greater the variability of the random error $\varepsilon$ (which is measured by its variance $\sigma^2$), the greater will be the errors in the estimation of the model parameters $\beta_0$ and $\beta_1$, and in the error of prediction when $\hat{y}$ is used to predict $y$ for some value of $x$. Consequently, you should not be surprised, as we proceed through this chapter, to find that $\sigma^2$ appears in the formulas for all confidence intervals and test statistics that we will be using.

In most practical situations, $\sigma^2$ will be unknown and we must use our data to estimate its value. The best (proof omitted) estimate $s^2$ of $\sigma^2$ is obtained by dividing the sum of squares of the deviations of the $y$ values from the prediction line,

$$SSE = \sum (y_i - \hat{y}_i)^2$$

by the number of degrees of freedom associated with this quantity. We use 2 df to estimate the $y$-intercept and slope in the straight-line model, leaving $(n - 2)$ df for the error variance estimation.

---

ESTIMATION OF $\sigma^2$ FOR A (FIRST-ORDER) STRAIGHT-LINE MODEL

$$s^2 = \frac{SSE}{\text{Degrees of freedom for error}} = \frac{SSE}{n - 2}$$

where

$$SSE = \sum (y_i - \hat{y}_i)^2 = SS_{yy} - \hat{\beta}_1 SS_{xy}$$

$$SS_{yy} = \sum (y_i - \bar{y})^2 = \sum y_i^2 - \frac{\left(\sum y_i\right)^2}{n}$$

*Warning:* When performing these calculations, you may be tempted to round the calculated values of $SS_{yy}$, $\hat{\beta}_1$, and $SS_{xy}$. Be certain to carry at least six significant figures for each of these quantities to avoid substantial errors in the calculation of the SSE.

---

In the drug reaction time example, we previously calculated SSE = 1.10 for the least squares line $\hat{y} = -.1 + .7x$. Recalling that there were $n = 5$ data points, we have $n - 2 = 5 - 2 = 3$ df for estimating $\sigma^2$. Thus,

$$s^2 = \frac{SSE}{n-2} = \frac{1.10}{3} = .367$$

is the estimated variance, and

$$s = \sqrt{.367} = .61$$

is the estimated standard deviation of $\varepsilon$.

You may be able to obtain an intuitive feeling for $s$ by recalling the interpretation given to a standard deviation in Chapter 2 and remembering that the least squares line estimates the mean value of $y$ for a given value of $x$. Since $s$ measures the spread of the distribution of $y$ values about the least squares line, we should not be surprised to find that most of the observations lie within $2s$ or $2(.61) = 1.22$ of the least squares line. For this simple example (only five data points), all five data points fall within $2s$ of the least squares line. In Section 9.8 we will use $s$ to evaluate the error of prediction when the least squares line is used to predict a value of $y$ to be observed for a given value of $x$.

EXERCISES

Learning the mechanics

**9.16.** Suppose that you fit a least squares line to nine data points and the calculated value of SSE is .217.

    a. Find $s^2$, the estimator of $\sigma^2$ (the variance of the random error $\varepsilon$).
    b. What is the largest deviation that you might expect between any one of the nine points and the least squares line?

**9.17.** Calculate SSE and $s^2$ for the least squares lines found in:

    a. Exercise 9.8
    b. Exercise 9.9
    c. Exercise 9.10
    d. Exercise 9.11
    e. Exercise 9.12
    f. Exercise 9.13
    g. Exercise 9.14
    h. Exercise 9.15

Applying the concepts

**9.18.** An electronics dealer believes that there is a linear relationship between the number of hours of quadraphonic programming on a city's FM stations and sales of quadraphonic systems. Records for the dealer's sales during the last 6 months and the amount of quadraphonic programming for the corresponding months are:

| MONTH | AVERAGE AMOUNT OF QUADRAPHONIC PROGRAMMING $x$, hours | NUMBER OF QUADRAPHONIC SYSTEMS SOLD $y$ |
|---|---|---|
| 1 | 33.6 | 7 |
| 2 | 36.3 | 10 |
| 3 | 38.7 | 13 |
| 4 | 36.6 | 11 |
| 5 | 39.0 | 14 |
| 6 | 38.4 | 18 |

a. Fit a least squares line to the data.
b. Plot the data and graph the least squares line as a check.
c. Calculate the SSE and $s^2$.

9.19. A company keeps extensive records on its new salespeople on the premise that sales should increase with experience. A random sample of seven new salespeople produced the data on experience and sales shown in the table.

| MONTHS ON JOB $x$ | MONTHLY SALES $y$, thousands of dollars |
|---|---|
| 2 | 2.4 |
| 4 | 7.0 |
| 8 | 11.3 |
| 12 | 15.0 |
| 1 | .8 |
| 5 | 3.7 |
| 9 | 12.0 |

a. Fit a least squares line to the data.
b. Plot the data and graph the least squares line.
c. Predict the sales that a new salesperson would be expected to generate after 6 months on the job. After 9 months.
d. Calculate the SSE and $s^2$.

## 9.5 ASSESSING THE UTILITY OF THE MODEL: MAKING INFERENCES ABOUT THE SLOPE $\beta_1$

Now that we have specified the probability distribution of $\varepsilon$ and found an estimate of the variance $\sigma^2$, we are ready to proceed to making statistical inferences about the model's utility for predicting the response, $y$. This is step 4 in our regression modeling procedure.

Refer again to the data of Table 9.1 and suppose that the reaction times are completely unrelated to the percentage of drug in the bloodstream. What could be said about the values of $\beta_0$ and $\beta_1$ in the hypothesized probabilistic model

$$y = \beta_0 + \beta_1 x + \varepsilon$$

if $x$ contributes no information for the prediction of $y$? The implication is that the mean of $y$, i.e., the deterministic part of the model $E(y) = \beta_0 + \beta_1 x$, does not change as $x$ changes. In the straight-line model, this means that the true slope, $\beta_1$, is equal to zero (see Figure 9.7). Therefore, to test the null hypothesis that $x$ contributes no information for the prediction of $y$ against the alternative hypothesis that these variables are linearly related with a slope differing from zero, we test

$$H_0: \quad \beta_1 = 0 \qquad H_a: \quad \beta_1 \neq 0$$

**FIGURE 9.7**
**GRAPHING**
**THE MODEL**
$y = \beta_0 + \varepsilon \quad (\beta_1 = 0)$

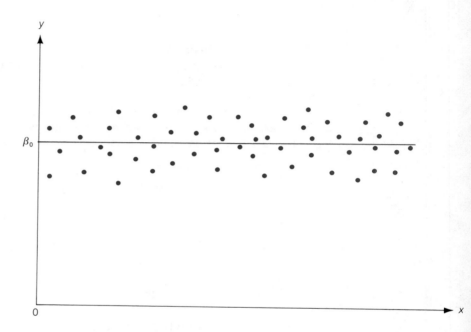

If the data support the alternative hypothesis, we will conclude that $x$ does contribute information for the prediction of $y$ using the straight-line model [although the true relationship between $E(y)$ and $x$ could be more complex than a straight line]. Thus, to some extent, this is a test of the utility of the hypothesized model.

The appropriate test statistic is found by considering the sampling distribution of $\hat{\beta}_1$, the least squares estimator of the slope $\beta_1$.

---

**SAMPLING DISTRIBUTION OF $\hat{\beta}_1$**

If we make the four assumptions about $\varepsilon$ (see Section 9.3), the sampling distribution of the least squares estimator $\hat{\beta}_1$ of the slope will be normal, with mean $\beta_1$ (the true slope) and standard deviation

$$\sigma_{\hat{\beta}_1} = \frac{\sigma}{\sqrt{SS_{xx}}} \qquad \text{(See Figure 9.8.)}$$

---

FIGURE 9.8
SAMPLING
DISTRIBUTION OF $\hat{\beta}_1$

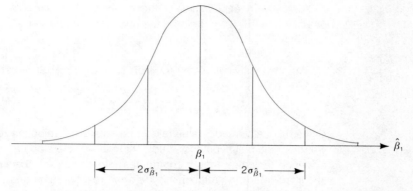

Since $\sigma$ will usually be unknown, the appropriate test statistic will be a Student's $t$ statistic formed as follows:

$$t = \frac{\hat{\beta}_1 - \text{Hypothesized value of } \beta_1}{s_{\hat{\beta}_1}} \qquad \text{where} \quad s_{\hat{\beta}_1} = \frac{s}{\sqrt{SS_{xx}}}$$

Thus,

$$t = \frac{\hat{\beta}_1 - 0}{s/\sqrt{SS_{xx}}}$$

Note that we have substituted the estimator $s$ for $\sigma$, and then formed $s_{\hat{\beta}_1}$ by dividing $s$ by $\sqrt{SS_{xx}}$. The number of degrees of freedom associated with this $t$ statistic is the same as the number of degrees of freedom associated with $s$. Recall that this will be $(n - 2)$ df when the hypothesized model is a straight line (see Section 9.4).

The setup of our test of the utility of the model is summarized in the box.

---

**A TEST OF MODEL UTILITY**

| One-tailed test | Two-tailed test |
|---|---|
| $H_0$: $\beta_1 = 0$ | $H_0$: $\beta_1 = 0$ |
| $H_a$: $\beta_1 < 0$ | $H_a$: $\beta_1 \neq 0$ |
| (or $H_a$: $\beta_1 > 0$) | |

Test statistic:                            Test statistic:

$$t = \frac{\hat{\beta}_1}{s_{\hat{\beta}_1}} = \frac{\hat{\beta}_1}{s/\sqrt{SS_{xx}}} \qquad\qquad t = \frac{\hat{\beta}_1}{s_{\hat{\beta}_1}} = \frac{\hat{\beta}_1}{s/\sqrt{SS_{xx}}}$$

Rejection region:   $t < -t_\alpha$         Rejection region:   $t < -t_{\alpha/2}$

                 (or   $t > t_\alpha$                     or   $t > t_{\alpha/2}$

                 when   $H_a$: $\beta_1 > 0$)

where $t_\alpha$ and $t_{\alpha/2}$ are based on $(n - 2)$ degrees of freedom.

Assumptions:   The four assumptions about $\varepsilon$ listed in Section 9.3.

---

For the drug reaction time example, we will choose $\alpha = .05$ and, since $n = 5$, $t$ will be based on $n - 2 = 3$ df and the rejection region will be

$$t < -t_{.025} = -3.182 \qquad \text{or} \qquad t > t_{.025} = 3.182$$

We previously calculated $\hat{\beta}_1 = .7$, $s = .61$, and $SS_{xx} = 10$. Thus,

$$t = \frac{\hat{\beta}_1}{s/\sqrt{SS_{xx}}} = \frac{.7}{.61/\sqrt{10}} = \frac{.7}{.19} = 3.7$$

Since this calculated $t$ value falls in the upper-tail rejection region (see Figure 9.9), we reject the null hypothesis and conclude that the slope $\beta_1$ is not zero. The sample evidence indicates that $x$ contributes information for the prediction of $y$ using a linear model for the relationship between reaction time and the amount of the drug in the bloodstream.

**FIGURE 9.9**
REJECTION REGION
AND CALCULATED
$t$ VALUE FOR TESTING
$H_0$: $\beta_1 = 0$
VERSUS $H_a$: $\beta_1 \neq 0$

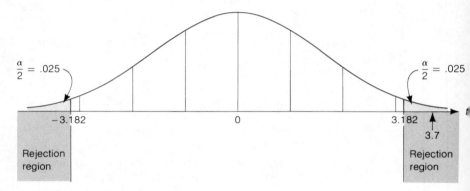

What conclusion can be drawn if the calculated $t$ value does not fall in the rejection region? We know from previous discussions of the philosophy of hypothesis testing that such a $t$ value does *not* lead us to accept the null hypothesis. That is, we do not conclude that $\beta_1 = 0$. Additional data might indicate that $\beta_1$ differs from zero, or a more complex relationship may exist between $x$ and $y$, requiring the fitting of a model other than the straight-line model.

Another way to make inferences about the slope $\beta_1$ is to estimate it using a confidence interval. This interval is formed as shown in the box.

---

A 100(1 − $\alpha$)PERCENT CONFIDENCE INTERVAL FOR THE SLOPE $\beta_1$

$$\hat{\beta}_1 \pm t_{\alpha/2}\, s_{\hat{\beta}_1}$$

where

$$s_{\hat{\beta}_1} = \frac{s}{\sqrt{SS_{xx}}}$$

and $t_{\alpha/2}$ is based on $(n - 2)$ degrees of freedom.

Assumptions: The four assumptions about $\varepsilon$ listed in Section 9.3.

---

For the drug reaction time example, $t_{\alpha/2}$ will be based on $(n - 2) = 3$ degrees of freedom. Therefore, a 95% confidence interval for the slope $\beta_1$, the expected change in reaction time for a 1% increase in the amount of drug in the bloodstream, is

$$\hat{\beta}_1 \pm t_{.025}s_{\hat{\beta}_1} = .7 \pm 3.182\left(\frac{s}{\sqrt{SS_{xx}}}\right)$$

$$= .7 \pm 3.182\left(\frac{.61}{\sqrt{10}}\right) = .7 \pm .61$$

Thus, the interval estimate of the slope parameter $\beta_1$ is .09 to 1.31.

Since all the values in this interval are positive, it appears that $\beta_1$ is positive and that the mean of $y$, $E(y)$, increases as $x$ increases. However, the rather large width of the confidence interval reflects the small number of data points (and, consequently, a lack of information) in the experiment. We would expect a narrower interval if the sample size were increased.

EXERCISES

Learning the mechanics

**9.20.** Give the values of $\hat{\beta}_1$, $s^2$, $SS_{xx}$, and the degrees of freedom associated with $s^2$ for each of the following exercises:

a. Exercise 9.9  b. Exercise 9.10  c. Exercise 9.11
d. Exercise 9.12  e. Exercise 9.13  f. Exercise 9.14
g. Exercise 9.15

**9.21.** For each of the following exercises, determine whether there is sufficient evidence to indicate that $\beta_1$ is greater than 0. Test using $\alpha = .05$.

a. Exercise 9.9  b. Exercise 9.12

**9.22.** For each of the following exercises, determine whether there is sufficient evidence to indicate that $\beta_1$ is less than 0. Test using $\alpha = .05$.

a. Exercise 9.10  b. Exercise 9.11

**9.23.** For each of the following exercises, determine whether there is sufficient evidence to indicate that $\beta_1$ differs from 0. Test using $\alpha = .05$.

a. Exercise 9.13  b. Exercise 9.14  c. Exercise 9.15

Applying the concepts

**9.24.** A breeder of thoroughbred horses wishes to model the relationship between the gestation period and the length of life of a horse. The breeder believes that the two variables may follow a linear trend. The information in the table on page 364 was supplied to the breeder from various thoroughbred stables across the state. (Note that the horse has the greatest variation of gestation period of any species, due to seasonal and feed factors.)

| HORSE | GESTATION PERIOD $x$, days | LIFE LENGTH $y$, years |
|-------|------------------|-------------|
| 1 | 416 | 24 |
| 2 | 279 | 25.5 |
| 3 | 298 | 20 |
| 4 | 307 | 21.5 |
| 5 | 356 | 22 |
| 6 | 403 | 23.5 |
| 7 | 265 | 21 |

a. Fit a least squares line to the data. Plot the data points and graph the least squares line as a check on your calculations.

b. Do the data provide sufficient evidence to support the horse breeder's hypothesis? Test using $\alpha = .05$.

c. Find a 90% confidence interval for $\beta_1$. Interpret this interval.

**9.25.** A group of children, ranging from 10 to 12 years of age, were administered a verbal test in order to study the relationship between the number of words used and the silence interval before response. The tester believes that a linear relationship exists between the two variables. Each subject was asked a series of questions, and the total number of words used in answering was recorded. The time (in seconds) before the subject responded to each question was also recorded. The data for eight children are given below:

| SUBJECT | TOTAL WORDS $y$ | TOTAL SILENCE TIME $x$, seconds |
|---------|------------|-------------------|
| 1 | 61 | 23 |
| 2 | 70 | 37 |
| 3 | 42 | 38 |
| 4 | 52 | 25 |
| 5 | 91 | 17 |
| 6 | 63 | 21 |
| 7 | 71 | 42 |
| 8 | 55 | 16 |

a. Write a simple linear probabilistic model relating total words to total silence time, and use the least squares method to estimate the deterministic part of the model.

b. Does $x$ contribute information for the prediction of $y$? Test the null hypothesis, slope $\beta_1 = 0$, against the alternative hypothesis, $\beta_1 \neq 0$. Use $\alpha = .05$. Interpret the results of the test.

**9.26.** A local brewery is interested in determining whether a linear relationship exists between the amount it spends on television advertising and total sales. The data listed in the table are available.

| MONTH | SALES Thousands of dollars | TELEVISION ADVERTISING EXPENDITURES Thousands of dollars |
|---|---|---|
| January | 50 | 0.5 |
| February | 90 | 0.9 |
| March | 30 | 0.4 |
| April | 90 | 0.7 |
| May | 91 | 1.1 |
| June | 95 | 0.75 |
| July | 95 | 0.8 |

a.  Find the least squares line for the given data. Plot the data on a scatter-gram and graph the line as a check on your calculations.

b.  Letting $\alpha = .05$, test the null hypothesis that $\beta_1 = 0$. What alternative hypothesis would you select for this test? Draw appropriate conclusions concerning the adequacy of a linear model to describe the relationship between sales and television advertising expenditures.

c.  Construct a 90% confidence interval for the slope parameter in the hypothesized linear model.

d.  Interpret the confidence interval and explain what it tells you about the relationship between sales and television advertising expenditure.

**9.27.**  A large car rental agency sells its cars after using them for a year. Among the records kept for each car are mileage and maintenance costs for the year. To evaluate the performance of a particular car model in terms of maintenance costs, the agency wants to use a 95% confidence interval to estimate the mean increase in maintenance costs for each additional 1,000 miles driven. Assume the relationship between maintenance cost and miles driven is linear. Use the data in the table to accomplish the objective of the rental agency.

| CAR | MILES DRIVEN $x$, thousands | MAINTENANCE COST $y$, dollars |
|---|---|---|
| 1 | 54 | 326 |
| 2 | 27 | 159 |
| 3 | 29 | 202 |
| 4 | 32 | 200 |
| 5 | 28 | 181 |
| 6 | 36 | 217 |

**9.28.**  In an investigation into the possibility of dysfunction in monocular depth perception in schizophrenics, subjects were asked to align images at different distances from a light source. Eleven schizophrenic patients were used in the experiment. The distance and perception score recorded for each subject are listed in the table on page 366. (Better depth perception is reflected by higher perception scores.)

| SUBJECT | DISTANCE FROM LIGHT $x$, feet | PERCEPTION SCORE $y$ |
|---------|-------------------------------|----------------------|
| 1 | 6 | 22 |
| 2 | 10 | 16 |
| 3 | 4 | 25 |
| 4 | 5 | 26 |
| 5 | 10 | 18 |
| 6 | 6 | 18 |
| 7 | 12 | 13 |
| 8 | 8 | 17 |
| 9 | 4 | 26 |
| 10 | 8 | 22 |
| 11 | 12 | 19 |

a. Construct a scattergram for the data.

b. Assuming the relationship between the variables is linear, use the least squares method to estimate the $y$-intercept and slope of the line.

c. Plot the least squares line on your scattergram.

d. Test the adequacy of the model by testing the null hypothesis, $H_0: \beta_1 = 0$, against the alternative hypothesis, $H_a$: $\beta_1 < 0$, i.e., that the average perception score decreases as the distance of the subject from the light increases. Use $\alpha = .05$.

9.29.    Buyers are often influenced by bulk advertising of a particular product. For example, suppose you have a product that sells for 25¢. If it is advertised at 2 for 50¢, 3 for 75¢, or 4 for $1, some people may think they are getting a bargain. To test this theory, a store manager advertised an item for equal periods of time at five different bulk rates and observed the sales volumes listed in the table. Do the data provide sufficient evidence to indicate that average sales increase as the number in the bulk increases?

| ADVERTISED NUMBER IN BULK SALE $x$ | VOLUME SOLD $y$ |
|------------------------------------|-----------------|
| 1 | 27 |
| 2 | 36 |
| 3 | 34 |
| 4 | 63 |
| 5 | 52 |

## 9.6
## CORRELATION: A MEASURE OF THE UTILITY OF THE MODEL

The claim is often made that the number of cigarettes smoked and the incidence of lung cancer are "highly correlated." Another popular belief is that the crime rate and the unemployment rate are "correlated." Some people even believe that the Dow Jones Industrial Average and the lengths of fashionable skirts are "correlated." In this section we will discuss the concept of correlation.

A numerical descriptive measure of the correlation between two variables $x$ and $y$ is provided by the Pearson product moment coefficient of correlation, $r$.

---

**DEFINITION 9.2**

The sample Pearson product moment coefficient of correlation, $r$, is defined as

$$r = \frac{SS_{xy}}{\sqrt{SS_{xx}SS_{yy}}}$$

It is a measure of the strength of the linear relationship between two random variables $x$ and $y$.

---

Note that the computational formula for the correlation coefficient $r$ given in Definition 9.2 involves the same quantities that were used in computing the least squares prediction equation. In fact, since the numerators of the expressions for $\hat{\beta}_1$ and $r$ are identical, you can see that $r = 0$ when $\hat{\beta}_1 = 0$ (the case where $x$ contributes no information for the prediction of $y$) and that $r$ will be positive when the slope is positive and negative when the slope is negative. Unlike $\hat{\beta}_1$, $r$ is scaleless and will assume a value between $-1$ and $+1$, regardless of the units of $x$ and $y$.

A value of $r$ near or equal to zero implies little or no linear relationship between $y$ and $x$. In contrast, the closer $r$ becomes to 1 or $-1$, the stronger the linear relationship between $y$ and $x$. And, if $r = 1$ or $r = -1$, all the sample points fall exactly on the least squares line. Positive values of $r$ imply a positive linear relationship between $y$ and $x$; i.e., $y$ increases as $x$ increases. Negative values of $r$ imply a negative linear relationship between $y$ and $x$, i.e., $y$ decreases as $x$ increases. Each of these situations is portrayed in Figure 9.10 (page 368).

We will demonstrate how to calculate the coefficient of correlation, $r$, using the data in Table 9.1 for the drug reaction time example. The quantities needed to calculate $r$ are $SS_{xy}$, $SS_{xx}$, and $SS_{yy}$. The first two quantities have been calculated previously and are repeated here for convenience.

$$SS_{xy} = 7 \qquad SS_{xx} = 10$$

$$SS_{yy} = \Sigma\, y^2 - \frac{(\Sigma\, y)^2}{n} = 26 - \frac{(10)^2}{5} = 26 - 20 = 6$$

We now find the coefficient of correlation:

$$r = \frac{SS_{xy}}{\sqrt{SS_{xx}SS_{yy}}} = \frac{7}{\sqrt{(10)(6)}} = \frac{7}{\sqrt{60}}$$
$$= .904$$

The fact that $r$ is positive and near 1 in value indicates that the reaction time tends to increase as the amount of drug in the bloodstream increases, *for this sample of five subjects*. This is the same conclusion we reached when the calculated value of the least squares slope was found to be positive.

**FIGURE 9.10**
**VALUES OF r AND**
**THEIR IMPLICATIONS**

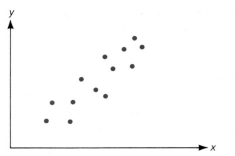

(a)  Positive r: y increases as x increases

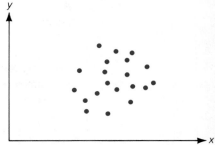

(b)  r near zero: little or no linear
relationship between y and x

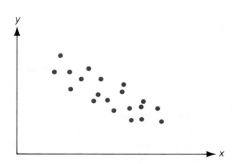

(c)  Negative r: y decreases as x increases

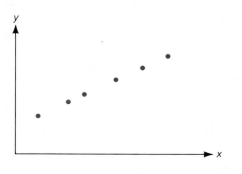

(d)  r = 1: a perfect positive linear
relationship between y and x

(e)  r = −1: a perfect negative linear
relationship between y and x

EXAMPLE 9.1      A firm wants to know the correlation between the size of its sales force and its yearly
sales revenue. The records for the past 10 years are examined, and the results listed
in Table 9.5 are obtained. Calculate the coefficient of correlation, r, for the data.

| | | |
|---|---|---|
| **TABLE 9.5** **SALES VERSUS NUMBER OF SALESPEOPLE** | | |

| YEAR | NUMBER OF SALESPEOPLE | SALES |
|------|------------------------|-------|
| | $x$ | $y$, hundred thousand dollars |
| 1972 | 15 | 1.35 |
| 1973 | 18 | 1.63 |
| 1974 | 24 | 2.33 |
| 1975 | 22 | 2.41 |
| 1976 | 25 | 2.63 |
| 1977 | 29 | 2.93 |
| 1978 | 30 | 3.41 |
| 1979 | 32 | 3.26 |
| 1980 | 35 | 3.63 |
| 1981 | 38 | 4.15 |

**Solution**

We need to calculate $SS_{xy}$, $SS_{xx}$, and $SS_{yy}$.

$$SS_{xy} = \Sigma\, xy - \frac{(\Sigma\, x)(\Sigma\, y)}{n} = 800.62 - \frac{(268)(27.73)}{10} = 57.456$$

$$SS_{xx} = \Sigma\, x^2 - \frac{(\Sigma\, x)^2}{n} = 7{,}668 - \frac{(268)^2}{10} = 485.6$$

$$SS_{yy} = \Sigma\, y^2 - \frac{(\Sigma\, y)^2}{n} = 83.8733 - \frac{(27.73)^2}{10} = 6.97801$$

Then, the coefficient of correlation is

$$r = \frac{SS_{xy}}{\sqrt{SS_{xx}SS_{yy}}} = \frac{57.456}{\sqrt{(485.6)(6.97801)}} = \frac{57.456}{58.211} = .99$$

Thus, the size of the sales force and sales revenue are very highly correlated—at least over the past 10 years. The implication is that a strong positive linear relationship exists between these variables (see Figure 9.11). We must be careful, however, not to jump to any unwarranted conclusions. For instance, the firm may be tempted to conclude that the best thing it can do to increase sales is to hire a large number of new salespeople, that is, that there is a causal relationship between the two variables. However, high correlation does not imply causality. The fact is, many things have probably contributed both to the increase in the size of the sales force and to the increase in sales revenue. The firm's expertise has undoubtedly grown, the economy has inflated (so that 1981 dollars are not worth as much as 1972 dollars), and perhaps the scope of products and services sold by the firm has widened. We must be careful not to infer a causal relationship on the basis of high sample correlation. The only safe conclusion when a high correlation is observed in the sample data is that a linear trend may exist between $x$ and $y$.

Keep in mind that the correlation coefficient $r$ measures the linear correlation between $x$ values and $y$ values in the sample, and that a similar linear coefficient of correlation exists for the population from which the data points were selected. The population correlation coefficient is denoted by the symbol $\rho$ (rho). As you might expect, $\rho$ is

FIGURE 9.11
SCATTERGRAM FOR
EXAMPLE 9.1

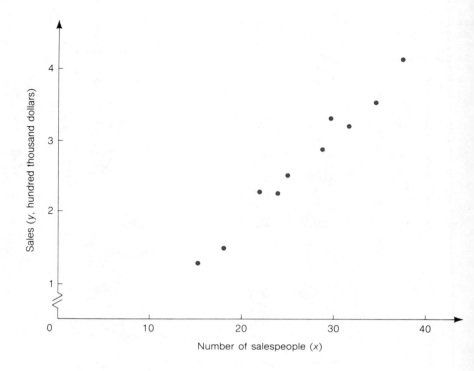

estimated by the corresponding sample statistic, $r$. Or, rather than estimating $\rho$, we might want to test the hypothesis $H_0$: $\rho = 0$ against $H_a$: $\rho \neq 0$, i.e., test the hypothesis that $x$ contributes no information for the prediction of $y$ using the straight-line model against the alternative that the two variables are at least linearly related.

However, we have already performed this *identical* test in Section 9.5 when we tested $H_0$: $\beta_1 = 0$ against $H_a$: $\beta_1 \neq 0$. When we tested the null hypothesis $H_0$: $\beta_1 = 0$ in connection with the drug reaction time example, the data led to a rejection of the null hypothesis at the $\alpha = .05$ level. This implies that the null hypothesis of a zero linear correlation between the two variables (drug and reaction time) can also be rejected at the $\alpha = .05$ level. The only real difference between the least squares slope $\hat{\beta}_1$ and the coefficient of correlation $r$ is the measurement scale. Therefore, the information they provide about the utility of the least squares model is to some extent redundant. For this reason, we will use the slope to make inferences about the existence of a positive or negative linear relationship between two variables.

**9.7
COEFFICIENT
OF
DETERMINATION**

Another way to measure the contribution of $x$ in predicting $y$ is to consider how much the errors of prediction of $y$ were reduced by using the information provided by $x$. If you do not use $x$, the best prediction for any value of $y$ would be $\bar{y}$, and the sum of squares of the deviations of the $y$ values about $\bar{y}$ is the familiar

$$SS_{yy} = \Sigma (y - \bar{y})^2$$

On the other hand, if we use $x$ to predict $y$, the sum of squares of the deviations of the $y$ values about the least squares line is

$$SSE = \Sigma\,(y - \hat{y})^2$$

Then, the reduction in the sum of squares of deviations that can be attributed to $x$, expressed as a proportion of $SS_{yy}$, is

$$\frac{SS_{yy} - SSE}{SS_{yy}}$$

It can be shown that this quantity is equal to the square of the simple linear coefficient of correlation.

---

**DEFINITION 9.3**

The square of the coefficient of correlation is called the **coefficient of determination.** It represents the proportion of the sum of squares of deviations of the $y$ values about their mean that can be attributed to a linear relationship between $y$ and $x$.

$$r^2 = \frac{SS_{yy} - SSE}{SS_{yy}} = 1 - \frac{SSE}{SS_{yy}}$$

---

Note that $r^2$ is always between 0 and 1, because $r$ is between $-1$ and $+1$. Thus, $r^2 = .60$ means that 60% of the sum of squares of deviations of the $y$ values about their mean is attributable to the linear relation between $y$ and $x$.

**EXAMPLE 9.2**

Calculate the coefficient of determination for the drug reaction time example using the formula given in the box. The data are repeated in Table 9.6 for convenience.

**TABLE 9.6**

| AMOUNT OF DRUG $x$, % | REACTION TIME $y$, seconds |
|:---:|:---:|
| 1 | 1 |
| 2 | 1 |
| 3 | 2 |
| 4 | 2 |
| 5 | 4 |

**Solution**

From previous calculations,

$$SS_{yy} = 6$$

and

$$SSE = \Sigma\,(y - \hat{y})^2 = 1.10$$

Then, the coefficient of determination is given by

$$r^2 = \frac{SS_{yy} - SSE}{SS_{yy}} = \frac{6.0 - 1.1}{6.0} = \frac{4.9}{6.0}$$

$$= .82$$

[In Section 9.6, we calculated $r = .904$. Now we have $r^2 = (.904)^2 = .82$.] So we know that the use of the amount of drug in the blood, $x$, to predict $y$ with the least squares line

$$\hat{y} = -.1 + .7x$$

accounts for 82% of the total sum of squares of deviations of the five sample $y$ values about their mean.

CASE STUDY 9.1

PREDICTING THE
UNITED STATES
CRIME INDEX

Reporting and analyzing crime rates is an important function of many law enforcement agencies. David Heaukulani (1975) comments, "The simple linear regression analysis remains one of the most useful tools for crime prediction." He demonstrates this point by fitting a straight-line model to predict the annual value of the United States crime index as a function of the United States population for each year. The data for the years 1963–1973 are given in Table 9.7, and the least squares line is shown in Figure 9.12.

TABLE 9.7
UNITED STATES
POPULATION AND
CRIME INDEX

| YEAR | POPULATION $x$ | ACTUAL CRIME INDEX $y$ |
|------|------------|------------|
| 1963 | 188,531,000 | 2,259,081 |
| 1964 | 191,334,000 | 2,604,426 |
| 1965 | 193,818,000 | 2,780,015 |
| 1966 | 195,857,000 | 3,243,400 |
| 1967 | 197,864,000 | 3,802,273 |
| 1968 | 199,861,000 | 4,466,573 |
| 1969 | 201,921,000 | 4,989,747 |
| 1970 | 203,184,772 | 5,568,197 |
| 1971 | 206,256,000 | 5,995,211 |
| 1972 | 208,232,000 | 5,891,924 |
| 1973 | 209,851,000 | 8,638,375 |

Heaukulani reports that the correlation is $r = .94$ ($r^2 = .88$), which provides sufficient evidence at $\alpha = .01$ to indicate that the size of the population is useful for predicting the crime index using the straight-line model. Heaukulani concludes in part that

> most agencies are using too narrow a span of statistical measurement to evaluate the overall crime picture. Year-by-year comparisons and percent analyses do not take average fluctuations into consideration. A lack of knowledge about statistical principles on the part of laymen, especially those in the news media, leads to a distortion or an invalid representation of the

FIGURE 9.12
LEAST SQUARES LINE
RELATING
CRIME INDEX TO
POPULATION SIZE
(1963–1973)

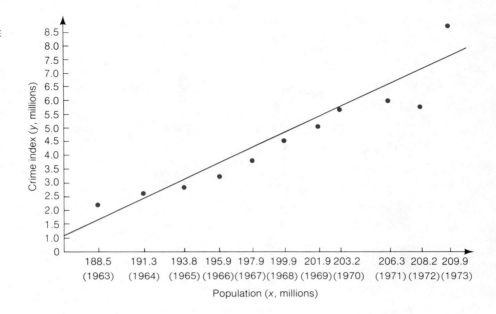

Population (x, millions)

crime figures. Any increase is reported either as "bucking a trend" or "in line with the rise in crime."

If we expect to make any progress in solving the crime problem, we must evaluate the crime index objectively. We have to accept the fact that the crime index will probably increase from year to year. We should be willing to accept an average amount of increase and a set maximum limit. If the crime index exceeds the limit, only then should we become concerned and attempt to examine the factors that may have been responsible for the excess deviation.

Note that the 1973 crime rate of 8.6 million greatly exceeded the predicted value of less than 7 million. Perhaps a special effort should be made to determine the causes of this "excess deviation."

EXERCISES

Learning the mechanics

**9.30.** Calculate the values of $r$ and $r^2$ for each of the following exercises:

| | | |
|---|---|---|
| a. Exercise 9.8 | b. Exercise 9.9 | c. Exercise 9.10 |
| d. Exercise 9.11 | e. Exercise 9.12 | f. Exercise 9.13 |
| g. Exercise 9.14 | h. Exercise 9.15 | |

Applying the concepts

**9.31.** When student seating in a large lecture course is by personal choice, is seat location correlated to a student's grade? In particular, do students who choose seats in the front of the classroom tend to obtain better grades than those who choose the back? The grades of ten randomly selected statistics students are given at the top of the next page along with their row location (number from the front of the classroom).

| Grade: | 93 | 68 | 89 | 98 | 66 | 86 | 73 | 55 | 80 | 71 |
|--------|----|----|----|----|----|----|----|----|----|----|
| Row:   | 7  | 1  | 4  | 3  | 25 | 12 | 9  | 30 | 13 | 27 |

a.  Calculate $r$ and $r^2$.

b.  Do the data provide sufficient evidence to indicate that students sitting in the front of the classroom tend to receive better grades? Test using $\alpha = .05$.*

**9.32.**  Find the correlation coefficient and the coefficient of determination for the sample data listed in the table and interpret your results.

| YEAR | NUMBER OF EIGHTEEN HOLE AND LARGER GOLF COURSES IN THE UNITED STATES | UNITED STATES DIVORCE RATE PER 1,000 POPULATION |
|------|------|------|
| 1960 | 2,725 | 2.2 |
| 1965 | 3,769 | 2.5 |
| 1970 | 4,845 | 3.5 |
| 1972 | 5,385 | 4.1 |
| 1975 | 6,282 | 4.8 |

Source: United States Bureau of the Census, *Statistical Abstract of the United States 1976.*

**9.33.**  Is the maximal oxygen uptake, a measure often used by physiologists to indicate an individual's state of cardiovascular fitness, related to the performance of distance runners? Six long-distance runners submitted to treadmill tests for determination of their maximal oxygen uptake. These results, along with each runner's best mile time (in seconds), are shown in the table.

| ATHLETE | MAXIMAL OXYGEN UPTAKE Milliliters/kilogram | MILE TIME Seconds |
|---------|------|------|
| 1 | 63.3 | 241.5 |
| 2 | 60.1 | 249.8 |
| 3 | 53.6 | 246.1 |
| 4 | 58.8 | 232.4 |
| 5 | 67.5 | 237.2 |
| 6 | 62.5 | 238.4 |

a.  Calculate $r$ and $r^2$.

b.  Do the data provide sufficient evidence to indicate that mile time is negatively correlated with maximal oxygen uptake? Test $H_0$: $\rho = 0$ against the alternative hypothesis, $H_a$: $\rho < 0$, using $\alpha = .05$.*

*Recall (Section 9.6) that the test of $H_0$: $\rho = 0$ is equivalent to the test of $H_0$: $\beta_1 = 0$.

**9.34.** Is there a correlation between the amount of education received by people living in central cities and that received by people living in the urban fringes surrounding the cities? To determine whether a correlation exists, a sociologist compared the percentages of people with 4 years of high school education or more for the two groups.

PERCENTAGES OF PEOPLE WITH 4 YEARS
OF HIGH SCHOOL EDUCATION OR MORE

| CITY | CENTRAL CITY $y$ | URBAN FRINGE $x$ |
|------|------------------|------------------|
| Baltimore | 28.2 | 42.3 |
| Boston | 44.6 | 55.8 |
| Chicago | 35.3 | 53.9 |
| Cleveland | 30.1 | 55.5 |
| Dallas | 48.9 | 56.4 |
| Detroit | 34.4 | 47.5 |
| Houston | 45.2 | 50.1 |
| Los Angeles | 53.4 | 53.4 |
| Milwaukee | 39.7 | 54.4 |
| New Orleans | 33.3 | 44.6 |
| New York | 36.4 | 48.7 |
| Philadelphia | 30.7 | 48.0 |
| St. Louis | 26.3 | 43.3 |
| San Francisco | 49.4 | 57.9 |
| Washington | 47.8 | 67.5 |

Source: Computed from United States Bureau of the Census, *U.S. Census of Population: 1960, General Social and Economic Characteristics,* and *U.S. Census of Population and Housing: 1960, Census Tracts* (Washington, D.C.: Government Printing Office, 1961).

a. Find the correlation coefficient for the data.

b. Find the coefficient of determination and explain its meaning in terms of this problem.

c. Is there sufficient evidence to indicate a nonzero correlation between $x$ and $y$? Test using $\alpha = .05$. *

**9.35.** In analyzing the costs of construction, labor and material costs are two basic components. Changes in the component costs, of course, will lead to changes in total construction costs.

a. Use the data in the table on page 376 to find a measure of the importance of the materials component. Do this by determining the fraction of variability of the construction cost index that can be explained by a linear relationship between the construction cost index and the material cost index.

*Recall (Section 9.6) that the test of $H_0$: $\rho = 0$ is equivalent to the test of $H_0$: $\beta_1 = 0$.

| MONTH | CONSTRUCTION COST*<br>y | INDEX OF ALL<br>CONSTRUCTION MATERIALS†<br>x |
|---|---|---|
| January | 193.2 | 180.0 |
| February | 193.1 | 181.7 |
| March | 193.6 | 184.1 |
| April | 195.1 | 185.3 |
| May | 195.6 | 185.7 |
| June | 198.1 | 185.9 |
| July | 200.9 | 187.7 |
| August | 202.7 | 189.6 |

*Source: United States Department of Commerce, Bureau of the Census.

†Source: United States Department of Labor, Bureau of Labor Statistics. Tables were given in Tables E-1 (p. 43) and E-2 (p. 44), respectively, in *Construction Review*, United States Department of Commerce, Oct. 1976, 22 (8).

b. Do the data provide sufficient evidence to indicate a nonzero correlation between $y$ and $x$?

9.36. Data on monthly sales, $y$, price per unit during the month, $x_1$, and amount spent on advertising, $x_2$, for a product are shown in the table for a 5 month period. Based on this sample, which variable—price or advertising expenditure—appears to provide more information about sales? Explain.

| MONTH | TOTAL<br>MONTHLY<br>SALES<br>y, thousands of dollars | PRICE<br>PER UNIT<br>$x_1$, dollars | AMOUNT SPENT<br>ON ALL FORMS<br>OF ADVERTISING<br>$x_2$, hundreds of dollars |
|---|---|---|---|
| June | 40 | 0.85 | 6.0 |
| July | 50 | 0.76 | 5.0 |
| August | 55 | 0.75 | 8.0 |
| September | 30 | 1.00 | 7.5 |
| October | 45 | 0.80 | 5.5 |

## 9.8
## USING THE MODEL FOR ESTIMATION AND PREDICTION

If we are satisfied that a useful model has been found to describe the relationship between reaction time and amount of drug in the bloodstream, we are ready for step 5 in our regression modeling procedure, using the model for estimation and prediction.

The most common uses of a probabilistic model for making inferences can be divided into two categories. The first is the use of the model for estimating the mean value of $y$, $E(y)$, for a specific value of $x$. For our drug reaction time example, we may want to estimate the mean response time for "all" people whose blood contains 4% of the drug. The second use of the model entails predicting a particular $y$ value for a given $x$. That is, we may want to predict the reaction time for a particular person who possesses 4% of the drug in their bloodstream.

In the first case, we are attempting to estimate the mean value of $y$ for a very large number of experiments at the given $x$ value. In the second case, we are trying to predict the outcome of a single experiment at the given $x$ value. Which of these model uses, estimating the mean value of $y$ or predicting an individual value of $y$ (for the same value of $x$) can be accomplished with the greater accuracy?

Before answering this question, we first consider the problem of choosing an estimator (or predictor) of the mean (or individual) $y$ value. We will use the least squares prediction equation

$$\hat{y} = \hat{\beta}_0 + \hat{\beta}_1 x$$

both to estimate the mean value of $y$ and to predict a particular value of $y$ for a given value of $x$. For our example, we found

$$\hat{y} = -.1 + .7x$$

so that the estimated mean reaction time for all people when $x = 4$ (drug is 4% of blood content) is

$$\hat{y} = -.1 + .7(4) = 2.7 \text{ seconds}$$

The identical value is used to predict the $y$ value when $x = 4$. That is, both the estimated mean and the predicted value of $y$ are $\hat{y} = 2.7$ when $x = 4$, as shown in Figure 9.13.

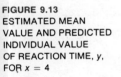

**FIGURE 9.13**
**ESTIMATED MEAN VALUE AND PREDICTED INDIVIDUAL VALUE OF REACTION TIME, $y$, FOR $x = 4$**

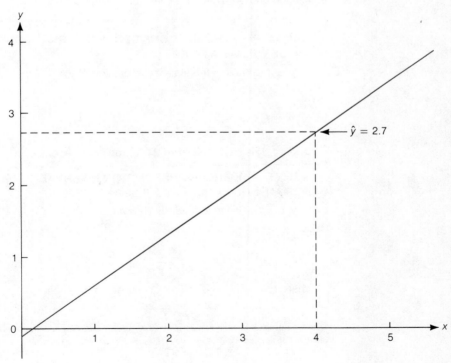

The difference between these two model uses lies in the relative accuracy of the estimate and the prediction. These accuracies are best measured by the repeated sampling errors of the least squares line when it is used as an estimator and as a predictor, respectively. These errors are given in the box.

---

SAMPLING ERRORS FOR THE ESTIMATOR OF THE MEAN OF $y$ AND THE PREDICTOR OF AN INDIVIDUAL $y$

**1.** The standard deviation of the sampling distribution of the estimator $\hat{y}$ of the mean value of $y$ at a fixed $x$ is

$$\sigma_{\hat{y}} = \sigma \sqrt{\frac{1}{n} + \frac{(x - \bar{x})^2}{SS_{xx}}}$$

where $\sigma$ is the standard deviation of the random error $\varepsilon$.

**2.** The standard deviation of the prediction error for the predictor $\hat{y}$ of an individual $y$ value at a fixed $x$ is

$$\sigma_{(y - \hat{y})} = \sigma \sqrt{1 + \frac{1}{n} + \frac{(x - \bar{x})^2}{SS_{xx}}}$$

where $\sigma$ is the standard deviation of the random error $\varepsilon$.

---

The true value of $\sigma$ will rarely be known, so we estimate $\sigma$ by $s$ and calculate the estimation and prediction intervals as shown below.

---

A $100(1 - \alpha)$ PERCENT CONFIDENCE INTERVAL FOR THE MEAN VALUE OF $y$ AT A FIXED $x$

$$\hat{y} \pm t_{\alpha/2} \text{ (Estimated standard deviation of } \hat{y})$$

or

$$\hat{y} \pm t_{\alpha/2} \, s \sqrt{\frac{1}{n} + \frac{(x - \bar{x})^2}{SS_{xx}}}$$

where $t_{\alpha/2}$ is based on $(n - 2)$ degrees of freedom.

---

A $100(1 - \alpha)$ PERCENT PREDICTION INTERVAL*
FOR AN INDIVIDUAL $y$ AT A FIXED $x$

$$\hat{y} \pm t_{\alpha/2} \text{ [Estimated standard deviation of } (y - \hat{y})]$$

or

$$\hat{y} \pm t_{\alpha/2} \, s \sqrt{1 + \frac{1}{n} + \frac{(x - \bar{x})^2}{SS_{xx}}}$$

where $t_{\alpha/2}$ is based on $(n - 2)$ degrees of freedom.

---

*The term prediction interval is used when the interval formed is intended to enclose the value of a random variable. The term confidence interval is reserved for estimation of population parameters (such as the mean).

EXAMPLE 9.3    Find a 95% confidence interval for the mean reaction time when the concentration of the drug in the bloodstream is 4%.

Solution    For a 4% concentration, $x = 4$ and the confidence interval for the mean value of $y$ is

$$\hat{y} \pm t_{\alpha/2}\, s\sqrt{\frac{1}{n} + \frac{(x - \bar{x})^2}{SS_{xx}}} = \hat{y} \pm t_{.025}\, s\sqrt{\frac{1}{5} + \frac{(4 - \bar{x})^2}{SS_{xx}}}$$

where $t_{.025}$ is based on $n - 2 = 5 - 2 = 3$ degrees of freedom. Recall that $\hat{y} = 2.7$, $s = .61$, $\bar{x} = 3$, and $SS_{xx} = 10$. From Table IV in the Appendix, $t_{.025} = 3.182$. Thus, we have

$$2.7 \pm (3.182)(.61)\sqrt{\frac{1}{5} + \frac{(4 - 3)^2}{10}} = 2.7 \pm (3.182)(.61)(.55)$$

$$= 2.7 \pm 1.1$$

Therefore, when the percentage of drug in the bloodstream is 4%, the 95% confidence interval for the mean reaction time for all possible subjects is 1.6–3.8 seconds. Note that we used a small amount of data (small sample size) for purposes of illustration in fitting the least squares line and that the interval would probably be narrower if more information had been obtained from a larger sample.

EXAMPLE 9.4    Predict the reaction time for the next performance of the experiment for a subject with a drug concentration of 4%. Use a 95% prediction interval.

Solution    To predict the response time for an individual subject for whom $x = 4$, we calculate the 95% prediction interval as

$$\hat{y} \pm t_{\alpha/2}\, s\sqrt{1 + \frac{1}{n} + \frac{(x - \bar{x})^2}{SS_{xx}}} = 2.7 \pm (3.182)(.61)\sqrt{1 + \frac{1}{5} + \frac{(4 - 3)^2}{10}}$$

$$= 2.7 \pm (3.182)(.61)(1.14)$$

$$= 2.7 \pm 2.2$$

Therefore, we predict that the reaction time for this individual will fall in the interval from .5 second to 4.9 seconds. As in the case for the confidence interval for the mean value of $y$, the prediction interval for $y$ is quite large. This is because we have chosen a simple example (only five data points) to fit the least squares line. The width of the prediction interval could be reduced by using a larger number of data points.

A comparison of the confidence interval for the mean value of $y$ and the prediction interval for a future value of $y$ for 4% drug concentration ($x = 4$) is illustrated in

FIGURE 9.14
A 95% CONFIDENCE
INTERVAL FOR MEAN
REACTION TIME AND
A PREDICTION
INTERVAL
FOR REACTION
TIME WHEN x = 4

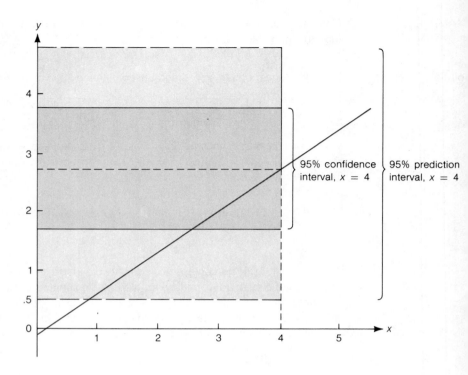

95% confidence interval, x = 4    95% prediction interval, x = 4

FIGURE 9.15
ERROR OF ESTIMATING
THE MEAN VALUE
OF y FOR A GIVEN
VALUE OF x

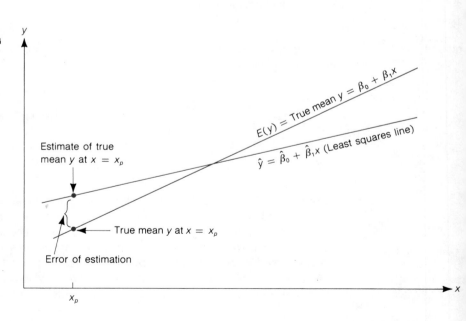

$E(y)$ = True mean $y = \beta_0 + \beta_1 x$

$\hat{y} = \hat{\beta}_0 + \hat{\beta}_1 x$ (Least squares line)

Estimate of true mean y at $x = x_p$

True mean y at $x = x_p$

Error of estimation

$x_p$

Figure 9.14. It is important to note that the prediction interval for an individual value of $y$ will always be wider than the corresponding confidence interval for the mean value of $y$. You can see this by examining the formulas for the two intervals and you can see it in Figure 9.14.

The error in estimating the mean value of $y$, $E(y)$, for a given value of $x$, say $x_p$, is the vertical distance between the least squares line and the true line of means, $E(y) = \beta_0 + \beta_1 x$. This error, shown in Figure 9.15, will take its smallest value when $x = \bar{x}$. The farther $x$ lies from $\bar{x}$, the larger will be the error of estimation. In contrast, the error in predicting some future value of $y$ is the sum of the two errors—the error of estimating the mean of $y$, $E(y)$, shown in Figure 9.15, plus the random error $\varepsilon$ that is a component of the value of $y$ to be predicted (see Figure 9.16). Consequently, the error of predicting a particular value of $y$ will usually be larger than the error of estimating the mean value of $y$ for a given value of $x$. As the sample size $n$ is increased, both the confidence interval and the prediction interval will decrease. The least squares line will get closer and closer to the line of means, $E(y) = \beta_0 + \beta_1 x$, and the $100(1 - \alpha)\%$ prediction interval for $y$ will decrease to $\hat{y} \pm z_{\alpha/2}s$.

IGURE 9.16
RROR OF PREDICTING
A FUTURE VALUE
OF $y$ FOR A GIVEN
VALUE OF $x$

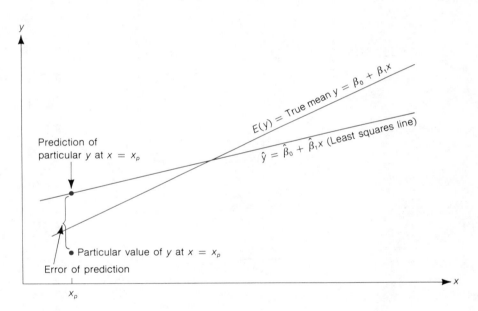

Be careful not to use the least squares prediction equation to estimate the mean value of $y$ or to predict a particular value of $y$ for values of $x$ that fall outside the *range* of the values of $x$ contained in your sample data. The model might provide a good fit to the data over the range of $x$ values contained in the sample but a very poor fit for values of $x$ outside this region (see Figure 9.17). Failure to heed this warning may lead to errors of estimation and prediction that are much larger than expected.

FIGURE 9.17
THE DANGER OF USING
A MODEL TO PREDICT
OUTSIDE THE RANGE
OF THE SAMPLE
VALUES OF x

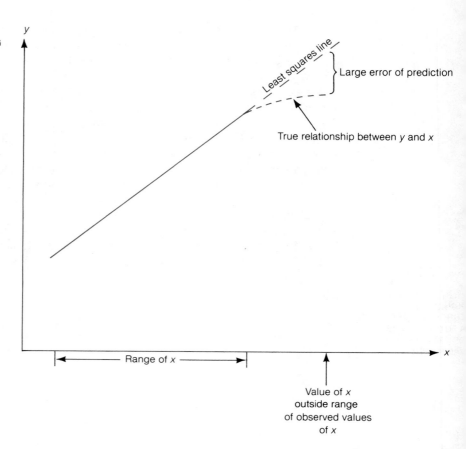

EXERCISES

Learning the mechanics

**9.37.** The data from Exercise 9.10 are repeated below.

| $x$ | $-2$ | $-1$ | $0$ | $1$ | $2$ |
|---|---|---|---|---|---|
| $y$ | $4$ | $3$ | $3$ | $1$ | $-1$ |

   a. Find a 90% confidence interval for the mean value of $y$ when $x = -1$.

   b. Find a 90% prediction interval for $y$ when $x = -1$.

[*Note:* Both intervals in parts a and b are wide because of the small number of data points.]

**9.38.** In fitting a least squares line to $n = 22$ data points, the following quantities were computed:

$$\begin{aligned}
SS_{xx} &= 25 & \bar{x} &= 2 \\
SS_{yy} &= 17 & \bar{y} &= 3 \\
SS_{xy} &= 20 &
\end{aligned}$$

a. Find the least squares line.
b. Graph the least squares line.
c. Calculate SSE.
d. Calculate $s^2$.
e. Find a 95% confidence interval for the mean value of $y$ when $x = 1$.
f. Find a 95% prediction interval for $y$ when $x = 1.5$.
g. Find a 95% confidence interval for the mean value of $y$ when $x = 0$.

### Applying the concepts

**9.39.** Certain dosages of a new drug developed to reduce a smoker's reliance on tobacco may reduce one's pulse rate to dangerously low levels. To investigate the drug's effect on pulse rate, different dosages of the drug were administered to six randomly selected patients, and 30 minutes later the decrease in each patient's pulse rate was recorded.

| PATIENT | DOSAGE $x$, cubic centimeters | DECREASE IN PULSE RATE $y$, beats/minute |
|---------|--------|---------------------|
| 1 | 2.0 | 15 |
| 2 | 1.5 | 9 |
| 3 | 3.0 | 18 |
| 4 | 2.5 | 16 |
| 5 | 4.0 | 23 |
| 6 | 3.0 | 20 |

a. Is there evidence of a linear relationship between drug dosage and change in pulse rate? Test at $\alpha = .10$.
b. Find a 95% confidence interval for the slope $\beta_1$.
c. Find a 99% confidence interval for the mean decrease in pulse rate corresponding to a dosage of 3.5 cubic centimeters.
d. Find a 99% prediction interval for the decrease in pulse rate corresponding to a dosage of 3.5 cubic centimeters.

**9.40.** Does a linear relationship exist between the Consumer Price Index (CPI) and the Dow Jones Industrial Average (DJA)? A random sample of 10 months selected from the past several years produced the following corresponding DJA and CPI data:

| DJA $y$ | CPI $x$ |
|---------|---------|
| 660 | 13.0 |
| 638 | 14.2 |
| 639 | 13.7 |
| 597 | 15.1 |
| 702 | 12.6 |
| 650 | 13.8 |
| 579 | 15.7 |
| 570 | 16.0 |
| 725 | 11.3 |
| 738 | 10.4 |

a. Find the least squares line relating the DJA, $y$, to the CPI, $x$.

b. Do the data provide sufficient evidence to indicate that $x$ contributes information for the prediction of $y$? Test using $\alpha = .05$.

c. Find a 95% confidence interval for $\beta_1$ and interpret your result.

d. Suppose you want to estimate the value of the DJA when the CPI is at 15.0. Should you calculate a 95% prediction interval for a particular value of the DJA, or a 95% confidence interval for the mean value of the DJA? Explain the difference.

e. Calculate both intervals considered in part d when the CPI is 15.0.

**9.41.** Will the national 55 mile per hour highway speed limit provide a substantial savings in fuel? To investigate the relationship between automobile gasoline consumption and driving speed, a small economy car was driven twice over the same stretch of an interstate freeway at each of six different speeds. The numbers of miles per gallon measured for each of the twelve trips are shown below:

| MILES PER HOUR | 50 | 55 | 60 |
|---|---|---|---|
| MILES PER GALLON | 34.8, 33.6 | 34.6, 34.1 | 32.8, 31.9 |

| MILES PER HOUR | 65 | 70 | 75 |
|---|---|---|---|
| MILES PER GALLON | 32.6, 30.0 | 31.6, 31.8 | 30.9, 31.7 |

a. Fit a least squares line to the data.

b. Is there sufficient evidence to conclude there is a linear relationship between speed and gasoline consumption?

c. Construct a 90% confidence interval for $\beta_1$.

d. Construct a 95% confidence interval for miles per gallon when the speed is 72 miles per hour.

e. Construct a 95% prediction interval for miles per gallon when the speed is 58 miles per hour.

**9.42.** In planning for an initial orientation meeting with new accounting majors, the chairperson of the Accounting Department wants to emphasize the importance of doing well in the major courses in order to get better-paying jobs after graduation. To support this point, the chairperson plans to show that there is a strong positive correlation between starting salaries for recent accounting graduates and their grade-point averages in the major courses. Records for seven of last year's accounting graduates are selected at random and are given in the table.

| GRADE-POINT AVERAGE IN MAJOR COURSES $x$ | STARTING SALARY $y$, thousands of dollars |
|---|---|
| 2.58 | 11.5 |
| 3.27 | 13.8 |
| 3.85 | 14.5 |
| 3.50 | 14.2 |
| 3.33 | 13.5 |
| 2.89 | 11.6 |
| 2.23 | 10.6 |

a. Find the least squares prediction equation.

b. Plot the data and graph the line as a check on your calculations.

c. Find the values of $r$ and $r^2$ and interpret these values.

d. Find a 95% prediction interval for a graduate whose grade-point average is 3.2.

e. What is the mean starting salary for graduates with grade-point averages equal to 3.0? Use a 95% confidence interval.

**9.43.** Explain why for a given $x$ value, the prediction interval for an individual $y$ value will always be wider than the confidence interval for a mean value of $y$.

**9.44.** Explain why the confidence interval for the mean value of $y$ for a given $x$ value gets wider the farther $x$ is from $\bar{x}$. What are the implications of this phenomenon for estimation and prediction?

**9.9
SIMPLE LINEAR
REGRESSION:
AN EXAMPLE**

In the previous sections we have presented the basic elements necessary to fit and use a straight-line regression model. In this final section we will assemble these elements by applying them in an example.

Suppose a fire insurance company wants to relate the amount of fire damage in major residential fires to the distance between the residence and the nearest fire station. The study is to be conducted in a large suburb of a major city; a sample of fifteen recent fires in this suburb is selected. The amount of damage, $y$, and the distance, $x$, between the fire and the nearest fire station are recorded for each fire. The results are given in Table 9.8 on page 386.

**Step 1.** First, we hypothesize a model to relate fire damage, $y$, to the distance from the nearest fire station, $x$. We will hypothesize a straight-line probabilistic model:

$$y = \beta_0 + \beta_1 x + \varepsilon$$

**Step 2.** Next, we use the data to estimate the unknown parameters in the deterministic component of the hypothesized model. We make some preliminary calculations:

$$SS_{xx} = \Sigma x^2 - \frac{(\Sigma x)^2}{n} = 196.16 - \frac{(49.2)^2}{15}$$

$$= 196.160 - 161.376 = 34.784$$

TABLE 9.8
FIRE DAMAGE DATA

| DISTANCE FROM FIRE STATION $x$, miles | FIRE DAMAGE $y$, thousands of dollars |
|---|---|
| 3.4 | 26.2 |
| 1.8 | 17.8 |
| 4.6 | 31.3 |
| 2.3 | 23.1 |
| 3.1 | 27.5 |
| 5.5 | 36.0 |
| 0.7 | 14.1 |
| 3.0 | 22.3 |
| 2.6 | 19.6 |
| 4.3 | 31.3 |
| 2.1 | 24.0 |
| 1.1 | 17.3 |
| 6.1 | 43.2 |
| 4.8 | 36.4 |
| 3.8 | 26.1 |

$$SS_{yy} = \Sigma\, y^2 - \frac{(\Sigma\, y)^2}{n} = 11{,}376.48 - \frac{(396.2)^2}{15}$$

$$= 11{,}376.480 - 10{,}464.963 = 911.517$$

$$SS_{xy} = \Sigma\, xy - \frac{(\Sigma\, x)(\Sigma\, y)}{n} = 1{,}470.65 - \frac{(49.2)(396.2)}{15}$$

$$= 1{,}470.650 - 1{,}299.536 = 171.114$$

Then the least squares estimates of the slope $\beta_1$ and intercept $\beta_0$ are

$$\hat{\beta}_1 = \frac{SS_{xy}}{SS_{xx}} = \frac{171.114}{34.784} = 4.919$$

$$\hat{\beta}_0 = \bar{y} - \hat{\beta}_1 \bar{x} = \frac{396.2}{15} - 4.919\left(\frac{49.2}{15}\right)$$

$$= 26.413 - (4.919)(3.28) = 26.413 - 16.134$$

$$= 10.279$$

And the least squares equation is

$$\hat{y} = 10.279 + 4.919x$$

This prediction equation is graphed in Figure 9.18, along with a plot of the data points.

**Step 3.** Now, we specify the probability distribution of the random error component $\varepsilon$. The assumptions about the distribution will be identical to those listed in Section 9.3. Although we know these assumptions are not completely satisfied (they rarely

FIGURE 9.18
LEAST SQUARES
MODEL FOR THE
FIRE DAMAGE DATA

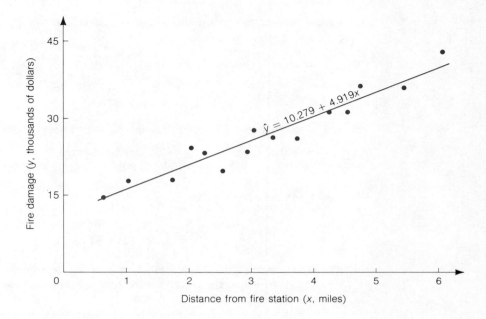

are for any practical problem), we are willing to assume they are approximately satisfied for this example. We have to estimate the variance, $\sigma^2$, of $\varepsilon$, so we calculate

$$\text{SSE} = \Sigma\,(y - \hat{y})^2 = \text{SS}_{yy} - \hat{\beta}_1\text{SS}_{xy}$$

where the last expression represents a shortcut formula for SSE. Thus,

$$\text{SSE} = 911.517 - (4.919)(171.114)$$
$$= 911.517 - 841.709766 = 69.807234^*$$

To estimate $\sigma^2$, we divide SSE by the degrees of freedom available for error, $n - 2$. Thus,

$$s^2 = \frac{\text{SSE}}{n - 2} = \frac{69.807234}{15 - 2} = 5.3698$$
$$s = \sqrt{5.3698} = 2.32$$

**Step 4.** We can now check the utility of the hypothesized model, that is, whether $x$ really contributes information for the prediction of $y$ using the straight-line model. First test the null hypothesis that the slope $\beta_1$ is zero, i.e., that there is no linear relationship between fire damage and the distance from the nearest fire station,

---

*The values for $\text{SS}_{yy}$, $\hat{\beta}_1$, and $\text{SS}_{xy}$ used to calculate SSE are exact for this example. For other problems where rounding is necessary, at least six significant figures should be carried for these quantities. Otherwise, the calculated value of SSE may be substantially in error.

against the alternative hypothesis that fire damage increases as the distance increases, i.e., $H_a$: $\beta_1 > 0$. We test

$$H_0: \quad \beta_1 = 0 \qquad H_a: \quad \beta_1 > 0$$

Test statistic: $\quad t = \dfrac{\hat{\beta}_1 - 0}{s_{\hat{\beta}_1}} = \dfrac{\hat{\beta}_1}{s/\sqrt{\text{SS}_{xx}}}$

Assumptions: The same assumptions made about $\varepsilon$ in Section 9.3

Rejection region: For $\alpha = .05$, we will reject $H_0$ if $t > t_\alpha$, where, for $n - 2 = 13$ df, we find $t_{.05} = 1.771$.

We then calculate the $t$ statistic:

$$t = \frac{\hat{\beta}_1}{s_{\hat{\beta}_1}} = \frac{\hat{\beta}_1}{s/\sqrt{\text{SS}_{xx}}} = \frac{4.919}{2.32/\sqrt{34.784}} = \frac{4.919}{.393} = 12.5$$

This large $t$ value leaves little doubt that mean fire damage and distance between the fire and the fire station are at least linearly related, with mean fire damage increasing as the distance increases.

We gain additional information about the relationship by forming a confidence interval for the slope $\beta_1$. A 95% confidence interval is

$$\hat{\beta}_1 \pm t_{.025} s_{\hat{\beta}_1} = 4.919 \pm (2.160)(.393)$$
$$= 4.919 \pm .849 = (4.070, 5.768)$$

We estimate that the interval from \$4,070 to \$5,768 encloses the mean increase ($\beta_1$) in fire damage per additional mile distance from the fire station.

Another measure of the utility of the model is the coefficient of correlation $r$. We have

$$r = \frac{\text{SS}_{xy}}{\sqrt{\text{SS}_{xx}\text{SS}_{yy}}} = \frac{171.114}{\sqrt{(34.784)(911.517)}}$$
$$= \frac{171.114}{178.062} = .96$$

The high correlation confirms our conclusion that $\beta_1$ is greater than zero; it appears that fire damage and distance from the fire station are positively correlated.

The coefficient of determination is

$$r^2 = (.96)^2 = .92$$

which implies that 92% of the sum of squares of deviations of the $y$ values about $\bar{y}$ is explained by the distance $x$ between the fire and the fire station. All signs point to a strong relationship between $x$ and $y$.

**Step 5.** We are now prepared to use the least squares model. Suppose the insurance company wants to predict the fire damage if a major residential fire were to occur 3.5 miles from the nearest fire station. The predicted value is

$$\hat{y} = \hat{\beta}_0 + \hat{\beta}_1 x = 10.279 + (4.919)(3.5)$$
$$= 10.279 + 17.216 = 27.5$$

(we round to the nearest tenth to be consistent with the units of the original data in Table 9.8). If we want a 95% prediction interval, we calculate

$$\hat{y} \pm t_{.025} s \sqrt{1 + \frac{1}{n} + \frac{(x - \bar{x})^2}{SS_{xx}}} = 27.5 \pm (2.16)(2.32)\sqrt{1 + \frac{1}{15} + \frac{(3.5 - 3.28)^2}{34.784}}$$

$$= 27.5 \pm (2.16)(2.32)\sqrt{1.0681}$$
$$= 27.5 \pm 5.2 = (22.3, 32.7)$$

The model yields a 95% prediction interval of \$22,300 to \$32,700 for fire damage in a major residential fire 3.5 miles from the nearest station.

*One caution before closing:* We would not use this prediction model to make predictions for homes less than .7 mile or more than 6.1 miles from the nearest fire station. A look at the data in Table 9.8 reveals that all the x values fall between .7 and 6.1. It is dangerous to use the model to make predictions outside the region in which the sample data fall. A straight line might not provide a good model for the relationship between the mean value of y and the value of x when stretched over a wider range of x values.

## SUMMARY

We have introduced an extremely useful tool in this chapter—the **method of least squares** for fitting a prediction equation to a set of data. This procedure, along with associated statistical tests and estimations, is called a **regression analysis**. In five steps we showed how to use sample data to build a model relating a dependent variable y to a single independent variable x.

**1.** The first step is to hypothesize a **probabilistic model**. In this chapter, we confined our attention to the straight-line model, $y = \beta_0 + \beta_1 x + \varepsilon$.

**2.** The second step is to use the method of least squares to estimate the unknown parameters in the **deterministic component**, $\beta_0 + \beta_1 x$. The least squares estimates yield a model $\hat{y} = \hat{\beta}_0 + \hat{\beta}_1 x$ with a sum of squared errors (SSE) that is smaller than that produced by any other straight-line model.

**3.** The third step is to specify the probability distribution of the **random error component**, $\varepsilon$.

**4.** The fourth step is to assess the utility of the hypothesized model. Included here are making inferences about the slope $\beta_1$, calculating the **coefficient of correlation**, r, and calculating the **coefficient of determination**, $r^2$.

**5.** Finally, if we are satisfied with the model, we are prepared to use it. We used the model to estimate the mean y value, $E(y)$, for a given x value and to predict an individual y value for a specific value of x.

Learning the mechanics

**9.45.** In fitting a least squares line to $n = 15$ data points, the following quantities were computed:

$$SS_{xx} = 50 \qquad \bar{x} = 1.3$$
$$SS_{yy} = 25 \qquad \bar{y} = 27$$
$$SS_{xy} = -30$$

a. Find the least squares line.
b. Graph the least squares line.
c. Calculate SSE.
d. Calculate $s^2$.
e. Find a 90% confidence interval for $\beta_1$. Interpret this estimate.
f. Find a 90% confidence interval for the mean value of $y$ when $x = 1.8$.
g. Find a 90% prediction interval for $y$ when $x = 1.8$.

**9.46.** Consider the following ten data points:

| x | 3 | 5 | 6 | 4 | 3 | 7 | 6 | 5 | 4 | 7 |
|---|---|---|---|---|---|---|---|---|---|---|
| y | 4 | 3 | 2 | 1 | 2 | 3 | 3 | 5 | 4 | 2 |

a. Plot the data on a scattergram.
b. Calculate the values of $r$ and $r^2$.
c. Is there sufficient evidence to indicate that $x$ and $y$ are linearly correlated? Test at the $\alpha = .10$ level of significance.

Applying the concepts

**9.47.** A study was conducted to determine whether the final grade in an introductory sociology course was related to a student's performance on a verbal ability test administered before college entrance. The verbal test scores and final grades for a random sample of ten students are shown in the table.

| STUDENT | VERBAL ABILITY TEST SCORE<br>x | FINAL INTRODUCTORY SOCIOLOGY GRADE<br>y |
|---|---|---|
| 1 | 39 | 65 |
| 2 | 43 | 78 |
| 3 | 21 | 52 |
| 4 | 64 | 82 |
| 5 | 57 | 92 |
| 6 | 47 | 89 |
| 7 | 28 | 73 |
| 8 | 75 | 98 |
| 9 | 34 | 56 |
| 10 | 52 | 75 |

a. Assuming a linear relationship exists between verbal scores and final grades, find the least squares line relating $y$ to $x$.

b. Plot the data points and graph the least squares line.

c. Do the data provide sufficient evidence to indicate that a positive correlation exists between verbal score and final grade? (Use $\alpha = .01$)

d. Find a 95% confidence interval for the slope $\beta_1$.

e. Predict a student's final grade in the introductory course when his or her verbal test score is 50. (Use a 90% prediction interval.)

f. Find a 95% confidence interval for the mean final grade for students scoring 35 on the college entrance verbal exam.

**9.48.** A large supermarket chain has its own store brand for many grocery items. These tend to be priced lower than other brands. For a particular item, the chain wants to study the effect of varying the price for the major competing brand on the sales of the store brand item, while the prices for the store brand and all other brands are held fixed. The experiment is conducted at one of the chain's stores over a 7 week period and the results are shown in the table.

| WEEK | MAJOR COMPETITOR'S PRICE $x$, cents | STORE BRAND SALES $y$ |
|---|---|---|
| 1 | 37 | 122 |
| 2 | 32 | 107 |
| 3 | 29 | 99 |
| 4 | 35 | 110 |
| 5 | 33 | 113 |
| 6 | 31 | 104 |
| 7 | 35 | 116 |

a. Find the least squares line relating store brand sales, $y$, to major competitor's price, $x$.

b. Plot the data and graph the line as a check on your calculations.

c. Does $x$ contribute information for the prediction of $y$?

d. Calculate $r$ and $r^2$ and interpret their values.

e. Find a 90% confidence interval for mean store brand sales when the competitor's price is 33¢.

f. Suppose you were to set the competitor's price at 33¢. Find a 90% prediction interval for next week's sales.

**9.49.** As part of the first-year evaluation for new salespeople, a large food-processing firm projects the second-year sales for each salesperson based on his or her sales for the first year.

| FIRST-YEAR SALES $x$, thousands of dollars | SECOND-YEAR SALES $y$, thousands of dollars |
|---|---|
| 75.2 | 99.3 |
| 91.7 | 125.7 |
| 100.3 | 136.1 |
| 64.2 | 108.6 |
| 81.8 | 102.0 |
| 110.2 | 153.7 |
| 77.3 | 108.8 |
| 80.1 | 105.4 |

a.   Use the data in the table on eight salespeople for this firm to fit a simple linear prediction model for second-year sales based on the first year's sales. Assume the data have been adjusted in terms of a base year to discount inflation effects.

b.   Plot the data and graph the line as a check on your calculations.

c.   Do the data provide sufficient information to indicate that $x$ contributes information for the prediction of $y$?

d.   Calculate $r^2$ and interpret its value.

e.   If a salesperson has first-year sales of $90,000, find a 90% prediction interval for the second year's sales.

**9.50.**   The data shown in the table give the consumer purchases of oranges, $y$, and the price per box for an 8 month period.

| MONTH | UNITED STATES CONSUMER PURCHASES OF ORANGES $y$, thousands of boxes | PRICE PER BOX $x$, dollars |
|---|---|---|
| October | 6,100 | 6.20 |
| November | 6,800 | 6.05 |
| December | 8,400 | 6.20 |
| January | 6,000 | 8.10 |
| February | 5,800 | 8.70 |
| March | 5,500 | 8.75 |
| April | 4,400 | 9.40 |
| May | 4,000 | 9.75 |

a.   Construct a scattergram of the data.

b.   Find the least squares line for the data and plot it on your scattergram. The least squares line may be viewed as an estimate of the short-run demand function for oranges.

c.   Define $\beta_1$ in the context of this problem.

d.   Test the hypothesis that the price per box of oranges contributes no information for the prediction of the number of boxes consumed when a linear model of short-run demand is used (let $\alpha = .05$). Draw the appropriate conclusions.

e. Find a 90% confidence interval for $\beta_1$. Interpret your results.

f. Find the coefficient of correlation for the given data.

g. Find the coefficient of determination for the linear model of demand you constructed in part b. Interpret your result.

h. Find a 90% prediction interval for the number of boxes of oranges that will be consumed if the price per box is $8.00.

i. Find a 95% confidence interval for the expected number of boxes that will be consumed at a price of $8.00 per box.

**9.51.** In placing a weekly order, a concessionaire that provides services at a baseball stadium must know what size crowd is expected during the coming week in order to know how much food, etc., to order. Since advanced ticket sales give an indication of expected attendance, food needs might be predicted on the basis of the advanced sales.

| HOT DOGS PURCHASED DURING WEEK $y$, thousands | ADVANCED TICKET SALES FOR WEEK $x$, thousands |
|---|---|
| 39.1 | 54.0 |
| 35.9 | 48.1 |
| 20.8 | 28.8 |
| 42.4 | 62.4 |
| 46.0 | 64.4 |
| 40.7 | 59.5 |
| 29.9 | 42.3 |

a. Use the data in the table from 7 previous weeks of home games to develop a simple linear model for hot dogs purchased as a function of advanced ticket sales.

b. Plot the data and graph the line as a check on your calculations.

c. Do the data provide sufficient information to indicate that advanced ticket sales provide information for the prediction of hot dog demand?

d. Calculate $r^2$ and interpret its value.

e. Find a 90% confidence interval for the mean number of hot dogs purchased when the advanced ticket sales equal 50,000.

f. If the advanced ticket sales this week equal 55,000, find a 90% prediction interval for the number of hot dogs that will be purchased this week at the game.

**9.52.** At temperatures approaching absolute zero (273 degrees below zero Celsius), helium exhibits traits that defy many laws of conventional physics. An experiment has been conducted with helium in solid form at various temperatures near absolute zero. The solid helium is placed in a dilution refrigerator along with a solid impure substance, and the fraction (in weight) of the impurity passing through the solid helium is recorded. (This phenomenon of solids passing directly through solids is known as *quantum tunnelling*.) The data are given in the table on page 394.

| TEMPERATURE $x$, °C | PROPORTION OF IMPURITY PASSING THROUGH HELIUM $y$ |
|---|---|
| −262.0 | .315 |
| −265.0 | .202 |
| −256.0 | .204 |
| −267.0 | .620 |
| −270.0 | .715 |
| −272.0 | .935 |
| −272.4 | .957 |
| −272.7 | .906 |
| −272.8 | .985 |
| −272.9 | .987 |

a.  Fit a least squares line to the data.

b.  Test the null hypothesis, $H_0$: $\beta_1 = 0$, against the alternative hypothesis, $H_a$: $\beta_1 < 0$, at the $\alpha = .01$ level of significance.

c.  Compute $r^2$ and interpret your results.

d.  Find a 95% prediction interval for the percentage of the solid impurity passing through solid helium at $-273°C$. (Note that this value of $x$ is outside the experimental region, where use of the model for prediction may be dangerous.)

**9.53.**  A certain manufacturer evaluates the sales potential for a product in a new marketing area by selecting several stores within the area to sell the product on a trial basis for a 1 month period. The sales figures for the trial period are then used to project sales for the entire area. [*Note:*  The same number of trial stores are used each time.]

| TOTAL SALES DURING TRIAL PERIOD $x$, hundreds of dollars | TOTAL SALES FOR FIRST MONTH FOR ENTIRE AREA $y$, hundreds of dollars |
|---|---|
| 16.8 | 48.2 |
| 14.0 | 46.8 |
| 18.3 | 54.3 |
| 22.1 | 59.7 |
| 14.9 | 48.3 |
| 23.2 | 67.5 |

a.  Use the data in the table to develop a simple linear model for predicting first-month sales for the entire area based on sales during the trial period.

b.  Plot the data and graph the line as a check on your calculations.

c.  Do the data provide sufficient evidence to indicate that total sales during the trial period contribute information for predicting total sales during the first month?

d.  Use a 90% prediction interval to predict total sales for the first month for the entire area if the trial sales equal $2,000.

**9.54.** The management of a manufacturing firm is considering the possibility of setting up its own market research department rather than continuing to use the services of a market research firm. The management wants to know what salary should be paid to a market researcher, based on years of experience. An independent consultant checks with several other firms in the area and obtains the information shown in the table on market researchers.

| ANNUAL SALARY <br> $y$, thousands of dollars | EXPERIENCE <br> $x$, years |
|:---:|:---:|
| 21.3 | 2 |
| 21.2 | 1.5 |
| 30.0 | 11 |
| 34.1 | 15 |
| 30.4 | 9 |
| 26.9 | 6 |

a. Fit a least squares line to the data.

b. Plot the data and graph the line as a check on your calculations.

c. Calculate $r$ and $r^2$. Explain how these values measure the utility of the model.

d. Estimate the mean annual salary of a market researcher with 8 years of experience. Use a 90% confidence interval.

e. Predict the salary of a market researcher with 7 years of experience using a 90% prediction interval.

**9.55.** A study was conducted to determine whether there is a linear relationship between the breaking strength, $y$, of wooden beams and the specific gravity, $x$, of the wood. Ten randomly selected beams of the same cross-sectional dimensions were stressed until they broke. The breaking strengths and the density of the wood are shown below for each of the ten beams:

| BEAM | SPECIFIC GRAVITY <br> $x$ | STRENGTH <br> $y$ |
|:---:|:---:|:---:|
| 1 | .499 | 11.14 |
| 2 | .558 | 12.74 |
| 3 | .604 | 13.13 |
| 4 | .441 | 11.51 |
| 5 | .550 | 12.38 |
| 6 | .528 | 12.60 |
| 7 | .418 | 11.13 |
| 8 | .480 | 11.70 |
| 9 | .406 | 11.02 |
| 10 | .467 | 11.41 |

a. Fit the model $y = \beta_0 + \beta_1 x + \varepsilon$.

b.  Test $H_0$: $\beta_1 = 0$ against the alternative hypothesis, $H_a$: $\beta_1 \neq 0$.

c.  Estimate the mean strength for beams with specific gravity .590 using a 90% confidence interval.

**9.56.**  Although the income tax system is structured so that people with higher incomes should pay a higher percentage of their incomes in taxes, there are many loopholes and tax shelters available for individuals with higher incomes. A sample of seven individual 1980 tax returns gave the data listed in the table on income and percent taxes paid.

| INDIVIDUAL | GROSS INCOME<br>x, thousands of dollars | TAXES PAID<br>y, percentage of total income |
|:---:|:---:|:---:|
| 1 | 35.8 | 16.7 |
| 2 | 80.2 | 21.4 |
| 3 | 14.9 | 15.2 |
| 4 | 7.3 | 10.1 |
| 5 | 9.1 | 12.2 |
| 6 | 150.7 | 19.6 |
| 7 | 25.9 | 17.3 |

a.  Fit a least squares line to the data.

b.  Plot the data and graph the line as a check on your calculations.

c.  Calculate $r$ and $r^2$ and interpret each.

d.  Find a 90% confidence interval for the mean percent taxes paid for individuals with gross incomes of $70,000.

**9.57.**  The data in the table were collected to calibrate a new instrument for measuring interocular pressure. The interocular pressure for each of ten glaucoma patients was measured by the new instrument and by a standard, reliable, but more time-consuming method.

| PATIENT | RELIABLE METHOD<br>x | NEW INSTRUMENT<br>y |
|:---:|:---:|:---:|
| 1 | 20.2 | 20.0 |
| 2 | 16.7 | 17.1 |
| 3 | 17.1 | 17.2 |
| 4 | 26.3 | 25.1 |
| 5 | 22.2 | 22.0 |
| 6 | 21.8 | 22.1 |
| 7 | 19.1 | 18.9 |
| 8 | 22.9 | 22.2 |
| 9 | 23.5 | 24.0 |
| 10 | 17.0 | 18.1 |

a. Fit a least squares line to the data.

b. Calculate $r$ and $r^2$. Interpret each of these quantities.

c. Predict the pressure measured by the new instrument when the reliable method gives a reading of 20.0. Use a 90% prediction interval.

**9.58.** The data in the table give the mileages per gallon obtained by a test automobile when using gasolines of varying levels of octane.

| MILEAGE<br>$y$, miles per gallon | OCTANE<br>$x$ |
|---|---|
| 13.0 | 89 |
| 13.2 | 93 |
| 13.0 | 87 |
| 13.6 | 90 |
| 13.3 | 89 |
| 13.8 | 95 |
| 14.1 | 100 |
| 14.0 | 98 |

a. Calculate $r$ and $r^2$.

b. Do the data provide sufficient evidence to indicate a correlation between octane level and miles per gallon for the test automobile?

**ON YOUR OWN . . .**

There are many dependent variables in all areas of research which are the subject of regression modeling efforts. We list five such variables below:

**1.** Crime rate in various communities

**2.** Daily maximum temperature in your town

**3.** Grade-point average of students who have completed one academic year at your college

**4.** Gross National Product of the United States

**5.** Points scored by your favorite football team in any one game

Choose one of these dependent variables which is of particular interest to you or choose some other dependent variable for which you want to construct a prediction model. There may be a large number of independent variables that should be included in a prediction equation for the dependent variable you choose. List three potentially important independent variables, $x_1$, $x_2$, and $x_3$, that you think might be (individually) strongly related to your dependent variable. Next, obtain ten data values, each of which consists of a measure of your dependent variable, $y$, and the corresponding values of $x_1$, $x_2$, and $x_3$.

**a.** Use the least squares formulas given in this chapter to fit three straight-line models—one for each independent variable—for predicting $y$.

**b.** Interpret the sign of the estimated slope coefficient $\hat{\beta}_1$ in each case, and test the utility of the model by testing $H_0$: $\beta_1 = 0$ against $H_a$: $\beta_1 \neq 0$. What assumptions must be satisfied to assure the validity of these tests?

**c.** Calculate the coefficient of determination, $r^2$, for each model. Which of the independent variables predicts $y$ best for the ten sampled sets of data? Is this variable necessarily best in general (i.e., for the entire population)? Explain.

**REFERENCES**

Graybill, F. *Theory and application of the linear model.* North Scituate, Mass.: Duxbury, 1976. Chapter 5.

Heaukulani, D. "The normal distribution of crime." *Journal of Police Science and Administration,* 1975, *3*(3), 312–318.

Mendenhall, W. *Introduction to linear models and the design and analysis of experiments.* Belmont, Ca.: Wadsworth, 1968.

Mendenhall, W., & McClave, J. T. *A second course in business statistics: regression analysis.* San Francisco: Dellen Publishing Co., 1981.

Neter, J., & Wasserman, W. *Applied linear statistical models.* Homewood, Ill.: Richard Irwin, 1974. Chapters 2–6.

Younger, M. S. *A handbook for linear regression.* North Scituate, Mass.: Duxbury, 1979.

# APPENDIX

**TABLE I**
**RANDOM NUMBERS**

| ROW \ COLUMN | 1 | 2 | 3 | 4 | 5 | 6 | 7 | 8 | 9 | 10 | 11 | 12 | 13 | 14 |
|---|---|---|---|---|---|---|---|---|---|---|---|---|---|---|
| 1 | 10480 | 15011 | 01536 | 02011 | 81647 | 91646 | 69179 | 14194 | 62590 | 36207 | 20969 | 99570 | 91291 | 90700 |
| 2 | 22368 | 46573 | 25595 | 85393 | 30995 | 89198 | 27982 | 53402 | 93965 | 34095 | 52666 | 19174 | 39615 | 99505 |
| 3 | 24130 | 48360 | 22527 | 97265 | 76393 | 64809 | 15179 | 24830 | 49340 | 32081 | 30680 | 19655 | 63348 | 58629 |
| 4 | 42167 | 93093 | 06243 | 61680 | 07856 | 16376 | 39440 | 53537 | 71341 | 57004 | 00849 | 74917 | 97758 | 16379 |
| 5 | 37570 | 39975 | 81837 | 16656 | 06121 | 91782 | 60468 | 81305 | 49684 | 60672 | 14110 | 06927 | 01263 | 54613 |
| 6 | 77921 | 06907 | 11008 | 42751 | 27756 | 53498 | 18602 | 70659 | 90655 | 15053 | 21916 | 81825 | 44394 | 42880 |
| 7 | 99562 | 72905 | 56420 | 69994 | 98872 | 31016 | 71194 | 18738 | 44013 | 48840 | 63213 | 21069 | 10634 | 12952 |
| 8 | 96301 | 91977 | 05463 | 07972 | 18876 | 20922 | 94595 | 56869 | 69014 | 60045 | 18425 | 84903 | 42508 | 32307 |
| 9 | 89579 | 14342 | 63661 | 10281 | 17453 | 18103 | 57740 | 84378 | 25331 | 12566 | 58678 | 44947 | 05585 | 56941 |
| 10 | 85475 | 36857 | 53342 | 53988 | 53060 | 59533 | 38867 | 62300 | 08158 | 17983 | 16439 | 11458 | 18593 | 64952 |
| 11 | 28918 | 69578 | 88231 | 33276 | 70997 | 79936 | 56865 | 05859 | 90106 | 31595 | 01547 | 85590 | 91610 | 78188 |
| 12 | 63553 | 40961 | 48235 | 03427 | 49626 | 69445 | 18663 | 72695 | 52180 | 20847 | 12234 | 90511 | 33703 | 90322 |
| 13 | 09429 | 93969 | 52636 | 92737 | 88974 | 33488 | 36320 | 17617 | 30015 | 08272 | 84115 | 27156 | 30613 | 74952 |
| 14 | 10365 | 61129 | 87529 | 85689 | 48237 | 52267 | 67689 | 93394 | 01511 | 26358 | 85104 | 20285 | 29975 | 89868 |
| 15 | 07119 | 97336 | 71048 | 08178 | 77233 | 13916 | 47564 | 81056 | 97735 | 85977 | 29372 | 74461 | 28551 | 90707 |
| 16 | 51085 | 12765 | 51821 | 51259 | 77452 | 16308 | 60756 | 92144 | 49442 | 53900 | 70960 | 63990 | 75601 | 40719 |
| 17 | 02368 | 21382 | 52404 | 60268 | 89368 | 19885 | 55322 | 44819 | 01188 | 65255 | 64835 | 44919 | 05944 | 55157 |
| 18 | 01011 | 54092 | 33362 | 94904 | 31273 | 04146 | 18594 | 29852 | 71585 | 85030 | 51132 | 01915 | 92747 | 64951 |
| 19 | 52162 | 53916 | 46369 | 58586 | 23216 | 14513 | 83149 | 98736 | 23495 | 64350 | 94738 | 17752 | 35156 | 35749 |
| 20 | 07056 | 97628 | 33787 | 09998 | 42698 | 06691 | 76988 | 13602 | 51851 | 46104 | 88916 | 19509 | 25625 | 58104 |
| 21 | 48663 | 91245 | 85828 | 14346 | 09172 | 30168 | 90229 | 04734 | 59193 | 22178 | 30421 | 61666 | 99904 | 32812 |
| 22 | 54164 | 58492 | 22421 | 74103 | 47070 | 25306 | 76468 | 26384 | 58151 | 06646 | 21524 | 15227 | 96909 | 44592 |
| 23 | 32639 | 32363 | 05597 | 24200 | 13363 | 38005 | 94342 | 28728 | 35806 | 06912 | 17012 | 64161 | 18296 | 22851 |
| 24 | 29334 | 27001 | 87637 | 87308 | 58731 | 00256 | 45834 | 15398 | 46557 | 41135 | 10367 | 07684 | 36188 | 18510 |
| 25 | 02488 | 33062 | 28834 | 07351 | 19731 | 92420 | 60952 | 61280 | 50001 | 67658 | 32586 | 86679 | 50720 | 94953 |
| 26 | 81525 | 72295 | 04839 | 96423 | 24878 | 82651 | 66566 | 14778 | 76797 | 14780 | 13300 | 87074 | 79666 | 95725 |
| 27 | 29676 | 20591 | 68086 | 26432 | 46901 | 20849 | 89768 | 81536 | 86645 | 12659 | 92259 | 57102 | 80428 | 25280 |
| 28 | 00742 | 57392 | 39064 | 66432 | 84673 | 40027 | 32832 | 61362 | 98947 | 96067 | 64760 | 64584 | 96096 | 98253 |
| 29 | 05366 | 04213 | 25669 | 26422 | 44407 | 44048 | 37937 | 63904 | 45766 | 66134 | 75470 | 66520 | 34693 | 90449 |
| 30 | 91921 | 26418 | 64117 | 94305 | 26766 | 25940 | 39972 | 22209 | 71500 | 64568 | 91402 | 42416 | 07844 | 69618 |
| 31 | 00582 | 04711 | 87917 | 77341 | 42206 | 35126 | 74087 | 99547 | 81817 | 42607 | 43808 | 76655 | 62028 | 76630 |
| 32 | 00725 | 69884 | 62797 | 56170 | 86324 | 88072 | 76222 | 36086 | 84637 | 93161 | 76038 | 65855 | 77919 | 88006 |
| 33 | 69011 | 65795 | 95876 | 55293 | 18988 | 27354 | 26575 | 08625 | 40801 | 59920 | 29841 | 80150 | 12777 | 48501 |
| 34 | 25976 | 57948 | 29888 | 88604 | 67917 | 48708 | 18912 | 82271 | 65424 | 69774 | 33611 | 54262 | 85963 | 03547 |

| | | | | | | | | | | | | | | |
|---|---|---|---|---|---|---|---|---|---|---|---|---|---|---|
| 35 | 09763 | 83473 | 73577 | 12908 | 30883 | 18317 | 28290 | 35797 | 05998 | 41688 | 34952 | 37888 | 38917 | 88050 |
| 36 | 91576 | 42595 | 27958 | 30134 | 04024 | 86385 | 29880 | 99730 | 55536 | 84855 | 29080 | 09250 | 79656 | 73211 |
| 37 | 17955 | 56349 | 90999 | 49127 | 20044 | 59931 | 06115 | 20542 | 18059 | 02008 | 73708 | 83517 | 36103 | 42791 |
| 38 | 46503 | 18584 | 18845 | 49618 | 02304 | 51038 | 20655 | 58727 | 28168 | 15475 | 56942 | 53389 | 20562 | 87338 |
| 39 | 92157 | 89634 | 94824 | 78171 | 84610 | 82834 | 09922 | 25417 | 44137 | 48413 | 25555 | 21246 | 35509 | 20468 |
| 40 | 14577 | 62765 | 35605 | 81263 | 39667 | 47358 | 56873 | 56307 | 61607 | 49518 | 89656 | 20103 | 77490 | 18062 |
| 41 | 98427 | 07523 | 33362 | 64270 | 01638 | 92477 | 66969 | 98420 | 04880 | 45585 | 46565 | 04102 | 46880 | 45709 |
| 42 | 34914 | 63976 | 88720 | 82765 | 34476 | 17032 | 87589 | 40836 | 32427 | 70002 | 70663 | 88863 | 77775 | 69348 |
| 43 | 70060 | 28277 | 39475 | 46473 | 23219 | 53416 | 94970 | 25832 | 69975 | 94884 | 19661 | 72828 | 00102 | 66794 |
| 44 | 53976 | 54914 | 06990 | 67245 | 68350 | 82948 | 11398 | 42878 | 80287 | 88267 | 47363 | 46634 | 06541 | 97809 |
| 45 | 76072 | 29515 | 40980 | 07391 | 58745 | 25774 | 22987 | 80059 | 39911 | 96189 | 41151 | 14222 | 60697 | 59583 |
| 46 | 90725 | 52210 | 83974 | 29992 | 65831 | 38857 | 50490 | 83765 | 55657 | 14361 | 31720 | 57375 | 56228 | 41546 |
| 47 | 64364 | 67412 | 33339 | 31926 | 14883 | 24413 | 59744 | 92351 | 97473 | 89286 | 35931 | 04110 | 23726 | 51900 |
| 48 | 08962 | 00358 | 31662 | 25388 | 61642 | 34072 | 81249 | 35648 | 56891 | 69352 | 48373 | 45578 | 78547 | 81788 |
| 49 | 95012 | 68379 | 93526 | 70765 | 10592 | 04542 | 76463 | 54328 | 02349 | 17247 | 28865 | 14777 | 62730 | 92277 |
| 50 | 15664 | 10493 | 20492 | 38391 | 91132 | 21999 | 59516 | 81652 | 27195 | 48223 | 46751 | 22923 | 32261 | 85653 |
| 51 | 16408 | 81899 | 04153 | 53381 | 79401 | 21438 | 83035 | 92350 | 36693 | 31238 | 59649 | 91754 | 72772 | 02338 |
| 52 | 18629 | 81953 | 05520 | 91962 | 04739 | 13092 | 97662 | 24822 | 94730 | 06496 | 35090 | 04822 | 86774 | 98289 |
| 53 | 73115 | 35101 | 47498 | 87637 | 99016 | 71060 | 88824 | 71013 | 18735 | 20286 | 23153 | 72924 | 35165 | 43040 |
| 54 | 57491 | 16703 | 23167 | 49323 | 45021 | 33132 | 12544 | 41035 | 80780 | 45393 | 44812 | 12515 | 98931 | 91202 |
| 55 | 30405 | 83946 | 23792 | 14422 | 15059 | 45799 | 22716 | 19792 | 09983 | 74353 | 68668 | 30429 | 70735 | 25499 |
| 56 | 16631 | 35006 | 85900 | 98275 | 32388 | 52390 | 16815 | 69298 | 82732 | 38480 | 73817 | 32523 | 41961 | 44437 |
| 57 | 96773 | 20206 | 42559 | 78985 | 05300 | 22164 | 24369 | 54224 | 35083 | 19687 | 11052 | 91491 | 60383 | 19746 |
| 58 | 38935 | 64202 | 14349 | 82674 | 66523 | 44133 | 00697 | 35552 | 35970 | 19124 | 63318 | 29686 | 03387 | 59846 |
| 59 | 31624 | 76384 | 17403 | 53363 | 44167 | 64486 | 64758 | 75366 | 76554 | 31601 | 12614 | 33072 | 60332 | 92325 |
| 60 | 78919 | 19474 | 23632 | 27889 | 47914 | 02584 | 37680 | 20801 | 72152 | 39339 | 34806 | 08930 | 85001 | 87820 |
| 61 | 03931 | 33309 | 57047 | 74211 | 63445 | 17361 | 62825 | 39908 | 05607 | 91284 | 68833 | 25570 | 38818 | 46920 |
| 62 | 74426 | 33278 | 43972 | 10119 | 89917 | 15665 | 52872 | 73823 | 73144 | 88662 | 88970 | 74492 | 51805 | 99378 |
| 63 | 09066 | 00903 | 20795 | 95452 | 92648 | 45454 | 09552 | 88815 | 16553 | 51125 | 79375 | 97596 | 16296 | 66092 |
| 64 | 42238 | 12426 | 87025 | 14267 | 20979 | 04508 | 64535 | 31355 | 86064 | 29472 | 47689 | 05974 | 52468 | 16834 |
| 65 | 16153 | 08002 | 26504 | 41744 | 81959 | 65642 | 74240 | 56302 | 00033 | 67107 | 77510 | 70625 | 28725 | 34191 |
| 66 | 21457 | 40742 | 29820 | 96783 | 29400 | 21840 | 15035 | 34537 | 33310 | 06116 | 95240 | 15957 | 16572 | 06004 |
| 67 | 21581 | 57802 | 02050 | 89728 | 17937 | 37621 | 47075 | 42080 | 97403 | 48626 | 68995 | 43805 | 33386 | 21597 |
| 68 | 55612 | 78095 | 83197 | 33732 | 05810 | 24813 | 86902 | 60397 | 16489 | 03264 | 88525 | 42786 | 05269 | 92532 |
| 69 | 44657 | 66999 | 99324 | 51281 | 84463 | 60563 | 79312 | 93454 | 68876 | 25471 | 93911 | 25650 | 12682 | 73572 |
| 70 | 91340 | 84979 | 46949 | 81973 | 37949 | 61023 | 43997 | 15263 | 80644 | 43942 | 89203 | 71795 | 99533 | 50501 |
| 71 | 91227 | 21199 | 31935 | 27022 | 84067 | 05462 | 35216 | 14486 | 29891 | 68607 | 41867 | 14951 | 91696 | 85065 |
| 72 | 50001 | 38140 | 66321 | 19924 | 72163 | 09538 | 12151 | 06878 | 91903 | 18749 | 34405 | 56087 | 82790 | 70925 |

*(continued)*

**TABLE I**
CONTINUED

| ROW \ COLUMN | 1 | 2 | 3 | 4 | 5 | 6 | 7 | 8 | 9 | 10 | 11 | 12 | 13 | 14 |
|---|---|---|---|---|---|---|---|---|---|---|---|---|---|---|
| 73 | 65390 | 05224 | 72958 | 28609 | 81406 | 39147 | 25549 | 48542 | 42627 | 45233 | 57202 | 94617 | 23772 | 07896 |
| 74 | 27504 | 96131 | 83944 | 41575 | 10573 | 08619 | 64482 | 73923 | 36152 | 05184 | 94142 | 25299 | 84387 | 34925 |
| 75 | 37169 | 94851 | 39117 | 89632 | 00959 | 16487 | 65536 | 49071 | 39782 | 17095 | 02330 | 74301 | 00275 | 48280 |
| 76 | 11508 | 70225 | 51111 | 38351 | 19444 | 66499 | 71945 | 05422 | 13442 | 78675 | 84081 | 66938 | 93654 | 59894 |
| 77 | 37449 | 30362 | 06694 | 54690 | 04052 | 53115 | 62757 | 95348 | 78662 | 11163 | 81651 | 50245 | 34971 | 52924 |
| 78 | 46515 | 70331 | 85922 | 38329 | 57015 | 15765 | 97161 | 17869 | 45349 | 61796 | 66345 | 81073 | 49106 | 79860 |
| 79 | 30986 | 81223 | 42416 | 58353 | 21532 | 30502 | 32305 | 86482 | 05174 | 07901 | 54339 | 58861 | 74818 | 46942 |
| 80 | 63798 | 64995 | 46583 | 09785 | 44160 | 78128 | 83991 | 42865 | 92520 | 83531 | 80377 | 35909 | 81250 | 54238 |
| 81 | 82486 | 84846 | 99254 | 67632 | 43218 | 50076 | 21361 | 64816 | 51202 | 88124 | 41870 | 52689 | 51275 | 83556 |
| 82 | 21885 | 32906 | 92431 | 09060 | 64297 | 51674 | 64126 | 62570 | 26123 | 05155 | 59194 | 52799 | 28225 | 85762 |
| 83 | 60336 | 98782 | 07408 | 53458 | 13564 | 59089 | 26445 | 29789 | 85205 | 41001 | 12535 | 12133 | 14645 | 23541 |
| 84 | 43937 | 46891 | 24010 | 25560 | 86355 | 33941 | 25786 | 54990 | 71899 | 15475 | 95434 | 98227 | 21824 | 19585 |
| 85 | 97656 | 63175 | 89303 | 16275 | 07100 | 92063 | 21942 | 18611 | 47348 | 20203 | 18534 | 03862 | 78095 | 50136 |
| 86 | 03299 | 01221 | 05418 | 38982 | 55758 | 92237 | 26759 | 86367 | 21216 | 98442 | 08303 | 56613 | 91511 | 75928 |
| 87 | 79626 | 06486 | 03574 | 17668 | 07785 | 76020 | 79924 | 25651 | 83325 | 88428 | 85076 | 72811 | 22717 | 50585 |
| 88 | 85636 | 68335 | 47539 | 03129 | 65651 | 11977 | 02510 | 26113 | 99447 | 68645 | 34327 | 15152 | 55230 | 93448 |
| 89 | 18039 | 14367 | 61337 | 06177 | 12143 | 46609 | 32989 | 74014 | 64708 | 00533 | 35398 | 58408 | 13261 | 47908 |
| 90 | 08362 | 15656 | 60627 | 36478 | 65648 | 16764 | 53412 | 09013 | 07832 | 41574 | 17639 | 82163 | 60859 | 75567 |
| 91 | 79556 | 29068 | 04142 | 16268 | 15387 | 12856 | 66227 | 38358 | 22478 | 73373 | 88732 | 09443 | 82558 | 05250 |
| 92 | 92608 | 82674 | 27072 | 32534 | 17075 | 27698 | 98204 | 63863 | 11951 | 34648 | 88022 | 56148 | 34925 | 57031 |
| 93 | 23982 | 25835 | 40055 | 67006 | 12293 | 02753 | 14827 | 23235 | 35071 | 99704 | 37543 | 11601 | 35503 | 85171 |
| 94 | 09915 | 96306 | 05908 | 97901 | 28395 | 14186 | 00821 | 80703 | 70426 | 75647 | 76310 | 88717 | 37890 | 40129 |
| 95 | 59037 | 33300 | 26695 | 62247 | 69927 | 76123 | 50842 | 43834 | 86654 | 70959 | 79725 | 93872 | 28117 | 19233 |
| 96 | 42488 | 78077 | 69882 | 61657 | 34136 | 79180 | 97526 | 43092 | 04098 | 73571 | 80799 | 76536 | 71255 | 64239 |
| 97 | 46764 | 86273 | 63003 | 93017 | 31204 | 36692 | 40202 | 35275 | 57306 | 55543 | 53203 | 18098 | 47625 | 88684 |
| 98 | 03237 | 45430 | 55417 | 63282 | 90816 | 17349 | 88298 | 90183 | 36600 | 78406 | 06216 | 95787 | 42579 | 90730 |
| 99 | 86591 | 81482 | 52667 | 61582 | 14972 | 90053 | 89534 | 76036 | 49199 | 43716 | 97548 | 04379 | 46370 | 28672 |
| 100 | 38534 | 01715 | 94964 | 87288 | 65680 | 43772 | 39560 | 12918 | 86537 | 62738 | 19636 | 51132 | 25739 | 56947 |

Source: Abridged from W. H. Beyer, Ed., *CRC Standard Mathematical Tables*, 24th ed.(Cleveland: The Chemical Rubber Company), 1976. Reproduced by permission of the publisher.

**TABLE II**

**BINOMIAL PROBABILITIES**

Tabulated values are $\sum_{x=0}^{k} p(x)$. *(Computations are rounded at the third decimal place.)*

**a.** $n = 5$

| k \ p | 0.01 | 0.05 | 0.10 | 0.20 | 0.30 | 0.40 | 0.50 | 0.60 | 0.70 | 0.80 | 0.90 | 0.95 | 0.99 |
|---|---|---|---|---|---|---|---|---|---|---|---|---|---|
| 0 | .951 | .774 | .590 | .328 | .168 | .078 | .031 | .010 | .002 | .000 | .000 | .000 | .000 |
| 1 | .999 | .977 | .919 | .737 | .528 | .337 | .188 | .087 | .031 | .007 | .000 | .000 | .000 |
| 2 | 1.000 | .999 | .991 | .942 | .837 | .683 | .500 | .317 | .163 | .058 | .009 | .001 | .000 |
| 3 | 1.000 | 1.000 | 1.000 | .993 | .969 | .913 | .812 | .663 | .472 | .263 | .081 | .023 | .001 |
| 4 | 1.000 | 1.000 | 1.000 | 1.000 | .998 | .990 | .969 | .922 | .832 | .672 | .410 | .226 | .049 |

**b.** $n = 6$

| k \ p | 0.01 | 0.05 | 0.10 | 0.20 | 0.30 | 0.40 | 0.50 | 0.60 | 0.70 | 0.80 | 0.90 | 0.95 | 0.99 |
|---|---|---|---|---|---|---|---|---|---|---|---|---|---|
| 0 | .941 | .735 | .531 | .262 | .118 | .047 | .016 | .004 | .001 | .000 | .000 | .000 | .000 |
| 1 | .999 | .967 | .886 | .655 | .420 | .233 | .109 | .041 | .011 | .002 | .000 | .000 | .000 |
| 2 | 1.000 | .998 | .984 | .901 | .744 | .544 | .344 | .179 | .070 | .017 | .001 | .000 | .000 |
| 3 | 1.000 | 1.000 | .999 | .983 | .930 | .821 | .656 | .456 | .256 | .099 | .016 | .002 | .000 |
| 4 | 1.000 | 1.000 | 1.000 | .998 | .989 | .959 | .891 | .767 | .580 | .345 | .114 | .033 | .001 |
| 5 | 1.000 | 1.000 | 1.000 | 1.000 | .999 | .996 | .984 | .953 | .882 | .738 | .469 | .265 | .059 |

**c.** $n = 7$

| k \ p | 0.01 | 0.05 | 0.10 | 0.20 | 0.30 | 0.40 | 0.50 | 0.60 | 0.70 | 0.80 | 0.90 | 0.95 | 0.99 |
|---|---|---|---|---|---|---|---|---|---|---|---|---|---|
| 0 | .932 | .698 | .478 | .210 | .082 | .028 | .008 | .002 | .000 | .000 | .000 | .000 | .000 |
| 1 | .998 | .956 | .850 | .577 | .329 | .159 | .063 | .019 | .004 | .000 | .000 | .000 | .000 |
| 2 | 1.000 | .996 | .974 | .852 | .647 | .420 | .227 | .096 | .029 | .005 | .000 | .000 | .000 |
| 3 | 1.000 | 1.000 | .997 | .967 | .874 | .710 | .500 | .290 | .126 | .033 | .003 | .000 | .000 |
| 4 | 1.000 | 1.000 | 1.000 | .995 | .971 | .904 | .773 | .580 | .353 | .148 | .026 | .004 | .000 |
| 5 | 1.000 | 1.000 | 1.000 | 1.000 | .996 | .981 | .937 | .841 | .671 | .423 | .150 | .044 | .002 |
| 6 | 1.000 | 1.000 | 1.000 | 1.000 | 1.000 | .998 | .992 | .972 | .918 | .790 | .522 | .302 | .068 |

**d.** $n = 8$

| k \ p | 0.01 | 0.05 | 0.10 | 0.20 | 0.30 | 0.40 | 0.50 | 0.60 | 0.70 | 0.80 | 0.90 | 0.95 | 0.99 |
|---|---|---|---|---|---|---|---|---|---|---|---|---|---|
| 0 | .923 | .663 | .430 | .168 | .058 | .017 | .004 | .001 | .000 | .000 | .000 | .000 | .000 |
| 1 | .997 | .943 | .813 | .503 | .255 | .106 | .035 | .009 | .001 | .000 | .000 | .000 | .000 |
| 2 | 1.000 | .994 | .962 | .797 | .552 | .315 | .145 | .050 | .011 | .001 | .000 | .000 | .000 |
| 3 | 1.000 | 1.000 | .995 | .944 | .806 | .594 | .363 | .174 | .058 | .010 | .000 | .000 | .000 |
| 4 | 1.000 | 1.000 | 1.000 | .990 | .942 | .826 | .637 | .406 | .194 | .056 | .005 | .000 | .000 |
| 5 | 1.000 | 1.000 | 1.000 | .999 | .989 | .950 | .855 | .685 | .448 | .203 | .038 | .006 | .000 |
| 6 | 1.000 | 1.000 | 1.000 | 1.000 | .999 | .991 | .965 | .894 | .745 | .497 | .187 | .057 | .003 |
| 7 | 1.000 | 1.000 | 1.000 | 1.000 | 1.000 | .999 | .996 | .983 | .942 | .832 | .570 | .337 | .077 |

*(continued)*

**TABLE II**
**CONTINUED**

**e.**  $n = 9$

| p<br>k | 0.01 | 0.05 | 0.10 | 0.20 | 0.30 | 0.40 | 0.50 | 0.60 | 0.70 | 0.80 | 0.90 | 0.95 | 0.99 |
|---|---|---|---|---|---|---|---|---|---|---|---|---|---|
| 0 | .914 | .630 | .387 | .134 | .040 | .010 | .002 | .000 | .000 | .000 | .000 | .000 | .000 |
| 1 | .997 | .929 | .775 | .436 | .196 | .071 | .020 | .004 | .000 | .000 | .000 | .000 | .000 |
| 2 | 1.000 | .992 | .947 | .738 | .463 | .232 | .090 | .025 | .004 | .000 | .000 | .000 | .000 |
| 3 | 1.000 | .999 | .992 | .914 | .730 | .483 | .254 | .099 | .025 | .003 | .000 | .000 | .000 |
| 4 | 1.000 | 1.000 | .999 | .980 | .901 | .733 | .500 | .267 | .099 | .020 | .001 | .000 | .000 |
| 5 | 1.000 | 1.000 | 1.000 | .997 | .975 | .901 | .746 | .517 | .270 | .086 | .008 | .001 | .000 |
| 6 | 1.000 | 1.000 | 1.000 | 1.000 | .996 | .975 | .910 | .768 | .537 | .262 | .053 | .008 | .000 |
| 7 | 1.000 | 1.000 | 1.000 | 1.000 | 1.000 | .996 | .980 | .929 | .804 | .564 | .225 | .071 | .003 |
| 8 | 1.000 | 1.000 | 1.000 | 1.000 | 1.000 | 1.000 | .998 | .990 | .960 | .866 | .613 | .370 | .086 |

**f.**  $n = 10$

| p<br>k | 0.01 | 0.05 | 0.10 | 0.20 | 0.30 | 0.40 | 0.50 | 0.60 | 0.70 | 0.80 | 0.90 | 0.95 | 0.99 |
|---|---|---|---|---|---|---|---|---|---|---|---|---|---|
| 0 | .904 | .599 | .349 | .107 | .028 | .006 | .001 | .000 | .000 | .000 | .000 | .000 | .000 |
| 1 | .996 | .914 | .736 | .376 | .149 | .046 | .011 | .002 | .000 | .000 | .000 | .000 | .000 |
| 2 | 1.000 | .988 | .930 | .678 | .383 | .167 | .055 | .012 | .002 | .000 | .000 | .000 | .000 |
| 3 | 1.000 | .999 | .987 | .879 | .650 | .382 | .172 | .055 | .011 | .001 | .000 | .000 | .000 |
| 4 | 1.000 | 1.000 | .998 | .967 | .850 | .633 | .377 | .166 | .047 | .006 | .000 | .000 | .000 |
| 5 | 1.000 | 1.000 | 1.000 | .994 | .953 | .834 | .623 | .367 | .150 | .033 | .002 | .000 | .000 |
| 6 | 1.000 | 1.000 | 1.000 | .999 | .989 | .945 | .828 | .618 | .350 | .121 | .013 | .001 | .000 |
| 7 | 1.000 | 1.000 | 1.000 | 1.000 | .998 | .988 | .945 | .833 | .617 | .322 | .070 | .012 | .000 |
| 8 | 1.000 | 1.000 | 1.000 | 1.000 | 1.000 | .998 | .989 | .954 | .851 | .624 | .264 | .086 | .004 |
| 9 | 1.000 | 1.000 | 1.000 | 1.000 | 1.000 | 1.000 | .999 | .994 | .972 | .893 | .651 | .401 | .096 |

**g.**  $n = 15$

| p<br>k | 0.01 | 0.05 | 0.10 | 0.20 | 0.30 | 0.40 | 0.50 | 0.60 | 0.70 | 0.80 | 0.90 | 0.95 | 0.99 |
|---|---|---|---|---|---|---|---|---|---|---|---|---|---|
| 0 | .860 | .463 | .206 | .035 | .005 | .000 | .000 | .000 | .000 | .000 | .000 | .000 | .000 |
| 1 | .990 | .829 | .549 | .167 | .035 | .005 | .000 | .000 | .000 | .000 | .000 | .000 | .000 |
| 2 | 1.000 | .964 | .816 | .398 | .127 | .027 | .004 | .000 | .000 | .000 | .000 | .000 | .000 |
| 3 | 1.000 | .995 | .944 | .648 | .297 | .091 | .018 | .002 | .000 | .000 | .000 | .000 | .000 |
| 4 | 1.000 | .999 | .987 | .836 | .515 | .217 | .059 | .009 | .001 | .000 | .000 | .000 | .000 |
| 5 | 1.000 | 1.000 | .998 | .939 | .722 | .403 | .151 | .034 | .004 | .000 | .000 | .000 | .000 |
| 6 | 1.000 | 1.000 | 1.000 | .982 | .869 | .610 | .304 | .095 | .015 | .001 | .000 | .000 | .000 |
| 7 | 1.000 | 1.000 | 1.000 | .996 | .950 | .787 | .500 | .213 | .050 | .004 | .000 | .000 | .000 |
| 8 | 1.000 | 1.000 | 1.000 | .999 | .985 | .905 | .696 | .390 | .131 | .018 | .000 | .000 | .000 |
| 9 | 1.000 | 1.000 | 1.000 | 1.000 | .996 | .966 | .849 | .597 | .278 | .061 | .002 | .000 | .000 |
| 10 | 1.000 | 1.000 | 1.000 | 1.000 | .999 | .991 | .941 | .783 | .485 | .164 | .013 | .001 | .000 |
| 11 | 1.000 | 1.000 | 1.000 | 1.000 | 1.000 | .998 | .982 | .909 | .703 | .352 | .056 | .005 | .000 |
| 12 | 1.000 | 1.000 | 1.000 | 1.000 | 1.000 | 1.000 | .996 | .973 | .873 | .602 | .184 | .036 | .000 |
| 13 | 1.000 | 1.000 | 1.000 | 1.000 | 1.000 | 1.000 | 1.000 | .995 | .965 | .833 | .451 | .171 | .010 |
| 14 | 1.000 | 1.000 | 1.000 | 1.000 | 1.000 | 1.000 | 1.000 | 1.000 | .995 | .965 | .794 | .537 | .140 |

**h.** $n = 20$

| k \ p | 0.01 | 0.05 | 0.10 | 0.20 | 0.30 | 0.40 | 0.50 | 0.60 | 0.70 | 0.80 | 0.90 | 0.95 | 0.99 |
|---|---|---|---|---|---|---|---|---|---|---|---|---|---|
| 0 | .818 | .358 | .122 | .012 | .001 | .000 | .000 | .000 | .000 | .000 | .000 | .000 | .000 |
| 1 | .983 | .736 | .392 | .069 | .008 | .001 | .000 | .000 | .000 | .000 | .000 | .000 | .000 |
| 2 | .999 | .925 | .677 | .206 | .035 | .004 | .000 | .000 | .000 | .000 | .000 | .000 | .000 |
| 3 | 1.000 | .984 | .867 | .411 | .107 | .016 | .001 | .000 | .000 | .000 | .000 | .000 | .000 |
| 4 | 1.000 | .997 | .957 | .630 | .238 | .051 | .006 | .000 | .000 | .000 | .000 | .000 | .000 |
| 5 | 1.000 | 1.000 | .989 | .804 | .416 | .126 | .021 | .002 | .000 | .000 | .000 | .000 | .000 |
| 6 | 1.000 | 1.000 | .998 | .913 | .608 | .250 | .058 | .006 | .000 | .000 | .000 | .000 | .000 |
| 7 | 1.000 | 1.000 | 1.000 | .968 | .772 | .416 | .132 | .021 | .001 | .000 | .000 | .000 | .000 |
| 8 | 1.000 | 1.000 | 1.000 | .990 | .887 | .596 | .252 | .057 | .005 | .000 | .000 | .000 | .000 |
| 9 | 1.000 | 1.000 | 1.000 | .997 | .952 | .755 | .412 | .128 | .017 | .001 | .000 | .000 | .000 |
| 10 | 1.000 | 1.000 | 1.000 | .999 | .983 | .872 | .588 | .245 | .048 | .003 | .000 | .000 | .000 |
| 11 | 1.000 | 1.000 | 1.000 | 1.000 | .995 | .943 | .748 | .404 | .113 | .010 | .000 | .000 | .000 |
| 12 | 1.000 | 1.000 | 1.000 | 1.000 | .999 | .979 | .868 | .584 | .228 | .032 | .000 | .000 | .000 |
| 13 | 1.000 | 1.000 | 1.000 | 1.000 | 1.000 | .994 | .942 | .750 | .392 | .087 | .002 | .000 | .000 |
| 14 | 1.000 | 1.000 | 1.000 | 1.000 | 1.000 | .998 | .979 | .874 | .584 | .196 | .011 | .000 | .000 |
| 15 | 1.000 | 1.000 | 1.000 | 1.000 | 1.000 | 1.000 | .994 | .949 | .762 | .370 | .043 | .003 | .000 |
| 16 | 1.000 | 1.000 | 1.000 | 1.000 | 1.000 | 1.000 | .999 | .984 | .893 | .589 | .133 | .016 | .000 |
| 17 | 1.000 | 1.000 | 1.000 | 1.000 | 1.000 | 1.000 | 1.000 | .996 | .965 | .794 | .323 | .075 | .001 |
| 18 | 1.000 | 1.000 | 1.000 | 1.000 | 1.000 | 1.000 | 1.000 | .999 | .992 | .931 | .608 | .264 | .017 |
| 19 | 1.000 | 1.000 | 1.000 | 1.000 | 1.000 | 1.000 | 1.000 | 1.000 | .999 | .988 | .878 | .642 | .182 |

Continued

**TABLE II**
CONTINUED

**i.** $n = 25$

| k \ p | 0.01 | 0.05 | 0.10 | 0.20 | 0.30 | 0.40 | 0.50 | 0.60 | 0.70 | 0.80 | 0.90 | 0.95 | 0.99 |
|---|---|---|---|---|---|---|---|---|---|---|---|---|---|
| 0 | .778 | .277 | .072 | .004 | .000 | .000 | .000 | .000 | .000 | .000 | .000 | .000 | .000 |
| 1 | .974 | .642 | .271 | .027 | .002 | .000 | .000 | .000 | .000 | .000 | .000 | .000 | .000 |
| 2 | .998 | .873 | .537 | .098 | .009 | .000 | .000 | .000 | .000 | .000 | .000 | .000 | .000 |
| 3 | 1.000 | .966 | .764 | .234 | .033 | .002 | .000 | .000 | .000 | .000 | .000 | .000 | .000 |
| 4 | 1.000 | .993 | .902 | .421 | .090 | .009 | .000 | .000 | .000 | .000 | .000 | .000 | .000 |
| 5 | 1.000 | .999 | .967 | .617 | .193 | .029 | .002 | .000 | .000 | .000 | .000 | .000 | .000 |
| 6 | 1.000 | 1.000 | .991 | .780 | .341 | .074 | .007 | .000 | .000 | .000 | .000 | .000 | .000 |
| 7 | 1.000 | 1.000 | .998 | .891 | .512 | .154 | .022 | .001 | .000 | .000 | .000 | .000 | .000 |
| 8 | 1.000 | 1.000 | 1.000 | .953 | .677 | .274 | .054 | .004 | .000 | .000 | .000 | .000 | .000 |
| 9 | 1.000 | 1.000 | 1.000 | .983 | .811 | .425 | .115 | .013 | .000 | .000 | .000 | .000 | .000 |
| 10 | 1.000 | 1.000 | 1.000 | .994 | .902 | .586 | .212 | .034 | .002 | .000 | .000 | .000 | .000 |
| 11 | 1.000 | 1.000 | 1.000 | .998 | .956 | .732 | .345 | .078 | .006 | .000 | .000 | .000 | .000 |
| 12 | 1.000 | 1.000 | 1.000 | 1.000 | .983 | .846 | .500 | .154 | .017 | .000 | .000 | .000 | .000 |
| 13 | 1.000 | 1.000 | 1.000 | 1.000 | .994 | .922 | .655 | .268 | .044 | .002 | .000 | .000 | .000 |
| 14 | 1.000 | 1.000 | 1.000 | 1.000 | .998 | .966 | .788 | .414 | .098 | .006 | .000 | .000 | .000 |
| 15 | 1.000 | 1.000 | 1.000 | 1.000 | 1.000 | .987 | .885 | .575 | .189 | .017 | .000 | .000 | .000 |
| 16 | 1.000 | 1.000 | 1.000 | 1.000 | 1.000 | .996 | .946 | .726 | .323 | .047 | .000 | .000 | .000 |
| 17 | 1.000 | 1.000 | 1.000 | 1.000 | 1.000 | .999 | .978 | .846 | .488 | .109 | .002 | .000 | .000 |
| 18 | 1.000 | 1.000 | 1.000 | 1.000 | 1.000 | 1.000 | .993 | .926 | .659 | .220 | .009 | .000 | .000 |
| 19 | 1.000 | 1.000 | 1.000 | 1.000 | 1.000 | 1.000 | .998 | .971 | .807 | .383 | .033 | .001 | .000 |
| 20 | 1.000 | 1.000 | 1.000 | 1.000 | 1.000 | 1.000 | 1.000 | .991 | .910 | .579 | .098 | .007 | .000 |
| 21 | 1.000 | 1.000 | 1.000 | 1.000 | 1.000 | 1.000 | 1.000 | .998 | .967 | .766 | .236 | .034 | .000 |
| 22 | 1.000 | 1.000 | 1.000 | 1.000 | 1.000 | 1.000 | 1.000 | 1.000 | .991 | .902 | .463 | .127 | .002 |
| 23 | 1.000 | 1.000 | 1.000 | 1.000 | 1.000 | 1.000 | 1.000 | 1.000 | .998 | .973 | .729 | .358 | .026 |
| 24 | 1.000 | 1.000 | 1.000 | 1.000 | 1.000 | 1.000 | 1.000 | 1.000 | 1.000 | .996 | .928 | .723 | .222 |

TABLE III
NORMAL CURVE AREAS

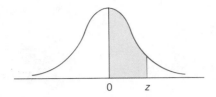

| z | .00 | .01 | .02 | .03 | .04 | .05 | .06 | .07 | .08 | .09 |
|---|---|---|---|---|---|---|---|---|---|---|
| 0.0 | .0000 | .0040 | .0080 | .0120 | .0160 | .0199 | .0239 | .0279 | .0319 | .0359 |
| 0.1 | .0398 | .0438 | .0478 | .0517 | .0557 | .0596 | .0636 | .0675 | .0714 | .0753 |
| 0.2 | .0793 | .0832 | .0871 | .0910 | .0948 | .0987 | .1026 | .1064 | .1103 | .1141 |
| 0.3 | .1179 | .1217 | .1255 | .1293 | .1331 | .1368 | .1406 | .1443 | .1480 | .1517 |
| 0.4 | .1554 | .1591 | .1628 | .1664 | .1700 | .1736 | .1772 | .1808 | .1844 | .1879 |
| 0.5 | .1915 | .1950 | .1985 | .2019 | .2054 | .2088 | .2123 | .2157 | .2190 | .2224 |
| 0.6 | .2257 | .2291 | .2324 | .2357 | .2389 | .2422 | .2454 | .2486 | .2517 | .2549 |
| 0.7 | .2580 | .2611 | .2642 | .2673 | .2704 | .2734 | .2764 | .2794 | .2823 | .2852 |
| 0.8 | .2881 | .2910 | .2939 | .2967 | .2995 | .3023 | .3051 | .3078 | .3106 | .3133 |
| 0.9 | .3159 | .3186 | .3212 | .3238 | .3264 | .3289 | .3315 | .3340 | .3365 | .3389 |
| 1.0 | .3413 | .3438 | .3461 | .3485 | .3508 | .3531 | .3554 | .3577 | .3599 | .3621 |
| 1.1 | .3643 | .3665 | .3686 | .3708 | .3729 | .3749 | .3770 | .3790 | .3810 | .3830 |
| 1.2 | .3849 | .3869 | .3888 | .3907 | .3925 | .3944 | .3962 | .3980 | .3997 | .4015 |
| 1.3 | .4032 | .4049 | .4066 | .4082 | .4099 | .4115 | .4131 | .4147 | .4162 | .4177 |
| 1.4 | .4192 | .4207 | .4222 | .4236 | .4251 | .4265 | .4279 | .4292 | .4306 | .4319 |
| 1.5 | .4332 | .4345 | .4357 | .4370 | .4382 | .4394 | .4406 | .4418 | .4429 | .4441 |
| 1.6 | .4452 | .4463 | .4474 | .4484 | .4495 | .4505 | .4515 | .4525 | .4535 | .4545 |
| 1.7 | .4554 | .4564 | .4573 | .4582 | .4591 | .4599 | .4608 | .4616 | .4625 | .4633 |
| 1.8 | .4641 | .4649 | .4656 | .4664 | .4671 | .4678 | .4686 | .4693 | .4699 | .4706 |
| 1.9 | .4713 | .4719 | .4726 | .4732 | .4738 | .4744 | .4750 | .4756 | .4761 | .4767 |
| 2.0 | .4772 | .4778 | .4783 | .4788 | .4793 | .4798 | .4803 | .4808 | .4812 | .4817 |
| 2.1 | .4821 | .4826 | .4830 | .4834 | .4838 | .4842 | .4846 | .4850 | .4854 | .4857 |
| 2.2 | .4861 | .4864 | .4868 | .4871 | .4875 | .4878 | .4881 | .4884 | .4887 | .4890 |
| 2.3 | .4893 | .4896 | .4898 | .4901 | .4904 | .4906 | .4909 | .4911 | .4913 | .4916 |
| 2.4 | .4918 | .4920 | .4922 | .4925 | .4927 | .4929 | .4931 | .4932 | .4934 | .4936 |
| 2.5 | .4938 | .4940 | .4941 | .4943 | .4945 | .4946 | .4948 | .4949 | .4951 | .4952 |
| 2.6 | .4953 | .4955 | .4956 | .4957 | .4959 | .4960 | .4961 | .4962 | .4963 | .4964 |
| 2.7 | .4965 | .4966 | .4967 | .4968 | .4969 | .4970 | .4971 | .4972 | .4973 | .4974 |
| 2.8 | .4974 | .4975 | .4976 | .4977 | .4977 | .4978 | .4979 | .4979 | .4980 | .4981 |
| 2.9 | .4981 | .4982 | .4982 | .4983 | .4984 | .4984 | .4985 | .4985 | .4986 | .4986 |
| 3.0 | .4987 | .4987 | .4987 | .4988 | .4988 | .4989 | .4989 | .4989 | .4990 | .4990 |

Source: Abridged from Table I of A. Hald, *Statistical Tables and Formulas* (New York: John Wiley & Sons, Inc.), 1952. Reproduced by permission of A. Hald and the publisher.

**TABLE IV**
**CRITICAL VALUES OF *t***

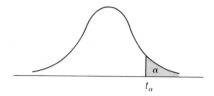

| DEGREES OF FREEDOM | $t_{.100}$ | $t_{.050}$ | $t_{.025}$ | $t_{.010}$ | $t_{.005}$ |
|---|---|---|---|---|---|
| 1 | 3.078 | 6.314 | 12.706 | 31.821 | 63.65 |
| 2 | 1.886 | 2.920 | 4.303 | 6.965 | 9.92 |
| 3 | 1.638 | 2.353 | 3.182 | 4.541 | 5.84 |
| 4 | 1.533 | 2.132 | 2.776 | 3.747 | 4.60 |
| 5 | 1.476 | 2.015 | 2.571 | 3.365 | 4.03 |
| 6 | 1.440 | 1.943 | 2.447 | 3.143 | 3.70 |
| 7 | 1.415 | 1.895 | 2.365 | 2.998 | 3.49 |
| 8 | 1.397 | 1.860 | 2.306 | 2.896 | 3.35 |
| 9 | 1.383 | 1.833 | 2.262 | 2.821 | 3.25 |
| 10 | 1.372 | 1.812 | 2.228 | 2.764 | 3.16 |
| 11 | 1.363 | 1.796 | 2.201 | 2.718 | 3.10 |
| 12 | 1.356 | 1.782 | 2.179 | 2.681 | 3.05 |
| 13 | 1.350 | 1.771 | 2.160 | 2.650 | 3.01 |
| 14 | 1.345 | 1.761 | 2.145 | 2.624 | 2.97 |
| 15 | 1.341 | 1.753 | 2.131 | 2.602 | 2.94 |
| 16 | 1.337 | 1.746 | 2.120 | 2.583 | 2.921 |
| 17 | 1.333 | 1.740 | 2.110 | 2.567 | 2.898 |
| 18 | 1.330 | 1.734 | 2.101 | 2.552 | 2.878 |
| 19 | 1.328 | 1.729 | 2.093 | 2.539 | 2.861 |
| 20 | 1.325 | 1.725 | 2.086 | 2.528 | 2.845 |
| 21 | 1.323 | 1.721 | 2.080 | 2.518 | 2.831 |
| 22 | 1.321 | 1.717 | 2.074 | 2.508 | 2.819 |
| 23 | 1.319 | 1.714 | 2.069 | 2.500 | 2.807 |
| 24 | 1.318 | 1.711 | 2.064 | 2.492 | 2.797 |
| 25 | 1.316 | 1.708 | 2.060 | 2.485 | 2.787 |
| 26 | 1.315 | 1.706 | 2.056 | 2.479 | 2.779 |
| 27 | 1.314 | 1.703 | 2.052 | 2.473 | 2.771 |
| 28 | 1.313 | 1.701 | 2.048 | 2.467 | 2.763 |
| 29 | 1.311 | 1.699 | 2.045 | 2.462 | 2.756 |
| $\infty$ | 1.282 | 1.645 | 1.960 | 2.326 | 2.576 |

Source: From M. Merrington, "Table of Percentage Points of the *t*-Distribution," *Biometrika*, 1941, *32*, 300. Reproduced by permission of the *Biometrika* Trustees.

TABLE V
CRITICAL VALUES OF THE *F* DISTRIBUTION, $\alpha = .10$

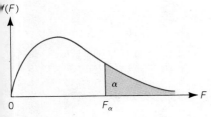

| $v_1$ | NUMERATOR DEGREES OF FREEDOM | | | | | | | | |
|---|---|---|---|---|---|---|---|---|---|
| $v_2$ | 1 | 2 | 3 | 4 | 5 | 6 | 7 | 8 | 9 |
| 1 | 39.86 | 49.50 | 53.59 | 55.83 | 57.24 | 58.20 | 58.91 | 59.44 | 59.86 |
| 2 | 8.53 | 9.00 | 9.16 | 9.24 | 9.29 | 9.33 | 9.35 | 9.37 | 9.38 |
| 3 | 5.54 | 5.46 | 5.39 | 5.34 | 5.31 | 5.28 | 5.27 | 5.25 | 5.24 |
| 4 | 4.54 | 4.32 | 4.19 | 4.11 | 4.05 | 4.01 | 3.98 | 3.95 | 3.94 |
| 5 | 4.06 | 3.78 | 3.62 | 3.52 | 3.45 | 3.40 | 3.37 | 3.34 | 3.32 |
| 6 | 3.78 | 3.46 | 3.29 | 3.18 | 3.11 | 3.05 | 3.01 | 2.98 | 2.96 |
| 7 | 3.59 | 3.26 | 3.07 | 2.96 | 2.88 | 2.83 | 2.78 | 2.75 | 2.72 |
| 8 | 3.46 | 3.11 | 2.92 | 2.81 | 2.73 | 2.67 | 2.62 | 2.59 | 2.56 |
| 9 | 3.36 | 3.01 | 2.81 | 2.69 | 2.61 | 2.55 | 2.51 | 2.47 | 2.44 |
| 10 | 3.29 | 2.92 | 2.73 | 2.61 | 2.52 | 2.46 | 2.41 | 2.38 | 2.35 |
| 11 | 3.23 | 2.86 | 2.66 | 2.54 | 2.45 | 2.39 | 2.34 | 2.30 | 2.27 |
| 12 | 3.18 | 2.81 | 2.61 | 2.48 | 2.39 | 2.33 | 2.28 | 2.24 | 2.21 |
| 13 | 3.14 | 2.76 | 2.56 | 2.43 | 2.35 | 2.28 | 2.23 | 2.20 | 2.16 |
| 14 | 3.10 | 2.73 | 2.52 | 2.39 | 2.31 | 2.24 | 2.19 | 2.15 | 2.12 |
| 15 | 3.07 | 2.70 | 2.49 | 2.36 | 2.27 | 2.21 | 2.16 | 2.12 | 2.09 |
| 16 | 3.05 | 2.67 | 2.46 | 2.33 | 2.24 | 2.18 | 2.13 | 2.09 | 2.06 |
| 17 | 3.03 | 2.64 | 2.44 | 2.31 | 2.22 | 2.15 | 2.10 | 2.06 | 2.03 |
| 18 | 3.01 | 2.62 | 2.42 | 2.29 | 2.20 | 2.13 | 2.08 | 2.04 | 2.00 |
| 19 | 2.99 | 2.61 | 2.40 | 2.27 | 2.18 | 2.11 | 2.06 | 2.02 | 1.98 |
| 20 | 2.97 | 2.59 | 2.38 | 2.25 | 2.16 | 2.09 | 2.04 | 2.00 | 1.96 |
| 21 | 2.96 | 2.57 | 2.36 | 2.23 | 2.14 | 2.08 | 2.02 | 1.98 | 1.95 |
| 22 | 2.95 | 2.56 | 2.35 | 2.22 | 2.13 | 2.06 | 2.01 | 1.97 | 1.93 |
| 23 | 2.94 | 2.55 | 2.34 | 2.21 | 2.11 | 2.05 | 1.99 | 1.95 | 1.92 |
| 24 | 2.93 | 2.54 | 2.33 | 2.19 | 2.10 | 2.04 | 1.98 | 1.94 | 1.91 |
| 25 | 2.92 | 2.53 | 2.32 | 2.18 | 2.09 | 2.02 | 1.97 | 1.93 | 1.89 |
| 26 | 2.91 | 2.52 | 2.31 | 2.17 | 2.08 | 2.01 | 1.96 | 1.92 | 1.88 |
| 27 | 2.90 | 2.51 | 2.30 | 2.17 | 2.07 | 2.00 | 1.95 | 1.91 | 1.87 |
| 28 | 2.89 | 2.50 | 2.29 | 2.16 | 2.06 | 2.00 | 1.94 | 1.90 | 1.87 |
| 29 | 2.89 | 2.50 | 2.28 | 2.15 | 2.06 | 1.99 | 1.93 | 1.89 | 1.86 |
| 30 | 2.88 | 2.49 | 2.28 | 2.14 | 2.05 | 1.98 | 1.93 | 1.88 | 1.85 |
| 40 | 2.84 | 2.44 | 2.23 | 2.09 | 2.00 | 1.93 | 1.87 | 1.83 | 1.79 |
| 60 | 2.79 | 2.39 | 2.18 | 2.04 | 1.95 | 1.87 | 1.82 | 1.77 | 1.74 |
| 120 | 2.75 | 2.35 | 2.13 | 1.99 | 1.90 | 1.82 | 1.77 | 1.72 | 1.68 |
| $\infty$ | 2.71 | 2.30 | 2.08 | 1.94 | 1.85 | 1.77 | 1.72 | 1.67 | 1.63 |

DENOMINATOR DEGREES OF FREEDOM

(continued)

**TABLE V**
CONTINUED

| $v_2$ | \multicolumn{10}{c}{NUMERATOR DEGREES OF FREEDOM $v_1$} |
|---|---|---|---|---|---|---|---|---|---|---|
|  | 10 | 12 | 15 | 20 | 24 | 30 | 40 | 60 | 120 | $\infty$ |
| 1 | 60.19 | 60.71 | 61.22 | 61.74 | 62.00 | 62.26 | 62.53 | 62.79 | 63.06 | 63.33 |
| 2 | 9.39 | 9.41 | 9.42 | 9.44 | 9.45 | 9.46 | 9.47 | 9.47 | 9.48 | 9.49 |
| 3 | 5.23 | 5.22 | 5.20 | 5.18 | 5.18 | 5.17 | 5.16 | 5.15 | 5.14 | 5.13 |
| 4 | 3.92 | 3.90 | 3.87 | 3.84 | 3.83 | 3.82 | 3.80 | 3.79 | 3.78 | 3.76 |
| 5 | 3.30 | 3.27 | 3.24 | 3.21 | 3.19 | 3.17 | 3.16 | 3.14 | 3.12 | 3.10 |
| 6 | 2.94 | 2.90 | 2.87 | 2.84 | 2.82 | 2.80 | 2.78 | 2.76 | 2.74 | 2.72 |
| 7 | 2.70 | 2.67 | 2.63 | 2.59 | 2.58 | 2.56 | 2.54 | 2.51 | 2.49 | 2.47 |
| 8 | 2.54 | 2.50 | 2.46 | 2.42 | 2.40 | 2.38 | 2.36 | 2.34 | 2.32 | 2.29 |
| 9 | 2.42 | 2.38 | 2.34 | 2.30 | 2.28 | 2.25 | 2.23 | 2.21 | 2.18 | 2.16 |
| 10 | 2.32 | 2.28 | 2.24 | 2.20 | 2.18 | 2.16 | 2.13 | 2.11 | 2.08 | 2.06 |
| 11 | 2.25 | 2.21 | 2.17 | 2.12 | 2.10 | 2.08 | 2.05 | 2.03 | 2.00 | 1.97 |
| 12 | 2.19 | 2.15 | 2.10 | 2.06 | 2.04 | 2.01 | 1.99 | 1.96 | 1.93 | 1.90 |
| 13 | 2.14 | 2.10 | 2.05 | 2.01 | 1.98 | 1.96 | 1.93 | 1.90 | 1.88 | 1.85 |
| 14 | 2.10 | 2.05 | 2.01 | 1.96 | 1.94 | 1.91 | 1.89 | 1.86 | 1.83 | 1.80 |
| 15 | 2.06 | 2.02 | 1.97 | 1.92 | 1.90 | 1.87 | 1.85 | 1.82 | 1.79 | 1.76 |
| 16 | 2.03 | 1.99 | 1.94 | 1.89 | 1.87 | 1.84 | 1.81 | 1.78 | 1.75 | 1.72 |
| 17 | 2.00 | 1.96 | 1.91 | 1.86 | 1.84 | 1.81 | 1.78 | 1.75 | 1.72 | 1.69 |
| 18 | 1.98 | 1.93 | 1.89 | 1.84 | 1.81 | 1.78 | 1.75 | 1.72 | 1.69 | 1.66 |
| 19 | 1.96 | 1.91 | 1.86 | 1.81 | 1.79 | 1.76 | 1.73 | 1.70 | 1.67 | 1.63 |
| 20 | 1.94 | 1.89 | 1.84 | 1.79 | 1.77 | 1.74 | 1.71 | 1.68 | 1.64 | 1.61 |
| 21 | 1.92 | 1.87 | 1.83 | 1.78 | 1.75 | 1.72 | 1.69 | 1.66 | 1.62 | 1.59 |
| 22 | 1.90 | 1.86 | 1.81 | 1.76 | 1.73 | 1.70 | 1.67 | 1.64 | 1.60 | 1.57 |
| 23 | 1.89 | 1.84 | 1.80 | 1.74 | 1.72 | 1.69 | 1.66 | 1.62 | 1.59 | 1.55 |
| 24 | 1.88 | 1.83 | 1.78 | 1.73 | 1.70 | 1.67 | 1.64 | 1.61 | 1.57 | 1.53 |
| 25 | 1.87 | 1.82 | 1.77 | 1.72 | 1.69 | 1.66 | 1.63 | 1.59 | 1.56 | 1.52 |
| 26 | 1.86 | 1.81 | 1.76 | 1.71 | 1.68 | 1.65 | 1.61 | 1.58 | 1.54 | 1.50 |
| 27 | 1.85 | 1.80 | 1.75 | 1.70 | 1.67 | 1.64 | 1.60 | 1.57 | 1.53 | 1.49 |
| 28 | 1.84 | 1.79 | 1.74 | 1.69 | 1.66 | 1.63 | 1.59 | 1.56 | 1.52 | 1.48 |
| 29 | 1.83 | 1.78 | 1.73 | 1.68 | 1.65 | 1.62 | 1.58 | 1.55 | 1.51 | 1.47 |
| 30 | 1.82 | 1.77 | 1.72 | 1.67 | 1.64 | 1.61 | 1.57 | 1.54 | 1.50 | 1.46 |
| 40 | 1.76 | 1.71 | 1.66 | 1.61 | 1.57 | 1.54 | 1.51 | 1.47 | 1.42 | 1.38 |
| 60 | 1.71 | 1.66 | 1.60 | 1.54 | 1.51 | 1.48 | 1.44 | 1.40 | 1.35 | 1.29 |
| 120 | 1.65 | 1.60 | 1.55 | 1.48 | 1.45 | 1.41 | 1.37 | 1.32 | 1.26 | 1.19 |
| $\infty$ | 1.60 | 1.55 | 1.49 | 1.42 | 1.38 | 1.34 | 1.30 | 1.24 | 1.17 | 1.00 |

DENOMINATOR DEGREES OF FREEDOM

TABLE VI

F-Table

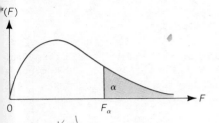

$f^*(F)$

Use for difference of two means w/ two independent small samples (Need for prelim. F-test)

$K-1$

| $\nu_1$ $\nu_2$ | NUMERATOR DEGREES OF FREEDOM | | | | | | | | |
|---|---|---|---|---|---|---|---|---|---|
| $n-K$ | 1 | 2 | 3 | 4 | 5 | 6 | 7 | 8 | 9 |
| 1 | 161.4 | 199.5 | 215.7 | 224.6 | 230.2 | 234.0 | 236.8 | 238.9 | 240.5 |
| 2 | 18.51 | 19.00 | 19.16 | 19.25 | 19.30 | 19.33 | 19.35 | 19.37 | 19.38 |
| 3 | 10.13 | 9.55 | 9.28 | 9.12 | 9.01 | 8.94 | 8.89 | 8.85 | 8.81 |
| 4 | 7.71 | 6.94 | 6.59 | 6.39 | 6.26 | 6.16 | 6.09 | 6.04 | 6.00 |
| 5 | 6.61 | 5.79 | 5.41 | 5.19 | 5.05 | 4.95 | 4.88 | 4.82 | 4.77 |
| 6 | 5.99 | 5.14 | 4.76 | 4.53 | 4.39 | 4.28 | 4.21 | 4.15 | 4.10 |
| 7 | 5.59 | 4.74 | 4.35 | 4.12 | 3.97 | 3.87 | 3.79 | 3.73 | 3.68 |
| 8 | 5.32 | 4.46 | 4.07 | 3.84 | 3.69 | 3.58 | 3.50 | 3.44 | 3.39 |
| 9 | 5.12 | 4.26 | 3.86 | 3.63 | 3.48 | 3.37 | 3.29 | 3.23 | 3.18 |
| 10 | 4.96 | 4.10 | 3.71 | 3.48 | 3.33 | 3.22 | 3.14 | 3.07 | 3.02 |
| 11 | 4.84 | 3.98 | 3.59 | 3.36 | 3.20 | 3.09 | 3.01 | 2.95 | 2.90 |
| 12 | 4.75 | 3.89 | 3.49 | 3.26 | 3.11 | 3.00 | 2.91 | 2.85 | 2.80 |
| 13 | 4.67 | 3.81 | 3.41 | 3.18 | 3.03 | 2.92 | 2.83 | 2.77 | 2.71 |
| 14 | 4.60 | 3.74 | 3.34 | 3.11 | 2.96 | 2.85 | 2.76 | 2.70 | 2.65 |
| 15 | 4.54 | 3.68 | 3.29 | 3.06 | 2.90 | 2.79 | 2.71 | 2.64 | 2.59 |
| 16 | 4.49 | 3.63 | 3.24 | 3.01 | 2.85 | 2.74 | 2.66 | 2.59 | 2.54 |
| 17 | 4.45 | 3.59 | 3.20 | 2.96 | 2.81 | 2.70 | 2.61 | 2.55 | 2.49 |
| 18 | 4.41 | 3.55 | 3.16 | 2.93 | 2.77 | 2.66 | 2.58 | 2.51 | 2.46 |
| 19 | 4.38 | 3.52 | 3.13 | 2.90 | 2.74 | 2.63 | 2.54 | 2.48 | 2.42 |
| 20 | 4.35 | 3.49 | 3.10 | 2.87 | 2.71 | 2.60 | 2.51 | 2.45 | 2.39 |
| 21 | 4.32 | 3.47 | 3.07 | 2.84 | 2.68 | 2.57 | 2.49 | 2.42 | 2.37 |
| 22 | 4.30 | 3.44 | 3.05 | 2.82 | 2.66 | 2.55 | 2.46 | 2.40 | 2.34 |
| 23 | 4.28 | 3.42 | 3.03 | 2.80 | 2.64 | 2.53 | 2.44 | 2.37 | 2.32 |
| 24 | 4.26 | 3.40 | 3.01 | 2.78 | 2.62 | 2.51 | 2.42 | 2.36 | 2.30 |
| 25 | 4.24 | 3.39 | 2.99 | 2.76 | 2.60 | 2.49 | 2.40 | 2.34 | 2.28 |
| 26 | 4.23 | 3.37 | 2.98 | 2.74 | 2.59 | 2.47 | 2.39 | 2.32 | 2.27 |
| 27 | 4.21 | 3.35 | 2.96 | 2.73 | 2.57 | 2.46 | 2.37 | 2.31 | 2.25 |
| 28 | 4.20 | 3.34 | 2.95 | 2.71 | 2.56 | 2.45 | 2.36 | 2.29 | 2.24 |
| 29 | 4.18 | 3.33 | 2.93 | 2.70 | 2.55 | 2.43 | 2.35 | 2.28 | 2.22 |
| 30 | 4.17 | 3.32 | 2.92 | 2.69 | 2.53 | 2.42 | 2.33 | 2.27 | 2.21 |
| 40 | 4.08 | 3.23 | 2.84 | 2.61 | 2.45 | 2.34 | 2.25 | 2.18 | 2.12 |
| 60 | 4.00 | 3.15 | 2.76 | 2.53 | 2.37 | 2.25 | 2.17 | 2.10 | 2.04 |
| 120 | 3.92 | 3.07 | 2.68 | 2.45 | 2.29 | 2.17 | 2.09 | 2.02 | 1.96 |
| $\infty$ | 3.84 | 3.00 | 2.60 | 2.37 | 2.21 | 2.10 | 2.01 | 1.94 | 1.88 |

DENOMINATOR DEGREES OF FREEDOM

*(continued)*

TABLE VI
CONTINUED

*(handwritten: ★ F-table — if tails in middle, right in middle, take average, otherwise, take closer one!)*

| $v_1$ / $v_2$ | 10 | 12 | 15 | 20 | 24 | 30 | 40 | 60 | 120 | ∞ |
|---|---|---|---|---|---|---|---|---|---|---|
| | | | | NUMERATOR DEGREES OF FREEDOM | | | | | | |
| 1 | 241.9 | 243.9 | 245.9 | 248.0 | 249.1 | 250.1 | 251.1 | 252.2 | 253.3 | 254.3 |
| 2 | 19.40 | 19.41 | 19.43 | 19.45 | 19.45 | 19.46 | 19.47 | 19.48 | 19.49 | 19.50 |
| 3 | 8.79 | 8.74 | 8.70 | 8.66 | 8.64 | 8.62 | 8.59 | 8.57 | 8.55 | 8.53 |
| 4 | 5.96 | 5.91 | 5.86 | 5.80 | 5.77 | 5.75 | 5.72 | 5.69 | 5.66 | 5.63 |
| 5 | 4.74 | 4.68 | 4.62 | 4.56 | 4.53 | 4.50 | 4.46 | 4.43 | 4.40 | 4.36 |
| 6 | 4.06 | 4.00 | 3.94 | 3.87 | 3.84 | 3.81 | 3.77 | 3.74 | 3.70 | 3.67 |
| 7 | 3.64 | 3.57 | 3.51 | 3.44 | 3.41 | 3.38 | 3.34 | 3.30 | 3.27 | 3.23 |
| 8 | 3.35 | 3.28 | 3.22 | 3.15 | 3.12 | 3.08 | 3.04 | 3.01 | 2.97 | 2.93 |
| 9 | 3.14 | 3.07 | 3.01 | 2.94 | 2.90 | 2.86 | 2.83 | 2.79 | 2.75 | 2.71 |
| 10 | 2.98 | 2.91 | 2.85 | 2.77 | 2.74 | 2.70 | 2.66 | 2.62 | 2.58 | 2.54 |
| 11 | 2.85 | 2.79 | 2.72 | 2.65 | 2.61 | 2.57 | 2.53 | 2.49 | 2.45 | 2.40 |
| 12 | 2.75 | 2.69 | 2.62 | 2.54 | 2.51 | 2.47 | 2.43 | 2.38 | 2.34 | 2.30 |
| 13 | 2.67 | 2.60 | 2.53 | 2.46 | 2.42 | 2.38 | 2.34 | 2.30 | 2.25 | 2.21 |
| 14 | 2.60 | 2.53 | 2.46 | 2.39 | 2.35 | 2.31 | 2.27 | 2.22 | 2.18 | 2.13 |
| 15 | 2.54 | 2.48 | 2.40 | 2.33 | 2.29 | 2.25 | 2.20 | 2.16 | 2.11 | 2.07 |
| 16 | 2.49 | 2.42 | 2.35 | 2.28 | 2.24 | 2.19 | 2.15 | 2.11 | 2.06 | 2.01 |
| 17 | 2.45 | 2.38 | 2.31 | 2.23 | 2.19 | 2.15 | 2.10 | 2.06 | 2.01 | 1.96 |
| 18 | 2.41 | 2.34 | 2.27 | 2.19 | 2.15 | 2.11 | 2.06 | 2.02 | 1.97 | 1.92 |
| 19 | 2.38 | 2.31 | 2.23 | 2.16 | 2.11 | 2.07 | 2.03 | 1.98 | 1.93 | 1.88 |
| 20 | 2.35 | 2.28 | 2.20 | 2.12 | 2.08 | 2.04 | 1.99 | 1.95 | 1.90 | 1.84 |
| 21 | 2.32 | 2.25 | 2.18 | 2.10 | 2.05 | 2.01 | 1.96 | 1.92 | 1.87 | 1.81 |
| 22 | 2.30 | 2.23 | 2.15 | 2.07 | 2.03 | 1.98 | 1.94 | 1.89 | 1.84 | 1.78 |
| 23 | 2.27 | 2.20 | 2.13 | 2.05 | 2.01 | 1.96 | 1.91 | 1.86 | 1.81 | 1.76 |
| 24 | 2.25 | 2.18 | 2.11 | 2.03 | 1.98 | 1.94 | 1.89 | 1.84 | 1.79 | 1.73 |
| 25 | 2.24 | 2.16 | 2.09 | 2.01 | 1.96 | 1.92 | 1.87 | 1.82 | 1.77 | 1.71 |
| 26 | 2.22 | 2.15 | 2.07 | 1.99 | 1.95 | 1.90 | 1.85 | 1.80 | 1.75 | 1.69 |
| 27 | 2.20 | 2.13 | 2.06 | 1.97 | 1.93 | 1.88 | 1.84 | 1.79 | 1.73 | 1.67 |
| 28 | 2.19 | 2.12 | 2.04 | 1.96 | 1.91 | 1.87 | 1.82 | 1.77 | 1.71 | 1.65 |
| 29 | 2.18 | 2.10 | 2.03 | 1.94 | 1.90 | 1.85 | 1.81 | 1.75 | 1.70 | 1.64 |
| 30 | 2.16 | 2.09 | 2.01 | 1.93 | 1.89 | 1.84 | 1.79 | 1.74 | 1.68 | 1.62 |
| 40 | 2.08 | 2.00 | 1.92 | 1.84 | 1.79 | 1.74 | 1.69 | 1.64 | 1.58 | 1.51 |
| 60 | 1.99 | 1.92 | 1.84 | 1.75 | 1.70 | 1.65 | 1.59 | 1.53 | 1.47 | 1.39 |
| 120 | 1.91 | 1.83 | 1.75 | 1.66 | 1.61 | 1.55 | 1.50 | 1.43 | 1.35 | 1.25 |
| ∞ | 1.83 | 1.75 | 1.67 | 1.57 | 1.52 | 1.46 | 1.39 | 1.32 | 1.22 | 1.00 |

DENOMINATOR DEGREES OF FREEDOM

*(handwritten: 2.315)*

TABLE VII

CRITICAL VALUES OF THE F DISTRIBUTION, α = .025

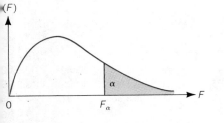

6.61904762

252

16.4195377

| $\nu_1$ | NUMERATOR DEGREES OF FREEDOM | | | | | | | | |
|---|---|---|---|---|---|---|---|---|---|
| $\nu_2$ | 1 | 2 | 3 | 4 | 5 | 6 | 7 | 8 | 9 |
| 1 | 647.8 | 799.5 | 864.2 | 899.6 | 921.8 | 937.1 | 948.2 | 956.7 | 963.3 |
| 2 | 38.51 | 39.00 | 39.17 | 39.25 | 39.30 | 39.33 | 39.36 | 39.37 | 39.39 |
| 3 | 17.44 | 16.04 | 15.44 | 15.10 | 14.88 | 14.73 | 14.62 | 14.54 | 14.47 |
| 4 | 12.22 | 10.65 | 9.98 | 9.60 | 9.36 | 9.20 | 9.07 | 8.98 | 8.90 |
| 5 | 10.01 | 8.43 | 7.76 | 7.39 | 7.15 | 6.98 | 6.85 | 6.76 | 6.68 |
| 6 | 8.81 | 7.26 | 6.60 | 6.23 | 5.99 | 5.82 | 5.70 | 5.60 | 5.52 |
| 7 | 8.07 | 6.54 | 5.89 | 5.52 | 5.29 | 5.12 | 4.99 | 4.90 | 4.82 |
| 8 | 7.57 | 6.06 | 5.42 | 5.05 | 4.82 | 4.65 | 4.53 | 4.43 | 4.36 |
| 9 | 7.21 | 5.71 | 5.08 | 4.72 | 4.48 | 4.32 | 4.20 | 4.10 | 4.03 |
| 10 | 6.94 | 5.46 | 4.83 | 4.47 | 4.24 | 4.07 | 3.95 | 3.85 | 3.78 |
| 11 | 6.72 | 5.26 | 4.63 | 4.28 | 4.04 | 3.88 | 3.76 | 3.66 | 3.59 |
| 12 | 6.55 | 5.10 | 4.47 | 4.12 | 3.89 | 3.73 | 3.61 | 3.51 | 3.44 |
| 13 | 6.41 | 4.97 | 4.35 | 4.00 | 3.77 | 3.60 | 3.48 | 3.39 | 3.31 |
| 14 | 6.30 | 4.86 | 4.24 | 3.89 | 3.66 | 3.50 | 3.38 | 3.29 | 3.21 |
| 15 | 6.20 | 4.77 | 4.15 | 3.80 | 3.58 | 3.41 | 3.29 | 3.20 | 3.12 |
| 16 | 6.12 | 4.69 | 4.08 | 3.73 | 3.50 | 3.34 | 3.22 | 3.12 | 3.05 |
| 17 | 6.04 | 4.62 | 4.01 | 3.66 | 3.44 | 3.28 | 3.16 | 3.06 | 2.98 |
| 18 | 5.98 | 4.56 | 3.95 | 3.61 | 3.38 | 3.22 | 3.10 | 3.01 | 2.93 |
| 19 | 5.92 | 4.51 | 3.90 | 3.56 | 3.33 | 3.17 | 3.05 | 2.96 | 2.88 |
| 20 | 5.87 | 4.46 | 3.86 | 3.51 | 3.29 | 3.13 | 3.01 | 2.91 | 2.84 |
| 21 | 5.83 | 4.42 | 3.82 | 3.48 | 3.25 | 3.09 | 2.97 | 2.87 | 2.80 |
| 22 | 5.79 | 4.38 | 3.78 | 3.44 | 3.22 | 3.05 | 2.93 | 2.84 | 2.76 |
| 23 | 5.75 | 4.35 | 3.75 | 3.41 | 3.18 | 3.02 | 2.90 | 2.81 | 2.73 |
| 24 | 5.72 | 4.32 | 3.72 | 3.38 | 3.15 | 2.99 | 2.87 | 2.78 | 2.70 |
| 25 | 5.69 | 4.29 | 3.69 | 3.35 | 3.13 | 2.97 | 2.85 | 2.75 | 2.68 |
| 26 | 5.66 | 4.27 | 3.67 | 3.33 | 3.10 | 2.94 | 2.82 | 2.73 | 2.65 |
| 27 | 5.63 | 4.24 | 3.65 | 3.31 | 3.08 | 2.92 | 2.80 | 2.71 | 2.63 |
| 28 | 5.61 | 4.22 | 3.63 | 3.29 | 3.06 | 2.90 | 2.78 | 2.69 | 2.61 |
| 29 | 5.59 | 4.20 | 3.61 | 3.27 | 3.04 | 2.88 | 2.76 | 2.67 | 2.59 |
| 30 | 5.57 | 4.18 | 3.59 | 3.25 | 3.03 | 2.87 | 2.75 | 2.65 | 2.57 |
| 40 | 5.42 | 4.05 | 3.46 | 3.13 | 2.90 | 2.74 | 2.62 | 2.53 | 2.45 |
| 60 | 5.29 | 3.93 | 3.34 | 3.01 | 2.79 | 2.63 | 2.51 | 2.41 | 2.33 |
| 120 | 5.15 | 3.80 | 3.23 | 2.89 | 2.67 | 2.52 | 2.39 | 2.30 | 2.22 |
| ∞ | 5.02 | 3.69 | 3.12 | 2.79 | 2.57 | 2.41 | 2.29 | 2.19 | 2.11 |

DENOMINATOR DEGREES OF FREEDOM

(continued)

**TABLE VII**
CONTINUED

*if balls in middle, right in take average*

| $\nu_2$ \ $\nu_1$ | 10 | 12 | 15 | 20 | 24 | 30 | 40 | 60 | 120 | ∞ |
|---|---|---|---|---|---|---|---|---|---|---|
| | | | | NUMERATOR DEGREES OF FREEDOM | | | | | | |
| 1 | 968.6 | 976.7 | 984.9 | 993.1 | 997.2 | 1001 | 1006 | 1010 | 1014 | 1018 |
| 2 | 39.40 | 39.41 | 39.43 | 39.45 | 39.46 | 39.46 | 39.47 | 39.48 | 39.49 | 39.50 |
| 3 | 14.42 | 14.34 | 14.25 | 14.17 | 14.12 | 14.08 | 14.04 | 13.99 | 13.95 | 13.90 |
| 4 | 8.84 | 8.75 | 8.66 | 8.56 | 8.51 | 8.46 | 8.41 | 8.36 | 8.31 | 8.26 |
| 5 | 6.62 | 6.52 | 6.43 | 6.33 | 6.28 | 6.23 | 6.18 | 6.12 | 6.07 | 6.02 |
| 6 | 5.46 | 5.37 | 5.27 | 5.17 | 5.12 | 5.07 | 5.01 | 4.96 | 4.90 | 4.85 |
| 7 | 4.76 | 4.67 | 4.57 | 4.47 | 4.42 | 4.36 | 4.31 | 4.25 | 4.20 | 4.14 |
| 8 | 4.30 | 4.20 | 4.10 | 4.00 | 3.95 | 3.89 | 3.84 | 3.78 | 3.73 | 3.67 |
| 9 | 3.96 | 3.87 | 3.77 | 3.67 | 3.61 | 3.56 | 3.51 | 3.45 | 3.39 | 3.33 |
| 10 | 3.72 | 3.62 | 3.52 | 3.42 | 3.37 | 3.31 | 3.26 | 3.20 | 3.14 | 3.08 |
| 11 | 3.53 | 3.43 | 3.33 | 3.23 | 3.17 | 3.12 | 3.06 | 3.00 | 2.94 | 2.88 |
| 12 | 3.37 | 3.28 | 3.18 | 3.07 | 3.02 | 2.96 | 2.91 | 2.85 | 2.79 | 2.72 |
| 13 | 3.25 | 3.15 | 3.05 | 2.95 | 2.89 | 2.84 | 2.78 | 2.72 | 2.66 | 2.60 |
| 14 | 3.15 | 3.05 | 2.95 | 2.84 | 2.79 | 2.73 | 2.67 | 2.61 | 2.55 | 2.49 |
| 15 | 3.06 | 2.96 | 2.86 | 2.76 | 2.70 | 2.64 | 2.59 | 2.52 | 2.46 | 2.40 |
| 16 | 2.99 | 2.89 | 2.79 | 2.68 | 2.63 | 2.57 | 2.51 | 2.45 | 2.38 | 2.32 |
| 17 | 2.92 | 2.82 | 2.72 | 2.62 | 2.56 | 2.50 | 2.44 | 2.38 | 2.32 | 2.25 |
| 18 | 2.87 | 2.77 | 2.67 | 2.56 | 2.50 | 2.44 | 2.38 | 2.32 | 2.26 | 2.19 |
| 19 | 2.82 | 2.72 | 2.62 | 2.51 | 2.45 | 2.39 | 2.33 | 2.27 | 2.20 | 2.13 |
| 20 | 2.77 | 2.68 | 2.57 | 2.46 | 2.41 | 2.35 | 2.29 | 2.22 | 2.16 | 2.09 |
| 21 | 2.73 | 2.64 | 2.53 | 2.42 | 2.37 | 2.31 | 2.25 | 2.18 | 2.11 | 2.04 |
| 22 | 2.70 | 2.60 | 2.50 | 2.39 | 2.33 | 2.27 | 2.21 | 2.14 | 2.08 | 2.00 |
| 23 | 2.67 | 2.57 | 2.47 | 2.36 | 2.30 | 2.24 | 2.18 | 2.11 | 2.04 | 1.97 |
| 24 | 2.64 | 2.54 | 2.44 | 2.33 | 2.27 | 2.21 | 2.15 | 2.08 | 2.01 | 1.94 |
| 25 | 2.61 | 2.51 | 2.41 | 2.30 | 2.24 | 2.18 | 2.12 | 2.05 | 1.98 | 1.91 |
| 26 | 2.59 | 2.49 | 2.39 | 2.28 | 2.22 | 2.16 | 2.09 | 2.03 | 1.95 | 1.88 |
| 27 | 2.57 | 2.47 | 2.36 | 2.25 | 2.19 | 2.13 | 2.07 | 2.00 | 1.93 | 1.85 |
| 28 | 2.55 | 2.45 | 2.34 | 2.23 | 2.17 | 2.11 | 2.05 | 1.98 | 1.91 | 1.83 |
| 29 | 2.53 | 2.43 | 2.32 | 2.21 | 2.15 | 2.09 | 2.03 | 1.96 | 1.89 | 1.81 |
| 30 | 2.51 | 2.41 | 2.31 | 2.20 | 2.14 | 2.07 | 2.01 | 1.94 | 1.87 | 1.79 |
| 40 | 2.39 | 2.29 | 2.18 | 2.07 | 2.01 | 1.94 | 1.88 | 1.80 | 1.72 | 1.64 |
| 60 | 2.27 | 2.17 | 2.06 | 1.94 | 1.88 | 1.82 | 1.74 | 1.67 | 1.58 | 1.48 |
| 120 | 2.16 | 2.05 | 1.94 | 1.82 | 1.76 | 1.69 | 1.61 | 1.53 | 1.43 | 1.31 |
| ∞ | 2.05 | 1.94 | 1.83 | 1.71 | 1.64 | 1.57 | 1.48 | 1.39 | 1.27 | 1.00 |

DENOMINATOR DEGREES OF FREEDOM

**TABLE VIII**
CRITICAL VALUES OF THE $F$ DISTRIBUTION, $\alpha = .01$

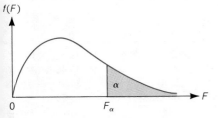

| $\nu_2$ \ $\nu_1$ | NUMERATOR DEGREES OF FREEDOM | | | | | | | | |
|---|---|---|---|---|---|---|---|---|---|
| | 1 | 2 | 3 | 4 | 5 | 6 | 7 | 8 | 9 |
| 1 | 4,052 | 4,999.5 | 5,403 | 5,625 | 5,764 | 5,859 | 5,928 | 5,982 | 6,022 |
| 2 | 98.50 | 99.00 | 99.17 | 99.25 | 99.30 | 99.33 | 99.36 | 99.37 | 99.39 |
| 3 | 34.12 | 30.82 | 29.46 | 28.71 | 28.24 | 27.91 | 27.67 | 27.49 | 27.35 |
| 4 | 21.20 | 18.00 | 16.69 | 15.98 | 15.52 | 15.21 | 14.98 | 14.80 | 14.66 |
| 5 | 16.26 | 13.27 | 12.06 | 11.39 | 10.97 | 10.67 | 10.46 | 10.29 | 10.16 |
| 6 | 13.75 | 10.92 | 9.78 | 9.15 | 8.75 | 8.47 | 8.26 | 8.10 | 7.98 |
| 7 | 12.25 | 9.55 | 8.45 | 7.85 | 7.46 | 7.19 | 6.99 | 6.84 | 6.72 |
| 8 | 11.26 | 8.65 | 7.59 | 7.01 | 6.63 | 6.37 | 6.18 | 6.03 | 5.91 |
| 9 | 10.56 | 8.02 | 6.99 | 6.42 | 6.06 | 5.80 | 5.61 | 5.47 | 5.35 |
| 10 | 10.04 | 7.56 | 6.55 | 5.99 | 5.64 | 5.39 | 5.20 | 5.06 | 4.94 |
| 11 | 9.65 | 7.21 | 6.22 | 5.67 | 5.32 | 5.07 | 4.89 | 4.74 | 4.63 |
| 12 | 9.33 | 6.93 | 5.95 | 5.41 | 5.06 | 4.82 | 4.64 | 4.50 | 4.39 |
| 13 | 9.07 | 6.70 | 5.74 | 5.21 | 4.86 | 4.62 | 4.44 | 4.30 | 4.19 |
| 14 | 8.86 | 6.51 | 5.56 | 5.04 | 4.69 | 4.46 | 4.28 | 4.14 | 4.03 |
| 15 | 8.68 | 6.36 | 5.42 | 4.89 | 4.56 | 4.32 | 4.14 | 4.00 | 3.89 |
| 16 | 8.53 | 6.23 | 5.29 | 4.77 | 4.44 | 4.20 | 4.03 | 3.89 | 3.78 |
| 17 | 8.40 | 6.11 | 5.18 | 4.67 | 4.34 | 4.10 | 3.93 | 3.79 | 3.68 |
| 18 | 8.29 | 6.01 | 5.09 | 4.58 | 4.25 | 4.01 | 3.84 | 3.71 | 3.60 |
| 19 | 8.18 | 5.93 | 5.01 | 4.50 | 4.17 | 3.94 | 3.77 | 3.63 | 3.52 |
| 20 | 8.10 | 5.85 | 4.94 | 4.43 | 4.10 | 3.87 | 3.70 | 3.56 | 3.46 |
| 21 | 8.02 | 5.78 | 4.87 | 4.37 | 4.04 | 3.81 | 3.64 | 3.51 | 3.40 |
| 22 | 7.95 | 5.72 | 4.82 | 4.31 | 3.99 | 3.76 | 3.59 | 3.45 | 3.35 |
| 23 | 7.88 | 5.66 | 4.76 | 4.26 | 3.94 | 3.71 | 3.54 | 3.41 | 3.30 |
| 24 | 7.82 | 5.61 | 4.72 | 4.22 | 3.90 | 3.67 | 3.50 | 3.36 | 3.26 |
| 25 | 7.77 | 5.57 | 4.68 | 4.18 | 3.85 | 3.63 | 3.46 | 3.32 | 3.22 |
| 26 | 7.72 | 5.53 | 4.64 | 4.14 | 3.82 | 3.59 | 3.42 | 3.29 | 3.18 |
| 27 | 7.68 | 5.49 | 4.60 | 4.11 | 3.78 | 3.56 | 3.39 | 3.26 | 3.15 |
| 28 | 7.64 | 5.45 | 4.57 | 4.07 | 3.75 | 3.53 | 3.36 | 3.23 | 3.12 |
| 29 | 7.60 | 5.42 | 4.54 | 4.04 | 3.73 | 3.50 | 3.33 | 3.20 | 3.09 |
| 30 | 7.56 | 5.39 | 4.51 | 4.02 | 3.70 | 3.47 | 3.30 | 3.17 | 3.07 |
| 40 | 7.31 | 5.18 | 4.31 | 3.83 | 3.51 | 3.29 | 3.12 | 2.99 | 2.89 |
| 60 | 7.08 | 4.98 | 4.13 | 3.65 | 3.34 | 3.12 | 2.95 | 2.82 | 2.72 |
| 120 | 6.85 | 4.79 | 3.95 | 3.48 | 3.17 | 2.96 | 2.79 | 2.66 | 2.56 |
| $\infty$ | 6.63 | 4.61 | 3.78 | 3.32 | 3.02 | 2.80 | 2.64 | 2.51 | 2.41 |

DENOMINATOR DEGREES OF FREEDOM

(continued)

**TABLE VIII**
CONTINUED

| $\nu_2$ \ $\nu_1$ | NUMERATOR DEGREES OF FREEDOM | | | | | | | | | |
|---|---|---|---|---|---|---|---|---|---|---|
| | 10 | 12 | 15 | 20 | 24 | 30 | 40 | 60 | 120 | $\infty$ |
| 1 | 6,056 | 6,106 | 6,157 | 6,209 | 6,235 | 6,261 | 6,287 | 6,313 | 6,339 | 6,366 |
| 2 | 99.40 | 99.42 | 99.43 | 99.45 | 99.46 | 99.47 | 99.47 | 99.48 | 99.49 | 99.50 |
| 3 | 27.23 | 27.05 | 26.87 | 26.69 | 26.60 | 26.50 | 26.41 | 26.32 | 26.22 | 26.13 |
| 4 | 14.55 | 14.37 | 14.20 | 14.02 | 13.93 | 13.84 | 13.75 | 13.65 | 13.56 | 13.46 |
| 5 | 10.05 | 9.89 | 9.72 | 9.55 | 9.47 | 9.38 | 9.29 | 9.20 | 9.11 | 9.02 |
| 6 | 7.87 | 7.72 | 7.56 | 7.40 | 7.31 | 7.23 | 7.14 | 7.06 | 6.97 | 6.88 |
| 7 | 6.62 | 6.47 | 6.31 | 6.16 | 6.07 | 5.99 | 5.91 | 5.82 | 5.74 | 5.65 |
| 8 | 5.81 | 5.67 | 5.52 | 5.36 | 5.28 | 5.20 | 5.12 | 5.03 | 4.95 | 4.86 |
| 9 | 5.26 | 5.11 | 4.96 | 4.81 | 4.73 | 4.65 | 4.57 | 4.48 | 4.40 | 4.31 |
| 10 | 4.85 | 4.71 | 4.56 | 4.41 | 4.33 | 4.25 | 4.17 | 4.08 | 4.00 | 3.91 |
| 11 | 4.54 | 4.40 | 4.25 | 4.10 | 4.02 | 3.94 | 3.86 | 3.78 | 3.69 | 3.60 |
| 12 | 4.30 | 4.16 | 4.01 | 3.86 | 3.78 | 3.70 | 3.62 | 3.54 | 3.45 | 3.36 |
| 13 | 4.10 | 3.96 | 3.82 | 3.66 | 3.59 | 3.51 | 3.43 | 3.34 | 3.25 | 3.17 |
| 14 | 3.94 | 3.80 | 3.66 | 3.51 | 3.43 | 3.35 | 3.27 | 3.18 | 3.09 | 3.00 |
| 15 | 3.80 | 3.67 | 3.52 | 3.37 | 3.29 | 3.21 | 3.13 | 3.05 | 2.96 | 2.87 |
| 16 | 3.69 | 3.55 | 3.41 | 3.26 | 3.18 | 3.10 | 3.02 | 2.93 | 2.84 | 2.75 |
| 17 | 3.59 | 3.46 | 3.31 | 3.16 | 3.08 | 3.00 | 2.92 | 2.83 | 2.75 | 2.65 |
| 18 | 3.51 | 3.37 | 3.23 | 3.08 | 3.00 | 2.92 | 2.84 | 2.75 | 2.66 | 2.57 |
| 19 | 3.43 | 3.30 | 3.15 | 3.00 | 2.92 | 2.84 | 2.76 | 2.67 | 2.58 | 2.49 |
| 20 | 3.37 | 3.23 | 3.09 | 2.94 | 2.86 | 2.78 | 2.69 | 2.61 | 2.52 | 2.42 |
| 21 | 3.31 | 3.17 | 3.03 | 2.88 | 2.80 | 2.72 | 2.64 | 2.55 | 2.46 | 2.36 |
| 22 | 3.26 | 3.12 | 2.98 | 2.83 | 2.75 | 2.67 | 2.58 | 2.50 | 2.40 | 2.31 |
| 23 | 3.21 | 3.07 | 2.93 | 2.78 | 2.70 | 2.62 | 2.54 | 2.45 | 2.35 | 2.26 |
| 24 | 3.17 | 3.03 | 2.89 | 2.74 | 2.66 | 2.58 | 2.49 | 2.40 | 2.31 | 2.21 |
| 25 | 3.13 | 2.99 | 2.85 | 2.70 | 2.62 | 2.54 | 2.45 | 2.36 | 2.27 | 2.17 |
| 26 | 3.09 | 2.96 | 2.81 | 2.66 | 2.58 | 2.50 | 2.42 | 2.33 | 2.23 | 2.13 |
| 27 | 3.06 | 2.93 | 2.78 | 2.63 | 2.55 | 2.47 | 2.38 | 2.29 | 2.20 | 2.10 |
| 28 | 3.03 | 2.90 | 2.75 | 2.60 | 2.52 | 2.44 | 2.35 | 2.26 | 2.17 | 2.06 |
| 29 | 3.00 | 2.87 | 2.73 | 2.57 | 2.49 | 2.41 | 2.33 | 2.23 | 2.14 | 2.03 |
| 30 | 2.98 | 2.84 | 2.70 | 2.55 | 2.47 | 2.39 | 2.30 | 2.21 | 2.11 | 2.01 |
| 40 | 2.80 | 2.66 | 2.52 | 2.37 | 2.29 | 2.20 | 2.11 | 2.02 | 1.92 | 1.80 |
| 60 | 2.63 | 2.50 | 2.35 | 2.20 | 2.12 | 2.03 | 1.94 | 1.84 | 1.73 | 1.60 |
| 120 | 2.47 | 2.34 | 2.19 | 2.03 | 1.95 | 1.86 | 1.76 | 1.66 | 1.53 | 1.38 |
| $\infty$ | 2.32 | 2.18 | 2.04 | 1.88 | 1.79 | 1.70 | 1.59 | 1.47 | 1.32 | 1.00 |

DENOMINATOR DEGREES OF FREEDOM

CRITICAL VALUES OF $\chi^2$

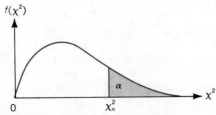

| DEGREES OF FREEDOM | $\chi^2_{.995}$ | $\chi^2_{.990}$ | $\chi^2_{.975}$ | $\chi^2_{.950}$ | $\chi^2_{.900}$ |
|---|---|---|---|---|---|
| 1 | 0.0000393 | 0.0001571 | 0.0009821 | 0.0039321 | 0.0157908 |
| 2 | 0.0100251 | 0.0201007 | 0.0506356 | 0.102587 | 0.210720 |
| 3 | 0.0717212 | 0.114832 | 0.215795 | 0.351846 | 0.584375 |
| 4 | 0.206990 | 0.297110 | 0.484419 | 0.710721 | 1.063623 |
| 5 | 0.411740 | 0.554300 | 0.831211 | 1.145476 | 1.61031 |
| 6 | 0.675727 | 0.872085 | 1.237347 | 1.63539 | 2.20413 |
| 7 | 0.989265 | 1.239043 | 1.68987 | 2.16735 | 2.83311 |
| 8 | 1.344419 | 1.646482 | 2.17973 | 2.73264 | 3.48954 |
| 9 | 1.734926 | 2.087912 | 2.70039 | 3.32511 | 4.16816 |
| 10 | 2.15585 | 2.55821 | 3.24697 | 3.94030 | 4.86518 |
| 11 | 2.60321 | 3.05347 | 3.81575 | 4.57481 | 5.57779 |
| 12 | 3.07382 | 3.57056 | 4.40379 | 5.22603 | 6.30380 |
| 13 | 3.56503 | 4.10691 | 5.00874 | 5.89186 | 7.04150 |
| 14 | 4.07468 | 4.66043 | 5.62872 | 6.57063 | 7.78953 |
| 15 | 4.60094 | 5.22935 | 6.26214 | 7.26094 | 8.54675 |
| 16 | 5.14224 | 5.81221 | 6.90766 | 7.96164 | 9.31223 |
| 17 | 5.69724 | 6.40776 | 7.56418 | 8.67176 | 10.0852 |
| 18 | 6.26481 | 7.01491 | 8.23075 | 9.39046 | 10.8649 |
| 19 | 6.84398 | 7.63273 | 8.90655 | 10.1170 | 11.6509 |
| 20 | 7.43386 | 8.26040 | 9.59083 | 10.8508 | 12.4426 |
| 21 | 8.03366 | 8.89720 | 10.28293 | 11.5913 | 13.2396 |
| 22 | 8.64272 | 9.54249 | 10.9823 | 12.3380 | 14.0415 |
| 23 | 9.26042 | 10.19567 | 11.6885 | 13.0905 | 14.8479 |
| 24 | 9.88623 | 10.8564 | 12.4011 | 13.8484 | 15.6587 |
| 25 | 10.5197 | 11.5240 | 13.1197 | 14.6114 | 16.4734 |
| 26 | 11.1603 | 12.1981 | 13.8439 | 15.3791 | 17.2919 |
| 27 | 11.8076 | 12.8786 | 14.5733 | 16.1513 | 18.1138 |
| 28 | 12.4613 | 13.5648 | 15.3079 | 16.9279 | 18.9392 |
| 29 | 13.1211 | 14.2565 | 16.0471 | 17.7083 | 19.7677 |
| 30 | 13.7867 | 14.9535 | 16.7908 | 18.4926 | 20.5992 |
| 40 | 20.7065 | 22.1643 | 24.4331 | 26.5093 | 29.0505 |
| 50 | 27.9907 | 29.7067 | 32.3574 | 34.7642 | 37.6886 |
| 60 | 35.5346 | 37.4848 | 40.4817 | 43.1879 | 46.4589 |
| 70 | 43.2752 | 45.4418 | 48.7576 | 51.7393 | 55.3290 |
| 80 | 51.1720 | 53.5400 | 57.1532 | 60.3915 | 64.2778 |
| 90 | 59.1963 | 61.7541 | 65.6466 | 69.1260 | 73.2912 |
| 100 | 67.3276 | 70.0648 | 74.2219 | 77.9295 | 82.3581 |

(continued)

**TABLE IX**
**CONTINUED**

| DEGREES OF FREEDOM | $\chi^2_{.100}$ | $\chi^2_{.050}$ | $\chi^2_{.025}$ | $\chi^2_{.010}$ | $\chi^2_{.005}$ |
|---|---|---|---|---|---|
| 1 | 2.70554 | 3.84146 | 5.02389 | 6.63490 | 7.87944 |
| 2 | 4.60517 | 5.99147 | 7.37776 | 9.21034 | 10.5966 |
| 3 | 6.25139 | 7.81473 | 9.34840 | 11.3449 | 12.8381 |
| 4 | 7.77944 | 9.48773 | 11.1433 | 13.2767 | 14.8602 |
| 5 | 9.23635 | 11.0705 | 12.8325 | 15.0863 | 16.7496 |
| 6 | 10.6446 | 12.5916 | 14.4494 | 16.8119 | 18.5476 |
| 7 | 12.0170 | 14.0671 | 16.0128 | 18.4753 | 20.2777 |
| 8 | 13.3616 | 15.5073 | 17.5346 | 20.0902 | 21.9550 |
| 9 | 14.6837 | 16.9190 | 19.0228 | 21.6660 | 23.5893 |
| 10 | 15.9871 | 18.3070 | 20.4831 | 23.2093 | 25.1882 |
| 11 | 17.2750 | 19.6751 | 21.9200 | 24.7250 | 26.7569 |
| 12 | 18.5494 | 21.0261 | 23.3367 | 26.2170 | 28.2995 |
| 13 | 19.8119 | 22.3621 | 24.7356 | 27.6883 | 29.8194 |
| 14 | 21.0642 | 23.6848 | 26.1190 | 29.1413 | 31.3193 |
| 15 | 22.3072 | 24.9958 | 27.4884 | 30.5779 | 32.8013 |
| 16 | 23.5418 | 26.2962 | 28.8454 | 31.9999 | 34.2672 |
| 17 | 24.7690 | 27.5871 | 30.1910 | 33.4087 | 35.7185 |
| 18 | 25.9894 | 28.8693 | 31.5264 | 34.8053 | 37.1564 |
| 19 | 27.2036 | 30.1435 | 32.8523 | 36.1908 | 38.5822 |
| 20 | 28.4120 | 31.4104 | 34.1696 | 37.5662 | 39.9968 |
| 21 | 29.6151 | 32.6705 | 35.4789 | 38.9321 | 41.4010 |
| 22 | 30.8133 | 33.9244 | 36.7807 | 40.2894 | 42.7956 |
| 23 | 32.0069 | 35.1725 | 38.0757 | 41.6384 | 44.1813 |
| 24 | 33.1963 | 36.4151 | 39.3641 | 42.9798 | 45.5585 |
| 25 | 34.3816 | 37.6525 | 40.6465 | 44.3141 | 46.9278 |
| 26 | 35.5631 | 38.8852 | 41.9232 | 45.6417 | 48.2899 |
| 27 | 36.7412 | 40.1133 | 43.1944 | 46.9630 | 49.6449 |
| 28 | 37.9159 | 41.3372 | 44.4607 | 48.2782 | 50.9933 |
| 29 | 39.0875 | 42.5569 | 45.7222 | 49.5879 | 52.3356 |
| 30 | 40.2560 | 43.7729 | 46.9792 | 50.8922 | 53.6720 |
| 40 | 51.8050 | 55.7585 | 59.3417 | 63.6907 | 66.7659 |
| 50 | 63.1671 | 67.5048 | 71.4202 | 76.1539 | 79.4900 |
| 60 | 74.3970 | 79.0819 | 83.2976 | 88.3794 | 91.9517 |
| 70 | 85.5271 | 90.5312 | 95.0231 | 100.425 | 104.215 |
| 80 | 96.5782 | 101.879 | 106.629 | 112.329 | 116.321 |
| 90 | 107.565 | 113.145 | 118.136 | 124.116 | 128.299 |
| 100 | 118.498 | 124.342 | 129.561 | 135.807 | 140.169 |

# ANSWERS TO SELECTED EXERCISES

CHAPTER 2

**2.1.** **a.** Quantitative    **b.** Qualitative    **c.** Qualitative    **d.** Quantitative
**2.2.** **a.** Qualitative    **b.** Quantitative    **c.** Qualitative    **d.** Qualitative
**2.7.** **a.** 7.5 to 9.5    **b.** .15    **c.** .20    **d.** 20    **2.8.** **b.** .15    **2.10.** **b.** .65
**2.11.** **a.** 5    **b.** 50    **c.** .02    **d.** 12.47
**2.12.** **a.** 2.5, 3.5, 4    **b.** 3.08, 3.0, 3    **c.** 53.8, 50, 50
**2.13.** **a.** Mean is smaller    **b.** Mean is larger    **c.** They are equal
**2.14.** **a.** 2, 2    **b.** 6, 2    **c.** 0, 0    **d.** $-27.7$, 0    **2.15.** **c.** 81.15, 83, 83
**2.16.** **a.** Right    **b.** Left    **c.** Right    **d.** Symmetric    **e.** Right    **f.** Left
**2.17.** **a.** Mean 15,325; median 11,450; modes 9,000, 10,000, 12,000, 17,000
**b.** Mean 11,352.63; median 11,400; modes 9,000, 10,000, 12,000, 17,000; mean is most sensitive
**2.18.** **a.** 78.15, 84, 88    **b.** Median    **c.** 83.06, 86, 88; mean is most sensitive
**2.19.** **a.** No    **b.** 110    **2.20.** **a.** Mean    **b.** Median
**2.21.** **a.** 3, 1.07, 1.03    **b.** 4, 2, 1.41    **c.** 14, 25.69, 5.07    **d.** 5, 1.62, 1.27
**2.22.** **a.** 10.57, 3.25    **b.** 6.41, 2.53    **c.** 0, 0
**2.23.** **a.** 4, 2.2, 1.48    **b.** 96, 2,856.25, 53.44    **c.** 96, 1,998.25, 44.70
**2.24.** **a.** 4, 2.3, 1.52    **b.** 4, 2.3, 1.52    **c.** 4, 2.3, 1.52    **d.** No effect    **2.25.** First; less variability
**2.26.** **b.** 10, 10    **c.** 26.4, 10    **d.** Variance
**2.27.** **a.** Perhaps very few    **b.** At least $\frac{3}{4}$    **c.** At least $\frac{8}{9}$
**2.28.** **a.** Approx. 68%    **b.** Approx. 95%    **c.** Approx. all
**2.29.** **a.** 8.24, 3.36, 1.83    **b.** 72%, 96%, 100%    **d.** 7, $\frac{7}{4} = 1.75$
**2.30.** **a.** 54.9, 94.99, 9.75    **b.** Perhaps very few, at least 75%, at least 89%    **c.** 60%, 97%, 100%
**2.31.** **a.** At most 25%    **b.** Approx. 2.5%    **2.32.** Do not buy. At most 555 are at least 40 feet tall.
**2.33.** Approx. none of them; yes.    **2.34.** .025; yes    **2.35.** **a.** At least $\frac{8}{9}$    **b.** Approx. .84    **c.** Yes

**2.36.** **a.** 9.1% are in $1 \pm .98$ **b.** 76.2% are in $1 \pm .99$ **c.** 89.3% are in $1 \pm .99$
**2.37.** **a.** $z = 2$ **b.** $z = .5$ **c.** $z = 0$ **d.** $z = -2.5$
**e.** Sample; population; population; sample **f.** Above by 2, above by .5, equal to the mean, below by 2.5
**2.38.** **a.** 25% above, 75% below **b.** 50% above, 50% below **c.** 80% above, 20% below
**d.** 16% above, 84% below **2.39.** **a.** $z = 2$ (most above) **b.** $z = -3$ (most below) **c.** $z = -2$
**d.** $z = 1.67$ **2.40.** **a.** $-.4, -3, -2.2, .8, 2.4, 0$
**b.** 83(.4 below), 70(3 below), 74(2.2 below), 89(.8 above), 97(2.4 above), 85(equals mean)
**2.41.** None; 81.5%; 85
**2.42.** **b.** 90% of the test scores are below 660 **c.** 94% of the test scores are below your score
**2.43.** **a.** $z = -3$ **b.** Approx. 0% **c.** 90 or above ($z = 2$)
**2.44.** **a.** $z = -3$ **b.** Yes **c.** Yes, since $z = -1.5$
**2.46.** **a.** $-1, 1, 2$ **b.** $-2, 2, 4$ **c.** 1, 3, 4 **d.** .1, .3, .4 **2.47.** 5
**2.48.** **a.** 2.67, 14.27, 3.78 **b.** 6, 3.33, 1.83 **c.** 12.86, 57.81, 7.60
**2.49.** **b.** 50.8, 57.06, 7.55
**c.** 43.25 to 58.35 (perhaps very few), 35.70 to 65.90 (at least 75%), 28.15 to 73.45 (at least 89%)
**d.** 63.3%, 100%, 100% **2.50.** **a.** $-1, 0, 9$ **b.** 25, 25, 16 **c.** 77.5, 80, 73
**2.51.** **a.** 2.58, .982, .991 **b.** .057, .0013, .036
**2.52.** **a.** Qualitative **b.** Quantitative **c.** Quantitative **d.** Qualitative **2.53.** 10, 32, 5.66
**2.54.** Decrease; smaller; smaller
**2.55.** **b.** 30, 4.83, 2.20 **c.** Perhaps very few; at least 75%; at least 89% **d.** 70%, 96.7%, 100%
**2.56.** No ($z = 1.90$) **2.57.** **a.** 11, 2.75 **b.** 14.73, 11.21, 3.35 **c.** All
**2.58.** **a.** 58.24, 65.44, 8.09 **b.** Perhaps very few; at least 75%; at least 89% **c.** 64%, 96%, 100%
**2.59.** **a.** 16.48, 15, 14 **b.** 22, 5.5 **c.** 33.93, 5.82 **d.** All
**2.60.** **a.** 68% **b.** 47.5% **c.** 16% **2.61.** 52 minutes
**2.62.** **a.** 18.2, 13.64, 3.69 **b.** 20.2, 13.64, 3.69
**2.63.** **a.** 98.67, 330.24, 18.17 **b.** 49.33, 82.56, 9.09
**2.64.** **a.** Approx. none **b.** No **c.** Yes ($z = -1.33$) **2.65.** 13.8, 11, 13.29, 3.65
**2.66.** **a.** At most 25% **b.** Approx. 2.5%
**2.67.** 3,100 to 3,150; 3,075 to 3,175; 3,050 to 3,200; perhaps very few; at least $\frac{3}{4}$; at least $\frac{8}{9}$
**2.68.** **a.** 23.76, 6.61, 2.57 **b.** Perhaps very few; at least 75%; at least 89% **c.** 68%, 96%, 100%

## CHAPTER 3

**3.1.** $A, .55; B, .65; C, .75$ **3.2.** **a.** $\{HHH, HHT, HTH, THH, HTT, THT, TTH, TTT\}$
**b.** Each simple event has probability $\frac{1}{8}$ **c.** $\frac{7}{8}, \frac{3}{8}, \frac{1}{2}$ **3.3.** **a.** $\frac{4}{9}$ **b.** $\frac{3}{9}$ **c.** 0
**3.4.** **a.** (1, 1), (1, 2), (1, 3), (1, 4), (1, 5), (1, 6)
(2, 1), (2, 2), (2, 3), (2, 4), (2, 5), (2, 6)
(3, 1), (3, 2), (3, 3), (3, 4), (3, 5), (3, 6)
(4, 1), (4, 2), (4, 3), (4, 4), (4, 5), (4, 6)
(5, 1), (5, 2), (5, 3), (5, 4), (5, 5), (5, 6)
(6, 1), (6, 2), (6, 3), (6, 4), (6, 5), (6, 6)
**b.** $\frac{1}{36}, \frac{1}{2}, \frac{1}{6}, \frac{11}{36}, \frac{1}{6}$ **3.5.** $A, \frac{1}{12}; B, \frac{1}{4}; C, \frac{1}{2}; D, \frac{1}{2}$ **3.6.** $\frac{1}{10}, \frac{6}{10}, \frac{3}{10}$ **3.8.** .81, .19, .18
**3.9.** **a.** $\frac{1}{3}$ **b.** $\frac{1}{3}$ **c.** $\frac{1}{6}$ **d.** $\frac{1}{6}$ **3.10.** **a.** $\frac{1}{4}$ **b.** $\frac{1}{2}$ **c.** 0
**3.11** $\frac{3}{4}$, no, the probability of being odd man out remains $\frac{1}{4}$
**3.12.** **a.** (W, R), (W, D), (W, I), (B, R), (B, D), (B, I) **b.** Sample space **c.** .30 **d.** .35 **e.** .55
**f.** .12
**3.13.** **a.** (H, T, L), (H, T, 220), (H, T, 100), (H, L, 220), (H, L, 100), (H, 220, 100), (T, L, 220), (T, L, 100), (T, 220, 100), (L, 220, 100) **b.** $\frac{3}{10}$
**3.14.** **a.** 18 simple events **b.** $\frac{1}{18}$ **c.** $\frac{1}{3}$

**3.15.** **a.** $A$: {$HHH, HHT, HTH, THH, HTT, THT, TTH$}; $B$: {$HHH, HTT, THT, TTH$}; $A \cup B$: Same as $A$; $A'$: {$TTT$};

$AB$: Same as $B$ **b.** $^7\!/_8, ^1\!/_2, ^7\!/_8, ^1\!/_8, ^1\!/_2$

**3.16.** **a.** $A$: {(1, 6), (2, 5), (3, 4), (4, 3), (5, 2), (6, 1)};

$B$: {(1, 4), (2, 4), (3, 4), (4, 1), (4, 2), (4, 3), (4, 4), (4, 5), (4, 6), (5, 4), (6, 4)}; $AB$: {(3, 4), (4, 3)};

$A \cup B$: {(1, 4), (1, 6), (2, 4), (2, 5), (3, 4), (4, 1), (4, 2), (4, 3), (4, 4), (4, 5), (4, 6), (5, 2), (5, 4), (6, 1), (6, 4)};

$A'$: The 30 simple events not in $A$ **b.** $^1\!/_6, ^{11}\!/_{36}, ^1\!/_{18}, ^{15}\!/_{36}, ^5\!/_6$ **3.17.** $^4\!/_{10}, ^2\!/_{10}, ^3\!/_{10}, ^5\!/_{10}, ^9\!/_{10}, ^7\!/_{10}, ^5\!/_{10}, ^1\!/_{10}$

**3.18.** **a.** {$G_1P_1, G_2P_1$} **b.** $S$ **c.** {$P_1P_2$} **d.** $^5\!/_6, ^3\!/_6, ^2\!/_6, 1, ^1\!/_6$ **3.19.** 0, 0, .40, .85, .75

**3.20.** **a.** {11, 13, 15, 17, 29, 31, 33, 35}

**b.** {1, 2, 3, 4, 5, 6, 7, 8, 9, 10, 11, 13, 15, 17, 19, 20, 21, 22, 23, 24, 25, 26, 27, 28, 29, 31, 33, 35}

**c.** $^{18}\!/_{38}, ^{18}\!/_{38}, ^8\!/_{38}, ^{28}\!/_{38}, ^{18}\!/_{38}$ **d.** {11, 13, 15, 17} **e.** $^4\!/_{38}$

**f.** All simple events except {00, 0, 30, 32, 34, 36} **g.** $^{32}\!/_{38}$

**3.21.** $^3\!/_4, ^2\!/_4, ^2\!/_4, 1, ^2\!/_3$ **3.22.** **a.** .60, .40, .30 **b.** .30, 0, 0 **c.** .75, 0, 0, .75

**3.23.** $^1\!/_3, 0, ^1\!/_9, ^1\!/_9, ^8\!/_{12}$ **3.24.** **a.** .16, .64, .16 **b.** .25, 1

**3.25.** **a.** $^1\!/_{15}$ **b.** $^2\!/_5$ **c.** $^{14}\!/_{15}$ **d.** $^3\!/_7$ **e.** $^5\!/_{14}$

**3.26.** **a.** $^{25}\!/_{44}$ **b.** $^6\!/_{25}$ **c.** $^{15}\!/_{20}$ **d.** 1

**3.27.** **a.** .10 **b.** .67 **c.** .56

**3.28.** **a.** .30 **b.** $^{10}\!/_{52} \approx .192$ **c.** $^{22}\!/_{30} \approx .733$ **d.** $^{40}\!/_{48} \approx .833$

**3.29.** **a.** $^7\!/_8, ^3\!/_8, ^3\!/_8, ^1\!/_2, ^3\!/_8, ^3\!/_8, 0, 0$ **b.** $^7\!/_8, 1, ^6\!/_8, ^3\!/_7, ^3\!/_4, 0$ **c.** No; no **d.** No; no **e.** No; yes

**3.30.** **a.** Yes; $P(A|B) = P(A) = ^1\!/_2$ **b.** No; $AB$: {(3, 6), (4, 5), (5, 4), (5, 6), (6, 3), (6, 5)}

**3.31.** **a.** .5 **b.** 0 **c.** 0 **d.** No; $P(A) \neq P(A|B)$ **3.32.** **a.** $^1\!/_6, ^1\!/_2, ^4\!/_6$ **b.** $^1\!/_5, ^2\!/_5$

**3.33.** **a.** $^{24}\!/_{50}$ **b.** $^{16}\!/_{50}$ **c.** $^{32}\!/_{50}$ **d.** $^8\!/_{50}$ **e.** $^{16}\!/_{50}$ **f.** $^1\!/_2$ **g.** No; $P(A|B) \neq P(A)$

**h.** No; $AB$ is nonempty

**3.34.** **a.** $^1\!/_5$ **b.** $^5\!/_{29}$ **c.** $^{276}\!/_{435}$ **3.35.** **a.** .3 **b.** .58 **c.** .7 **d.** .3

**3.36.** **b.** .15 **c.** .50 **3.37.** .891 **3.38.** **a.** No; yes **b.** Yes **c.** Yes; no

**3.39.** $1 - (.95)^{10} \approx .40$, independence **3.40.** **a.** .028 **b.** More than 10% need repair in the first year.

**3.41.** **a.** .4 **b.** .3 **c.** .375 **d.** No **3.42.** **a.** $^8\!/_{36}$ **b.** $^1\!/_{36}$ **c.** $^{11}\!/_{36}$

**3.43.** .04, .38

**3.44.** **a.** (1, $H$), (1, $T$)

(2, 1), (2, 2), (2, 3), (2, 4), (2, 5), (2, 6)

(3, $H$), (3, $T$)

(4, 1), (4, 2), (4, 3), (4, 4), (4, 5), (4, 6)

(5, $H$), (5, $T$)

(6, 1), (6, 2), (6, 3), (6, 4), (6, 5), (6, 6)

**b.** Probability of a simple event with odd number first is $^1\!/_{12}$; otherwise, $^1\!/_{36}$

**c.** $^1\!/_4, ^1\!/_2$

**d.** $A'$ contains all simple events except (1, $H$), (3, $H$), and (5, $H$);

$B'$: {(2, 1), (2, 2), (2, 3), (2, 4), (2, 5), (2, 6), (4, 1), (4, 2), (4, 3), (4, 4), (4, 5), (4, 6), (6, 1), (6, 2), (6, 3), (6, 4), (6, 5), (6, 6)}; $AB$: {(1, $H$), (3, $H$), (5, $H$)};

$A \cup B$: {(1, $H$), (1, $T$), (3, $H$), (3, $T$), (5, $H$), (5, $T$)}

**e.** $^3\!/_4, ^1\!/_2, ^1\!/_4, ^1\!/_2, ^1\!/_2, 1$ **f.** No; no

**3.45.** **a.** 0, .1, .8, 1, .7, .3, $^1\!/_3$, 0 **b.** No; yes **c.** No; no

**3.46.** **a.** No; $P(AB) = .03$ **b.** .3, .1 **c.** .37 **3.47.** **a.** $^{34}\!/_{60} \approx .567$ **b.** $^4\!/_{34} \approx .118$

**3.48.** **a.** $BC$: {$HHH$} **b.** $A \cup B$: {$HHH, HHT, HTH, THH, THT$} **c.** $^1\!/_8$ **d.** $^5\!/_8$ **e.** $^5\!/_8$ **f.** $^5\!/_8$

**3.49.** **a.** .64, .32, .04 **b.** .72, .22, .06 **c.** Dependent

**3.51.** **a.** .5, .5, .3, .15, .2, .35 **b.** 1 **c.** .05 **d.** .05 **e.** 0 **f.** .45 **g.** .35

**3.53.** **a.** .15, .85, .70, .15, .05, .10 **b.** 1 **c.** .08 **d.** .85 **e.** 0 **f.** .87

**3.54.** $^{15}\!/_{36}, (^{21}\!/_{36})(^{15}\!/_{36}), (^{21}\!/_{36})^2(^{15}\!/_{36}), (^{21}\!/_{36})^{n-1}(^{15}\!/_{36})$ **3.55.** **a.** $(^1\!/_2)^5$ **b.** $5(^1\!/_2)^5$ **3.56.** **b.** 18%

**3.57.** **a.** $^6\!/_{30}$ **b.** $^6\!/_{29} \approx .207$ **c.** $(^4\!/_5)(^{23}\!/_{29}) \approx .634$ **3.58.** .79 **3.59.** $^{.7425}\!/_{.7475} \approx .993$

**3.60.** **a.** $^{48}\!/_{1,326} \approx .036$ **b.** $(^{48}\!/_{1,326})(^{1,192}\!/_{1,225}) \approx .035$ **3.61.** .046, .747

**3.62.** **a.** There are 20 simple events **b.** $^1\!/_{20}$ **c.** $^1\!/_{20}$ **d.** $^1\!/_2$

**4.1.** **a.** Discrete    **b.** Discrete    **c.** Continuous    **d.** Continuous    **e.** Discrete    **f.** Discrete
**4.2.** **a.** Discrete    **b.** Discrete    **c.** Continuous    **d.** Continuous    **e.** Continuous
**f.** Discrete
**4.3.** **a.** Discrete    **b.** Continuous    **c.** Continuous    **d.** Discrete    **e.** Continuous
**f.** Continuous
**4.4.** **a.** $-4, 0, 1, 3$    **b.** 1    **c.** .6    **d.** 0    **4.5.** **a.** .30    **b.** .35    **c.** .85

**4.6.** **a.**

| Simple event | HHH | HHT | HTH | THH | HTT | THT | TTH | TTT |
|---|---|---|---|---|---|---|---|---|
| $x$ | 3 | 2 | 2 | 2 | 1 | 1 | 1 | 0 |

**b.**

| $x$ | 0 | 1 | 2 | 3 |
|---|---|---|---|---|
| $p(x)$ | $\frac{1}{8}$ | $\frac{3}{8}$ | $\frac{3}{8}$ | $\frac{1}{8}$ |

**d.** $\frac{1}{2}$

**4.7.** **a.**

| $x$ | 2 | 3 | 4 | 5 | 6 | 7 | 8 | 9 | 10 | 11 | 12 |
|---|---|---|---|---|---|---|---|---|---|---|---|
| $p(x)$ | $\frac{1}{36}$ | $\frac{2}{36}$ | $\frac{3}{36}$ | $\frac{4}{36}$ | $\frac{5}{36}$ | $\frac{6}{36}$ | $\frac{5}{36}$ | $\frac{4}{36}$ | $\frac{3}{36}$ | $\frac{2}{36}$ | $\frac{1}{36}$ |

**b.** $\frac{15}{36}$    **c.** $\frac{21}{36}$    **d.** $\frac{18}{36}, \frac{18}{36}$    **e.** $\frac{6}{36}$

**4.8.**

| $x$ | 1 | 2 | 3 |
|---|---|---|---|
| $p(x)$ | $\frac{6}{10}$ | $\frac{3}{10}$ | $\frac{1}{10}$ |

**4.9.** $\frac{7}{8}$

**4.10.** **a.**

| $x$ | 0 | 1 | 2 | 3 |
|---|---|---|---|---|
| $p(x)$ | .3446 | .4408 | .1879 | .0267 |

**c.** .6554

**4.11.** **a.**

| $x$ | 0 | 1 | 2 |
|---|---|---|---|
| $p(x)$ | .01 | .05 | .94 |

**b.** No

**4.12.** **a.** Distribution I    **b.** $P(x > 5 \text{ for I}) = .08$; $P(x > 5 \text{ for II}) = .73$
**c.** $P(x \le 3 \text{ for I}) = .61$; $P(x \le 3 \text{ for II}) = .08$
**4.13.** **a.** 10    **b.** 140    **c.** 11.83    **4.14.** **a.** 4.80    **b.** 12.56    **c.** 3.54
**4.15.** **a.** 0, 3.96, 1.99    **c.** .90    **4.16.** **a.** $E(x) = 1$ for both    **b.** The first    **c.** 1, .6; 1, .2
**4.17.** 1.29    **4.18.** **a.** $81.25    **b.** 195.3125    **c.** .90    **4.19.** $E(x) = 11,500$; yes
**4.20.** No, expected cost = $79    **4.21.** $0.25    **4.22.** **a.** 114.4    **b.** 1,000.6    **c.** 31.6
**4.23.** **a.** 6    **b.** 1    **c.** 10    **d.** .00032    **e.** 20    **f.** 1    **g.** .0036
**4.24.** **a.** .375    **b.** .001    **c.** .49    **d.** .1536    **e.** .441    **f.** .42

**4.25.** **a.**

| $x$ | 0 | 1 | 2 | 3 | 4 | 5 | 6 |
|---|---|---|---|---|---|---|---|
| $p(x)$ | .016 | .094 | .234 | .312 | .234 | .094 | .016 |

**b.** $\mu = 3$, $\sigma^2 = 1.5$    **d.** .968

**4.26.** **b.**

| $x$ | 0 | 1 | 2 | 3 | 4 | 5 |
|---|---|---|---|---|---|---|
| $p(x)$ | .168 | .360 | .309 | .132 | .028 | .002 |

**4.27.** **a.** 4, 3.2, 1.79 **b.** 16, 3.2, 1.79 **c.** 50, 25, 5 **d.** 20, 16, 4 **e.** 10, 8, 2.83
**f.** 5, 4.95, 2.22
**4.28.** **a.** .194 **b.** .403 **c.** .812 **d.** .000 **e.** .998 **f.** .000
**4.29.** **a.** 5, 2.5, 1.58 **c.** .978
**4.30.** **a.** No **b.** Yes **c.** No **d.** For part b, $\mu = 4$, $\sigma^2 = 3.2$, $\sigma = 1.79$
**4.31.** **a.** 7.2 **b.** $(.4)^{12} \approx .000017$ **c.** Yes **4.32.** .0005063 **4.33.** .32
**4.34.** **a.** .590 **b.** .410 **4.35.** .098 **4.36.** **a.** 152 **b.** $\mu \pm 2\sigma$ is about 140 to 164
**4.37.** **a.** .009 **c.** Yes **4.38.** $\mu = .5$, $\sigma = .707$, no
**4.39.** **a.** .2 **b.** 4 **c.** .196 **d.** No
**4.40.** **a.** $\mu = 48$, $\sigma = 6.50$ **b.** No, 31 is more than 2 standard deviations below the mean **c.** No
**4.41.** **a.** .000 **b.** Drug seems effective **4.42.** 15
**4.43.** **a.** 0, .000, .000, .098, .873, 1 **c.** 0, .000, .000, .234, .966, 1
**4.44.** **a.** .4772 **b.** .4332 **c.** .4989 **d.** .1915 **e.** .4772 **f.** .4332 **g.** .4989
**h.** .1915
**4.45.** **a.** .4772 **b.** .4332 **c.** .4989 **d.** .1915
**4.46.** **a.** .6826 **b.** .9544 **c.** .9361 **d.** .6104 **e.** .9846 **f.** .9949
**4.47.** **a.** .1587 **b.** .0228 **c.** .0368 **d.** .2240 **e.** .5000 **f.** .8384
**4.48.** **a.** 0 **b.** 1.96 **c.** −1.96 **d.** 2.00 **e.** 2.06 **f.** −1.75
**4.49.** **a.** −2.17 **b.** 2.17 **c.** 1.96 **d.** 1.645 **e.** 1.00 **f.** 2.81
**4.50.** **a.** .9573 **b.** .7964 **c.** .0019 **d.** .9554 **e.** .0788 **f.** 1.000 **g.** 0.000
**h.** .9010
**4.51.** **a.** −3.01 **b.** 3.01 **c.** 1.66 **d.** .94 **e.** 1.30 **f.** 1.80
**4.52.** **a.** 1 **b.** −1 **c.** 0 **d.** −2.5 **e.** 3
**4.53.** **a.** −1 **b.** 0 **c.** 1.5 **d.** −4 **e.** 4 **f.** .4
**4.54.** **a.** −2.5 **b.** 0 **c.** 1.25 **d.** −1.25 **e.** −5 **f.** 10
**4.55.** **a.** .3413 **b.** .4772 **c.** .0655 **d.** .7675 **e.** .0526 **f.** .8830
**4.56.** **a.** .5 **b.** .0749 **c.** .8189 **d.** .0431 **e.** .9932 **f.** .0068
**4.57.** **a.** 40 **b.** 22.36 **c.** 51.52 **d.** 25.195 **e.** 28.48 **f.** 47.56 **g.** 60.97
**4.58.** **a.** .9544 **b.** .0228 **c.** .1587 **d.** .8185 **e.** .1498 **f.** .9974
**4.59.** **a.** .0150 **b.** $\mu \pm 2\sigma$ is 5.1 to 7.5 **4.60.** **a.** .0048 **b.** Yes **4.61.** .0019
**4.62.** 2.22% **4.63.** **a.** .0228 **b.** .9835 **c.** .00052 **d.** Mean of 4.5 is too high
**4.64.** 5.068 **4.65.** 398.8 **4.66.** 473 **4.67.** **a.** 7.667 **b.** 45.62%
**4.68.** **a.** .586; approx. .5793 **b.** .846; approx. .8461 **c.** .120; approx. .1170 **d.** .725; approx. .7280
**4.69.** **a.** Approx. 1.0 **b.** Approx. .0031 **c.** Approx. .5557
**4.71.** **a.** .9738 **b.** .0162 **c.** .0000 **d.** .0314 **e.** .9292
**4.72.** **a.** .9838 **b.** .1314 **4.73.** Approx. .0559 **4.74.** Approx. 0
**4.75.** **a.** Approx. .1762 **b.** Not likely; $z = 3.42$ **4.76.** **a.** Approx. .9808 **b.** Approx. 0
**4.77.** **a.** Approx. .0885 **b.** Approx. .7123
**4.78.** **a.** Approx. .011 **b.** True percentage affected is smaller than 45%
**4.79.** **a.** Approx. .0516 **b.** Approx. .8324 **4.80.** **a.** Binomial **b.** $\mu = 30$, $\sigma = 4.58$ **c.** 0
**4.81.** **a.** Approx. .4681 **b.** Approx. .0436 **c.** Approx. .9822
**4.82.** **a.** Approx. .0559 **b.** Yes
**4.83.** **a.** 16.2, 18.76, 4.33 **b.** .4 **c.** 7.54 to 24.86 **d.** 1
**4.84.** **a.** .243 **b.** .131 **c.** .36 **d.** .157 **e.** .128 **f.** .121
**4.85.** **a.** .124 **b.** .245 **c.** .755 **d.** .975 **e.** .927 **f.** 12, 4.8, 2.19 **g.** .963
**4.86.** **a.** .4750 **b.** .9500 **c.** .9010 **d.** .0388 **e.** .1658 **f.** .9218
**4.87.** **a.** .7257 **b.** .2743 **c.** .9938 **d.** .3983 **e.** .5957 **f.** .1762
**4.88.** **a.** .05 **b.** .44 **c.** 0 **d.** 1.59 **e.** −1.04 **f.** 2.36
**4.89.** **a.** .3085 **b.** .1587 **c.** .1359 **d.** .6915 **e.** 0 **f.** .9938
**4.90.** **a.** 75 **b.** 96.33 **c.** 50.07 **d.** 93 **e.** 63.48 **f.** 67.44

**4.91. a.** Approx. .9441 **b.** Approx. .5557 **c.** Approx. .0031 **d.** Approx. .8426
**e.** Approx. .1114 **f.** Approx. .1118
**4.92. a.** .599 **b.** .988 **4.93. a.** .027 **b.** .376 **c.** No **4.94. a.** .126 **b.** .057
**4.95. a.** .033 **b.** .985 **4.96.** .401 **4.97.** 22,250
**4.98. a.** .53 **b.** $1,240 **c.** $240 **d.** $\sigma^2 = 312,400; \sigma = 558.93$
**4.99. a.** .021 **b.** 292 **4.100. a.** .4019 **b.** .1608 **d.** $50,000
**4.101. a.** .343 **b.** .902 **c.** .004
**4.102. a.** $A$: 4.6; $B$: 3.7 **b.** $A$: $46,000; $B$: $55,500 **c.** $A$: $\sigma^2 = 1.34, \sigma = 1.16$; $B$: $\sigma^2 = 1.21, \sigma = 1.1$
**d.** $A$: .95; $B$: .95
**4.103. a.** .657 **b.** .027 **4.104.** .1715 **4.105. a.** .346 **b.** .683
**4.106. a.** 5, 4 **b.** .617 **c.** .006 **4.107. a.** .006 **4.108.** .059 **4.109.** 4.664%
**4.110.** .8315
**4.111. a.** .8790 **b.** .7967 **c.** About 163 **4.112. a.** .0456 **b.** .9082 **c.** .0023
**4.113 a.** .2033 **b.** $12,800 **4.114.** .0721 **4.115. a.** .8850 **b.** .0304
**4.116. a.** .0087 **b.** 22.7 **4.117. a.** .0853 **b.** .1858 **4.118. a.** .0122 **b.** .0062
**4.119. a.** .1469 **b.** .0216 **4.120. a.** Normal **c.** $\bar{x} = 9.9825; s = .1960$
**4.121. a.** .0548 **b.** .6006 **c.** .3446 **d.** $6,503.80 **4.122.** .1056
**4.123.** $(.9938)^{100} = .5369$; approx. 1.0 **4.124. a.** Approx. 1.0 **b.** Approx. .8830
**4.125. a.** Approx. .1056 **b.** Approx. .0465 **c.** Approx. .9539
**4.126. a.** Approx. 1.0 **b.** Approx. .6618 **4.127.** $P(x \geq 400$ when $p = .2) \approx 0$
**4.128.** Approx. .1446 **4.129. a.** .4880 **b.** .1093 **c.** .8926

## CHAPTER 5

**5.1. a. and b.**

| Samples | (0, 0) | (0, 2) (2, 0) | (0, 4) (4, 0) | (0, 6) (6, 0) | (2, 2) | (2, 4) (4, 2) | (2, 6) (6, 2) | (4, 4) | (4, 6) (6, 4) | (6, 6) |
|---------|--------|---------------|---------------|---------------|--------|---------------|---------------|--------|---------------|--------|
| $\bar{x}$ | 0 | 1 | 2 | 3 | 2 | 3 | 4 | 4 | 5 | 6 |

**c.** $\frac{1}{16}$ **d.**

| $\bar{x}$ | 0 | 1 | 2 | 3 | 4 | 5 | 6 |
|-----------|---|---|---|---|---|---|---|
| $p(\bar{x})$ | $\frac{1}{16}$ | $\frac{2}{16}$ | $\frac{3}{16}$ | $\frac{4}{16}$ | $\frac{3}{16}$ | $\frac{2}{16}$ | $\frac{1}{16}$ |

**5.3. a.**

| $\bar{x}$ | 1 | 1.5 | 2 | 2.5 | 3 | 3.5 | 4 | 4.5 | 5 |
|-----------|---|-----|---|-----|---|-----|---|-----|---|
| $p(\bar{x})$ | .04 | .12 | .17 | .20 | .20 | .14 | .08 | .04 | .01 |

**b.** .05 **c.** No

**5.6. a.** $\mu = 2, \sigma^2 = \frac{14}{3}$ **b.**

| $\bar{x}$ | 0 | $\frac{1}{2}$ | 1 | $\frac{5}{2}$ | 3 | 5 |
|-----------|---|---------------|---|---------------|---|---|
| $p(\bar{x})$ | $\frac{1}{9}$ | $\frac{2}{9}$ | $\frac{1}{9}$ | $\frac{2}{9}$ | $\frac{2}{9}$ | $\frac{1}{9}$ |

**c.** $E(\bar{x}) = 2 = \mu$

**d.**

| $s^2$ | 0 | $\frac{1}{2}$ | 8 | $\frac{25}{2}$ |
|-------|---|---------------|---|----------------|
| $p(s^2)$ | $\frac{3}{9}$ | $\frac{2}{9}$ | $\frac{2}{9}$ | $\frac{2}{9}$ |

**e.** $E(s^2) = \frac{42}{9} = \frac{14}{3} = \sigma^2$

**5.7. a.** $\mu = 5$ **b.**

| $\bar{x}$ | 2 | $\frac{8}{3}$ | $\frac{10}{3}$ | 4 | $\frac{13}{3}$ | 5 | $\frac{17}{3}$ | $\frac{20}{3}$ | $\frac{22}{3}$ | 9 |
|-----------|---|---------------|----------------|---|----------------|---|----------------|----------------|----------------|---|
| $p(\bar{x})$ | $\frac{1}{27}$ | $\frac{3}{27}$ | $\frac{3}{27}$ | $\frac{1}{27}$ | $\frac{3}{27}$ | $\frac{6}{27}$ | $\frac{3}{27}$ | $\frac{3}{27}$ | $\frac{3}{27}$ | $\frac{1}{27}$ |

$E(\bar{x}) = \frac{135}{27} = 5 = \mu$

**5.7. c.**

| $m$ | 2 | 4 | 9 |
|---|---|---|---|
| $p(m)$ | $7/27$ | $13/27$ | $7/27$ |

$E(m) = 129/27 = 4.778 < \mu = 5$

**d.** The sample mean, $\bar{x}$; it is unbiased

**5.8. a.** $\mu = 1$  **b.**

| $\bar{x}$ | 0 | $1/3$ | $2/3$ | 1 | $4/3$ | $5/3$ | 2 |
|---|---|---|---|---|---|---|---|
| $p(\bar{x})$ | $1/27$ | $3/27$ | $6/27$ | $7/27$ | $6/27$ | $3/27$ | $1/27$ |

**c.**

| $m$ | 0 | 1 | 2 |
|---|---|---|---|
| $p(m)$ | $7/27$ | $13/27$ | $7/27$ |

**d.** $E(\bar{x}) = 27/27 = 1 = \mu$; $E(m) = 27/27 = 1 = \mu$  **e.** Variance of $\bar{x}$ is $2/9 \approx .222$; variance of $m$ is $14/27 \approx .519$

**f.** $\bar{x}$, smaller variance

**5.10. a.** $\mu_{\bar{x}} = 50, \sigma_{\bar{x}} = \sqrt{70/10} \approx 2.646$  **b.** $\mu_{\bar{x}} = 50, \sigma_{\bar{x}} = \sqrt{70/25} \approx 1.673$

**c.** $\mu_{\bar{x}} = 50, \sigma_{\bar{x}} = \sqrt{70/100} \approx .837$  **d.** $\mu_{\bar{x}} = 50, \sigma_{\bar{x}} = \sqrt{70/70} = 1$  **e.** $\mu_{\bar{x}} = 50, \sigma_{\bar{x}} = \sqrt{70/1000} \approx .265$

**f.** $\mu_{\bar{x}} = 50, \sigma_{\bar{x}} = \sqrt{70/400} \approx .418$

**5.11. a.** 10, 2  **b.** 20, 1  **c.** 50, 30  **d.** 30, 20

**5.12. a.** $\mu = 4, \sigma^2 = 12.5, \sigma = \sqrt{12.5} \approx 3.536$

**b.**

| $\bar{x}$ | 1 | $3/2$ | 2 | $5/2$ | 3 | $11/2$ | 6 | $13/2$ | 10 |
|---|---|---|---|---|---|---|---|---|---|
| $p(\bar{x})$ | $1/16$ | $2/16$ | $3/16$ | $2/16$ | $1/16$ | $2/16$ | $2/16$ | $2/16$ | $1/16$ |

**c.** $\mu_{\bar{x}} = 64/16 = 4 = \mu$; $\sigma_{\bar{x}}^2 = 6.25, \sigma_{\bar{x}} = \sqrt{6.25} = 2.5 = \sqrt{12.5}/\sqrt{2} = \sigma/\sqrt{n}$

**5.13.** No; to be sure, sample must be large.  **5.14. a.** .5  **b.** .6271  **c.** .0019  **d.** .0202

**5.15. a.** .9974  **b.** .0823  **c.** .5118  **d.** .9871

**5.16. a.** Largest: $\mu + 2\sigma_{\bar{x}} = 100.70$; smallest: $\mu - 2\sigma_{\bar{x}} = 99.30$  **b.** $3\sigma_{\bar{x}} = 1.05$  **c.** No

**5.18. a.** $\mu_{\bar{x}} = 1.3, \sigma_{\bar{x}} = 1.7/\sqrt{50} \approx .24$  **b.** Yes; large random sample  **c.** .1056  **d.** .0062

**5.19. a.** It is approximately normal, $\mu_{\bar{x}} = 100, \sigma_{\bar{x}} = 15/\sqrt{200} \approx 1.06$  **b.** .0023

**c.** Claim seems incorrect, since a sample mean less than 97 is so rare.

**d.** $P(\bar{x} \leq 98.5) \approx .0793$; not as rare; not as much evidence to doubt claim

**5.20. a.** .0013  **b.** Employees are using fewer sick days (using rare event approach).

**5.21. a.** $\mu_{\bar{x}} = 6, \sigma_{\bar{x}} = 2.5/\sqrt{50} \approx .35$  **b.** .5222  **c.** .0793  **d.** Same mean, $\sigma_{\bar{x}} = 2.5/\sqrt{100} = .25$

**5.22. a.** $\mu = 5/3$

**b.**

| Sample | (0, 0, 0) | (0, 0, 2)<br>(0, 2, 0)<br>(2, 0, 0) | (0, 0, 3)<br>(0, 3, 0)<br>(3, 0, 0) | (0, 2, 2)<br>(2, 0, 2)<br>(2, 2, 0) | (0, 3, 3)<br>(3, 0, 3)<br>(3, 3, 0) | (0, 2, 3),(0, 3, 2)<br>(2, 0, 3),(2, 3, 0)<br>(3, 0, 2),(3, 2, 0) | (2, 2, 2) | (2, 2, 3)<br>(2, 3, 2)<br>(3, 2, 2) | (2, 3, 3)<br>(3, 2, 3)<br>(3, 3, 2) | (3, 3, 3) |
|---|---|---|---|---|---|---|---|---|---|---|
| Mean, $\bar{x}$ | 0 | $2/3$ | 1 | $4/3$ | 2 | $5/3$ | 2 | $7/3$ | $8/3$ | 3 |
| Median, $m$ | 0 | 0 | 0 | 2 | 3 | 2 | 2 | 2 | 3 | 3 |

**c.**

| $\bar{x}$ | 0 | $2/3$ | 1 | $4/3$ | $5/3$ | 2 | $7/3$ | $8/3$ | 3 |
|---|---|---|---|---|---|---|---|---|---|
| $p(\bar{x})$ | $1/27$ | $3/27$ | $3/27$ | $3/27$ | $6/27$ | $4/27$ | $3/27$ | $3/27$ | $1/27$ |

**d.** $E(\bar{x}) = 45/27 = 5/3 = \mu$; $E(m) = 47/27 > \mu = 5/3$

| $m$ | 0 | 2 | 3 |
|---|---|---|---|
| $p(m)$ | $7/27$ | $13/27$ | $7/27$ |

**5.23. a.** $\mu_{\bar{x}} = 120$, $\sigma_{\bar{x}} = \sqrt{410/75} \approx 2.34$    **b.** Approx. normal    **c.** .1949    **d.** .8835    **e.** .9756
**f.** .1251    **5.24. a.** .5    **b.** .0418    **c.** .0749    **d.** .8729

**5.25. a.**

| $\bar{x}$ | 1 | 1.5 | 2 | 2.5 | 3 | 3.5 | 4 | 4.5 | 5 | 5.5 | 6 |
|---|---|---|---|---|---|---|---|---|---|---|---|
| $p(\bar{x})$ | $\frac{1}{36}$ | $\frac{2}{36}$ | $\frac{3}{36}$ | $\frac{4}{36}$ | $\frac{5}{36}$ | $\frac{6}{36}$ | $\frac{5}{36}$ | $\frac{4}{36}$ | $\frac{3}{36}$ | $\frac{2}{36}$ | $\frac{1}{36}$ |

**b.** $\frac{33}{36}$

**5.27. b.** 4.68    **c.** 7.931    **d.** 4.78, 1.642    **e.** 4.68, 1.172 (means rounded to two decimals)
**f.** $\bar{x}$
**5.28. b.** 4.84, 1.313    **5.29. b.** 4.68, .8270    **5.30. a.** .9887    **b.** Normal distribution for $\bar{x}$
**5.31. a.** Approx. normal, $\mu_{\bar{x}} = 400$, $\sigma_{\bar{x}} = 11.86$    **b.** .0174    **c.** .5    **5.32. a.** .2327    **b.** .2215
**5.33. a.** .5    **b.** .0023    **c.** Normal distribution for $\bar{x}$    **5.34. b.** $\mu_y = .5$, $\sigma_y = .29/\sqrt{n}$    **c.** Normal
**5.35.** .383    **5.36.** .9772
**5.37. a.** 45 samples; $\{(1, 2), (1, 3), \ldots, (9, 10)\}$    **b.** $\{(1, 4), (1, 5), \ldots, (8, 10)\}$; $\frac{21}{45}$
**c.** $p(0) = \frac{3}{45}$, $p(1) = \frac{21}{45}$, $p(2) = \frac{21}{45}$
**5.38. a.** Approx. normal, $\mu_{\bar{x}} = 45$, $\sigma_{\bar{x}} = .258$    **b.** .8508    **c.** .000    **5.39. a.** .0082    **b.** None
**5.40. a.** 1.342    **b.** Approx. .0026
**5.41. a.** Approx. normal, $\mu = 7{,}500$, $\sigma = 246.48$    **b.** .9576    **c.** .0075


# CHAPTER 6

**6.1. a.** (71.96, 78.04)    **b.** (72.45, 77.55)    **c.** (71.00, 79.00)
**6.2. a.** (38.48, 41.52)    **b.** (39.24, 40.76)    **c.** (46.08, 53.92)    **d.** (42.16, 57.84)
**6.3. a.** (4.43, 11.17)    **b.** (5.65, 9.95)
**6.4. a.** (12.592, 13.808)    **c.** (12.4, 14.0)    **d.** Increases
**6.6. a.** Yes, for a large random sample
**6.7. a.** (12.8, 14.4)    **b.** (13.2, 14.0)    **c.** Part a, width is 1.6; part b, width is .8; width is half as large
**6.8.** (18.12, 25.88)    **6.9.** (8.9, 10.3)    **6.10.** (.67, .77)    **6.11.** (18.575, 21.025)
**6.12.** (31.79, 33.41)
**6.15. a.** .025    **b.** .05    **c.** .005    **d.** .10    **e.** .10    **f.** .01
**6.16. a.** $z = 1.38$, do not reject $H_0$    **b.** $z = 1.39$, reject $H_0$    **c.** $z = -3.16$, reject $H_0$
**6.17. a.** $z = 1.57$, do not reject $H_0$    **b.** $z = 2.83$, reject $H_0$    **c.** $z = -2.03$, reject $H_0$
**6.18. a.** $z = 3.86$, reject $H_0$    **b.** $z = -1.08$, do not reject $H_0$    **c.** $z = -1.59$, reject $H_0$
**6.19. a.** $z = -1.84$, reject $H_0$    **b.** $z = -1.84$, do not reject $H_0$    **6.21.** No
**6.23. a.** $H_0$: $\mu = 43$, $H_a$: $\mu > 43$    **b.** $z = 2.92$, reject $H_0$
**6.24. a.** $H_0$: $\mu = 62{,}000$, $H_a$: $\mu < 62{,}000$    **b.** $z = -2.60$, reject $H_0$
**6.25.** $z = -4.34$; yes    **6.26. a.** .166    **b.** .0853    **c.** $z = 2.36$; yes
**6.27.** $z = 2.0$; yes    **6.28.** $z = 3.53$; yes
**6.29. a.** .0427    **b.** .0064    **c.** .0885    **d.** .0808    **e.** .1190    **f.** .0016
**6.31. a.** .0854    **b.** .0128    **c.** .1498    **d.** .0198    **e.** .1118    **f.** .1970
**6.33.** $p$ value $= .0392$    **6.34. a.** $z = -1.06$; no; $p$ value $= .2892$    **6.35.** $p$ value $= .0018$
**6.36.** $p$ value $= .1230$    **6.37.** $p$ value $= .0228$
**6.38. a.** Reject $H_0$ if $t > 2.093$ or $t < -2.093$    **b.** Reject $H_0$ if $t > 2.492$
**c.** Reject $H_0$ if $t > 1.383$    **d.** Reject $H_0$ if $t < -2.681$
**e.** Reject $H_0$ if $t > 1.746$ or $t < -1.746$    **f.** Reject $H_0$ if $t < -1.895$
**6.39. a.** 1.761    **b.** 4.032    **c.** 2.947    **d.** 2.052    **e.** 1.703
**6.40. a.** $t = -1.79$, reject $H_0$    **b.** $t = -1.79$, do not reject $H_0$    **c.** (2.94, 6.66)
**d.** (3.37, 6.23)
**6.41. a.** $t = 3.56$, reject $H_0$    **b.** $t = 3.56$, do not reject $H_0$    **c.** $t = 3.56$, reject $H_0$    **d.** (3.12, 6.88)
**e.** (2.06, 7.94)

**6.42.** **a.** (51.13, 54.07)    **6.43.** **a.** $t = -2.01$; yes    **b.** Type II error    **c.** Type I error
**6.44.** (19.73, 19.99)    **6.45.** **b.** $t = -2.24$; no    **c.** $.05 < p$ value $< .10$
**6.46.** **a.** $t = .866$; no    **b.** $p$ value $> .10$    **6.47.** **a.** (9.1, 13.5)
**6.48.** **a.** $t = 3.57$; yes    **b.** $p$ value $< .005$    **6.49.** (4.03, 4.17)    **6.50.** $t = -4.19$; yes
**6.51.** (49.58, 49.82)    **6.52.** **a.** $t = 2.33$; yes    **6.53.** **a.** $t = -3.63$; yes    **b.** $p$ value $< .01$
**6.54.** $t = .264$; no    **6.55.** $(-.26, 13.02)$    **6.56.** (1.32, 7.02)    **6.57.** **a.** $t = -2.20$; no
**6.58.** **a.** (.33, .57)    **b.** (.35, .55)    **c.** $z = -.8$, do not reject $H_0$    **d.** $z = -.8$, do not reject $H_0$
**6.59.** **a.** $z = 2.5$, reject $H_0$    **b.** $z = 2.5$, reject $H_0$    **c.** $z = 1.40$, do not reject $H_0$    **d.** (.84, .96)
**e.** (.82, .98)
**6.60.** **b.** $z = -1.53$; no    **6.62.** (.04, .10)    **6.63.** (.58, .68)    **6.65.** **a.** $z = 2.5$; yes    **b.** .0062
**6.66.** **a.** $z = -1.51$; no    **b.** .0655    **6.67.** (.035, .075)    **6.68.** (.04, .08)
**6.69.** $z = -1.24$; no    **6.70.** $z = .98$; no    **6.71.** $z = -2.97$; yes    **6.72.** (.154, .218)
**6.73.** **a.** 683    **b.** 6,147    **c.** 4,733    **d.** 482
**6.74.** **a.** 5,991    **b.** 240    **c.** 2,305    **d.** 722    **e.** 601    **6.75.** 577
**6.76.** **a.** 1,421    **b.** 1,692    **6.77.** 68    **6.78.** **a.** .98, .784, .56, .392, .196
**6.79.** **a.** 68    **b.** Decrease    **6.80.** 666    **6.81.** 62    **6.82.** 135    **6.83.** 2,401
**6.84.** 55    **6.85.** 68    **6.86.** 322
**6.87.** **a.** (70.90, 74.30)    **b.** $t = -7.51$, reject $H_0$    **c.** $t = -7.51$, reject $H_0$    **d.** (69.78, 75.42)
**e.** 75
**6.88.** **a.** $z = -1.78$, reject $H_0$    **b.** $z = -1.78$, do not reject $H_0$    **c.** (.23, .35)    **d.** (.21, 37)
**e.** 549
**6.89.** **a.** (8.08, 8.32)    **b.** $z = -1.67$, do not reject $H_0$    **c.** $z = -3.35$, reject $H_0$
**6.90.** **a.** $z = -2.99$, reject $H_0$    **b.** $p$ value $= .0028$    **6.91.** $t = 3.51$; no
**6.92.** **a.** $t = 1.56$; yes    **b.** $.05 < p$ value $< .10$    **6.93.** **a.** $z = 6.13$; yes    **b.** $p$ value $\approx 0$
**6.94.** **a.** $t = 3.82$; yes    **6.95.** **a.** $z = 1.41$; no    **6.96.** Small
**6.97.** **a.** (10.555, 13.845)    **b.** 166    **6.98.** $z = 1.07$; no    **6.99.** 65
**6.100.** **a.** (.038, .112)    **b.** $z = 1.62$, reject $H_0$    **6.101.** 667    **6.102.** (3.27, 3.93)
**6.103.** $z = -1.67$; yes    **6.104.** (.82, .90)    **6.105.** **a.** (47.86, 60.14)    **b.** Maximum depth of snowfall
**6.106.** (.52, .66)    **6.107.** (.030, .220)    **6.108.** (.54, .58)    **6.109.** (24.86, 27.94)
**6.110.** (1.53, 2.47)


## CHAPTER 7

**7.1.** **a.** $(-.73, -.27)$    **b.** $z = -4.33$, reject $H_0$    **c.** $p$ value is near 0
**7.2.** **a.** $(-6.15, -2.05)$    **b.** $(-7.31, -.89)$    **c.** $z = -3.29$, reject $H_0$    **d.** $p$ value is near 0
**7.3.** **a.** $z = 1.67$, reject $H_0$    **b.** $z = 1.67$, do not reject $H_0$    **c.** $(-.37, 4.57)$
**7.4.** Yes, for large, random, and independent samples
**7.5.** **b.** $z = -3.76$; yes    **c.** $(-4.89, -0.91)$    **d.** Narrower
**7.6.** **a.** $z = 2.82$; yes    **b.** (9.21, 50.79)    **7.7.** $(-5.36, 12.16)$    **7.8.** $z = 2.69$; yes
**7.9.** $z = 3.43$; yes    **7.10.** $z = -2.09$; yes
**7.11.** **a.** $z = -4.22$; yes    **b.** $p$ value is near 0    **c.** $(-1.53, -.67)$
**7.12.** **a.** $z = 2.77$; yes    **c.** $p$ value $= .0028$
**7.14.** **a.** .56    **b.** $t = -1.89$; yes    **c.** $(-1.9, 0.0)$
**7.15.** **a.** $t = -1.41$; do not reject $H_0$    **b.** $(-6.17, 1.17)$    **7.18.** **a.** $t = 3.08$; yes    **b.** (4.45, 15.55)
**7.19.** $t = .82$; no    **7.20.** **a.** $t = 2.76$, reject $H_0$    **b.** $p$ value $= .01$    **7.21.** $t = 1.57$; no
**7.22.** **a.** $t = 2.10$; no    **b.** (.7, 7.3)    **7.23.** **a.** 88.622, 68.489    **b.** Yes    **c.** $t = .61$; no
**7.24.** **b.** $t = 1.54$; no    **c.** $p$ value $= .10$
**7.25.** **a.** 2, 4.4    **b.** $\mu_D = \mu_1 - \mu_2$    **c.** $(-.2, 4.2)$    **d.** $t = 2.34$, do not reject $H_0$
**7.26.** **a.** $t = -6.87$; yes    **b.** $(-5.32, -3.08)$    **7.27.** (3.04, 6.96)

**7.28.** **a.** $t = 3.73$; yes    **b.** $p$ value $= .02$    **c.** $(1.23, 4.11)$    **7.29.** $t = -2.58$; yes

**7.30.** $t = 3.66$; yes (with $\alpha \geq .05$)    **7.31.** **a.** $t = 2.98$; yes    **b.** $p$ value $= .025$

**7.32.** **a.** $t = 7.68$; yes    **c.** $(.28, .56)$    **7.33.** **a.** $t = .40$; no    **b.** $(-1.88, 2.72)$    **d.** No

**7.34.** **a.** No, same plants were used    **b.** $t = -5.88$; yes    **c.** $(-3.00, -1.40)$

**7.35.** **a.** $(-1,409.68, 4,441.34)$

**7.36.** **a.** 158    **b.** 54    **c.** 148    **7.37.** 68    **7.38.** $n = 34$

**7.39.** **a.**

| SOURCE | df | SS | MS | F |
|---|---|---|---|---|
| Treatments | 2 | 11.075 | 5.538 | 3.15 |
| Error | 7 | 12.301 | 1.757 | |
| Total | 9 | 23.376 | | |

**b.** Do not reject $H_0$    **c.** $2.54 \pm 1.92$
**d.** Smaller    **e.** $3.975 \pm 1.567$    **f.** 43

**7.40.** **a.**

| SOURCE | df | SS | MS | F |
|---|---|---|---|---|
| Treatments | 4 | 24.7 | 6.18 | 4.90 |
| Error | 30 | 37.7 | 1.26 | |
| Total | 34 | 62.4 | | |

**b.** 5    **c.** Yes    **d.** $t = -.667$; no
**e.** $-.4 \pm .987$    **f.** $3.7 \pm .698$

**7.41.** **a.** $F = 6.79$; yes    **b.** $21.8 \pm 11.01$    **c.** $5.1 \pm 11.68$

**7.42.** **a.** $F = 8.63$; yes    **b.** $p$ value is smaller than .01    **c.** $-.56 \pm .51$

**7.43.** **a.** $F = 4.55$; yes    **b.** $52.83 \pm 6.15$    **c.** $4.0 \pm 8.7$

**7.44.** **a.** $F = 3.168$; no    **b.** $473.3 \pm 22.93$    **c.** $37.05 \pm 30.76$

**7.45.** **a.** .37    **b.** $t^2 = (.61)^2 = .37$    **c.** Do not reject $H_0$

**7.46.** **a.** $t = .78$, do not reject $H_0$    **b.** $(-6.49, 11.49)$    **c.** $n_1 = n_2 = 225$

**7.47.** **a.** $t = 5.73$, reject $H_0$    **b.** $(1.96, 5.64)$

**7.49.** **a.** $F = 7.70$; yes    **b.** $2 \pm 1.4$    **c.** $11.4 \pm 1.2$    **7.50.** **a.** $t = 1.064$, do not reject $H_0$

**7.51.** $t = -4.02$, reject $H_0$    **7.52.** $z = .90$; no    **7.53.** $z = 1.68$; yes    **7.54.** $(-.22), 1.02)$; yes

**7.55.** $t = -.70$; no    **7.56.** $n_1 = n_2 = 193$    **7.57.** **a.** $t = 2.27$; yes    **7.58.** $t = 6.14$; yes

**7.59.** **a.** $t = -2.84$; yes    **b.** $(-76.94, -8.66)$    **7.60.** **a.** $F = 5.29$; yes    **c.** $p$ value is smaller than .01

**7.61.** **a.** $F = 10.00$; yes ($\alpha = .05$)    **b.** $2.0 \pm 11.5$    **c.** $70.5 \pm 7.04$

**7.62.** **a.** $F = 7.613$; yes    **b.** $8.067 \pm .4491$    **7.63.** **a.** $F = 7.79$; yes    **b.** $-5.65 \pm 3.25$

# CHAPTER 8

**8.1.** $(-.02, .24)$    **b.** $(-.176, -.004)$    **c.** $(-.12, .22)$

**8.2.** **a.** $z = -3.18$, reject $H_0$    **b.** $z = -3.18$, reject $H_0$    **c.** $z = -3.18$, reject $H_0$
**d.** $(-.15, -.05)$

**8.8.** $z = -1.67$; no (with $\alpha \leq .095$)    **8.9.** **a.** $z = 4$; yes    **b.** $(.12, .28)$    **8.10.** $(-.59, -.13)$

**8.11.** **a.** $z = -.86$; no    **b.** $p$ value $= .3898$

**8.12.** **a.** $(.17, .18)$    **b.** $(.07, .09)$    **c.** $(.09, .11)$    **8.13.** $z = 17.04$, reject $H_0$; yes, $B$

**8.14.** $(.13, .21)$    **8.15.** $z = 7.79$; yes    **8.17.** **a.** $z = .79$; do not reject $H_0$    **b.** $p$ value $= .2148$

**8.18.** **a.** 29,954    **b.** 542    **c.** 271

**8.19.** **a.** $n_1 = n_2 = 769$    **b.** Yes, $n_1 = n_2 = 385$ would be large enough    **8.20.** $n_1 = n_2 = 4,802$

**8.21.** $n_1 = 260, n_2 = 520$    **8.22.** **a.** $X^2 = 16.29$; yes    **b.** $(-.02, .17)$    **c.** $(.23, .38)$

**8.23.** $X^2 = 1.07$; no    **8.24.** $X^2 = 2.40$; no    **8.25.** $X^2 = 8.75$; yes

**8.26.** **a.** $X^2 = 66.25$; yes    **b.** $.04 \pm .014$    **8.27.** $X^2 = 1.35$; no    **8.28.** $X^2 = 14.406$; yes

**8.29.** **a.** $X^2 = 31.86$; yes    **b.** $.12 \pm .0947$

**8.30.** **a.** $z = -2.04$, reject $H_0$    **b.** $(-.196, -.004)$    **c.** $n_1 = n_2 = 18,248$    **8.31.** $X^2 = 54.14$; yes

**8.32.** $z = -1.30$, do not reject $H_0$

**8.33.** $(-.003, .065)$     **8.34.** $z = -2.62$; yes     **8.35.** $(-.13, .05)$     **8.36.** $n_1 = n_2 = 1{,}184$

**8.37.** $(-.13, .03)$     **8.38.** $z = -.54$; no     **8.39.** **a.** $X^2 = 84.89$; yes     **b.** $(.75, .82)$     **c.** $(.15, .27)$

**8.40.** $X^2 = 103.1$; yes     **8.41.** **a.** $X^2 = 46.25$; yes ($\alpha = .05$)     **b.** $.387$


# CHAPTER 9

**9.2.** **a.** $2, 2$     **b.** $1, 4$     **c.** $-1, 2$     **d.** $-1, -4$

**9.3.** **a.** $y = 2 + 2x$     **b.** $y = 4 + x$     **c.** $y = 2 - x$     **d.** $y = -4 - x$

**9.5.** **a.** $2, 3$     **b.** $1, 1$     **c.** $3, -2$     **d.** $5, 0$     **e.** $-2, 4$

**9.8.** **b.** $2, 3$     **c.** $\hat{y} = 3 + 2x$     **9.9.** **c.** $\hat{y} = .857x$ (without rounding off, $\hat{\beta}_0 = 0$)

**9.10.** **a.** $\hat{\beta}_0 = 2, \hat{\beta}_1 = -1.2$     **c.** $\hat{y} = .8$     **d.** $\hat{y} = 3.8$

**9.11.** **a.** $\hat{y} = 57.91 - .81x$     **c.** $9.22$

**9.12.** **b.** $\hat{\beta}_0 = -.125, \hat{\beta}_1 = 3.125$     **c.** $15.5$     **9.13.** **b.** $\hat{\beta}_0 = 24.45, \hat{\beta}_1 = 2.38$     **c.** $33.38$

**9.14.** $\hat{y} = .274 + .078x$     **9.15.** **a.** $\hat{y} = 133.1 - .2647x$     **9.16.** **a.** $.031$     **b.** $2s \approx .35$

**9.17.** **a.** $0, 0$     **b.** $1.143, .286$     **c.** $1.6, .533$     **d.** $7.18, 1.44$     **e.** $11.0, 3.667$

**f.** $52.48, 8.75$     **g.** $.039, .00978$     **h.** $2{,}235.9, 171.99$

**9.18.** **a.** $\hat{y} = -45.78 + 1.562x$     **c.** $19.60, 4.90$

**9.19.** **a.** $\hat{y} = -.246 + 1.315x$     **c.** $7.65, 11.59$     **d.** $12.44, 2.487$

**9.20.** **a.** $.857, .286, 17.5, 4$     **b.** $-1.2, .533, 10, 3$     **c.** $-.811, 1.44, 175.71, 5$

**d.** $3.125, 3.667, 12.8, 3$     **e.** $2.38, 8.75, 10.5, 6$     **f.** $.078, .00978, 18.833, 4$

**g.** $-.2647, 171.99, 9{,}043.333, 13$

**9.21.** **a.** $t = 6.71$; yes     **b.** $t = 5.84$; yes     **9.22.** **a.** $t = -5.20$; yes     **b.** $t = -8.98$; yes

**9.23.** **a.** $t = 2.61$; yes     **b.** $t = 3.42$; yes     **c.** $t = -1.92$; no

**9.24.** **a.** $\hat{y} = 18.89 + .0109x$     **b.** $t = .813$; no     **c.** $.01 \pm .027$

**9.25.** **a.** $\hat{y} = 70.8 - .28x$     **b.** $t = -.48$; no

**9.26.** **a.** $\hat{y} = 10.394 + 90.921x$     **b.** $H_a$: $\beta_1 > 0$; $t = 3.182$, reject $H_0$     **c.** $90.92 \pm 57.58$

**9.27.** $5.638 \pm 1.448$     **9.28.** **b.** $\hat{\beta}_0 = 29.423, \hat{\beta}_1 = -1.196$     **d.** $t = -4.295$, reject $H_0$

**9.29.** $t = 2.56$; yes ($\alpha = .05$)

**9.30.** **a.** $1, 1$     **b.** $.958, .92$     **c.** $-.949, .90$     **d.** $-.970, .94$     **e.** $.959, .92$     **f.** $.729, .53$

**g.** $.863, .75$     **h.** $-.470, .22$

**9.31.** **a.** $-.688, .473$     **b.** $t = -2.68$; yes     **9.32.** $r = .9836, r^2 = .9675$

**9.33.** **a.** $-.366, .134$     **b.** $t = -.79$; no     **9.34.** **a.** $.69$     **b.** $.476$     **c.** $t = 3.44$; yes

**9.35.** **a.** $.809$     **b.** $t = 5.04$; yes (at $\alpha = .05$)     **9.36.** $r_1 = -.9674, r_2 = -.1105$; $x_1$

**9.37.** **a.** $3.2 \pm .94$     **b.** $3.2 \pm 1.96$

**9.38.** **a.** $\hat{y} = 1.4 + .8x$     **c.** $1$     **d.** $.05$     **e.** $2.2 \pm .14$     **f.** $2.6 \pm .48$     **g.** $1.4 \pm .28$

**9.39.** **a.** $t = 6.96$; yes     **b.** $5.26 \pm 2.10$     **c.** $21.22 \pm 4.01$     **d.** $21.22 \pm 7.90$

**9.40.** **a.** $\hat{y} = 1079.848 - 31.668x$     **b.** $t = -16.40$; yes     **c.** $-31.67 \pm 4.45$

**e.** $604.83 \pm 9.95, 604.83 \pm 26.24$

**9.41.** **a.** $\hat{y} = 40.926 - .1343x$     **b.** $t = -4.21$; yes     **c.** $-.1343 \pm .0576$     **d.** $31.26 \pm .91$

**e.** $33.14 \pm 2.21$

**9.42.** **a.** $\hat{y} = 4.554 + 2.671x$     **c.** $r = .967, r^2 = .935$     **d.** $13.10 \pm 1.19$     **e.** $12.57 \pm .43$

**9.45.** **a.** $\hat{y} = 27.78 - .6x$     **c.** $7.0$     **d.** $.5385$     **e.** $-.6 \pm .184$     **f.** $26.70 \pm .35$

**g.** $26.70 \pm 1.35$

**9.46.** **b.** $-.1245, .016$     **c.** $t = -.35$; no

**9.47.** **a.** $\hat{y} = 40.7842 + .7656x$     **c.** $t = 4.38$, reject $H_0$     **d.** $.7656 \pm .4035$

**e.** $79.06 \pm 17.03$     **f.** $67.58 \pm 7.74$

**9.48.** **a.** $\hat{y} = 22.32 + 2.650x$     **c.** $t = 6.280$; yes     **d.** $.9421, .8875$     **e.** $109.8 \pm 2.155$

**f.** $109.8 \pm 6.088$

**9.49.** **a.** $\hat{y} = 18.22 + 1.166x$     **c.** $t = 4.967$; yes $(\alpha = .05)$     **d.** .8044     **e.** $123.2 \pm 19.09$

**9.50.** **b.** $\hat{y} = 11,898.8 - 763.1x$     **d.** $t = -3.944$, reject $H_0$     **e.** $-763.1 \pm 375.9$     **f.** $-.8495$
**g.** .7216     **h.** $5,793.9 \pm 1,609.3$     **i.** $5,793.9 \pm 677.2$

**9.51.** **a.** $\hat{y} = 2.084 + .6682x$     **c.** $t = 16.28$; yes     **d.** .9815     **e.** $35.49 \pm .979$
**f.** $38.83 \pm 2.766$

**9.52.** **a.** $\hat{y} = -13.4903 - .0528x$     **b.** $t = -6.84$, reject $H_0$     **c.** .8538     **d.** $.9320 \pm .3333$

**9.53.** **a.** $\hat{y} = 16.28 + 2.078x$     **c.** $t = 6.672$; yes $(\alpha = .05)$     **d.** $57.84 \pm 6.146$

**9.54.** **a.** $\hat{y} = 20.11 + .9720x$     **c.** .9815, .9634     **d.** $27.88 \pm .98$     **e.** $26.91 \pm 2.57$

**9.55.** **a.** $\hat{y} = 6.5143 + 10.8294x$     **b.** $t = 6.336$, reject $H_0$ $(\alpha = .05)$     **c.** $12.904 \pm .3594$

**9.56.** **a.** $\hat{y} = 13.48 + .0560x$     **c.** .7402, .5479     **d.** $17.40 \pm 2.474$

**9.57.** **a.** $\hat{y} = 2.76 + .866x$     **b.** .9853, .9707     **c.** $20.081 \pm 1.003$

**9.58.** **a.** .8914, .7946     **b.** $t = 4.8187$; yes $(\alpha = .05)$

# INDEX

Estimator (*continued*)
of the difference between binomial
proportions, 320
of the difference between
population means
large independent samples, 266
paired samples, 286
small independent samples, 276
of intercept in straight-line
regression, 350
of a population mean
large sample, 212
small sample, 239
of regression coefficient, 350
of slope in straight-line
regression, 350
unbiased, 184
Events, 67
complementary, 77
compound, 75
dependent, 95
independent, 95
intersection, 75
mutually exclusive, 92
simple, 64
union, 75
Expected value, 122
Experiment, 64

$F$ distribution, 300
$F$ statistic, 300
Factorial, 132
First-order model, 345, 357
Frequency, 15
Frequency distribution, 12
mound-shaped, 34
Frequency function, 142

Gosset, W. S., 233

Histogram, 12
Hypothesis, *see also* Tests of
hypotheses
alternative, 216
null, 216
one-tailed, 223
research, 216
two-tailed, 223

Independence, test for, 335
Independent events, 95
Independent sampling design, 299
Independent variables,
regression, 345
Inference, *see also* Confidence
interval; Tests of hypotheses
definition, 5
rare event approach, 44
Inferential statistics, 3
Intercept, straight-line
regression, 350
Intersection of events, 75
Interval estimate, 210

Least squares model, 349
formulas for straight-line
model, 350
Level
attained significance, 231
observed significance, 229
significance, 218
Lot acceptance sampling, 157

Mean
arithmetic, 21
binomial, 134
inferences, *see* Confidence
interval; Tests of hypotheses
normal, 144
population, 22
sample, 22
Mean square
error, analysis of variance, 300
treatments, analysis of
variance, 300
Measurement class, 14
Median, 23
Modal class, 26
Mode, 25
Model
deterministic, 343
first-order, 345, 357
probabilistic, 343
straight-line, 345
Multinomial experiment, 334
Multiplicative rule of probability, 94
Mutually exclusive events, 92

Normal approximation to
binomial, 156
Null hypothesis, 216, 222

Objective of statistics, 6
Observed significance level, 229
One-tailed statistical test, 223

$p$ value, 229
Paired difference experiment, 285
Parameters, 174
Pearson product moment coefficient
of correlation, 367
Percentile, 41
Population, 5
Prediction interval for an individual
value of the dependent variable
in the straight-line model, 378
Probabilistic model, 343
Probability, 63, 66
additive rule, 91
conditional, 83
of events, 69
multiplicative rule, 94
of simple events, 66
Probability density function, 142
Probability distribution, 118
binomial, 133, 156
chi square ($\chi^2$), 330
continuous random variable, 142
discrete random variable, 118
$F$, 300
normal, 144
normal approximation to
binomial, 156
$t$, 233

Qualitative data, 11
Quantitative data, 11

Random numbers, 104
Random sample, 103
Random variable, 113
binomial, 129
continuous, 115, 142
discrete, 115
normal, 143
Randomized block design, 285

Cover design:   John Williams
Technical art:   Boardworks/Reese Thornton
Production coordination:   Phyllis Niklas

| SYMBOL | DESCRIPTION | PAGE |
|---|---|---|
| $n!$ | $n$ factorial; product of first $n$ integers $(0! = 1)$ | 132 |
| $\binom{n}{x}$ | Number of simple events that have $x$ successes and $(n - x)$ failures in binomial experiment | 132 |
| $\nu_1$ | Numerator degrees of freedom for an $F$ statistic | 303 |
| $\nu_2$ | Denominator degrees of freedom for an $F$ statistic | 303 |
| $p$ | (1) Observed significance level for an hypothesis test | 229 |
| | (2) Probability of Success for the binomial distribution | 129 |
| $\hat{p}$ | (1) Sample estimator of binomial proportion $p$; proportion of successes in the sample | 245 |
| | (2) Pooled estimator of $p_1 = p_2 = p$ in test of equality of two population proportions | 320 |
| $p(x)$ | Probability distribution for a discrete random variable $x$ | 119 |
| $p_0$ | Hypothesized value for binomial probability | 247 |
| $p_1 - p_2$ | Difference between true proportions of success in two independent binomial experiments | 318 |
| $\hat{p}_1 - \hat{p}_2$ | Difference between sample proportions of success in two independent binomial samples; sample estimator of $(p_1 - p_2)$ | 318 |
| $P(A)$ | Probability of event $A$ | 68 |
| $P(A\|B)$ | Probability of event $A$ given that event $B$ occurs | 83 |
| $q$ | Probability of Failure for the binomial distribution | 129 |
| $r$ | Pearson product moment coefficient of correlation between two samples | 367 |
| $r^2$ | Coefficient of determination in simple linear regression | 371 |
| $r_i$ | Total count in row $i$ of a contingency table | 329 |
| $\rho$ (rho) | Pearson product moment coefficient of correlation between two populations | 369 |
| $s$ | Sample standard deviation | 31 |
| $s^2$ | Sample variance | 30 |
| $s_{\hat{\beta}_1}$ | Sample estimator of $\sigma_{\hat{\beta}_1}$ | 361 |
| $s_D$ | Sample standard deviation of differences in a paired difference experiment | 285 |
| $s_p^2$ | Pooled sample variance in two sample $t$ test | 274 |
| $S$ | Sample space | 65 |
| SS(Total) | Total sum of squares in an analysis of variance | 301 |
| $SS_{xx}$ | Sum of squares of the distances between $x$ measurements and their mean, i.e., $\Sigma (x - \bar{x})^2$ | 350 |
| $SS_{xy}$ | Sum of products of distances of $x$ and $y$ measurements from their means, i.e., $\Sigma (x - \bar{x})(y - \bar{y})$ | 350 |
| $SS_{yy}$ | Sum of squares of the distances between $y$ measurements and their mean, i.e., $\Sigma (y - \bar{y})^2$ | 357 |